Version 13

Design of Experiments Guide

"The real voyage of discovery consists not in seeking new landscapes, but in having new eyes."

Marcel Proust

JMP, A Business Unit of SAS
SAS Campus Drive
Cary, NC 27513

The correct bibliographic citation for this manual is as follows: SAS Institute Inc. 2016. *JMP® 13 Design of Experiments Guide*. Cary, NC: SAS Institute Inc.

JMP® 13 Design of Experiments Guide

ISBN 978-1-62960-469-5 (Hardcopy)
ISBN 978-1-62960-562-3 (EPUB)
ISBN 978-1-62960-563-0 (MOBI)

SAS Institute Inc., SAS Campus Drive, Cary, North Carolina 27513-2414.

September 2016

Technology License Notices

Contents

Design of Experiments Guide

12 Full Factorial Designs

13 Mixture Designs

21 Space-Filling Designs

22 Accelerated Life Test Designs

23 Nonlinear Designs

A Column Properties

Chapter 1

Learn about JMP

Documentation and Additional Resources

This chapter includes the following information:

- book conventions
- JMP documentation
- JMP Help
- additional resources, such as the following:
  - other JMP documentation
  - tutorials
  - indexes
  - Web resources
  - technical support options

Formatting Conventions

The following conventions help you relate written material to information that you see on your screen:

- Sample data table names, column names, pathnames, filenames, file extensions, and folders appear in Helvetica font.
- Code appears in `Lucida Sans Typewriter` font.
- Code output appears in *`Lucida Sans Typewriter`* italic font and is indented farther than the preceding code.
- **Helvetica bold** formatting indicates items that you select to complete a task:
  - buttons
  - check boxes
  - commands
  - list names that are selectable
  - menus
  - options
  - tab names
  - text boxes
- The following items appear in italics:
  - words or phrases that are important or have definitions specific to JMP
  - book titles
  - variables
  - script output
- Features that are for JMP Pro only are noted with the JMP Pro icon JMP PRO. For an overview of JMP Pro features, visit http://www.jmp.com/software/pro/.

Note: Special information and limitations appear within a Note.

Tip: Helpful information appears within a Tip.

JMP Documentation

JMP offers documentation in various formats, from print books and Portable Document Format (PDF) to electronic books (e-books).

- Open the PDF versions from the **Help > Books** menu.
- All books are also combined into one PDF file, called *JMP Documentation Library,* for convenient searching. Open the *JMP Documentation Library* PDF file from the **Help > Books** menu.
- You can also purchase printed documentation and e-books on the SAS website:

  http://www.sas.com/store/search.ep?keyWords=JMP

JMP Documentation Library

The following table describes the purpose and content of each book in the JMP library.

Document Title	Document Purpose	Document Content
Discovering JMP	If you are not familiar with JMP, start here.	Introduces you to JMP and gets you started creating and analyzing data.
Using JMP	Learn about JMP data tables and how to perform basic operations.	Covers general JMP concepts and features that span across all of JMP, including importing data, modifying columns properties, sorting data, and connecting to SAS.
Basic Analysis	Perform basic analysis using this document.	Describes these Analyze menu platforms: • Distribution • Fit Y by X • Tabulate • Text Explorer Covers how to perform bivariate, one-way ANOVA, and contingency analyses through Analyze > Fit Y by X. How to approximate sampling distributions using bootstrapping and how to perform parametric resampling with the Simulate platform are also included.

Document Title	Document Purpose	Document Content
Essential Graphing	Find the ideal graph for your data.	Describes these Graph menu platforms: • Graph Builder • Overlay Plot • Scatterplot 3D • Contour Plot • Bubble Plot • Parallel Plot • Cell Plot • Treemap • Scatterplot Matrix • Ternary Plot • Chart The book also covers how to create background and custom maps.
Profilers	Learn how to use interactive profiling tools, which enable you to view cross-sections of any response surface.	Covers all profilers listed in the Graph menu. Analyzing noise factors is included along with running simulations using random inputs.
Design of Experiments Guide	Learn how to design experiments and determine appropriate sample sizes.	Covers all topics in the DOE menu and the Specialized DOE Models menu item in the Analyze > Specialized Modeling menu.

Document Title	Document Purpose	Document Content
Fitting Linear Models	Learn about Fit Model platform and many of its personalities.	Describes these personalities, all available within the Analyze menu Fit Model platform: • Standard Least Squares • Stepwise • Generalized Regression • Mixed Model • MANOVA • Loglinear Variance • Nominal Logistic • Ordinal Logistic • Generalized Linear Model

Document Title	Document Purpose	Document Content
Predictive and Specialized Modeling	Learn about additional modeling techniques.	Describes these Analyze > Predictive Modeling menu platforms: • Modeling Utilities • Neural • Partition • Bootstrap Forest • Boosted Tree • K Nearest Neighbors • Naive Bayes • Model Comparison • Formula Depot Describes these Analyze > Specialized Modeling menu platforms: • Fit Curve • Nonlinear • Gaussian Process • Time Series • Matched Pairs Describes these Analyze > Screening menu platforms: • Response Screening • Process Screening • Predictor Screening • Association Analysis The platforms in the Analyze > Specialized Modeling > Specialized DOE Models menu are described in *Design of Experiments Guide*.

Document Title	Document Purpose	Document Content
Multivariate Methods	Read about techniques for analyzing several variables simultaneously.	Describes these Analyze > Multivariate Methods menu platforms: • Multivariate • Principal Components • Discriminant • Partial Least Squares Describes these Analyze > Clustering menu platforms: • Hierarchical Cluster • K Means Cluster • Normal Mixtures • Latent Class Analysis • Cluster Variables
Quality and Process Methods	Read about tools for evaluating and improving processes.	Describes these Analyze > Quality and Process menu platforms: • Control Chart Builder and individual control charts • Measurement Systems Analysis • Variability / Attribute Gauge Charts • Process Capability • Pareto Plot • Diagram

Document Title	Document Purpose	Document Content
Reliability and Survival Methods	Learn to evaluate and improve reliability in a product or system and analyze survival data for people and products.	Describes these Analyze > Reliability and Survival menu platforms: • Life Distribution • Fit Life by X • Cumulative Damage • Recurrence Analysis • Degradation and Destructive Degradation • Reliability Forecast • Reliability Growth • Reliability Block Diagram • Repairable Systems Simulation • Survival • Fit Parametric Survival • Fit Proportional Hazards
Consumer Research	Learn about methods for studying consumer preferences and using that insight to create better products and services.	Describes these Analyze > Consumer Research menu platforms: • Categorical • Multiple Correspondence Analysis • Multidimensional Scaling • Factor Analysis • Choice • MaxDiff • Uplift • Item Analysis
Scripting Guide	Learn about taking advantage of the powerful JMP Scripting Language (JSL).	Covers a variety of topics, such as writing and debugging scripts, manipulating data tables, constructing display boxes, and creating JMP applications.

Document Title	Document Purpose	Document Content
JSL Syntax Reference	Read about many JSL functions on functions and their arguments, and messages that you send to objects and display boxes.	Includes syntax, examples, and notes for JSL commands.

Note: The **Books** menu also contains two reference cards that can be printed: The *Menu Card* describes JMP menus, and the *Quick Reference* describes JMP keyboard shortcuts.

JMP Help

JMP Help is an abbreviated version of the documentation library that provides targeted information. You can open JMP Help in several ways:

- On Windows, press the F1 key to open the Help system window.
- Get help on a specific part of a data table or report window. Select the Help tool from the **Tools** menu and then click anywhere in a data table or report window to see the Help for that area.
- Within a JMP window, click the **Help** button.
- Search and view JMP Help on Windows using the **Help > Help Contents, Search Help**, and **Help Index** options. On Mac, select **Help > JMP Help**.
- Search the Help at http://jmp.com/support/help/ (English only).

Additional Resources for Learning JMP

In addition to JMP documentation and JMP Help, you can also learn about JMP using the following resources:

- Tutorials (see "Tutorials" on page 30)
- Sample data (see "Sample Data Tables" on page 30)
- Indexes (see "Learn about Statistical and JSL Terms" on page 30)
- Tip of the Day (see "Learn JMP Tips and Tricks" on page 30)
- Web resources (see "JMP User Community" on page 31)
- JMPer Cable technical publication (see "JMPer Cable" on page 31)
- Books about JMP (see "JMP Books by Users" on page 31)
- JMP Starter (see "The JMP Starter Window" on page 32)

- Teaching Resources (see "Sample Data Tables" on page 30)

Tutorials

You can access JMP tutorials by selecting **Help > Tutorials**. The first item on the **Tutorials** menu is **Tutorials Directory**. This opens a new window with all the tutorials grouped by category.

If you are not familiar with JMP, then start with the **Beginners Tutorial**. It steps you through the JMP interface and explains the basics of using JMP.

The rest of the tutorials help you with specific aspects of JMP, such as designing an experiment and comparing a sample mean to a constant.

Sample Data Tables

All of the examples in the JMP documentation suite use sample data. Select **Help > Sample Data Library** to open the sample data directory.

To view an alphabetized list of sample data tables or view sample data within categories, select **Help > Sample Data**.

Sample data tables are installed in the following directory:

On Windows: C:\Program Files\SAS\JMP\13\Samples\Data

On Macintosh: \Library\Application Support\JMP\13\Samples\Data

In JMP Pro, sample data is installed in the JMPPRO (rather than JMP) directory. In JMP Shrinkwrap, sample data is installed in the JMPSW directory.

To view examples using sample data, select **Help > Sample Data** and navigate to the Teaching Resources section. To learn more about the teaching resources, visit http://jmp.com/tools.

Learn about Statistical and JSL Terms

The **Help** menu contains the following indexes:

Statistics Index Provides definitions of statistical terms.

Scripting Index Lets you search for information about JSL functions, objects, and display boxes. You can also edit and run sample scripts from the Scripting Index.

Learn JMP Tips and Tricks

When you first start JMP, you see the Tip of the Day window. This window provides tips for using JMP.

To turn off the Tip of the Day, clear the **Show tips at startup** check box. To view it again, select **Help > Tip of the Day**. Or, you can turn it off using the Preferences window. See the *Using JMP* book for details.

Tooltips

JMP provides descriptive tooltips when you place your cursor over items, such as the following:

- Menu or toolbar options
- Labels in graphs
- Text results in the report window (move your cursor in a circle to reveal)
- Files or windows in the Home Window
- Code in the Script Editor

Tip: On Windows, you can hide tooltips in the JMP Preferences. Select **File > Preferences > General** and then deselect **Show menu tips**. This option is not available on Macintosh.

JMP User Community

The JMP User Community provides a range of options to help you learn more about JMP and connect with other JMP users. The learning library of one-page guides, tutorials, and demos is a good place to start. And you can continue your education by registering for a variety of JMP training courses.

Other resources include a discussion forum, sample data and script file exchange, webcasts, and social networking groups.

To access JMP resources on the website, select **Help > JMP User Community** or visit https://community.jmp.com/.

JMPer Cable

The JMPer Cable is a yearly technical publication targeted to users of JMP. The JMPer Cable is available on the JMP website:

http://www.jmp.com/about/newsletters/jmpercable/

JMP Books by Users

Additional books about using JMP that are written by JMP users are available on the JMP website:

http://www.jmp.com/en_us/software/books.html

The JMP Starter Window

The JMP Starter window is a good place to begin if you are not familiar with JMP or data analysis. Options are categorized and described, and you launch them by clicking a button. The JMP Starter window covers many of the options found in the Analyze, Graph, Tables, and File menus. The window also lists JMP Pro features and platforms.

- To open the JMP Starter window, select **View** (**Window** on the Macintosh) > **JMP Starter**.
- To display the JMP Starter automatically when you open JMP on Windows, select **File > Preferences > General**, and then select **JMP Starter** from the Initial JMP Window list. On Macintosh, select **JMP > Preferences > Initial JMP Starter Window**.

Technical Support

JMP technical support is provided by statisticians and engineers educated in SAS and JMP, many of whom have graduate degrees in statistics or other technical disciplines.

Many technical support options are provided at http://www.jmp.com/support, including the technical support phone number.

Chapter **2**

Introduction to DOE

Overview of Design of Experiment Platforms

The JMP DOE platforms help you to design experiments. Most of the platforms focus on constructing designs of various types. Other platforms support the design effort. This chapter gives a quick overview of each of the platforms found under the DOE menu.

Design Construction Platforms

Custom Design Constructs designs that fit a wide variety of settings. Custom designs tend to be more cost effective and flexible than approaches based exclusively on classical designs.

Custom designs accommodate various types of factors, constraints, and disallowed combinations. You can specify which effects are necessary to estimate and which are desirable to estimate, given the number of runs. You can specify a number of runs that matches the budget for your experimental situation. Custom designs also support hard-to-change and very-hard-to-change factors, allowing you to construct split-plot and related designs.

The Custom Design platform constructs many special design types:

- screening
- response surface
- mixture
- random block
- split-plot
- split-split-plot
- two-way split-plot

You can construct classical screening, response surface, and mixture designs using other platforms. However, the Custom Design platform gives you flexibility that is not available in the other platforms. Constructing designs for split-plot situations can be done only using the Custom Design platform.

Definitive Screening Design Constructs screening designs for continuous and two-level categorical factors. Definitive screening designs are useful if you suspect active interactions or curvature. Definitive screening designs enable you to identify the source of strong nonlinear effects while avoiding complete confounding between any effects up through the second order.

Definitive screening designs are small. They have roughly twice as many runs as there are factors. Continuous factors are set at three levels. Definitive screening designs support grouping runs into blocks. The number of blocks is user-specified.

Fit Definitive Screening Design Analyzes definitive screening designs using a methodology called *Effective Model Selection for DSDs*. This methodology takes advantage of the special structure of definitive screening designs.

Screening Design Constructs screening designs for continuous, discrete numeric, and categorical factors with an arbitrary number of levels. When standard designs exist, you have two options:

- Choose from a list of classical screening designs. These designs allow two-level continuous factors or two- or three-level categorical or discrete continuous factors.
- Generate a design that is orthogonal or nearly orthogonal for main effects. *Near-orthogonal designs* allow for categorical and discrete numeric factors with any number of levels, as well as two-level continuous factors. These designs focus on estimating main effects in the presence of negligible interactions.

For many screening situations, standard designs are not available. In these situations, you can construct near-orthogonal screening designs.

Response Surface Design Constructs designs that model a quadratic function of continuous factors. To fit the quadratic effects, response surface designs require three settings for each factor. JMP provides response surface designs for up to eight factors.

You can choose from a list of Central Composite or Box-Behnken designs. When appropriate, Central Composite designs that block orthogonally are included in the list. Various modifications to Central Composite designs are supported.

Full Factorial Design Constructs full factorial designs for any number of continuous or categorical factors, both with arbitrarily many levels. A full factorial design has a run at every combination of settings of the factors. Full factorial designs tend to be large. The number of runs equals the product of the numbers of factor levels.

Mixture Design Constructs designs that you use when factors are ingredients in a mixture. In a mixture experiment, a change in the proportion of one ingredient requires that one or more of the remaining ingredients change to maintain the sum. Choose from among several design types, including some classical mixture design approaches: optimal, simplex centroid, simplex lattice, extreme vertices, ABCD, and space filling. For optimal, extreme vertices, and space filling mixture designs, you can specify linear inequality constraints to limit the design space.

JMP PRO **Covering Array** Constructs combinatorial designs that you can use to test software, networks, and other systems. A strength *t* covering array has the property that every combination of levels of every *t* factors appears in at least one run. Covering arrays allow for any number of categorical factors, each with an arbitrary number of levels. Disallowed combinations can be specified.

Choice Design Constructs designs that you can use to compare prospective products. The factors in a choice design are product attributes. The design arranges product profiles, which are combinations of various attributes, in pairs or in groups of three or four. The experiment consists of having respondents indicate which profile in a pair of profiles that they prefer. You can generate a choice design that reflects prior information about the product attributes.

MaxDiff Constructs a design consisting of choice sets that can be presented to respondents as part of a MaxDiff study. Respondents report only the most and least preferred options from among a small set of choices. This forces respondents to rank options in terms of preference, which often results in rankings that are more definitive than rankings obtained using standard preference scales.

Space Filling Design Constructs designs for situations where the system of interest is deterministic or near-deterministic. A standard application involves creating a simpler surrogate model of a highly complex deterministic computer simulation model.

In a deterministic system, there is no variation. The goal is to minimize the difference between the fitted model and the true model (bias). Space-filling designs attempt to meet this goal either by spreading the design points out as far from each other as possible or by spacing the points evenly over the design region.

JMP provides seven space-filling design approaches. One of these approaches, the fast flexible filling design, accommodates categorical factors with any number of levels and supports linear constraints.

Accelerated Life Test Design Constructs and augments designs useful for testing products at extreme conditions which are intended to accelerate failure time. Use experimental results to predict reliability under normal operating conditions.

The life distribution can be lognormal or Weibull. Designs can include one or two accelerating factors. If there are two accelerating factors, you can choose to include their interaction. You can specify prior distributions for the acceleration model parameters. D-optimal and two types of I-optimal designs are available.

Nonlinear Design Constructs and augments designs that you use to fit models that are nonlinear in their parameters. You can construct a design using estimates from a model fit to existing data. You can also construct a design by applying prior knowledge if you do not have model-based estimates.

Taguchi Arrays Constructs designs that you use for signal-to-noise analysis. The designs are based on Taguchi's inner and outer array approach. Control factor settings constitute the inner array and noise factor settings form the outer array. The mean and signal-to-noise ratio are the responses of interest.

An alternative to using a Taguchi array is to construct a custom design that includes control factors, noise factors, and control-by-noise interactions. Such designs, called *combined arrays*, are generally more cost-effective and informative than Taguchi arrays.

Supporting Platforms

Evaluate Design Provides diagnostics for an existing experimental design. The Evaluate Design platform provides various ways for you to assess the strengths and limitations of your design. The platform can be used with any data table, not only designs created using JMP.

Several diagnostics are provided:

- power analysis
- prediction variance plots
- estimation efficiency for parameters
- the alias matrix, showing the bias structure for model effects
- a color map showing absolute correlations among effects
- design efficiency values

Compare Designs Compares two or three designs simultaneously to explore and evaluate their performance. Diagnostics show how the designs perform relative to each other and how they perform in an absolute sense.

Augment Design Adds runs to existing designs in such a way that the resulting design is optimal. Augment Design enables you to conduct experiments in an iterative fashion. You can replicate the design, add center points, create a fold-over design, add axial points, add points to create a space-filling design, or augment the design with a specified number of runs. You can group runs into blocks to distinguish the original runs from the augmented runs. You can add model effects that were not in the original model and specify estimability requirements for these effects.

Sample Size and Power Provides sample size and power calculations for a variety of testing situations: one or more sample means, a standard deviation, one or two proportions, counts per unit (Poisson mean), and sigma quality level. For these options, you specify two of three quantities to compute the third. These three quantities are the difference you want to detect, the sample size, and the power. If you supply only one of these values, a plot of the relationship between the other two values is provided.

You can compute the sample size required for a reliability test plan, where your goal is to estimate failure probabilities. You can also compute the sample size required for a reliability demonstration, where your goal is to demonstrate that a product meets or exceeds a specified standard.

Chapter 3

Starting Out with DOE

Example and Key Concepts

A designed experiment is a controlled set of tests designed to model and explore the relationship between factors and one or more responses. JMP includes a variety of tools that enable you to create efficient experimental designs that work for your situation. In particular, these classes of designs are available:

- The Custom Design platform customizes a design for your unique situation. It constructs designs that accommodate any number of factors of any type and factors that are difficult to change (split plot situations). You control the number of runs.
- The Definitive Screening Design platform constructs an innovative class of screening designs where main effects are not aliased with each other or with two-way interactions. These designs also allow estimation of quadratic terms.
- The Screening Design, Response Surface Design, Full Factorial Design, Mixture Design, and Taguchi Arrays platforms construct traditional experimental designs.
- The Covering Array, Choice Design, Space Filling Design, Accelerated Life Test Design, and Nonlinear Design platforms construct specialized design types.

The Evaluate Design and Augment Design platforms provide tools for evaluating and augmenting existing design. The Sample Size and Power platform addresses sample size and power calculations for specialized situations.

This chapter presents an example that illustrates the JMP approach to DOE. This chapter also discusses the framework for DOE, the workflow that supports design creation, and principles that are fundamental to DOE.

Figure 3.1 Example of a Profiler Plot

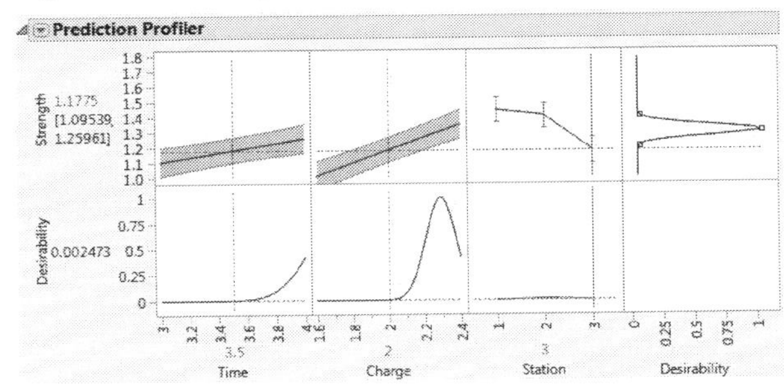

Overview of Experimental Design and the DOE Workflow

A six-step framework provides the structure for designing an experiment, running the experimental trials, and analyzing the results. Sound engineering and process knowledge is critical to all of these steps.

Figure 3.2 Framework for Experimental Design

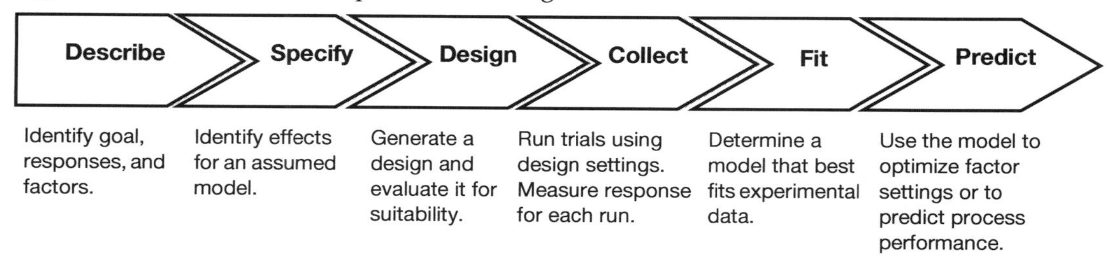

You perform the first three steps in the DOE platforms. The end result is a design that can be run in your work environment. For a detailed description of the workflow for these three steps, see "The DOE Workflow: Describe, Specify, Design" on page 54.

Describe Determine the goal of your experiment. Identify responses and factors.

Your goal might be to identify active factors, to find optimal factor settings, or to build a predictive model.

Specify Determine or specify an assumed model that you believe adequately describes the physical situation.

Your *assumed* model is an initial model that ideally contains all the effects that you want to estimate. In some platforms, you can explicitly build the model of interest. In others, the model is implicit in the choices that you make. For example, in the Screening Design platform, you might select a model with a given resolution. The resolution of the design determines which effects are confounded. Confounding of effects potentially leads to ambiguity about which effect is truly active.

Design Generate a design that is consistent with your assumed model. Evaluate this design to understand its strengths and limitations, and to ensure that it provides the information that you need, given your model and goals.

The Design Evaluation or Design Diagnostics outline in the design generation platform give you insight about the properties of your design.

The next step is the data collection phase, where the experiment is run under controlled conditions.

Collect Conduct each of the trials and record the response values.

After you run your experiment, scripts in the generated data table help you fit a model using platforms such as Fit Model and Screening. Depending on your goal, the model can help you identify active effects or find optimal settings.

Fit Fit your assumed model to the experimental data.

Use the JMP modeling platforms to fit and refine your model. In some situations, you might need to augment the design and perform additional runs to resolve model ambiguity.

Predict Use your refined model to address your experimental goals.

Determine which effects are active, find factor levels to optimize responses, or build a predictive model.

Designed experiments are typically used sequentially to construct process knowledge. A design strategy often begins with a screening design to narrow the list of potentially active factors. Then the identified factors are studied in designs that focus on building a better understanding of interactions and quadratic effects. Sometimes there is a need to augment a design to resolve ambiguities relating to the factors responsible for effects. The steps outlined in this section relate to conducting and analyzing a single experiment. However, you may require a sequence of experiments to achieve your goals.

The example in "The Coffee Strength Experiment" on page 39 explicitly illustrates the steps in the DOE workflow process. It also shows how to use a data table script to analyze your experimental data. Many examples in the *Design of Experiments Guide* illustrate both the workflow that supports a good design and the analysis of the experimental data from the study.

The Coffee Strength Experiment

This example contains the following processes:

• "Define the Study and Goals" on page 39
• "Create the Design" on page 41
• "Run the Experiment" on page 47
• "Analyze the Data" on page 48

Define the Study and Goals

Your employer is a local mid-size coffee roaster. You need to address the strength of individually brewed twelve ounce cups of coffee. Your goal is to determine which factors have an effect on coffee strength and to find optimal settings for those factors.

Response

The response is coffee Strength. It is measured as total dissolved solids, using a refractometer. The coffee is brewed using a single cup coffee dripper and measured five minutes after the liquid is released from the grounds.

Previous studies indicate that a strength reading of 1.3 is most desirable, though the strength is still acceptable if it falls between 1.2 and 1.4.

Factors

Four factors are identified for the study: Grind, Temperature, Time, and Charge. Coffee is brewed at three stations in the work area. To account for variation due to brewing location, Station is included in the study as a blocking factor. The following describes the factors:

- Grind is the coarseness of the grind. Grind is set at two levels, Medium and Coarse.
- Temperature is the temperature in degrees Fahrenheit of the water measured immediately before pouring it over the grounds. Temperature is set at 195 and 205 degrees Fahrenheit.
- Time is the brewing time in minutes. Time is set at 3 or 4 minutes.
- Charge is the amount of coffee placed in the cone filter, measured in grams of coffee beans per ounce of water. Charge is set at 1.6 and 2.4.
- Station is the location where the coffee is brewed. The three stations are labeled as 1, 2, and 3.

Table 3.1 summarizes information about the factors and their settings. The factors and levels are also given in the Coffee Factors.jmp sample data table, located in the Design Experiment folder.

Table 3.1 Factors and Range of Settings for Coffee Experiment

Factor	Role	Range of Settings
Grind	Categorical	Medium, Coarse
Temperature	Continuous	195 - 205
Time	Continuous	3 - 4
Charge	Continuous	1.6 - 2.4
Station	Blocking	1, 2, 3

Note the following:

- Grind is categorical with two levels.
- Temperature, Time, and Charge are continuous.
- Station is a blocking factor with three levels.

All factors can be varied and reset for each run. There are no hard-to-change factors for this experiment.

The apparatus used in running the coffee experiment is shown in Figure 3.3. This is the setup at one of the three brewing stations. The two other stations have the same type of equipment.

Figure 3.3 Coffee Experiment Apparatus

Number of Runs

Based on the resources and time available, you determine that you can conduct 12 runs in all. Since there are three stations, you conduct 4 runs at each station.

Create the Design

Create the design following the steps in the design workflow process outlined in "The DOE Workflow: Describe, Specify, Design" on page 54:

- "Define Responses and Factors" on page 55
- "Specify the Model" on page 56
- "Generate the Design" on page 57
- "Evaluate the Design" on page 58
- "Make the Table" on page 58

Define Responses and Factors

In the first outlines that appear, enter information about your response and factors.

Responses

1. Select **DOE > Custom Design.**
2. Double-click Y under Response Name and type Strength.

   Note that the default Goal is **Maximize.** Your goal is to find factor settings that enable you to brew coffee with a target strength of 1.3, within limits of 1.2 and 1.4.
3. Click on the default Goal of **Maximize** and change it to **Match Target.**

Figure 3.4 Selection of Match Target as the Goal

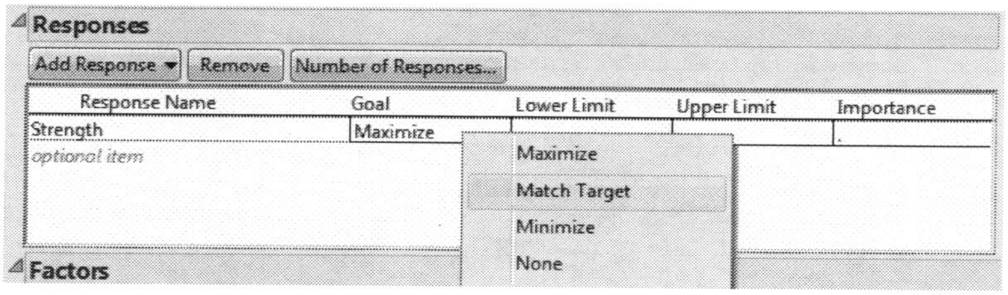

4. Click under **Lower Limit** and type 1.2.
5. Click under **Upper Limit** and type 1.4.
6. Leave the area under **Importance** blank.

   Because there is only one response, that response is given Importance 1 by default.

The completed Responses outline appears in Figure 3.5.

Factors

Enter factors either manually or from a pre-existing table that contains the factors and settings. If you are designing a new experiment, you must first enter the factors manually. Once you have saved the factors to a data table using the Save Factors option, you can load them using the saved table.

For this example, you can choose either option. See "Entering Factors Manually" on page 42 or see "Entering Factors Using Load Factors" on page 44.

Entering Factors Manually

1. Click **Add Factor > Categorical > 2 Level.**
2. Type Grind over the default Name of X1.

Note that Role is set to Categorical, as requested. The Changes attribute is set to Easy by default, indicating that **Grind** settings can be reset for every run.

3. Click the default Values, L1 and L2, and change them to Coarse and Medium.
4. Type 3 next to **Add N Factors.** Then click **Add Factor > Continuous.**
5. Type the following names and values over the default entries:
   – **Temperature** (195 and 205)
   – **Time** (3 and 4)
   – **Charge** (1.6 and 2.4)
6. Click **Add Factor > Blocking > 4 runs per block.**

   Recall that your run budget allows for 12 runs. You want to balance these runs among the three stations.
7. Type **Station** over the default Name of **X5**.

   Notice that Role is set to Blocking and that only one setting for Values appears. This is because JMP cannot determine the number of blocks until the desired number of runs is specified. Once you specify the Number of Runs in the Design Generation outline, JMP updates the number of levels for **Station** to what is required.

The completed Factors outline is shown in Figure 3.5.

Figure 3.5 Completed Responses and Factors Outlines

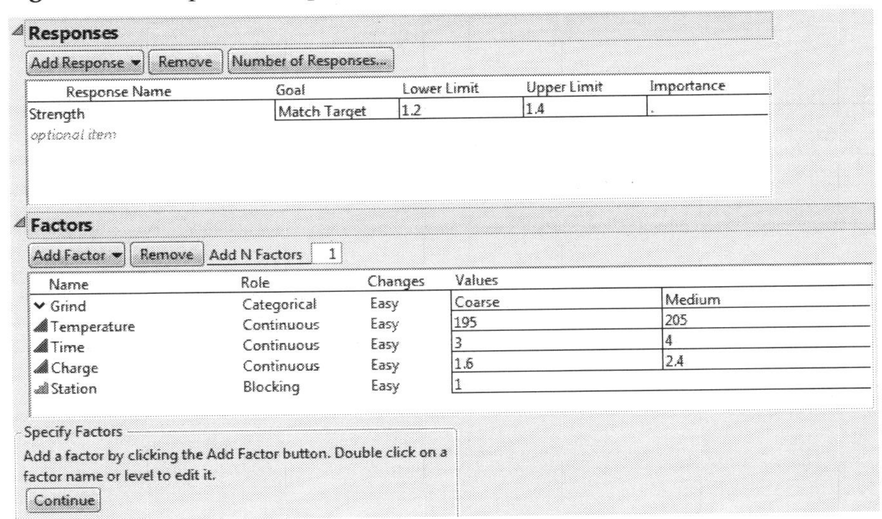

8. Click **Continue.**

   The following outlines are added to the Custom Design window:

   – Define Factor Constraints (not used in this example)

– Model
– Alias Terms
– Design Generation

Entering Factors Using Load Factors

To enter factors using a table containing factor information, proceed as follows:

1. From the Custom Design red triangle menu, select **Load Factors**.
2. Select **Help > Sample Data Library** and open Design Experiment/Coffee Factors.jmp.

   After loading the factors, the Custom Design window is updated. The following outlines are added to the Custom Design window:

   – Define Factor Constraints (not used in this example)
   – Model
   – Alias Terms
   – Design Generation

Define Factor Constraints

The Define Factor Constraints outline appears once you have entered your factors manually and clicked Continue, or once you have loaded the factors from the factor table. Adding factor constraints, if you have any, is part of the Responses and Factors step. Since there are no constraints on factor settings for this design, leave this outline unchanged.

Specify the Model

Model Outline

Figure 3.7 shows the Model outline. The Model outline is where you specify your assumed model, which contains the effects that you want to estimate. See "Specify" on page 38. The list that appears by default shows all main effects as **Necessary**, indicating that the design is capable of estimating all main effects. Because your main interest at this point is in the main effects of the factors, you do not add any effects to the Model outline.

Figure 3.6 Model Outline with Main Effects Only

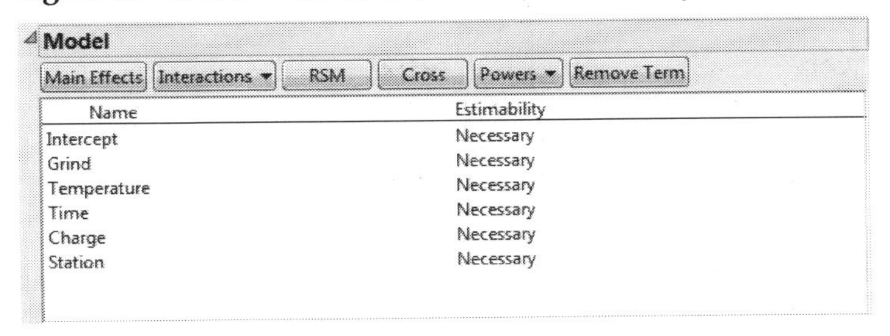

Steps to Duplicate Results (Optional)

Because the Custom Design algorithm begins with a random starting design, your design might differ from the one shown in Figure 3.8. To obtain a design with exactly the same runs, perform the following steps before generating your design:

1. From the Custom Design red triangle menu, select **Set Random Seed**.
2. Type 569534903.
3. Click **OK**.
4. From the Custom Design red triangle menu, select **Number of Starts**.
5. Type 100.
6. Click **OK**.

Note: Setting the Random Seed and Number of Starts reproduces the exact design shown in this example. However, the rows in the design table might be in a different order. In constructing a design on your own, these steps are not necessary.

Generate the Design

In the Design Generation outline, you can enter additional details about the structure and size of your design. The Default design is shown as having 12 runs. Recall that your design budget allows for 12 runs ("Number of Runs" on page 41).

Figure 3.7 Design Generation Outline

Design Generation

Number of Center Points: 0
Number of Replicate Runs: 0

Number of Runs:
Minimum 6
Default 12
User Specified 12
Make Design

1. Click **Make Design**.

   The Design and Design Evaluation outlines are added to the Custom Design window. The Output Options panel also appears.

   The Design outline shows the design (Figure 3.8). If you did not follow the steps in "Steps to Duplicate Results (Optional)" on page 45, your design might be different from the one in Figure 3.8. This is because the algorithm begins with a random starting design.

Figure 3.8 Design for Coffee Experiment

Design

Run	Grind	Temperature	Time	Charge	Station
1	Coarse	195	3	1.6	1
2	Medium	195	4	2.4	2
3	Medium	195	4	1.6	3
4	Medium	205	4	2.4	1
5	Coarse	205	3	2.4	2
6	Medium	205	3	2.4	3
7	Medium	205	3	1.6	1
8	Coarse	205	4	1.6	2
9	Coarse	205	4	1.6	3
10	Coarse	195	4	2.4	1
11	Medium	195	3	1.6	2
12	Coarse	195	3	2.4	3

Evaluate the Design

The Design Evaluation outline provides various ways to evaluate your design. This is an important topic, but for simplicity, it is not covered in the context of this example. See the "Evaluate Designs" chapter on page 419.

Make the Table

Specify the order of runs in your data table using the Output Options panel. The default selection, **Randomize within Blocks**, is appropriate. This selection arranges the runs in a random order for each Station.

Figure 3.9 Output Options

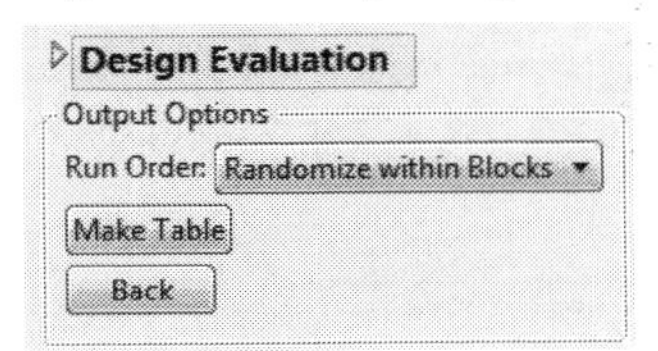

1. Click **Make Table.**

   The data table shown in Figure 3.10 opens. Keep in mind that, if you did not follow the steps in "Steps to Duplicate Results (Optional)" on page 45, your design table might be different. Your design table represents another optimal design.

Figure 3.10 Custom Design Table

Custom Design
Design: Custom Design
Criterion: D Optimal
▶ Model
▶ Evaluate Design
▶ DOE Dialog

Columns (6/0)
Grind *
Temperature *
Time *
Charge *
Station *
Strength *

	Grind	Temperature	Time	Charge	Station	Strength
1	Coarse	195	4	2.4	1	•
2	Medium	205	3	1.6	1	•
3	Medium	205	4	2.4	1	•
4	Coarse	195	3	1.6	1	•
5	Medium	195	4	2.4	2	•
6	Medium	195	3	1.6	2	•
7	Coarse	205	4	1.6	2	•
8	Coarse	205	3	2.4	2	•
9	Medium	195	4	1.6	3	•
10	Medium	205	3	2.4	3	•
11	Coarse	205	4	1.6	3	•
12	Coarse	195	3	2.4	3	•

Note the asterisks in the Columns panel to the right of the factors and response. These indicate column properties that have been saved to the columns in the data table. These column properties are used in the analysis of the data. For more information, see "Factors" on page 85 and "Factor Column Properties" on page 91.

Run the Experiment

At this point, you perform the experiment. At each Station, four runs are conducted in the order shown in the design table. Equipment and material are reset between runs. For example, if two consecutive runs require water at 195 degrees, separate 12-ounce batches of water are heated to 195 degrees after the heating container cools. The Strength measurements are recorded.

Your design and the experimental results for Strength are given in the Coffee Data.jmp sample data table (Figure 3.11), located in the Design Experiment folder.

Figure 3.11 Coffee Design with Strength Results

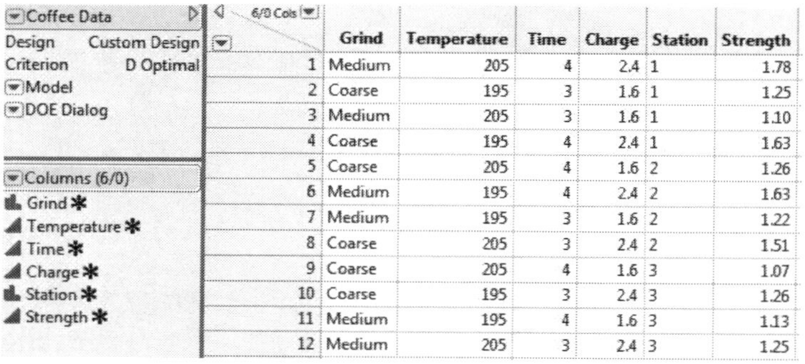

Coffee Data
Design Custom Design
Criterion D Optimal
Model
DOE Dialog

Columns (6/0)
Grind
Temperature
Time
Charge
Station
Strength

	Grind	Temperature	Time	Charge	Station	Strength
1	Medium	205	4	2.4	1	1.78
2	Coarse	195	3	1.6	1	1.25
3	Medium	205	3	1.6	1	1.10
4	Coarse	195	4	2.4	1	1.63
5	Coarse	205	4	1.6	2	1.26
6	Medium	195	4	2.4	2	1.63
7	Medium	195	3	1.6	2	1.22
8	Coarse	205	3	2.4	2	1.51
9	Coarse	205	4	1.6	3	1.07
10	Coarse	195	3	2.4	3	1.26
11	Medium	195	4	1.6	3	1.13
12	Medium	205	3	2.4	3	1.25

Analyze the Data

The Custom Design platform facilitates the task of data analysis by saving a Model script to the design table that it creates. See Figure 3.10. Run this script after you conduct your experiment and enter your data. The script opens a Fit Model window containing the effects that you specified in the Model outline of the Custom Design window.

Fit the Model

1. Select **Help > Sample Data Library** and open Design Experiment/Coffee Data.jmp.

   In the Table panel, notice the Model script created by Custom Design.

2. Click the green triangle next to the Model script.

   The Model Specification window shows the effects that you specified in the Model outline.

Figure 3.12 Model Specification Window

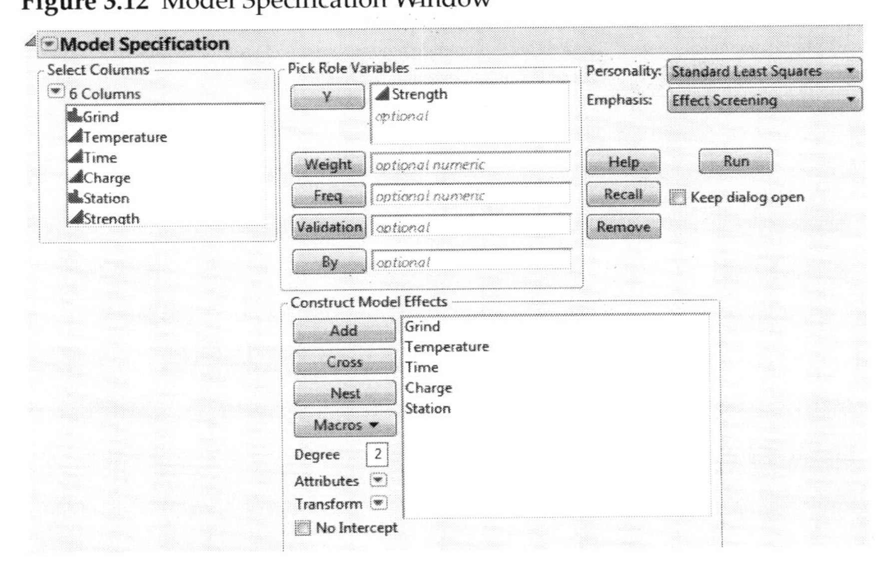

3. Select the **Keep dialog open** option.
4. Click **Run**.

Analyze the Model

The Effect Summary and Actual by Predicted Plot reports give high-level information about the model.

Figure 3.13 Effect Summary and Actual by Predicted Plot for Full Model

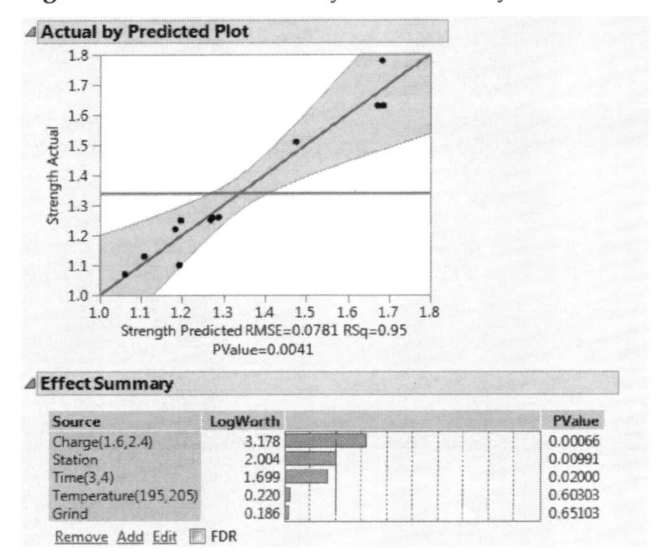

Note the following:

- The Actual by Predicted Plot shows no evidence of lack of fit.
- The model is significant, as indicated by the Actual by Predicted Plot. The notation **P = 0.0041**, shown below the plot, gives the significance level of the overall model test.
- The Effect Summary report shows that Charge, Station, and Time are significant at the 0.05 level.
- The Effect Summary report also shows that Temperature and Grind are not significant.

Reduce the Model

Because Temperature and Grind appear not to be active, they contribute random noise to the model. Refit the model without these effects to obtain more precise estimates of the model parameters associated with the active effects.

1. In the Model Specification window, select Temperature and Grind in the Construct Model Effects list.
2. Click **Remove.**
3. Change the **Emphasis** to **Effect Screening.**

   The Effect Screening emphasis presents reports (such as the Prediction Profiler) that are useful for analyzing experimental designs.
4. Click **Run.**

Figure 3.14 Partial Results for Reduced Model

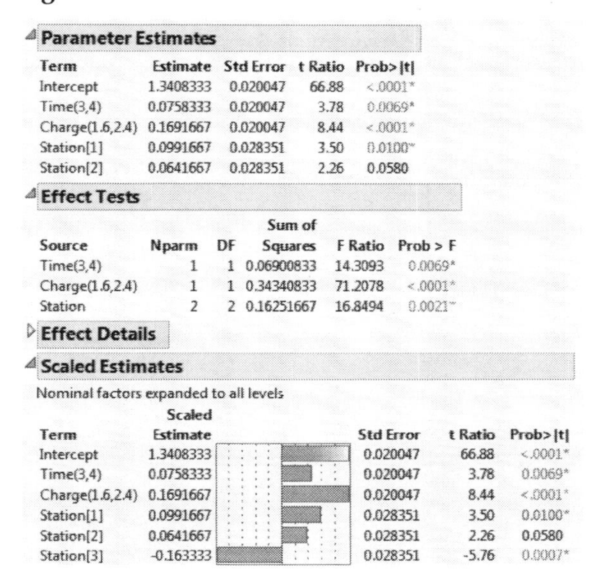

Parameter Estimates

Term	Estimate	Std Error	t Ratio	Prob>|t|
Intercept	1.3408333	0.020047	66.88	<.0001*
Time(3,4)	0.0758333	0.020047	3.78	0.0069*
Charge(1.6,2.4)	0.1691667	0.020047	8.44	<.0001*
Station[1]	0.0991667	0.028351	3.50	0.0100*
Station[2]	0.0641667	0.028351	2.26	0.0580

Effect Tests

Source	Nparm	DF	Sum of Squares	F Ratio	Prob > F
Time(3,4)	1	1	0.06900833	14.3093	0.0069*
Charge(1.6,2.4)	1	1	0.34340833	71.2078	<.0001*
Station	2	2	0.16251667	16.8494	0.0021*

Effect Details

Scaled Estimates

Nominal factors expanded to all levels

Term	Scaled Estimate	Std Error	t Ratio	Prob>|t|
Intercept	1.3408333	0.020047	66.88	<.0001*
Time(3,4)	0.0758333	0.020047	3.78	0.0069*
Charge(1.6,2.4)	0.1691667	0.020047	8.44	<.0001*
Station[1]	0.0991667	0.028351	3.50	0.0100*
Station[2]	0.0641667	0.028351	2.26	0.0580
Station[3]	-0.163333	0.028351	-5.76	0.0007*

Note the following:

- The Effect Tests report shows that all three effects remain significant.
- The Scaled Estimates report further indicates that the Station[1]and Station[3] means differ significantly from the average response of Strength.
- Note that the Estimates that appear in the Parameter Estimates report are identical to their counterparts in the Scaled Estimates report. This is because the effects are coded. See "Coding" on page 663 in the "Column Properties" appendix.
- The estimate of the Station[3] effect only appears in the Scaled Estimates report, where nominal factors are expanded to show estimates for all their levels.
- The Parameter Estimates report gives estimates for the model coefficients where the model is specified in terms of the coded effects.

Explore the Model

The Prediction Profiler appears at the bottom of the report.

Figure 3.15 Prediction Profiler

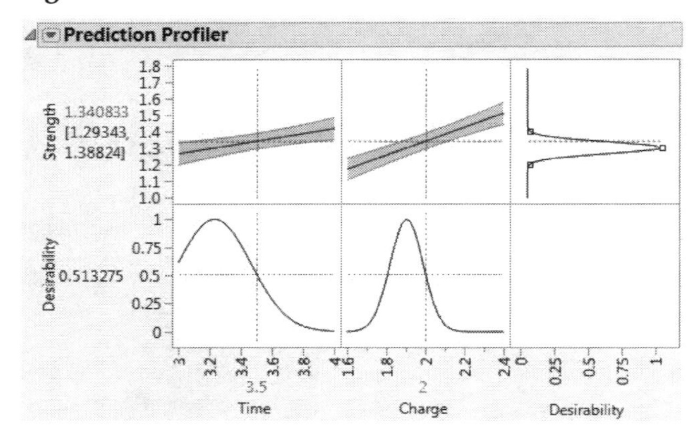

Recall that, in designing your experiment, you set a response Goal of Match Target with limits of 1.2 and 1.4. JMP uses this information to construct a desirability function to reflect your specifications. For more details, see "Factors" on page 85.

Note the following in Figure 3.15:

- The first two plots in the top row of the graph show how Strength varies for one of the factors, given the setting of the other factor. For example, when Charge is 2, the line in the plot for Time shows how predicted Strength changes with Time.
- The values to the left of the top row of plots give the Predicted Strength (in red) and a confidence interval for the mean Strength for the selected factor settings.
- The right-most plot in the top row shows the desirability function for Strength. The desirability function indicates that the target of 1.3 is most desirable. Desirability decreases as you move away from that target. Desirability is close to 0 at the limits of 1.2 and 1.4.
- The plots in the bottom row show the desirability trace for each factor at the setting of the other factor.
- The value to the left of the bottom row of plots gives the Desirability of the response value for the selected factor settings.

Explore various factor settings by dragging the red dashed vertical lines in the columns for Time and Charge. Since there are no interactions in the model, the profiler indicates that increasing Charge increases Strength. Also, Strength seems to be more sensitive to changes in Charge than to changes in Time.

Since Station is a blocking factor, it does not appear in the Prediction Profiler. However, you might like to see how predicted Strength varies by Station. To include Station in the Prediction Profiler, follow these steps:

1. From the Prediction Profiler red triangle menu, select **Reset Factor Grid**.

A Factor Settings window appears with columns for Time, Charge, and Station. Under Station, notice that the box corresponding to **Show** is not selected. This indicates that Station is not shown in the Prediction Profiler.

2. Select the box under Station in the row corresponding to **Show**.
3. Deselect the box under Station in the row corresponding to **Lock Factor Setting**.

Figure 3.16 Factor Settings Window

Factor	Time	Charge	Station
Current Value:	3.5	2	
Minimum Setting:	3	1.6	
Maximum Setting:	4	2.4	
Number of Plotted Points:	41	41	
Show	☑	☑	☑
Lock Factor Setting:	☐	☐	☐

OK Cancel

4. Click **OK**.

   Plots for Station appear in the Prediction Profiler.

5. Click in either plot above Station to insert a dashed red vertical line.
6. Move the dashed red vertical line to Station 1.

Figure 3.17 Prediction Profiler Showing Results for Station 1

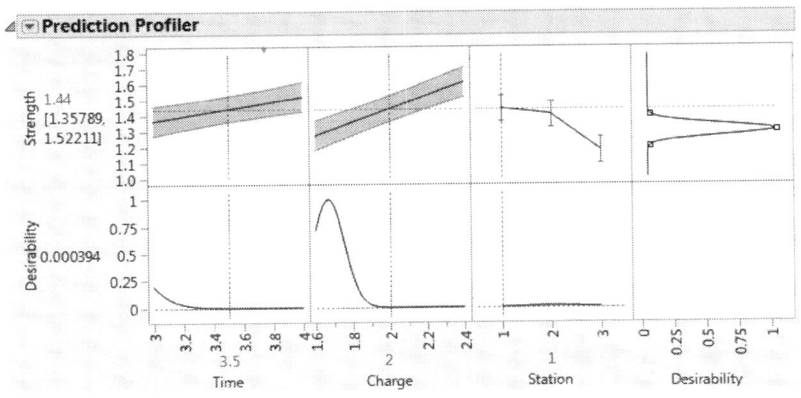

7. Move the dashed red vertical line to Station 3.

Figure 3.18 Prediction Profiler Showing Results for Station 3

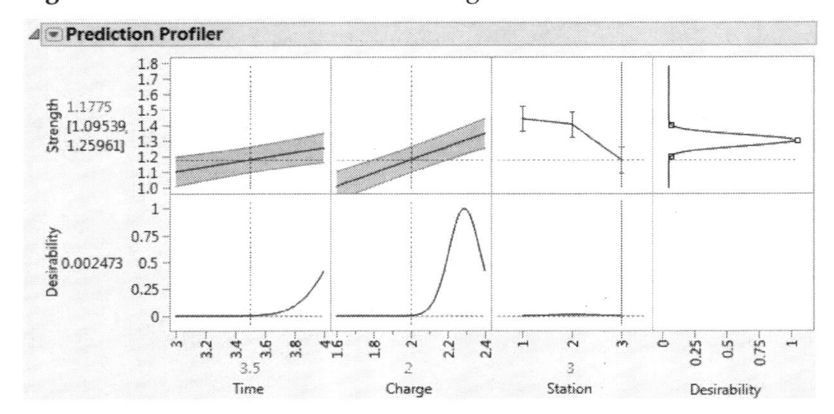

The predicted Strength in the center of the design region for Station 1 is 1.44. For Station 3, the predicted Strength is about 1.18. The magnitude of the difference indicates that you need to address Station variability. Better control of Station variation should lead to more consistent Strength. Once Station consistency is achieved, you can determine common optimal settings for Time and Charge.

The process that you used to construct the design for the coffee experiment followed the steps in the DOE workflow. The next section describes the DOE workflow in more detail.

The DOE Workflow: Describe, Specify, Design

The DOE platforms are structured as a series of steps that present the workflow that is intrinsic to designing experiments. Once you complete each step, you click **Continue** to move to the next step. The elements described in this section are common to nine of the design of experiments platforms. These are the platforms that are addressed in this section:

- Custom Design
- Definitive Screening Design
- Screening Design
- Response Surface Design
- Full Factorial Design
- Mixture Design
- JMP PRO Covering Array
- Space Filling Design
- Taguchi Arrays

Three special-purpose platforms differ substantially: Choice Designs, Accelerated Life Test Design, and Nonlinear Design. These three platforms are not addressed in this section.

This section describes the steps in the DOE workflow. It also discusses their implementation in the various design platforms.

1. "Define Responses and Factors" on page 55.
2. "Specify the Model" on page 56
3. "Generate the Design" on page 57
4. "Evaluate the Design" on page 58
5. "Make the Table" on page 58

Define Responses and Factors

In the Describe step of the experimental design framework:

- You identify the responses and factors of interest.
- You determine your goals for the experiment. Do you want to maximize the response, or hit a target? What is that target? Or do you simply want to identify which factors have an effect on the response?
- You identify factor settings that describe your experimental range or design space.

When they open, most of the JMP DOE platforms display outlines where you can list your responses and your factors. The Responses outline is common across platforms. There you insert your responses and additional information, such as the response goal, lower limit, upper limit, and importance.

The Factors outline varies across platforms. This is to accommodate the types of factors and specific design situations that each platform addresses. In certain platforms, once responses and factors are entered, a Define Factor Constraints outline appears after you click Continue. In this outline, you can constrain the values of the factors that are available for the design.

Figure 3.19 shows the Responses and Factors outline using the Custom Design platform for constructing the design in the Box Corrosion Split-Plot.jmp sample data table, located in the Design Experiment folder. Also shown is the Define Factor Constraints outline, which appears once you click Continue. The Define Constraints outline enables you to specify restrictions that your factor settings must satisfy.

Figure 3.19 Responses and Factors for Box Corrosion Split-Plot Experiment

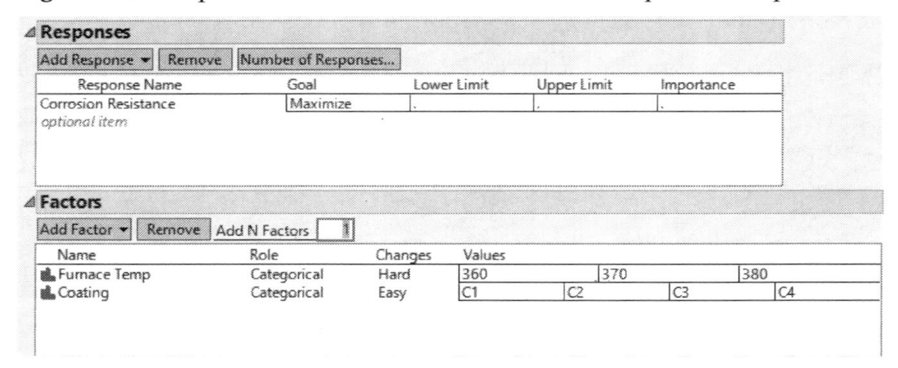

Specify the Model

Once you have completed filling in the Responses and Factors outlines, click the **Continue** button. This brings you to the next phase of design construction, where you either explicitly or implicitly choose an assumed model.

The Custom Design platform enables you to explicitly specify the model that you want to fit. The design that is generated is optimal for this model. The other design platforms do not allow you to explicitly specify your model. For example, in the screening platform, one option enables you to choose from a list of full factorial, fractional factorial, and Plackett-Burman designs. The aliasing relationships in these designs implicitly define the models that you can fit.

In Custom Design, when you click Continue after filling in the Responses and Factors, you see the Model outline. An example, for the design used in the Box Corrosion Split-Plot.jmp sample data table, is shown in Figure 3.20. The assumed model requires that the Furnace Temp and Coating main effects, and their interaction, be estimable. The design that is generated guarantees estimability of these effects.

In most other platforms, clicking Continue gives you a collection of designs to choose from. In Full Factorial, Continue takes you directly to Output Options, since the design is determined once the Factors outline is completed.

Figure 3.20 Model Outline for Box Corrosion Split-Plot Experiment

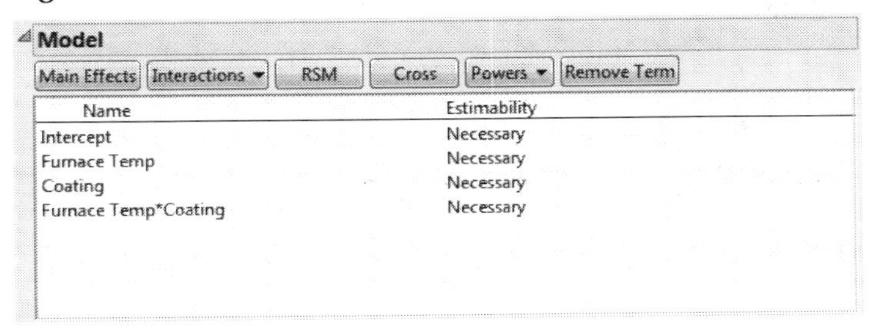

Generate the Design

Most of the DOE platforms give you some control over the size of the final design. In Custom Design, you can specify the number of runs and, when appropriate, the number of center points and replicate runs. In other platforms, you have various degrees of flexibility. Often you can specify the number of center points, replicate runs, or replicates of the design.

Once you have specified your options in terms of the number of runs, click Make Design. The DOE window is updated to show your design in a Design outline.

The Design outline for a 24-run custom design for the Box Corrosion Split-Plot.jmp experiment is shown in Figure 3.21. Because **Changes** for Furnace Temp was specified as **Hard**, a Whole Plots factor is constructed to represent the random blocks of settings for Furnace Temp.

Figure 3.21 Design Outline for Box Corrosion Split-Plot Experiment

Design

Run	Whole Plots	Furnace Temp	Coating
1	1	370	C2
2	1	370	C3
3	1	370	C1
4	1	370	C4
5	2	370	C3
6	2	370	C4
7	2	370	C2
8	2	370	C1
9	3	360	C3
10	3	360	C2
11	3	360	C4
12	3	360	C1
13	4	380	C4
14	4	380	C3
15	4	380	C2
16	4	380	C1
17	5	380	C4
18	5	380	C2
19	5	380	C1
20	5	380	C3
21	6	360	C3
22	6	360	C1
23	6	360	C4
24	6	360	C2

Evaluate the Design

When you click Make Design, in most platforms, a Design Evaluation outline appears. Here you can explore the design that you created in terms of the following: its power to detect effects, its prediction variance, its estimation efficiency, its aliasing relationships, the correlations between effects, and other design efficiency measures. The Design Evaluation outline for a Custom Design is shown in Figure 3.22. Design Evaluation is covered in the "Evaluate Designs" chapter on page 419.

For some platforms, other types of design diagnostics are appropriate. For example, Space Filling Design provides a Design Diagnostics outline with metrics specific to space-filling designs. Covering Array provides a Metrics outline with measures that are specific to coverage.

Figure 3.22 Design Evaluation Outline in Custom Design

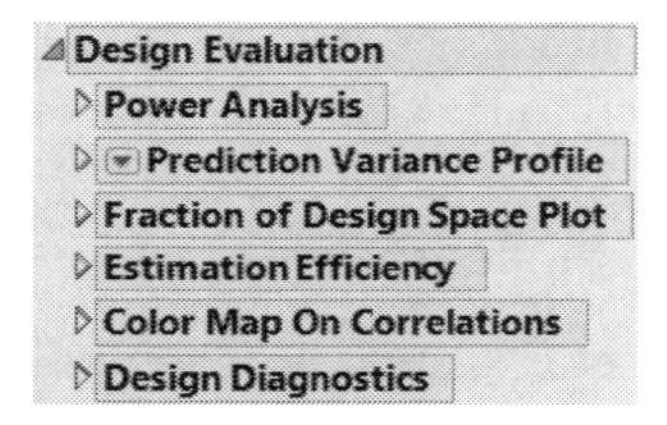

Make the Table

Most platforms provide an Output Options node or panel. Depending on the platform and the design, you can use the Output Options panel to specify additional design structure. For example, you can specify the number of runs, center points, replicates, or the order in which you want the design runs to appear in the generated data table.

The Output Options panel shown in Figure 3.23 is for the experiment in the Wine.jmp sample data table, located in the Design Experiment folder. In this example, you can choose various Run Order options and construct the design data table. Or, you can choose to go Back and restructure your design.

Figure 3.23 Output Options Panel for Wine Experiment

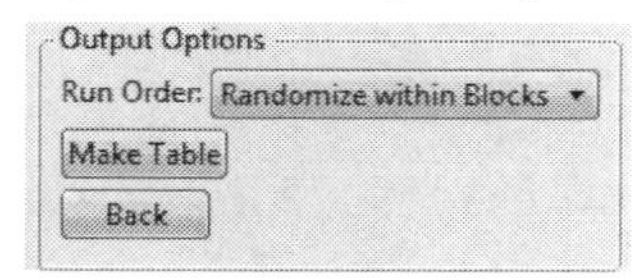

Principles and Guidelines for Experimental Design

Certain principles underlie the design of experiments and the analysis of experimental data. The principles of effect hierarchy, effect heredity, and effect sparsity relate primarily to model selection. These principles help you reduce the set of possible models in searching for a best model. For details, see Hamada and Wu, 1992, Wu and Hamada, 2009, and Goos and Jones, 2011.

Effect Hierarchy

In regression modeling, the principle of *effect hierarchy* maintains that main (first-order) effects tend to account for the largest amounts of variation in the response. Second-order effects, that is, interaction effects and quadratic terms, are next in terms of accounting for variation. Then come higher-order terms, in hierarchical order.

Here are the implications for modeling: main effects are more likely to be important than second-order effects; second-order effects are more likely to be important than third-order effects; and so on, for higher-order terms.

Effect Heredity

The principle of *effect heredity* relates to the inclusion in the model of lower-order components of higher-order effects. The motivation for this principle is observational evidence that factors with small main effects tend not to have significant interaction effects.

Strong effect heredity requires that all lower-order components of a model effect be included in the model. Suppose that a three-way interaction (ABC) is in the model. Then all of its component main effects and two-way interactions (A, B, C, AB, AC, BC) must also be in the model.

Weak effect heredity requires that only a sequence of lower-order components of a model effect be included. If a three-way interaction is in the model, then the model must contain one of the factors involved and one two-way interaction involving that factor. Suppose that the three-way interaction ABC is in the model. Then if B and BC are also in the model, the model satisfies weak effect heredity.

For continuous factors, effect heredity ensures that the model is invariant to changes in the location and scale of the factors.

Effect Sparsity

The principle of *effect sparsity* asserts that most of the variation in the response is explained by a relatively small number of effects. Screening designs, where many effects are studied, rely heavily on effect sparsity. Experience shows that the number of runs used in a screening design should be at least twice the number of effects that are likely to be significant.

Center Points, Replicate Runs, and Testing

Several DOE platforms enable you to add center points (for continuous factors), replicate runs, or full replicates of the design, to your design. Here is some background relative to adding design points.

Adding Center Points

Center points for continuous factors enable you to test for lack of fit due to nonlinear effects. Testing for lack of fit helps you determine whether the error variance estimate has been inflated due to a missing model term. This can be a wise investment of runs.

You can replicate runs solely at center points or you can replicate other design runs. JMP uses replicate runs to construct a model-independent error estimate (pure-error estimate).This pure-error estimate enables you to test for lack of fit.

Be aware that center points do not help you obtain more precise estimates of model effects. They enable you to test for evidence of curvature, but do not identify the responsible nonlinear effects.

To identify the source of curvature, you must set continuous factors at a minimum of three levels. Definitive screening designs are three-level designs with the ability to detect and identify any factors causing strong nonlinear effects on the response. For details, see Chapter 7, "Definitive Screening Designs".

Adding Replicate Runs

If your run budget allows, you can either replicate runs or distribute new runs optimally within the design space. Adding replicate runs adds precision for some estimates and improves the power of the lack of fit test. However, for a given run budget, adding replicate runs generally lowers the ability of the design to estimate model effects. You are not able to estimate as many terms as you could by distributing the runs optimally within the design space.

Testing for Lack of Fit

Designed experiments are typically constructed to require as few runs as possible, consistent with the goals of the experiment. With too few runs, only extremely large effects can be detected. For example, for a given effect, the *t*-test statistic is the ratio of the change in response means to their standard error. If there is only one error degree of freedom (df), then the critical value of the test exceeds 12. So, for such a nearly saturated design to detect an effect, it has to be very large.

A similar observation applies to the lack-of-fit test. The power of this test to detect lack-of-fit depends on the numbers of degrees of freedom in the numerator and denominator. If you have only 1 df of each kind, you need an *F* value that exceeds 150 to declare significance at the 0.05 level. If you have 2 df of each kind, then the *F* value must exceed 19. In order for the test to be significant in this second case, the lack-of-fit mean square must be 19 times larger than the pure error mean square. It is also true that the lack-of-fit test is sensitive to outliers.

For details about the Lack of Fit test, see the Standard Least Squares chapter in the *Fitting Linear Models* book.

Determining the Number of Runs

In industrial applications, each run is often very costly, so there is incentive to minimize the number of runs. To estimate the fixed effects of interest, you need only as many runs as there are terms in the model. To determine whether the effects are active, you need a reasonable estimate of the error variance. Unless you already have a good estimate of this variance, consider adding at least 4 runs to the number required to estimate the model terms.

Chapter 4

Custom Designs

Construct Designs That Meet Your Needs

Use the Custom Design platform to construct optimal designs that are custom built for your specific experimental setting. Generally, a custom design is more cost-effective than a design obtained using alternative methods. You can perform the following tasks:

- Enter factors of many different types.
- Specify constraints on the design space.
- Indicate which effects are necessary to estimate and which are desirable to estimate, if possible, given the number of runs.
- Specify a number of experimental runs that matches your budget.

The Custom Design platform constructs a wide variety of designs, including these special design types: Screening, Response Surface, Mixture, Random Block, Split Plot, Split-Split-Plot, and Two-Way Split Plot.

This chapter contains a detailed example of how to use the Custom Design platform, followed by information about the platform. See also the "Examples of Custom Designs" chapter on page 129.

Figure 4.1 Color Map for Absolute Correlations

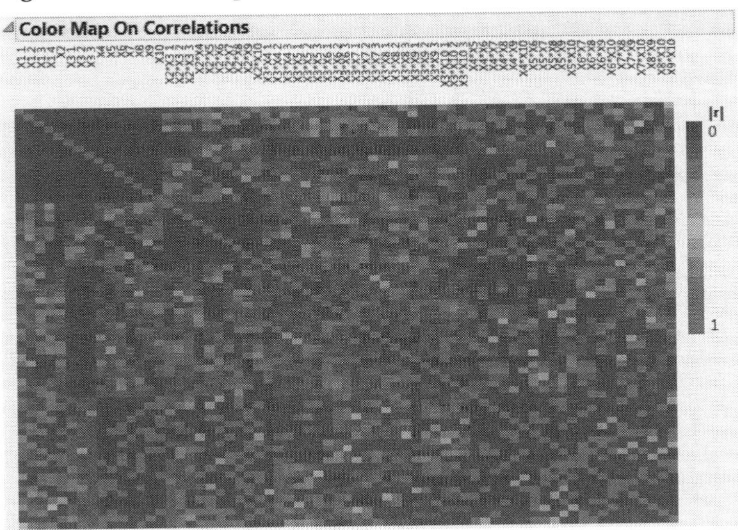

Overview of Custom Design

Use the Custom Design platform to construct an optimal design custom built for your specific experimental needs.

You can include a wide range of factor types, including the following:

- Continuous
- Discrete numeric (with any number of levels)
- Categorical (with any number of levels)
- Blocking (with an arbitrary number of runs per block)
- Covariate
- Mixture
- Constant
- Uncontrolled

Specify the Region of Operability

You can restrict your experimental region to reflect your operating conditions using linear factor constraints or disallowed combinations. In particular, restrictions can be specified for categorical, continuous, and discrete numeric factors. See "Define Factor Constraints" on page 92.

Specify Factors with Hard-to-Change Levels

For continuous, discrete numeric, categorical, and mixture factors, you can indicate two levels of difficult-to-change factors. These difficulty levels are represented by whole plots or whole plots and split plots. You can also specify hard-to-change covariates.

Specify the Effects of Primary Interest

You can explicitly specify your *assumed* model. Your assumed model is an initial model that ideally contains all the effects that you want to estimate. Your model can contain any combination of main effects, interactions, response surface effects, and polynomial effects (up to the fifth power). You can specify the effects for which estimability is necessary and those for which estimability is desired. Custom Design uses a Bayesian optimality approach to estimate effects whose estimability is desired, subject to the number of runs. See "Model" on page 95.

Specify the Number of Runs

The Custom Design platform enables you to specify the number of runs that matches the budget for your experimental situation. The platform indicates the minimum number of runs that can be used to estimate the required effects and provides a default number of runs. These values can serve as a guide for determining a feasible number of runs. See "Design Generation" on page 98.

Construct the Appropriate Design Type

Custom Design can construct a wide variety of design types. These include classical designs and random block designs. For examples of different design types, see the "Examples of Custom Designs" chapter on page 129.

Construct an Optimal Design

Given your specific requirements, the Custom Design platform constructs a design that is optimal. The algorithm supports several optimality criteria:

- D optimality
- I optimality
- Bayesian D and I optimality (using If Possible effects)
- Alias optimality

See "Optimality Criteria" on page 123.

Designs are constructed using the coordinate-exchange algorithm (Meyer and Nachtsheim, 1995). See "Coordinate-Exchange Algorithm" on page 127.

Example of a Custom Design

The following example describes a wine tasting experiment. Your employer grows two varieties of Pinot Noir grapes that can be processed in different ways. Your goal is to determine which factors affect the taste of Pinot Noir wine. Before the grapes are processed, you set up your experimental design. Once processed, the wine samples are aged for 12 months, then filtered and bottled. At this point, the wine samples are rated for quality by expert wine tasters.

Response

Most of your vineyard's product is sold to five large wine distributors. You arrange for a wine-tasting expert from each distributor to evaluate the wine samples for quality. To maximize the number of factors that you can study, you decide that each expert must rate eight different samples. This means that your design needs to have 40 wine samples, or runs.

The ratings follow a 0 – 20 scale, where 0 is the worst and 20 is the best. Rating, the variable consisting of the experts' ratings, is the response of interest. You want to identify the wine-related factors that maximize the response.

Blocking Factor

A blocking factor is used to account for variation that is not necessarily of direct interest. A blocking factor is particularly effective when observations taken at one factor level are expected to be more similar than observations at different levels. In your experiment, ratings by one expert are likely to have similar characteristics and to differ from ratings by a different expert. Yet, you are interested in which properties of the wine lead to high ratings by all experts.

Because each rater tastes eight wines, Rater is a blocking factor with eight runs per block. For this experiment, only these five raters are of concern. You are not interested in generalizing to a larger population of raters.

Process Factors

You have identified nine process factors for the study. These include the grape variety, the field on which the grapes were grown, and seven other factors related to processing. You can experiment with any combination of these factors. Also, the factors can be varied at will as part of the experiment. Relative to the experiment, these factors are all "Easy" to change. For information about specifying factor changes, see "Changes and Random Blocks" on page 89.

The factors and their levels appear in Table 4.1. Note that all of these factors are categorical. The factors and their levels are also given in the factor table Wine Factors.jmp in the Design Experiment folder of Sample Data.

To experiment with all possible combinations of these factors would require a staggering $4 \times 2^8 = 1024$ runs. However, in this example, you are able to construct a compelling design in only 40 runs.

Table 4.1 Process Factors and Levels for Wine Tasting Experiment

Factor	Levels
Variety	Bernard, Dijon
Field	1, 2, 3, 4
De-Stem	No, Yes
Yeast	Cultured, Wild
Temperature	High, Low
Press	Hard, Soft
Barrel Age	New, 2 Years
Barrel Seasoning	Air, Kiln
Filtering	No, Yes

Create the Design

Note: In order to introduce and describe the Custom Design outlines, this example works through the outlines in succession.

To create the custom design, follow the steps in these sections:

- "Responses" on page 67
- "Factors" on page 68
- "Model" on page 70
- "Alias Terms" on page 70
- "Duplicate Results (Optional)" on page 71
- "Design Generation" on page 71
- "Design" on page 71
- "Design Evaluation" on page 72
- "Output Options" on page 75

For information about the complete DOE workflow, see "The DOE Workflow: Describe, Specify, Design" on page 54 in the "Starting Out with DOE" chapter.

Responses

Add your response, the response Goal, and, if appropriate, the Lower Limit, Upper Limit, and Importance. Here, the response is Rating.

1. Select **DOE > Custom Design**.
2. Double-click Y under Response Name and type Rating.

   Note that the default Goal is Maximize. Because you want to maximize the taste rating, do not change the goal.
3. Click under Lower Limit and type 0.

   The least desirable rating is 0.
4. Click under Upper Limit and type 20.

   The most desirable rating is 20.
5. Leave the area under Importance blank.

   Because there is only one response, that response is given Importance 1 by default.

Figure 4.2 on page 69 shows the completed Responses outline.

Factors

Enter factors either manually or automatically using a pre-existing table that contains the factors and settings.

- If you are designing a new experiment, you must first enter the factors manually. See "Entering Factors Manually" on page 68.
- Once you have saved the factors using the Save Factors option, you can load them automatically using the saved table. See "Entering Factors Using Load Factors" on page 69.

Both methods add these four outlines to the Custom Design window: Define Factor Constraints, Model, Alias Terms, and Design Generation.

Entering Factors Manually

1. First, add the blocking factor, Rater. Click **Add Factor > Blocking > 8 runs per block**.
2. Type Rater over the default Name of X1.

   Note that Role is set to Blocking. Note also that only one setting for Values appears. This is because the number of blocks cannot be determined until the desired number of runs is specified. Once you specify the Number of Runs in the Design Generation outline, the number of levels for Rater updates to what is required.
3. Click **Add Factor > Categorical > 2 Level**.
4. Type Variety over the default Name of X2.

   Note that Role is set to Categorical, as requested, and that Changes is set to Easy by default.
5. Click L1 and L2 and change them to Bernard and Dijon.
6. Click **Add Factor > Categorical > 4 Level**.
7. Type Field over the default Name of X3.
8. Click L1, L2, L3, and L4, and change them to 1, 2, 3, and 4.
9. Click **Add Factor > Categorical > 2 Level**.
10. Type De-Stem over the default Name of X4.
11. Click L1 and L2 and change them to No and Yes.

    Add the rest of the factors as follows:
12. Type 6 next to **Add N Factors**, and then click **Add Factor > Categorical > 2 Level**.
13. Type the following names and values over the default ones:
    - Yeast (Cultured and Wild)
    - Temperature (High and Low)
    - Press (Hard and Soft)

– Barrel Age (New and Two Years)

– Barrel Seasoning (Air and Kiln)

– Filtering (No and Yes)

The completed Factors outline appears in Figure 4.2.

Figure 4.2 Completed Responses and Factors Outlines

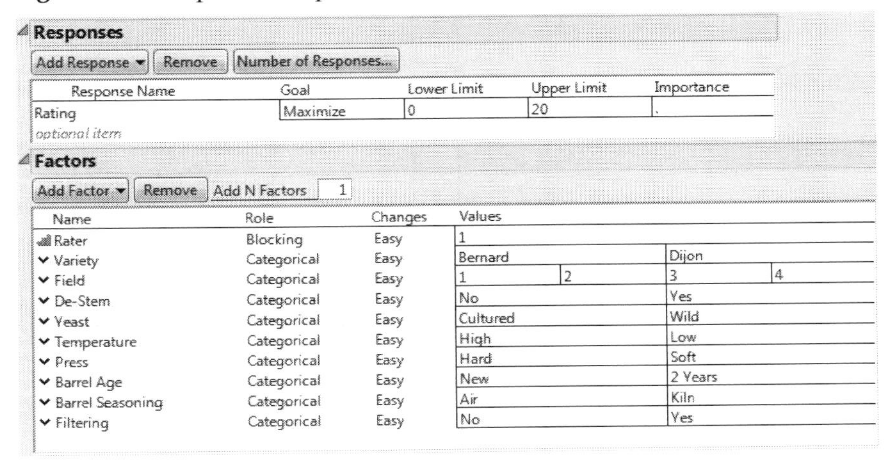

14. Click **Continue**.

The following outlines are added to the Custom Design window:

- Define Factor Constraints (not used in this example)
- Model
- Alias Terms
- Design Generation

Entering Factors Using Load Factors

To enter factors using a table containing factor information, proceed as follows:

1. Select **Help > Sample Data Library** and open Design Experiment/Wine Factors.jmp.
2. From the Custom Design red triangle menu, select **Load Factors**.

After loading the factors, the Custom Design window automatically updates. The following outlines are added to the Custom Design window:

- Define Factor Constraints (not used in this example)
- Model
- Alias Terms

- Design Generation

Model

The Model outline shows all main effects as Necessary, indicating that the design needs to be capable of estimating all main effects. For this example, your assumed model reflects your interest in main effects only. However, if you wanted to estimate other effects, you could add them to the Model outline. See "Model" on page 95.

Figure 4.3 Model Outline

Model

Main Effects | Interactions | RSM | Cross | Powers | Remove Term

Name	Estimability
Intercept	Necessary
Variety	Necessary
Field	Necessary
De-Stem	Necessary
Yeast	Necessary
Temperature	Necessary
Press	Necessary
Barrel Age	Necessary
Barrel Seasoning	Necessary
Filtering	Necessary
Rater	Necessary

Alias Terms

The Alias Terms outline specifies the effects to be shown in the Alias Matrix, which appears later. See "Alias Matrix" on page 74. The Alias Matrix shows the aliasing relationships between the Model terms and the effects listed in the Alias Terms outline. Open the Alias Terms outline node to verify that all two-factor interactions are listed.

Figure 4.4 Partial View of the Alias Terms Outline

Duplicate Results (Optional)

In the next step, you generate your design. Because the Custom Design algorithm begins with a random starting design, your design might differ from the one shown in Figure 4.5. If you want to obtain a design with exactly the same runs and run order, perform the following steps:

1. From the Custom Design red triangle menu, select **Set Random Seed**.
2. Type 100526291 (the random seed).
3. Click **OK**.
4. From the Custom Design red triangle menu, select **Number of Starts**.
5. Type 2.
6. Click **OK**.

Note: Setting the Random Seed and Number of Starts reproduces the exact results shown in this example. In constructing a design on your own, these steps are not necessary.

Proceed to the Design Generation section.

Design Generation

In the Design Generation outline, you can enter additional details about the structure and size of your design. In this example, the Default design shows 16 runs. But you have five raters, each of whom can sample eight wines. This means that you want a design with 40 runs. Change the number of runs as follows:

1. Under **Number of Runs**, type **40** in the **User Specified** box.

   Because you do not want to replicate runs, leave the **Number of Replicate Runs** set to **0**.
2. Click **Make Design**.

The Design and Design Evaluation outlines are added to the Custom Design window. The Output Options panel also appears.

Design

The Design outline shows the runs in the design that you have constructed. Later, you are able to randomize the order under Output Options. For now, verify that this design is appropriate for your experiment. For example, check that each of five Raters evaluates eight wines, that all necessary factors are shown, and that none of the settings represent infeasible combinations.

Figure 4.5 Design for Wine Experiment

Design

Run	Rater	Variety	Field	De-Stem	Yeast	Temperature	Press	Barrel Age	Barrel Seasoning	Filtering
1	1	Dijon	3	No	Wild	High	Soft	New	Kiln	Yes
2	2	Bernard	4	Yes	Cultured	Low	Hard	New	Kiln	No
3	3	Bernard	2	Yes	Wild	High	Hard	New	Air	No
4	4	Bernard	1	Yes	Wild	Low	Hard	2 Years	Air	Yes
5	5	Dijon	1	Yes	Wild	Low	Soft	New	Kiln	No
6	1	Dijon	2	Yes	Cultured	Low	Soft	2 Years	Air	Yes
7	2	Dijon	1	No	Cultured	High	Hard	New	Air	No
8	3	Dijon	4	No	Cultured	High	Soft	New	Air	Yes
9	4	Dijon	4	Yes	Wild	High	Hard	New	Kiln	Yes
10	5	Bernard	2	No	Wild	Low	Hard	New	Air	Yes
11	1	Bernard	3	Yes	Wild	High	Hard	New	Air	No
12	2	Dijon	1	Yes	Wild	High	Soft	2 Years	Air	Yes
13	3	Bernard	4	Yes	Wild	Low	Soft	2 Years	Kiln	No
14	4	Bernard	3	No	Cultured	Low	Soft	New	Air	No
15	5	Dijon	2	No	Cultured	High	Soft	2 Years	Kiln	No
16	1	Dijon	1	Yes	Cultured	Low	Hard	2 Years	Kiln	Yes
17	2	Bernard	2	No	Cultured	Low	Hard	2 Years	Air	No
18	3	Dijon	2	Yes	Cultured	High	Hard	New	Kiln	No
19	4	Bernard	1	No	Cultured	Low	Hard	New	Kiln	Yes
20	5	Bernard	3	Yes	Cultured	High	Hard	2 Years	Air	Yes
21	1	Bernard	4	No	Wild	High	Hard	2 Years	Kiln	No
22	2	Bernard	4	Yes	Cultured	High	Soft	2 Years	Kiln	Yes
23	3	Dijon	1	No	Wild	Low	Hard	2 Years	Kiln	No
24	4	Dijon	2	Yes	Wild	High	Soft	2 Years	Air	No
25	5	Bernard	1	No	Cultured	High	Soft	New	Kiln	No
26	1	Bernard	2	No	Wild	Low	Soft	New	Kiln	Yes
27	2	Dijon	3	No	Wild	High	Hard	New	Kiln	Yes
28	3	Bernard	1	No	Wild	High	Soft	2 Years	Air	Yes
29	4	Bernard	3	Yes	Cultured	High	Soft	2 Years	Kiln	No
30	5	Dijon	3	Yes	Wild	Low	Hard	2 Years	Kiln	No
31	1	Dijon	4	No	Cultured	Low	Hard	2 Years	Air	No
32	2	Bernard	2	Yes	Wild	Low	Soft	New	Kiln	Yes
33	3	Bernard	3	No	Cultured	Low	Soft	2 Years	Kiln	Yes
34	4	Dijon	2	No	Cultured	High	Hard	2 Years	Kiln	Yes
35	5	Dijon	4	Yes	Cultured	Low	Soft	New	Air	Yes
36	1	Bernard	1	Yes	Cultured	High	Soft	New	Air	No
37	2	Dijon	3	No	Wild	Low	Soft	2 Years	Air	No
38	3	Dijon	3	Yes	Cultured	Low	Hard	New	Air	Yes
39	4	Dijon	4	No	Wild	Low	Soft	New	Air	No
40	5	Bernard	4	No	Wild	High	Hard	2 Years	Air	Yes

Design Evaluation

The Design Evaluation outline provides different ways to evaluate your design.

Note: For details about the Design Evaluation outline, see the "Evaluate Designs" chapter on page 419.

For this example, open the Design Evaluation outline, and examine the Color Map on Correlations, the Alias Matrix, and Design Diagnostics.

Color Map on Correlations

The Color Map on Correlations shows the absolute value of the correlation between any two effects that appear in either the Model or the Alias Terms outline. (The colors shown in Figure 4.6 are the JMP default colors.)

Figure 4.6 Color Map on Correlations

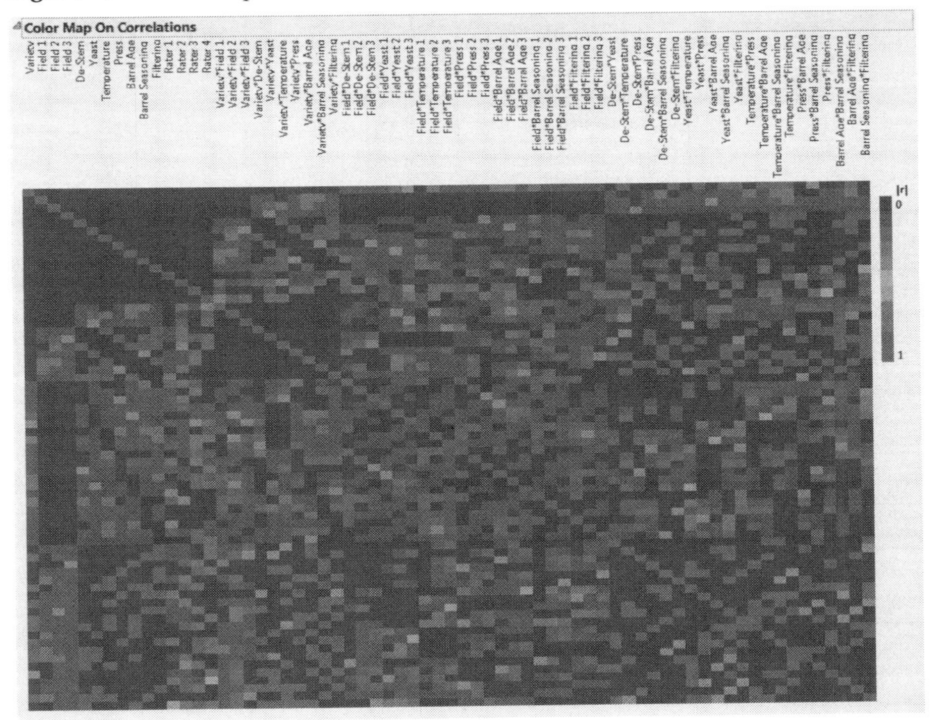

The main effects are represented by the 15 terms in the upper left corner of the map. The deep blue color corresponding to the correlations of main effects with other main effects indicate correlations of 0. This means that all main effects are orthogonal and can be estimated independently.

The only red in Figure 4.6 is on the main diagonal. The color indicates absolute correlations of one, reflecting that each term is perfectly correlated with itself. It follows that no main effect is completely confounded with any two-way interaction. In fact, the absolute values of the correlations of main effects with two-way interactions are fairly low. This means that estimates of main effects might be only slightly biased by the presence of active two-way interactions.

Tip: Position your mouse pointer over cells in the color map to see the absolute correlations between effects.

Alias Matrix

In the Alias Matrix, model effects are listed in the column on the left. For a given model effect, a column entry indicates the degree to which the column effect (if active) biases the estimate of the model effect.

Figure 4.7 Partial View of Alias Matrix

Alias Matrix

Effect	Variety*Field 1	Variety*Field 2	Variety*Field 3	Variety*De-Stem	Variety*Yeast	Variety*Temperature	Variety*Press	Variety*Barrel Age	Variety*Barrel Seasoning	Variety*Filtering
Intercept	0	0	0	0	0	0	0	0	0	0
Variety	0	0	0	0	0	0	0	0	0	0
Field 1	0	0	0	0.141	0.141	0	-0.14	0.283	0.283	-0.14
Field 2	0	0	0	0.082	-0.24	-0.33	0.082	0.327	0.163	-0.08
Field 3	0	0	0	-0.12	0.346	0.115	-0.12	-0.12	0.115	0.115
De-Stem	0.141	0.082	-0.12	0	0	-0.2	0	0	0	-0.2
Yeast	0.141	-0.24	0.346	0	0	0	-0.2	0	-0.2	0.2
Temperature	0	-0.33	0.115	-0.2	0	0	0	-0.2	0.2	0.2
Press	-0.14	0.082	-0.12	0	-0.2	0	0	0	0.4	0
Barrel Age	0.283	0.327	-0.12	0	0	-0.2	0	0	0	0.2
Barrel Seasoning	0.283	0.163	0.115	0	-0.2	0.2	0.4	0	0	0
Filtering	-0.14	-0.08	0.115	-0.2	0.2	0.2	0	0.2	0	0
Rater 1	0	0	0	-0.16	-0.16	0	-0.16	0.158	-0.16	0.316
Rater 2	-0.26	0.149	-0.21	-0.27	0.274	-0.37	-0.09	-0.09	-0.27	0
Rater 3	0.091	-0.05	0.075	0	-0.19	-0.06	-0.26	-0.26	0	0
Rater 4	0.354	-0.2	0.289	0	0.25	-0.25	0	0	0	0

For example, consider the model effect Barrel Seasoning. If Variety*Press is active, then the expected value of the estimate for the Barrel Seasoning effect differs from an unbiased estimate of that effect. The amount by which it differs is equal to 0.4 times the effect of Variety*Press. Therefore, what appears to be a significant Barrel Seasoning estimated effect could in reality be a significant Variety*Press effect.

Design Diagnostics

The Design Diagnostics outline provides information about the efficiency of the design. Efficiency measures compare your design to a theoretically optimal design, which might not exist. The efficiency values are ratios, expressed as percents, of the efficiency of your design to the efficiency of this optimal design. For details about the efficiency measures, see "Estimation Efficiency" on page 444 in the "Evaluate Designs" chapter.

Figure 4.8 Design Diagnostics Outline

Design Diagnostics

D Optimal Design	
D Efficiency	100
G Efficiency	100
A Efficiency	100
Average Variance of Prediction	0.4
Design Creation Time (seconds)	0.05

Notice that the D-, G-, and A-efficiency values are all 100%. Because your design is orthogonal for main effects, the design is optimal for the main effects model relative to all three efficiency criteria.

The first line in the Design Diagnostics outline indicates that your design was constructed to optimize the D-efficiency criterion. For more details, see the Optimality Criterion description in "Custom Design Options" on page 103. In this case, your design has D Efficiency of 100%.

Output Options

Specify the order of runs in your data table using the Output Options panel. The default selection, **Randomize within Blocks**, is appropriate for this example. Simply click **Make Table**.

A Custom Design table is created and opens, similar to the one in Figure 4.9.

Note: Your table might look different because the algorithm that creates it uses a random starting design. To obtain the precise table shown in Figure 4.9, follow the steps in "Duplicate Results (Optional)" on page 71.

Figure 4.9 Custom Design Table

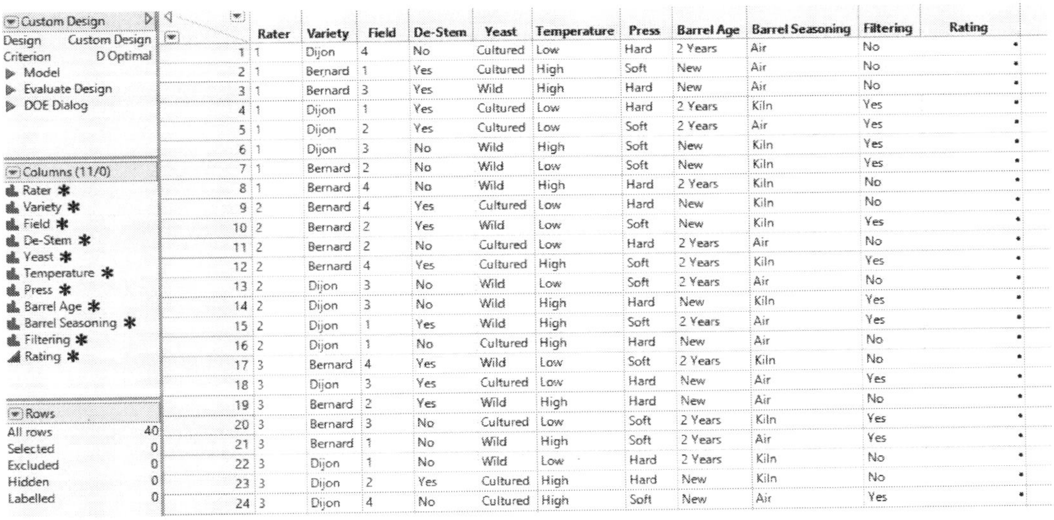

Custom Design
Design Custom Design
Criterion D Optimal
▶ Model
▶ Evaluate Design
▶ DOE Dialog

Columns (11/0)
Rater *
Variety *
Field *
De-Stem *
Yeast *
Temperature *
Press *
Barrel Age *
Barrel Seasoning *
Filtering *
Rating *

Rows
All rows 40
Selected 0
Excluded 0
Hidden 0
Labelled 0

	Rater	Variety	Field	De-Stem	Yeast	Temperature	Press	Barrel Age	Barrel Seasoning	Filtering	Rating
1	1	Dijon	4	No	Cultured	Low	Hard	2 Years	Air	No	•
2	1	Bernard	1	Yes	Cultured	High	Soft	New	Air	No	•
3	1	Bernard	3	Yes	Wild	High	Hard	New	Air	No	•
4	1	Dijon	1	Yes	Cultured	Low	Hard	2 Years	Kiln	Yes	•
5	1	Dijon	2	Yes	Cultured	Low	Soft	2 Years	Air	Yes	•
6	1	Dijon	3	No	Wild	High	Soft	New	Kiln	Yes	•
7	1	Bernard	2	No	Wild	Low	Soft	New	Kiln	Yes	•
8	1	Bernard	4	No	Wild	High	Hard	2 Years	Kiln	No	•
9	2	Bernard	4	Yes	Cultured	Low	Hard	New	Kiln	No	•
10	2	Bernard	2	Yes	Wild	Low	Soft	New	Kiln	Yes	•
11	2	Bernard	2	No	Cultured	Low	Hard	2 Years	Air	No	•
12	2	Bernard	4	Yes	Cultured	High	Soft	2 Years	Kiln	Yes	•
13	2	Dijon	3	No	Wild	Low	Soft	2 Years	Air	No	•
14	2	Dijon	3	No	Wild	High	Hard	New	Kiln	Yes	•
15	2	Dijon	1	Yes	Wild	High	Soft	2 Years	Air	Yes	•
16	2	Dijon	1	No	Cultured	High	Hard	New	Air	No	•
17	3	Bernard	4	Yes	Wild	Low	Soft	2 Years	Kiln	No	•
18	3	Dijon	3	Yes	Cultured	Low	Hard	New	Air	Yes	•
19	3	Bernard	2	Yes	Wild	High	Hard	New	Air	No	•
20	3	Bernard	3	No	Cultured	Low	Soft	2 Years	Kiln	Yes	•
21	3	Bernard	1	No	Wild	High	Soft	2 Years	Air	Yes	•
22	3	Dijon	1	No	Wild	Low	Hard	2 Years	Kiln	No	•
23	3	Dijon	2	Yes	Cultured	High	Hard	New	Kiln	No	•
24	3	Dijon	4	No	Cultured	High	Soft	New	Air	Yes	•

Note the following:

- In the Table panel, the Model, Evaluate Design, and DOE Dialog scripts are added during the design creation process. The Model script opens a Fit Model window containing the effects that you specified as Necessary in the Custom Design dialog. The DOE Dialog script re-creates the window used to generate the design table.
- In the Columns panel, the asterisks to the right of the factors and response indicate column properties that have been saved to the columns in the data table. These column properties are used in the analysis of the data. For details about column properties, see "Factors" on page 85 and "Factor Column Properties" on page 91.

Analyze the Data

Now you are ready to run your experiment, gather the Rating data, and insert the results in the Rating column of your Custom Design table.

1. Select **Help > Sample Data Library** and open Design Experiment/Wine Data.jmp.

   The Wine Data.jmp table is exactly the same as the Custom Design table shown in Figure 4.9, except that it contains your recorded experimental results.

2. In the Table panel, click the green triangle next to the **Model** script.

Figure 4.10 Fit Model Dialog for Wine Experiment

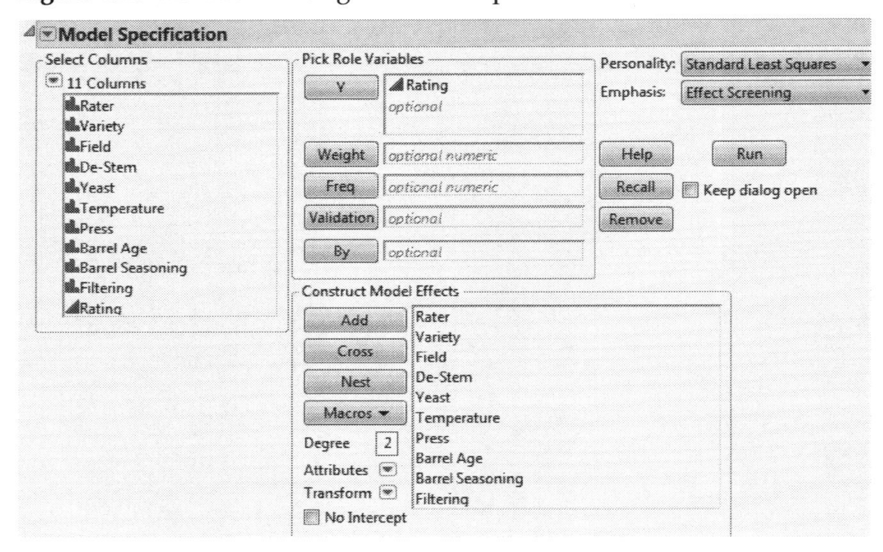

   Notice that Rater, the blocking factor, is added as a fixed effect, rather than as a random block effect. This is appropriate because the five raters were specifically chosen and are not a random sample from a larger population.

3. Click **Run**.

Interpret the Full Model Results

The results are shown below.

Figure 4.11 Partial Model Fit Results

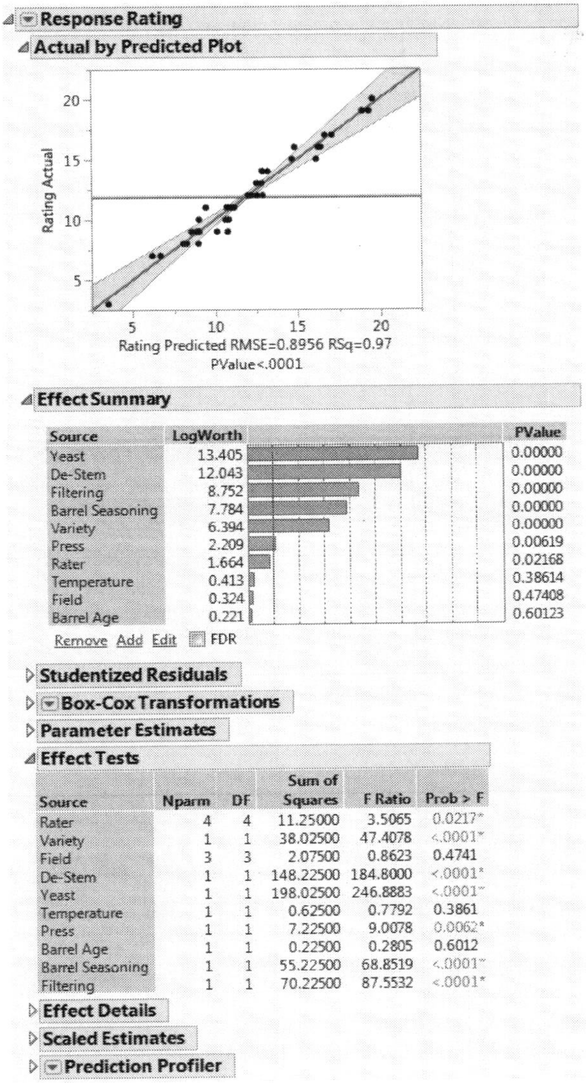

Response Rating

Actual by Predicted Plot

Rating Actual

Rating Predicted RMSE=0.8956 RSq=0.97
PValue<.0001

Effect Summary

Source	LogWorth	PValue
Yeast	13.405	0.00000
De-Stem	12.043	0.00000
Filtering	8.752	0.00000
Barrel Seasoning	7.784	0.00000
Variety	6.394	0.00000
Press	2.209	0.00619
Rater	1.664	0.02168
Temperature	0.413	0.38614
Field	0.324	0.47408
Barrel Age	0.221	0.60123

Remove Add Edit FDR

Studentized Residuals

Box-Cox Transformations

Parameter Estimates

Effect Tests

Source	Nparm	DF	Sum of Squares	F Ratio	Prob > F
Rater	4	4	11.25000	3.5065	0.0217*
Variety	1	1	38.02500	47.4078	<.0001*
Field	3	3	2.07500	0.8623	0.4741
De-Stem	1	1	148.22500	184.8000	<.0001*
Yeast	1	1	198.02500	246.8883	<.0001*
Temperature	1	1	0.62500	0.7792	0.3861
Press	1	1	7.22500	9.0078	0.0062*
Barrel Age	1	1	0.22500	0.2805	0.6012
Barrel Seasoning	1	1	55.22500	68.8519	<.0001*
Filtering	1	1	70.22500	87.5532	<.0001*

Effect Details

Scaled Estimates

Prediction Profiler

Note the following:

- The Actual by Predicted Plot shows no obvious evidence of lack of fit.
- The model is significant, as indicated by the Actual by Predicted Plot and by the P value beneath it.
- The Effect Tests report indicates that seven of the model terms are significant at the 0.05 level. Field, Temperature, and Barrel Age are not significant.

- The Effect Summary report lists these effects in decreasing order of significance. Larger LogWorth values correspond to smaller PValues and greater significance.

Reduce the Model

Reduce the model by removing the effects that you identified as inactive:

1. In the Effect Summary report, press the Control key and hold it as you select Temperature, Field, and Barrel Age.
2. Click **Remove**.

   The report updates to show the model fit with these three effects removed.

Interpret the Reduced Model Results

The Actual by Predicted Plot for the reduced model shows no lack of fit issues. The Effect Summary and the Effect Test report show that the remaining seven terms are significant at the 0.05 level.

Figure 4.12 shows the Prediction Profiler. Recall that you specified a response goal of Maximize, with lower and upper limits of 0 and 20. Setting these limits caused a Response Limits column property to be saved to the Rating column in the Custom Design table. The Prediction Profiler uses the Response Limits information to construct a Desirability function, which appears in the right-most plot in the top row in Figure 4.12. The bottom row displays Desirability traces.

The first six plots in the top row show traces of the predicted model. For each factor, the line in the plot shows how Rating varies when all other factors are set at the values defined by the red dashed vertical lines. By default, the profiler appears with categorical factors set at their low settings. By varying the settings for the factors, you can see how the predicted Rating for wines changes. Notice that a confidence interval is given for the mean predicted Rating.

Observe that Rater is not included among the factors shown in the profiler. This is because Rater is a block variable. You included Rater to explain variation, but Rater is not of direct interest in terms of optimizing process factor settings. The predicted Rating for a wine with the given settings is the average of the predicted ratings for that wine by all raters.

Figure 4.12 Profiler for Reduced Model

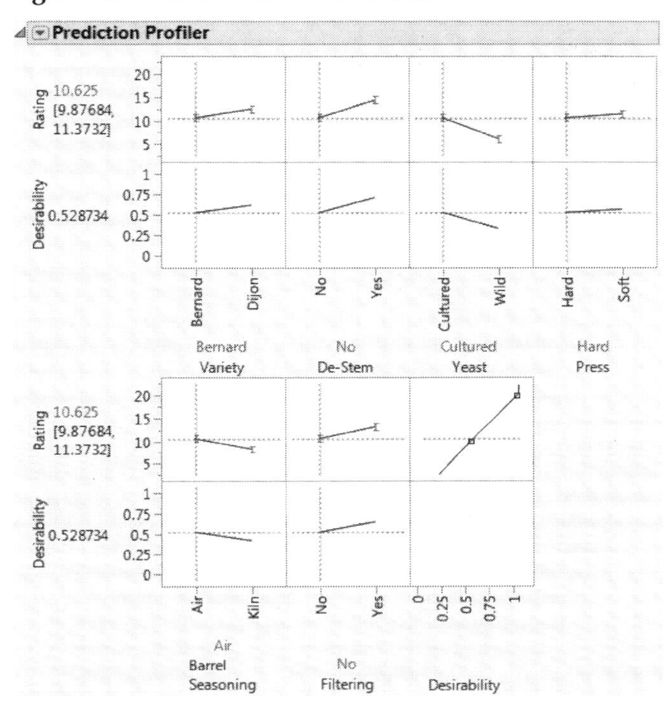

Optimize Factor Settings

You would like to identify settings that maximize Rating across raters.

1. From the Prediction Profiler red triangle menu, select **Optimization and Desirability > Maximize Desirability**.

   The red dashed vertical lines in the Prediction Profiler update to show optimal settings for each factor. The optimal settings result in a predicted rating of 19.925. In general, there can be different sets of factor settings that result in the same optimal value.

Figure 4.13 Prediction Profiler with Factor Settings Optimized

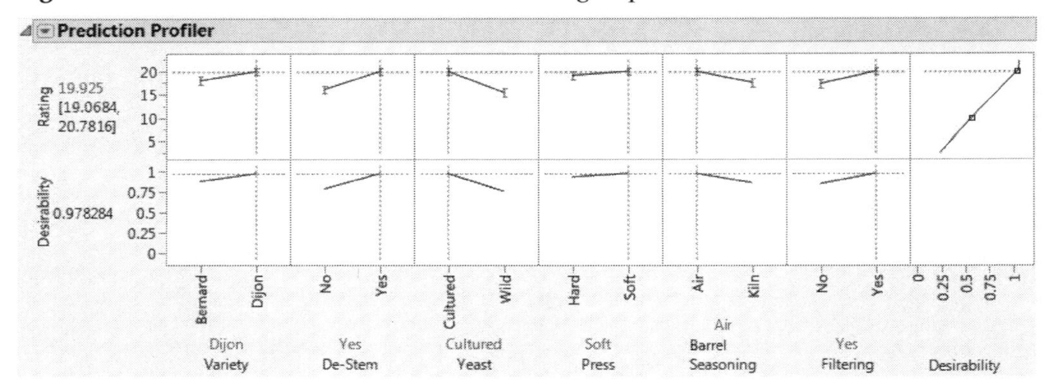

2. To see predicted ratings for all runs, save the Prediction Formula. From the Response Rating red triangle menu, select **Save Columns > Prediction Formula**.

   A column called Pred Formula Rating is added to the data table. Note that one of the runs, row 33, was given the maximum rating of 20 by Rater 5. The predicted rating for that run by Rater 5 is 19.550. But the row 33 trial was run at the optimal settings. The predicted value of 19.925 given for these settings in the Prediction Profiler is obtained by averaging the predicted ratings for that run over all five raters.

Lock a Factor Level

When you maximized desirability, you learned that the optimal rating is achieved with the Dijon variety of grapes. See Figure 4.13. Your manager points out that it would be cost-prohibitive to replant the fields that are growing Bernard grapes with young Dijon vines. Therefore, you need to find optimal process settings and the predicted rating for Bernard grapes.

1. In the Variety plot of the Prediction Profiler, drag the red dashed vertical line to Bernard.
2. Press Control and click in one of the Variety plots.

   The Factor Settings window appears.
3. Select **Lock Factor Setting** and click **OK**.
4. From the Prediction Profiler red triangle menu, select **Optimization and Desirability > Maximize Desirability**.

Figure 4.14 Prediction Profiler with Optimal Settings for Bernard Variety

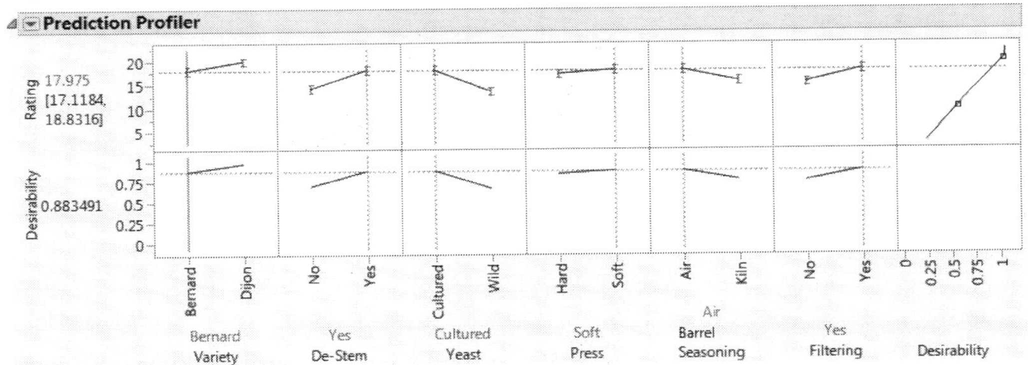

The optimal settings are unchanged because the model contains no interaction terms. The predicted rating at these settings is 17.975.

Profiler with Rater

If you want to see the Profiler traces for the levels of Rater, perform the following steps:

1. From the Prediction Profiler red triangle menu, select **Reset Factor Grid**.

   A Factor Settings window appears with columns for all of the factors, including **Rater**. The box under Rater and next to **Show** is not checked. This indicates that Rater is not shown in the Prediction Profiler.

2. Check the box under Rater in the row corresponding to **Show**.
3. Deselect the box under Rater in the row corresponding to **Lock Factor Setting**.
4. Click **OK**.

   The Profiler updates to show a plot for Rater.

5. Click in either plot above Rater.

Figure 4.15 Profiler for Reduced Model Showing Rater

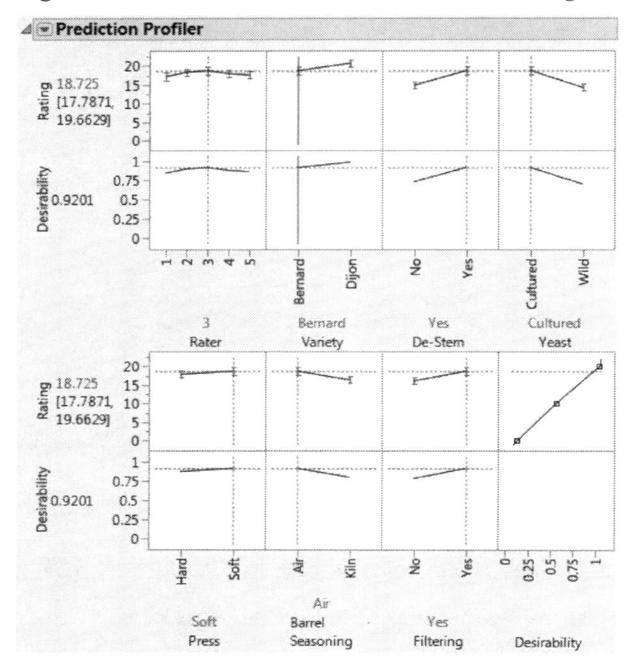

A dashed vertical red line appears. Drag this line to see the traces for each of the raters. Keep in mind that Variety is still locked at Bernard.To unlock Variety, press Control and click in one of the Variety plots. In the Factor Settings window that appears, deselect **Lock Factor Setting**.

Summary

In your wine tasting experiment, using only 40 runs, you have identified six (out of nine) factors that have an effect on ratings for Pinot Noir grapes. You found that you could achieve a predicted rating of 19.925 (out of a possible 20) at the optimal settings for those factors. You also identified optimal settings for both varieties of grapes.

In this section, you constructed a design using the outlines in the Custom Design window. The next section explains each outline and the design steps in more detail.

Custom Design Window

The Custom Design window updates as you work through the design steps. The outlines that appear, separated by buttons that update the window, follow the flow in Figure 4.16.

Figure 4.16 Custom Design Flow

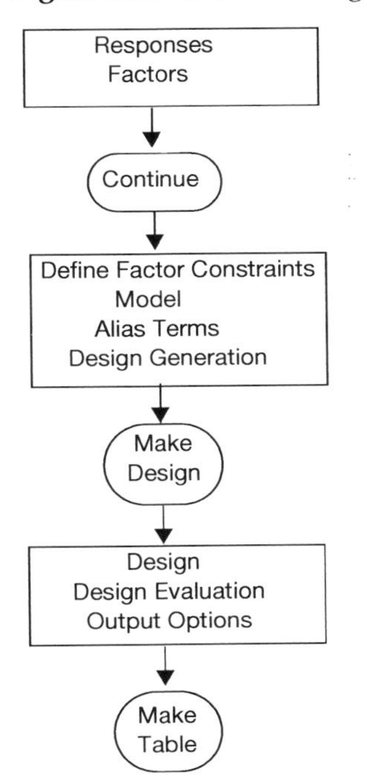

This section describes the outlines in the Custom Design window.

Responses

Use the Responses outline to specify one or more responses.

Tip: When you have completed the Responses outline, consider selecting **Save Responses** from the red triangle menu. This option saves the response names, goals, limits, and importance values in a data table that you can later reload in DOE platforms.

Figure 4.17 Responses Outline

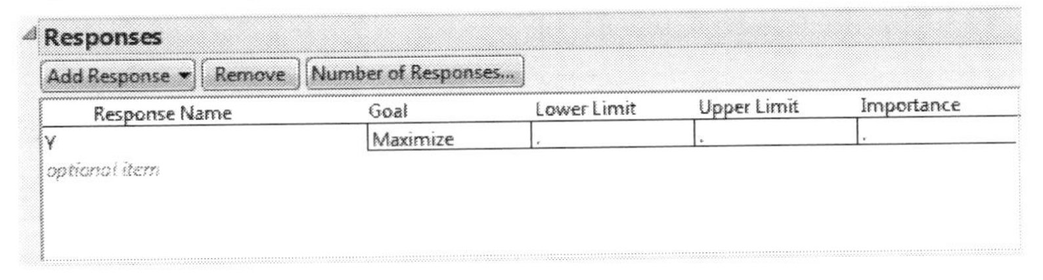

Add Response Enters a single response with a goal type of Maximize, Match Target, Minimize, or None. If you select Match Target, enter limits for your target value. If you select Maximize or Minimize, entering limits is not required but can be useful if you intend to use desirability functions.

Remove Removes the selected responses.

Number of Responses Enters additional responses so that the number that you enter is the total number of responses. If you have entered a response other than the default Y, the Goal for each of the additional responses is the Goal associated with the last response entered. Otherwise, the Goal defaults to Match Target. Click the Goal type in the table to change it.

The Responses outline contains the following columns:

Response Name The name of the response. When added, a response is given a default name of Y, Y2, and so on. To change this name, double-click it and enter the desired name.

Goal, Lower Limit, Upper Limit The Goal tells JMP whether you want to maximize your response, minimize your response, match a target, or that you have no response goal. JMP assigns a Response Limits column property, based on these specifications, to each response column in the design table. It uses this information to define a desirability function for each response. The Profiler and Contour Profiler use these desirability functions to find optimal factor settings. For further details, see the Profiler chapter in the *Profilers* book and "Response Limits" on page 653 in the "Column Properties" appendix.

– A Goal of Maximize indicates that the best value is the largest possible. If there are natural lower or upper bounds, you can specify these as the Lower Limit or Upper Limit.

– A Goal of Minimize indicates that the best value is the smallest possible. If there are natural lower or upper bounds, you can specify these as the Lower Limit or Upper Limit.

– A Goal of Match Target indicates that the best value is a specific target value. The default target value is assumed to be midway between the Lower Limit and Upper Limit.

– A Goal of None indicates that there is no goal in terms of optimization. No desirability function is constructed.

Note: If your target response is not midway between the Lower Limit and the Upper Limit, you can change the target after you generate your design table. In the data table, open the Column Info window for the response column (**Cols** > **Column Info**) and enter the desired target value.

Importance When you have several responses, the Importance values that you specify are used to compute an overall desirability function. These values are treated as weights for the responses. If there is only one response, then specifying the Importance is unnecessary because it is set to 1 by default.

Editing the Responses Outline

In the Responses outline, note the following:

- Double-click a response to edit the response name.
- Click the goal to change it.
- Click on a limit or importance weight to change it.
- For multiple responses, you might want to enter values for the importance weights.

Response Limits Column Property

The Goal, Lower Limit, Upper Limit, and Importance that you specify when you enter a response are used in finding optimal factor settings. For each response, the information is saved in the generated design data table as a Response Limits column property. JMP uses this information to define the desirability function. The desirability function is used in the Prediction Profiler to find optimal factor settings. For further details about the Response Limits column property and examples of its use, see "Response Limits" on page 653 in the "Column Properties" appendix.

If you do not specify a Lower Limit and Upper Limit, JMP uses the range of the observed data for the response to define the limits for the desirability function. Specifying the Lower Limit and Upper Limit gives you control over the specification of the desirability function. For more details about the construction of the desirability function, see the Profiler chapter in the *Profilers* book.

Factors

Add factors in the Factors outline.

Tip: When you have completed the Factors outline, consider selecting **Save Factors** from the red triangle menu. This saves the factor names, roles, changes, and values in a data table that you can later reload.

Figure 4.18 Factors Outline

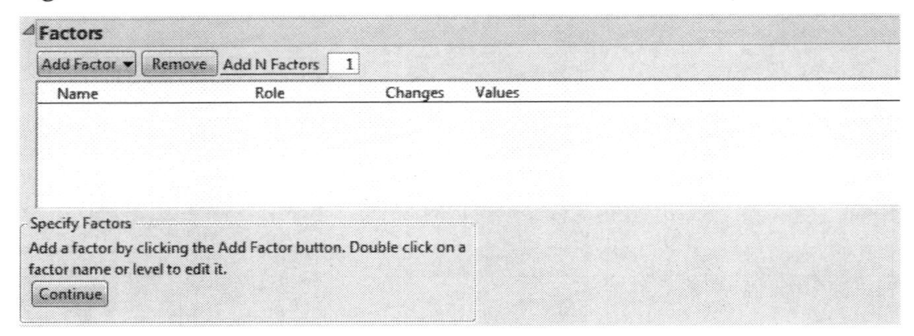

Add Factor Select the factor type. For details, see "Factor Types" on page 87.

Remove Removes the selected factors.

Note: If you attempt to remove all factors after clicking the Continue or Back button, one continuous factor remains. You can delete it after you add new factors.

Add N Factors Adds multiple factors. Enter the number of factors to add, click **Add Factor,** and then select the factor type. Repeat **Add N Factors** to add multiple factors of different types.

Factors Outline

The Factors outline contains the following columns:

Name The name of the factor. When added, a factor is given a default name of X1, X2, and so on. To change this name, double-click it and enter the desired name.

Role Specifies the Design Role of the factor. The Design Role column property for the factor is saved to the data table. This property ensures that the factor type is modeled appropriately.

Changes Indicates whether the factor levels are Easy, Hard, or Very Hard to change. Click on the default value of Easy to change it. When you specify factors as Hard or Very Hard to change, your design reflects these restrictions on randomization. A factor cannot be designated as Very Hard unless the Factors list contains a factor designated as Hard. The Factor Changes column property is saved to the data table. For more details, see "Changes and Random Blocks" on page 89.

Values The experimental settings for the factors. To insert Values, click on the default values and enter the desired values.

Editing the Factors Outline

In the Factors outline, note the following:

- To edit a factor name, double-click the factor name.
- Categorical factors have a down arrow to the left of the factor name. Click the arrow to add a level.
- To remove a factor level, click the value, click **Delete**, and click outside the text box.
- To modify the entry under Changes, click the value in the Changes column and select the appropriate entry.
- To edit a value, click the value in the Values column.

Factor Types

To choose a factor type, click **Add Factor** in Custom Design.

Note: A Design Role column property containing each factor's role is added to that factor's column in the design table that is generated. The Design Role column property ensures that the factor is modeled correctly.

Continuous Numeric data types only. A continuous factor is a factor that you can conceptually set to any value between the lower and upper limits you supply, given the limitations of your process and measurement system.

Discrete Numeric Numeric data types only. A discrete numeric factor can assume only a discrete number of values. These values have an implied order.

The default values for a discrete numeric factor with *k* levels, where $k > 2$, are the integers $1, 2, \ldots, k$. The default values for a discrete numeric factor with $k = 2$ levels are -1 and 1. Replace the default values with the settings that you plan to use in your experiment.

Note: Not all levels of a discrete numeric factor appear in the design. The levels that appear are determined by your specifications in the Model outline. If you need all levels to appear in your design, consider using the Screening Design platform.

In the assumed model, the effects for a discrete numeric factor with *k* levels include polynomial terms in that effect through order *k-1*. For *k* greater than 6, powers up to the 5th level are included. The Estimability for polynomial effects (powers of two or higher) is set to If Possible. This allows the algorithm to use the multiple levels as permitted by the run size. If the polynomial terms are not included, then a main effects only design is created. For more details about how discrete numeric factors are treated in the assumed model, see "Model" on page 95.

Fit Model treats a discrete numeric factor as a continuous predictor. The Model script that is saved to the design table does not contain any polynomial terms of order greater than two.

Categorical Either numeric or character data types. The data type in the resulting data table is categorical. The value ordering of the levels is the order of the values, as entered from left to right. This ordering is saved in the Value Ordering column property after the design data table is created.

Blocking Either numeric or character data types. A blocking factor is a special type of categorical factor that can enter the model only as a main effect. When you define a blocking factor, you specify the number of runs per block. The RunsPerBlock column property is saved to the design table. The Default run size always assumes that there are at least two blocks. If you specify a run size that is not an integer multiple of the number of runs per block, JMP tries to balance the design to the extent possible. In balancing the design, JMP ensures that there are at least two runs per block.

Covariate Either numeric or character data types. The values of a covariate factor are measurements on experimental units that are known in advance of an experiment. Covariate values are selected to ensure the optimality of the resulting design relative to the optimality criterion. See "Changes and Random Blocks" on page 89 and "Covariates with Hard-to-Change Levels" on page 122.

JMP obtains the covariate factors and their values from a data table that contains the measured covariates for the available experimental units. Make this data table your current data table. When you select Covariate, a list of columns in the current data table opens, and you select the columns containing covariates from this list.

In some situations, you may want to select a small set of design points from a larger set of candidate settings. For example, you may have multiple measurement columns (factors) for a large batch of units. You want to treat the measurements for each unit as a candidate run. From these candidate runs, you want to select a small but optimal collection for which you will measure a response. In this case, make the data table of all candidate runs the active table, select Add Factor > Covariate, and enter all of your measurement columns as covariates. Specify your desired run size. The Custom Design platform will identify an optimal collection of design settings.

Note: You cannot specify a Number of Runs or Number of Whole Plots that exceeds the number of rows in the covariate's data table.

Mixture Continuous factors that represent ingredients in a mixture. The values for a mixture factor must sum to a constant. By default, the values for all mixture factors sum to one. To set the sum of the mixture components to some other positive value, select **Advanced**

Options > Mixture Sum from the red triangle menu. The Mixture column property is saved to the data table.

Constant Either numeric or character data types. A constant factor is a factor whose values are fixed during an experiment. Constant factors are not included in the Model outline or in the Model script that is saved to the data table.

Uncontrolled Either numeric or character data types. An uncontrolled factor is one whose values cannot be controlled during production, but it is a factor that you want to include in the model. It is assumed that you can record the factor's value for each experimental run.

An empty column with a Continuous Modeling Type is created in the design table. You can change the column's Data Type and Modeling Type in the Column Info window if required. Enter your data in this column. Uncontrolled factors are included in the Model outline and the Model script that is saved to the data table.

Changes and Random Blocks

Specifying the relative difficulty of changing a factor from run to run is useful in industrial experimentation. It is often convenient to make several runs while keeping factors that are hard-to-change fixed at some setting. A Changes value of Hard results in a split-plot design. A Changes value of Very Hard results in a split-split-plot design or a two-way split-plot design.

You can set Changes for Continuous, Discrete Numeric, Categorical, and Mixture factors to Hard and Very Hard. To set a factor to Very Hard, the list must contain another factor that is set to Hard.

You can set Changes for a Covariate factor to Hard. In this case, all other covariates are also set to Hard and the remaining factors are set to Easy. The algorithm requires a combination of row exchange and coordinate exchange. For this reason, even moderately sized designs might take some time to generate.

For designs with Hard or Very Hard to change factors, Custom Design strives to find a design that is optimal, given your specified optimality criterion. See "Optimality Criteria" on page 123. For details about the methodology used to generate split-plot designs, see Jones and Goos (2007). For details relating to designs with hard-to-change covariates, see Jones and Goos (2015).

Figure 4.19 shows a split-split-plot scenario, using the factors from the Cheese Factors.jmp sample data table (located in the Design Experiment folder).

Figure 4.19 Factors and Design Generation Outline for a Split-Split-Plot Design

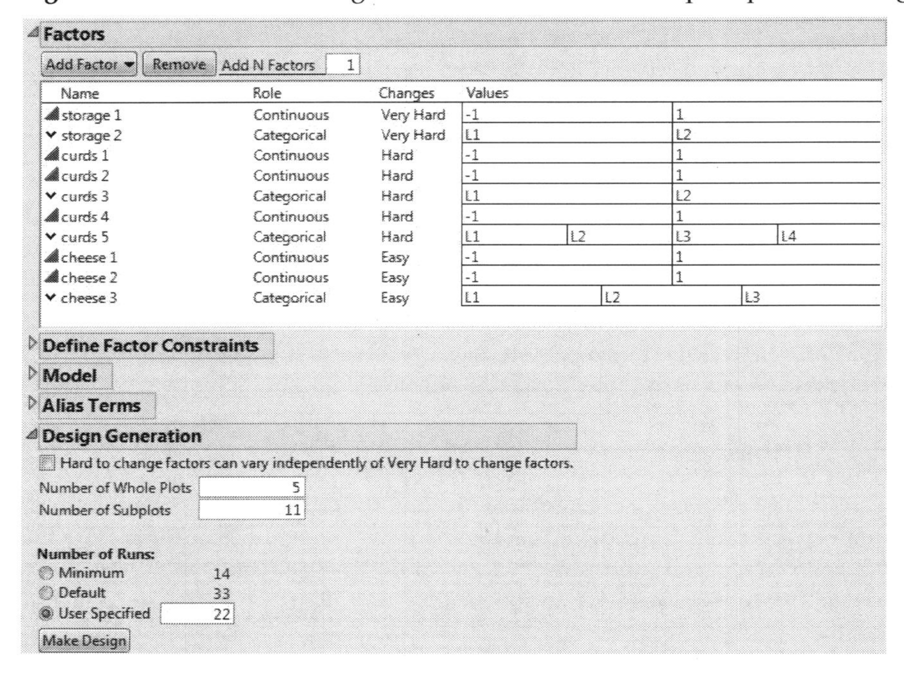

If you assign Changes as Hard for one or more factors, but no factors have Changes assigned as Very Hard, a categorical factor called Whole Plots is added to the design. This situation results in a split-plot design:

- Each level of Whole Plots corresponds to a block of constant settings of the hard-to-change factors.
- The Model script in the design table applies the Random Effect attribute to the factor Whole Plots.
- The factor Whole Plots is assigned the Design Role column property with a value of Random Block.

When you designate Changes as both Hard and Very Hard, categorical factors called Subplots and Whole Plots are added to the design. This situation results in a split-split-plot design:

- Each level of Subplots corresponds to a block of constant settings of the hard-to-change factors.
- Each level of Whole Plots corresponds to a block of constant settings of the very-hard-to-change factors.
- The Model script in the design table applies the Random Effect attribute to the Whole Plots and Subplots effects.

- The levels of the hard-to-change factor are assumed to be nested within the levels of the very-hard-to-change factor by default.
- In the design table, both of the factors Whole Plots and Subplots are assigned the Design Role column property with a value of Random Block.

To construct a two-way split-plot design, select the **Hard to change factors can vary independently of Very Hard to change factors** option under Design Generation. The option crosses the levels of the hard-to-change factor with the levels of the very-hard-to-change factor. See "Two-Way Split-Plot Designs" on page 118.

Use the Number of Whole Plots and Number of Subplots text boxes to specify values for the numbers of whole plots or subplots. These boxes are initialized to suggested numbers of whole plots and subplots. For information about how these values are obtained, see "Numbers of Whole Plots and Subplots" on page 122.

For more details and scenarios that illustrate random block split-plot, split-split-plot, and two-way split-plot designs, see "Designs with Randomization Restrictions" on page 113. For details about designs with hard-to-change covariates, see "Covariates with Hard-to-Change Levels" on page 122.

Factor Column Properties

For each factor, various column properties are saved to the data table. You can find details about these column properties and related examples in Appendix A, "Column Properties".

Design Role Each factor is given the Design Role column property. The Role that you specify in defining the factor determines the value of its Design Role column property. When you add a random block under Design Generation, that factor is assigned the Random Block value. The Design Role property reflects how the factor is intended to be used in modeling the experimental data. Design Role values are used in the Augment Design platform. For details, see "Design Role" on page 661 in the "Column Properties" appendix.

Factor Changes Each factor is assigned the Factor Changes column property. The value that you specify under Changes determines the value of its Factor Changes column property. The Factor Changes property reflects how the factor is used in modeling the experimental data. Factor Changes values are used in the Augment Design and Evaluate Design platforms. For details, see "Factor Changes" on page 676 in the "Column Properties" appendix.

Coding If the Role is Continuous, Discrete Numeric, a continuous Covariate, or Uncontrolled, the Coding column property for the factor is saved. This property transforms the factor values so that the low and high values correspond to –1 and +1, respectively. For details, see "Coding" on page 663 in the "Column Properties" appendix.

Value Ordering If the Role is Categorical or Blocking, the Value Ordering column property for the factor is saved. This property determines the order in which levels of the factor appear. For details, see "Value Ordering" on page 679 in the "Column Properties" appendix.

Mixture If the Role is Mixture, the Mixture column property for the factor is saved. This property indicates the limits for the factor and the mixture sum. It also enables you to choose the coding for the mixture factors. For details, see "Mixture" on page 671 in the "Column Properties" appendix.

RunsPerBlock For a blocking factor, indicates the maximum allowable number of runs in each block. When a Blocking factor is specified in the Factors outline, the RunsPerBlock column property is saved for that factor. For details, see "RunsPerBlock" on page 685 in the "Column Properties" appendix.

Define Factor Constraints

Note: If you are working in Covering Arrays, see the "Covering Arrays" chapter on page 541 for more information.

Use Define Factor Constraints to restrict the design space. Unless you have loaded a constraint or included one as part of a script, the **None** option is selected. To specify constraints, select one of the other options:

Specify Linear Constraints Specifies inequality constraints on linear combinations of factors. Only available for factors with a Role of Continuous or Mixture. See "Specify Linear Constraints".

Note: When you save a script for a design that involves a linear constraint, the script expresses the linear constraint as a *less than or equal to* inequality ($\leq$).

Use Disallowed Combinations Filter Defines sets of constraints based on restricting values of individual factors. You can define both AND and OR constraints. See "Use Disallowed Combinations Filter".

Use Disallowed Combinations Script Defines disallowed combinations and other constraints as Boolean JSL expressions in a script editor box. See "Use Disallowed Combinations Script".

Specify Linear Constraints

In cases where it is impossible to vary continuous factors independently over the design space, you can specify linear inequality constraints. Linear inequalities describe factor level settings that are allowed.

Click **Add** to enter one or more linear inequality constraints.

Add Adds a template for a linear expression involving all the continuous factors in your design. Enter coefficient values for the factors and select the direction of the inequality to reflect your linear constraint. Specify the constraining value in the box to the right of the inequality. To add more constraints, click **Add** again.

Note: The Add option is disabled if you have already constrained the design region by specifying a Sphere Radius.

Remove Last Constraint Removes the last constraint.

Check Constraints Checks the constraints for consistency. This option removes redundant constraints and conducts feasibility checks. A JMP alert appears if there is a problem. If constraints are equivalent to bounds on the factors, a JMP alert indicates that the bounds in the Factors outline have been updated.

Use Disallowed Combinations Filter

This option uses an adaptation of the Data Filter to facilitate specifying disallowed combinations. For detailed information about using the Data Filter, see the JMP Reports chapter in the *Using JMP* book.

Select factors from the Add Filter Factors list and click **Add**. Then specify the disallowed combinations by using the slider (for continuous factors) or by selecting levels (for categorical factors).

The red triangle options for the Add Filter Factors menu are those found in the Select Columns panel of many platform launch windows. See the Get Started chapter in the *Using JMP* book for additional details about the column selection menu.

When you click Add, the Disallowed Combinations control panel shows the selected factors and provides options for further control. Factors are represented as follows, based on their modeling types:

Continuous Factors For a continuous factor, a double-arrow slider that spans the range of factor settings appears. An expression that describes the range using an inequality appears above the slider. You can specify disallowed settings by dragging the slider arrows or by clicking on the inequality bounds in the expression and entering your desired constraints. In the slider, a solid blue highlight represents the disallowed values.

Categorical Factor For a categorical factor, the possible levels are displayed either as labeled blocks or, when the number of levels is large, as list entries. Select a level to disallow it. To select multiple levels, hold the Control key. The block or list entries are highlighted to indicate the levels that have been disallowed. When you add a categorical factor to the Disallowed Combinations panel, the number of levels of the categorical factor is given in parentheses following the factor name.

Disallowed Combinations Options

The control panel has the following controls:

Clear Clears all disallowed factor level settings that you have specified. This does not clear the selected factors.

Start Over Removes all selected factors and returns you to the initial list of factors.

AND Opens the Add Filter Factors list. Selected factors become an AND group. Any combination of factor levels specified within an AND group is disallowed.

To add a factor to an AND group later on, click the group's outline to see a highlighted rectangle. Select AND and add the factor.

To remove a single factor, select **Delete** from its red triangle menu.

OR Opens the Add Filter Factors list. Selected factors become a separate AND group. For AND groups separated by OR, a combination is disallowed if it is specified in at least one AND group.

Red Triangle Options for Factors

A factor can appear in several OR groups. An occurrence of the factor in a specific OR group is referred to as an *instance* of the factor.

Delete Removes the selected instance of the factor from the Disallowed Combinations panel.

Clear Selection Clears any selection for that instance of the factor.

Invert Selection Deselects the selected values and selects the values not previously selected for that instance of the factor.

Display Options Available only for categorical factors. Changes the appearance of the display. Options include:

– **Blocks Display** shows each level as a block.

– **List Display** shows each level as a member of a list.

– **Single Category Display** shows each level.

– **Check Box Display** adds a check box next to each value.

Find Available only for categorical factors. Provides a text box beneath the factor name where you can enter a search string for levels of the factor. Press the Enter key or click outside the text box to perform the search. Once **Find** is selected, the following Find options appear in the red triangle menu:

– **Clear Find** clears the results of the Find operation and returns the panel to its original state.

– **Match Case** uses the case of the search string to return the correct results.

– **Contains** searches for values that include the search string.

– **Does not contain** searches for values that do not include the search string.

– **Starts with** searches for values that start with the search string.

– **Ends with** searches for values that end with the search string.

Use Disallowed Combinations Script

Use this option to disallow particular combinations of factor levels using a JSL script. This option can be used with continuous factors or mixed continuous and categorical factors.

This option opens a script window where you insert a script that identifies the combinations that you want to disallow. The script must evaluate as a Boolean expression. When the expression evaluates as true, the specified combination is disallowed.

When forming the expression for a categorical factor, use the ordinal value of the level instead of the name of the level. If a factor's levels are high, medium, and low, specified in that order in the Factors outline, their associated ordinal values are 1, 2, and 3. For example, suppose that you have two continuous factors, X1 and X2, and a categorical factor X3 with three levels: L1, L2, and L3, in order. You want to disallow levels where the following holds:

$$e^{X_1} + 2X_2 < 0 \text{ and } X_3 = L2$$

Enter the expression `(Exp(X1) + 2*X2 < 0) & (X3 == 2)` into the script window.

Figure 4.20 Expression in Script Editor

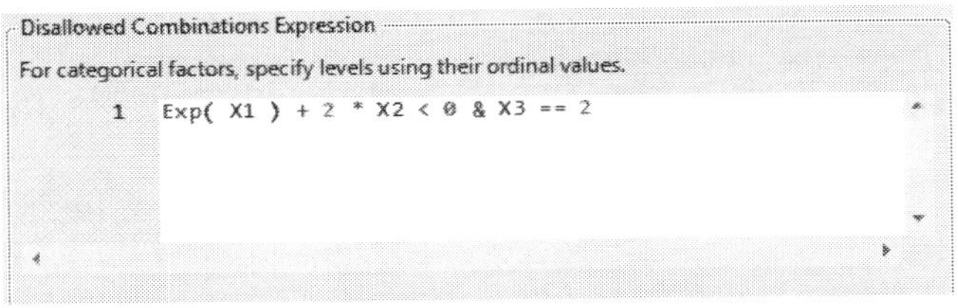

Disallowed Combinations Expression
For categorical factors, specify levels using their ordinal values.

```
1   Exp( X1 ) + 2 * X2 < 0 & X3 == 2
```

(In the figure, unnecessary parentheses were removed by parsing.) Notice that functions can be entered as part of the Boolean expression.

Model

Specify your assumed model (which contains all the effects that you want to estimate) in the Model outline. For each effect that you specify, you can designate that effect's Estimability. The Estimability value indicates whether it is Necessary to estimate that effect, or if you are content to estimate that effect If Possible.

When the Model outline opens, for most factors only the main effects appear. If you have entered a discrete numeric factor, polynomial terms also appear. The Estimability of second-and higher-order terms is set to If Possible. If you want to ensure that these terms are estimable, change their Estimability to Necessary.

Note: You can ensure that the estimability of discrete numeric polynomial terms is always set to Necessary. Select **File > Preferences > Platforms > DOE**. Check Discrete Numeric Powers Set to Necessary.

Figure 4.21 Model Outline

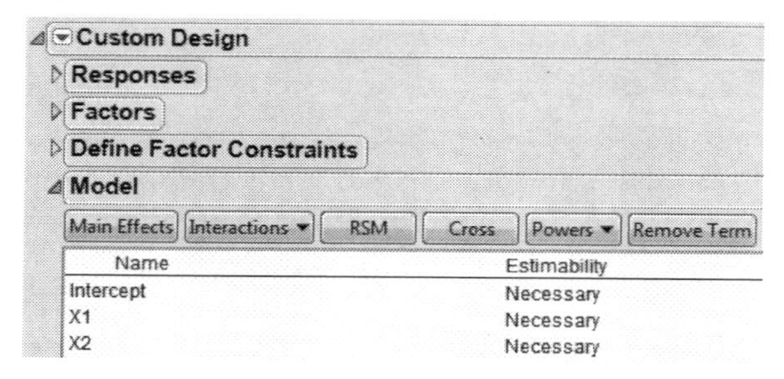

When you construct your design table, JMP saves a Model script to the data table. Except for discrete numeric factors, the Model script contains the effects shown in the Model outline. For a discrete numeric factor, the Model script contains only its main effect and quadratic term.

The Model outline contains the following buttons and fields:

Main Effects Adds main effects for all factors in the model, and polynomial terms for discrete numeric factors.

Interactions Adds interaction effects. If no factors are selected in the Factors outline, select 2nd, 3rd, 4th, or 5th to add all appropriate interactions up to that order. Add interactions up to a given order for specific factors by selecting the factor names in the Factors outline, selecting Interactions, and then specifying the appropriate order. Interactions between non-mixture and mixture factors, and interactions with blocking and constant factors, are not added.

RSM Adds interaction and quadratic terms up to the second order (response surface model terms) for continuous factors. Categorical factors are not included in RSM terms. Main effects for non-mixture factors that interact with all the mixture factors are removed.

Cross Adds specific interaction terms. Select factor names in the Factors outline and effect names in the Model outline. Click Cross to add the crossed terms to the Model outline.

Powers Adds polynomial terms. If no factor names are selected in the Factors outline, adds polynomial terms for all continuous factors. If factor names are selected in the Factors outline, adds polynomial terms for only those factors. Select 2nd, 3rd, 4th, or 5th to add polynomial terms of that order.

Scheffé Cubic Adds Scheffé cubic terms for all mixture factors. These terms are used to specify a mixture model with third-degree polynomial terms.

Remove Term Removes selected effects.

Name Name of the effect.

Estimability A designation of your need to estimate the effect.

– If Estimability is set to Necessary, the algorithm ensures that the effect is estimable.
– If Estimability is set to If Possible, the algorithm attempts to make that effect estimable, as permitted by the number of runs that you specify.

Except for polynomial terms for discrete numeric factors, all effects are specified as Necessary by default. Click an effect's Estimability value to change it.

Bayesian D-Optimality and Estimation of If Possible Effects

The Bayesian D-Optimal design approach obtains precise estimation of all Necessary terms while providing omnibus detectability (and some estimability) for If Possible terms. For more detail, see "Response Surface Experiments" on page 150 in the "Examples of Custom Designs" chapter and "Bayesian D-Optimality" on page 124.

Alias Terms

It is possible that effects *not* included in your assumed model are active. In the Alias Terms outline, add potentially active effects that are not in your assumed model but might bias the estimates of model terms. Once you generate your design, the Alias Matrix outline appears under Design Evaluation. The Alias Matrix entries represent the degree of bias imparted to model parameters by the effects that you specified in the Alias Terms outline. For details, see the "The Alias Matrix" on page 690 in the "Technical Details" appendix.

By default, the Alias Terms outline includes all two-way interaction effects that are not in your Model outline (with the exception of terms involving blocking factors). Add terms using the buttons. For a description of how to use these buttons to add effects to the Alias Terms table, see "Model" on page 95.

For example, suppose that you specify a design with three continuous factors. Your assumed model, specified in the Model outline, contains only those three main effects. You can afford only six runs. You want to see how estimates of the main effects might be biased by active two-way interactions and the three-way interaction.

The Alias Terms table includes all two-way interactions by default. You can add the three-way interaction by selecting **Interactions > 3rd**.

Figure 4.22 Alias Terms Outline

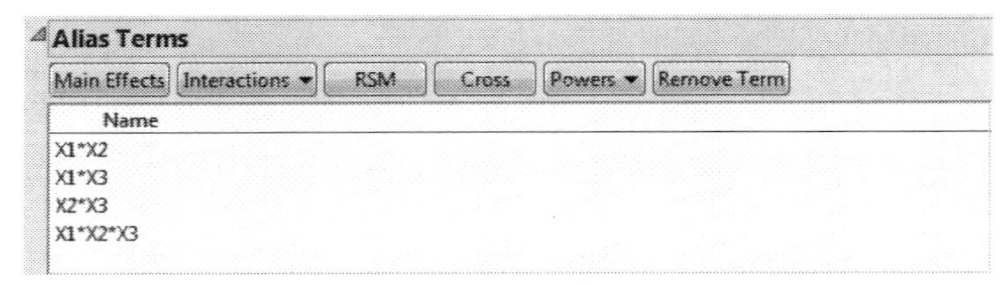

Once you specify six runs in the Design Generation outline and click **Make Design**, the Design Evaluation outline appears. Open the Design Evaluation outline and the Alias Matrix outline. See Figure 4.23.

Figure 4.23 Alias Matrix

Effect	X1*X2	X1*X3	X2*X3	X1*X2*X3
Intercept	0	0	0.5	-0.5
X1	0	0	-0.5	0.5
X2	0.5	-0.5	0	0
X3	-0.5	0.5	0	0

The Alias Matrix indicates that each main effect is partially aliased with two of the interactions. See "Alias Matrix" on page 446 in the "Evaluate Designs" chapter and "The Alias Matrix" on page 690 in the "Technical Details" appendix.

Design Generation

The Design Generation outline gives you choices relating to the size and structure of the design. Typically, the input area has two parts:

- Design structure options
- Number of runs options

Figure 4.24 Design Generation Outline

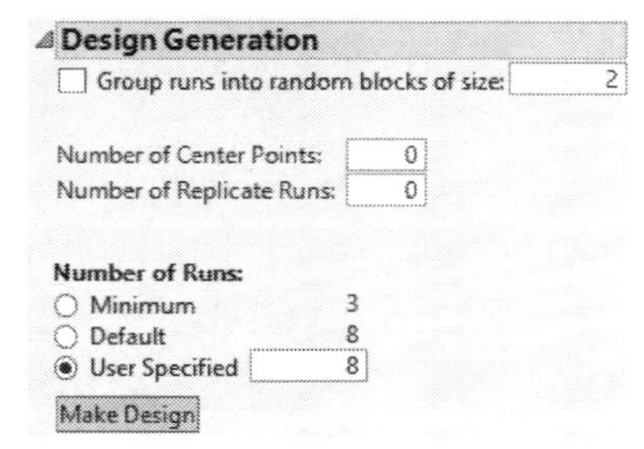

Design Structure Options

Group runs into random blocks of size (Not available if a blocking factor is specified) To construct a random block design, enter the number of runs that you want in each block. When you specify the sample size, a factor called Random Block is created. Its levels define blocks of a size that is consistent with the block size that you entered, given the specified number of runs. If the number of runs is an integer multiple of the block size, the block sizes equal your specified value.

Number of Whole Plots Appears when you specify a hard or very-hard-to-change factor. The factor Whole Plots corresponds to the very-hard-to-change factors (split-split-plot design), if there are any, otherwise to the hard-to-change factors (split-plot design). JMP suggests a value for the number of whole plots that maximizes the information about the coefficients in the model. Or, you can enter a value for the number of whole plots. For details, see "Numbers of Whole Plots and Subplots" on page 122.

Number of Subplots Appears when you specify a very-hard-to-change factor. The factor Subplots corresponds to the hard-to-change factors in the split-split-plot design. JMP suggests values for the number of whole plots and subplots that maximize the information about the coefficients in the model. Or, you can enter a value for the number of subplots. For details, see "Numbers of Whole Plots and Subplots" on page 122.

Hard to change factors can vary independently of Very Hard to change factors Select this option to create a strip-plot (also known as two-way split-plot or split block) design. This option creates a design where the hard-to-change factors are randomized within the levels of the very-hard-to-change factors. They are *not* nested within the very-hard-to-change factors.

Number of Center Points Appears only if the design contains factors with a Continuous or Mixture factor type. Specify how many additional runs you want to add as center points to the design. A center point is a run whose setting for each continuous factor is midway between the high and low settings. See "Center Points, Replicate Runs, and Testing" on page 60 in the "Starting Out with DOE" chapter.

If a design contains both continuous and other types of factors, center points might not be balanced relative to the levels of the other factors. Custom Design chooses the center points to maximize the D-, I-, or alias efficiency of the design.

Number of Replicate Runs Specify the number of replicate trials that you want to add to the design. This does not replicate the entire design, but chooses the optimal design points to replicate. See "Center Points, Replicate Runs, and Testing" on page 60 in the "Starting Out with DOE" chapter.

Number of Runs Options

Minimum A lower bound on the number of runs necessary to avoid failures in design generation. When you select Minimum, the resulting design is saturated. There are no degrees of freedom for error.

Note: If you select the Minimum number of runs, there will be no error term for testing. You will not be able to test parameter estimates. This choice is appropriate only when the cost of additional runs is prohibitive.

Default Suggests the number of runs. This value is based on heuristics for creating a balanced design with at least four runs more than the Minimum number of runs.

User Specified Specify the number of runs that you want. Enter that value into the Number of Runs text box. This option enables you to balance the cost of additional runs against the potential gain in information.

Number of Runs This is the only option that appears when a covariate factor with Changes set to Easy is specified. The number of runs shown is the number of rows in the data table associated with your covariate or covariates. You can specify a smaller number of runs. In that case, the covariate runs that are selected optimize the design criterion.

Make Design

Once you have completed the Design Generation outline, click **Make Design**. Custom Design generates the design, presents it in the Design outline, and provides evaluation information in the Design Evaluation outline. The Output Options panel also appears, allowing you to create the design table.

Note: Sometimes several designs can optimize the optimality criterion. When this is the case, the design algorithm might generate different designs when you click the **Back** and **Make Design** buttons repeatedly.

Design

The Design outline shows the runs for a design that is optimal, given the conditions that you have specified. The runs might not appear to be properly randomized. You can select Run Order options in the Output Options panel before generating your design table.

Design Evaluation

The Design Evaluation outline provides a number of ways to evaluate the properties of the generated design. Open the Design Evaluation outline to see the following options:

Power Analysis Enables you to explore your ability to detect effects of given sizes.

Prediction Variance Profile Shows the prediction variance over the range of factor settings.

Fraction of Design Space Plot Shows how much of the model prediction variance lies below (or above) a given value.

Prediction Variance Surface Shows a surface plot of the prediction variance for any two continuous factors.

Estimation Efficiency For each parameter, gives the fractional increase in the length of a confidence interval compared to that of an idealized (orthogonal) design, which might not exist. Also gives the relative standard error of the parameters.

Alias Matrix Gives coefficients that indicate the degree by which the model parameters are biased by effects that are potentially active, but not in the model. You specify the terms representing potentially active effects in the Alias Terms table. See "The Alias Matrix" on page 690 in the "Technical Details" appendix.

Color Map on Correlations Shows the absolute correlation between effects on a plot using an intensity scale.

Design Diagnostics Indicates the optimality criterion used to construct the design. Also gives efficiency measures for your design. See Optimality Criterion in "Custom Design Options" on page 103 and "Optimality Criteria" on page 123.

Note: The Design Diagnostics outline does not provide the following statistics when the model includes factors with Changes set to Hard or Very Hard or with Estimability set to If Possible: D Efficiency, G Efficiency, A Efficiency.

For more details about the Design Evaluation outline, see "Design Evaluation" on page 433 in the "Evaluate Designs" chapter.

Output Options

Use the Output Options panel to perform the following tasks:

- specify how you want the custom design data table to appear
- construct the design table
- return to a previous point in the Custom Design window

In most cases, the Output Options panel appears as shown in Figure 4.25.

Figure 4.25 Output Options Panel

The Output Options panel contains these options:

- "Run Order" on page 102
- "Make Table" on page 102

- "Back" on page 103

Run Order

The **Run Order** options determine the order of the runs in the design table. Choices include the following:

Keep the Same Rows in the design table are in the same order as in the Design outline.

Sort Left to Right Columns in the design table are sorted from left to right.

Randomize Rows in the design table are in random order.

Sort Right to Left Columns in the design table are sorted from right to left.

Randomize within Blocks Rows in the design table are in random order within the blocks.

Make Table

Click **Make Table** to construct the custom design data table. In the Custom Design table, the Table panel (in the upper left) can contain scripts, as appropriate given your design. The Model, Evaluate Design, and DOE Dialog scripts are always provided. To run a script, click the green triangle next to the script name.

Figure 4.26 Custom Design Table Showing Scripts

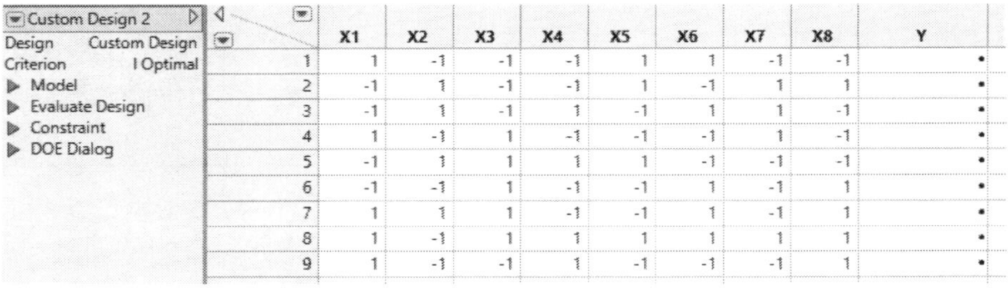

Custom Design 2
Design Custom Design
Criterion I Optimal
▶ Model
▶ Evaluate Design
▶ Constraint
▶ DOE Dialog

	X1	X2	X3	X4	X5	X6	X7	X8	Y
1	1	-1	-1	-1	1	1	-1	-1	•
2	-1	1	-1	-1	1	-1	1	1	•
3	-1	1	-1	1	-1	1	1	-1	•
4	1	-1	1	-1	-1	-1	1	-1	•
5	-1	1	1	1	1	-1	-1	-1	•
6	-1	-1	1	-1	-1	1	-1	1	•
7	1	1	1	-1	-1	1	-1	1	•
8	1	-1	1	1	1	1	1	1	•
9	1	-1	-1	1	-1	-1	-1	1	•

Possible scripts include the following:

Model Runs the **Analyze** > **Fit Model** platform. The model described by the script is determined by your choices in the Model outline and by the type of design.

Evaluate Design Runs the **DOE > Design Diagnostics > Evaluate Design** platform. The model described by the script is determined by your choices in the Model outline and by the type of design.

Constraint Shows model constraints that you entered in the Define Factor Constraints outline using the **Specify Linear Constraints** option.

Disallowed Combinations Shows model constraints that you entered in the Define Factor Constraints outline using the **Use Disallowed Combinations Filter** or the **Use Disallowed Combinations Script** options.

DOE Dialog Re-creates the Custom Design window that you used to generate the design table. The script also contains the random seed used to generate your design.

Back

The Back button takes you back to where you were before clicking Make Design. You can make changes to the previous outlines and regenerate the design.

Note: If you attempt to remove all factors after clicking the Back button, one continuous factor remains. You can delete the continuous factor after new factors are added.

Custom Design Options

This section describes the options available under the Custom Design red triangle menu.

Description of Options

The Custom Design red triangle menu contains the following options:

Save Responses Saves the information in the Responses panel to a new data table. You can then quickly load the responses and their associated information into most DOE windows. This option is helpful if you anticipate re-using the responses.

Load Responses Loads responses that you have saved using the Save Responses option.

Save Factors Saves the information in the Factors panel to a new data table. Each factor's column contains its levels. Other information is stored as column properties. You can then quickly load the factors and their associated information into most DOE windows.

Note: It is possible to create a factors table by entering data into an empty table, but remember to assign each column an appropriate Design Role. Do this by right-clicking on the column name in the data grid and selecting **Column Properties > Design Role**. In the Design Role area, select the appropriate role.

Load Factors Loads factors that you have saved using the Save Factors option.

Save Constraints (Unavailable for some platforms) Saves factor constraints that you have defined in the Define Factor Constraints or Linear Constraints outline into a data table, with a column for each constraint. You can then quickly load the constraints into most DOE windows.

In the constraint table, the first rows contain the coefficients for each factor. The last row contains the inequality bound. Each constraint's column contains a column property called ConstraintState that identifies the constraint as a "less than" or a "greater than" constraint. See "ConstraintState" on page 686 in the "Column Properties" appendix.

Load Constraints (Unavailable for some platforms) Loads factor constraints that you have saved using the Save Constraints option.

Set Random Seed Sets the random seed that JMP uses to control certain actions that have a random component. These actions include:

– simulating responses using the Simulate Responses option
– randomizing Run Order for design construction
– selecting a starting design for designs based on random starts

To reproduce a design or simulated responses, enter the random seed that generated them. For designs using random starts, set the seed before clicking Make Design. To control simulated responses or run order, set the seed before clicking Make Table.

Note: The random seed associated with a design is included in the DOE Dialog script that is saved to the design data table.

Simulate Responses Adds response values and a column containing a simulation formula to the design table. Select this option before you click Make Table.

When you click Make Table, the following occur:

– A set of simulated response values is added to each response column.
– For each response, a new a column that contains a simulation model formula is added to the design table. The formula and values are based on the model that is specified in the design window.
– A Model window appears where you can set the values of coefficients for model effects and specify one of three distributions: Normal, Binomial, or Poisson.
– A script called **DOE Simulate** is saved to the design table. This script re-opens the Model window, enabling you to re-simulate values or to make changes to the simulated response distribution.

Make selections in the Model window to control the distribution of simulated response values. When you click Apply, a formula for the simulated response values is saved in a new column called <Y> Simulated, where Y is the name of the response. Clicking Apply again updates the formula and values in <Y> Simulated.

For additional details, see "Simulate Responses" on page 106 in the "Custom Designs" chapter.

Note: JMP PRO You can use Simulate Responses to conduct simulation analyses using the JMP Pro Simulate feature. For information about Simulate and some DOE examples, see the Simulate chapter in the *Basic Analysis* book.

Save X Matrix Saves scripts called Moments Matrix and Model Matrix to the design data table. These scripts contain the moments and design matrices. See "Save X Matrix" on page 108.

Caution: For a design with nominal factors, the matrix in the Model Matrix script saved by the Save X Matrix option is *not* the coding matrix used in fitting the linear model. You can obtain the coding matrix used for fitting the model by selecting the option Save Columns > Save Coding Table in the Fit Model report that you obtain when you run the Model script.

Optimality Criterion Changes the design optimality criterion. The default criterion, **Recommended**, specifies D-optimality for all design types, unless you added quadratic effects using the RSM button in the Model outline. For more information about the D-, I-, and alias-optimal designs, see "Optimality Criteria" on page 123.

Note: You can set a preference to always use a given optimality criterion. Select File > Preferences > Platforms > DOE. Check Optimality Criterion and select your preferred criterion.

Number of Starts Enables you to specify the number of random starts used in constructing the design. See "Number of Starts" on page 110.

Design Search Time Maximum number of seconds spent searching for a design. The default search time is based on the complexity of the design. See "Design Search Time" on page 111 and "Number of Starts" on page 110.

If the iterations of the algorithm require more than a few seconds, a Computing Design progress window appears. If you click **Cancel** in the progress window, the calculation stops and gives the best design found at that point. The progress window also displays D-efficiency for D-optimal designs that do not include factors with Changes set to Hard or Very Hard or with Estimability set to If Possible.

Note: You can set a preference for Design Search Time. Select File > Preferences > Platforms > DOE. Check Design Search Time and enter the maximum number of seconds. In certain situations where more time is required, JMP extends the search time.

Sphere Radius Constrains the continuous factors in a design to a hypersphere. Specify the radius and click **OK**. Design points are chosen so that their distance from 0 equals the Sphere Radius. Select this option before you click Make Design.

Note: Sphere Radius constraints cannot be combined with constraints added using the Specify Linear Constraints option. Also, the option is not available when hard-to-change factors are included (split-plot designs).

Advanced Options > Mixture Sum Set the sum of the mixture factors to any positive value. Use this option to keep a component of a mixture constant throughout an experiment.

Advanced Options > Split Plot Variance Ratio Specify the ratio of the variance of the random whole plot and the subplot variance (if present) to the error variance. Before setting this value, you must define a hard-to-change factor for your split-plot design, or hard and very-hard-to-change factors for your split-split-plot design. Then you can enter one or two positive numbers for the variance ratios, depending on whether you have specified a split-plot or a split-split-plot design.

Advanced Options > Prior Parameter Variance (Available only when the Model outline is available) Specify the weights that are used for factors whose Estimability is set to If Possible. The option updates to show the default weights when you click Make Design. Enter a positive number for each of the terms for which you want to specify a weight. The value that you enter is the square root of the reciprocal of the prior variance. A larger value represents a smaller variance and therefore more prior information that the effect is not active.

Bayesian D- or I-optimality is used in constructing designs with If Possible factors. The default values used in the algorithm are 0 for Necessary terms, 4 for interactions involving If Possible terms, and 1 for If Possible terms. For more details, see "The Alias Matrix" on page 690 in the "Technical Details" appendix and "Optimality Criteria" on page 123.

Advanced Options > D Efficiency Weight Specify the relative importance of D-efficiency to alias optimality in constructing the design. Select this option to balance reducing the variance of the coefficients with obtaining a desirable alias structure. Values should be between 0 and 1. Larger values give more weight to D-Efficiency. The default value is 0.5. This option has an effect only when you select Make Alias Optimal Design as your Optimality Criterion.

For the definition of D-efficiency, see "Optimality Criteria" on page 123. For details about alias optimality, see "Alias Optimality" on page 127.

Advanced Options > Set Delta for Power Specify the difference in the mean response that you want to detect for model effects. See "Set Delta for Power" on page 112.

Save Script to Script Window Creates the script for the design that you specified in the Custom Design window and places it in an open script window.

Simulate Responses

When you click Make Table to create your design table, the Simulate Responses option does the following for each response:

- It adds random response values to the response column in your design table.
- It adds a new a column containing a simulation model formula to the design table. The formula and values are based on the model that is specified in the design window.

A Model window opens where you can specify parameter values and select a response distribution for simulation. When you click Apply in the Model window, each column containing a simulation model formula is updated.

Control Window

Figure 4.27 shows the Model window for a design with one continuous factor (X1) and one three-level categorical factor (X2). Notice that X2 is represented by two model terms.

Figure 4.27 Simulate Responses Control Window

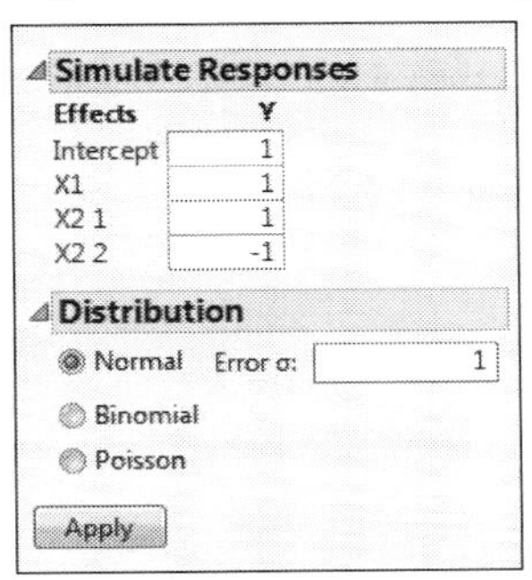

The initial window shows values for the coefficients of either 1 or -1, and a Normal distribution with error standard deviation equal to 1. If you have set Anticipated Coefficients as part of Power Analysis under Design Evaluation in the DOE window, then the default values in the Simulate Responses outline are the values that you specified as Anticipated Coefficients and Anticipated RMSE (Error Std) in the Power Analysis outline. If it is not possible to fit the model specified in the data table's Model script, the intercept and coefficients have default values of 0.

Simulate Responses

To specify a model for simulated values, do the following:

1. For each term in the list of Effects, enter coefficients for the linear model used to simulate the response values. These define a linear function, $L(\mathbf{x}, \beta) = \mathbf{x}'\beta$. See the Simulate Responses outline in Figure 4.27:
   – The vector **x** consists of the terms that define the effects listed under Effects.
   – The vector β is the vector of model coefficients that you specify under Y.
2. Under Distribution, select a response distribution.
3. Click **Apply**. A <Y> Simulated column containing simulated values and their formula is added to the design table, where Y is the name of the response column.

Distribution

Choose from one of the following distributions in the Simulate Responses window:

Normal Simulates values from a normal distribution. Enter a value for Error σ, the standard deviation of the normal error distribution. If you have designated factors to have Changes of Hard in the Factors outline, you can enter a value for Whole Plots σ, the whole plot error. If you have designated factors to have Changes of Hard and Very Hard, you can enter values for both the subplot and whole plot errors. When you click Apply, random values and a formula containing a random response vector based on the model are entered in the column <Y> Simulated.

Binomial Simulates values from a binomial distribution. Enter a value for N, the number of trials. Random integer values are generated according to a binomial distribution based on N trials with probability of success $1/(1 + exp(-L(\mathbf{x}, \beta))$. When you click Apply, random values and their formula are entered in the column <Y> Simulated. A column called N Trials that contains the value N is also added to the data table.

Poisson Simulates random integer values according to a Poisson distribution with parameter $exp((L(\mathbf{x}, \beta))$. When you click Apply, random values and their formula are entered in the column <Y> Simulated.

Note: You can set a preference to simulate responses every time you click Make Table. To do so, select **File > Preferences > Platforms > DOE**. Select **Simulate Responses**.

Save X Matrix

This option saves scripts called Moments Matrix and Model Matrix that contain the moments matrix and the model matrix. The moments matrix and the model matrix are used to calculate the Average Variance of Prediction, which appears in the Design Diagnostics section of the Design Evaluation outline. For details, see Goos and Jones (2011). If the design is a split-plot design, a V Inverse script is also saved. The V Inverse script contains the inverse of the covariance matrix of the responses.

Caution: For a design with nominal factors, the matrix in the Model Matrix script saved by the Save X Matrix option is *not* the coding matrix used in fitting the linear model. You can obtain the coding matrix used for fitting the model by selecting the option Save Columns > Save Coding Table in the Fit Model report that you obtain when you run the Model script.

Note: You can set a preference to always save the matrix script. Select **File > Preferences > Platforms > DOE**. Check Save X Matrix.

Model Matrix

The *model matrix* describes the design for the experiment. The model matrix has a row for each run and a column for each term of the model specified in the Model outline. For each run, the corresponding row of the model matrix contains the coded values of the model terms:

- Continuous terms are coded to range from -1 to 1.

- Nominal terms are coded by applying the Gram-Schmidt orthogonalization procedure to the coding for nominal effects that is used in fitting linear models.

Because of how nominal terms are coded for constructing optimal designs, when a design contains nominal factors, the model matrix coding differs from the coding used in fitting linear models. For information about the coding used for nominal terms in fitting linear models, see the Standard Least Squares chapter in the *Fitting Linear Models* book.

Moments Matrix

The *moments matrix* is dependent upon the model effects but is independent of the design. It is defined as follows:

$$\mathbf{M} = \int_R f(\mathbf{x})f(\mathbf{x})'d\mathbf{x}$$

where $f(\mathbf{x})$ denotes the model effects corresponding to factor combinations of the vector of factors, $\mathbf{x}$, and R denotes the design space. For additional details concerning moments and design matrices, see Goos and Jones (2011, pp 88-90) and Myers et al. (2009). Note that the moments matrix is called a matrix of region moments in Myers et al. (2009, p. 376).

Scripts

From the Custom Design red triangle menu, select **Save X Matrix**. After the design and the table are created, in the Custom Design table, the Moments Matrix and Model Matrix scripts, and if the design is a split plot, the V Inverse script, are saved as table properties.

- Select **Edit** from the red triangle next to either the Moments Matrix, Model Matrix, or V Inverse script. The script shows the corresponding matrix. You can copy this matrix into scripts that you write.
- When you run the Moments Matrix script, the log shows the number of rows in the moments matrix, called Moments.
- When you run the script Model Matrix, the log displays the number of rows in the model matrix, called X.
- When you run the script V Inverse, the log displays the number of rows in the inverse covariance matrix, called V Inverse.

Example

Follow these steps to illustrate these features:

Tip: To see the log, select **View > Log** (**Window > Log** on the Macintosh).

1. Select **DOE > Custom Design**.
2. Add 3 continuous factors and click **Continue**.

3. Click **Interactions > 2nd**.
4. From the Custom Design red triangle menu, select **Save X Matrix**.
5. Using the Default Number of Runs (12), click **Make Design** and then **Make Table**.
6. In the Table panel, right click the Moments Matrix script and select **Edit**.

   The script appears in a script window. The script shows the moments matrix, which is called **Moments**.

Figure 4.28 Moments Matrix Script

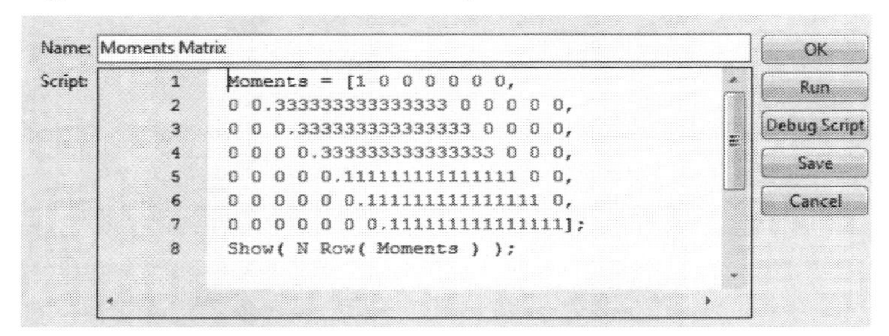

7. If it is not already open, select **View > Log** (**Window > Log** on the Macintosh).
8. In the Table panel, click the green triangle next to the Moments Matrix script.

   The number of rows appear in the log as `N Row(Moments)`=7.
9. In the Table panel, right click the Model Matrix script and select **Edit**.

   The script appears in a script window. The script shows the model matrix, which is called **X**.
10. Click **Run**.

    The number of rows appears in the log as `N Row(X)`=12.
11. To view the Model Matrix as a data table, add these lines to the script:

```
dt = New Table( "Model Matrix" );
dt << Set Matrix( X );
```

12. Click **Run**.

Number of Starts

The number of starts is the number of times that the coordinate-exchange algorithm initiates with a new design. See "Coordinate-Exchange Algorithm" on page 127. You can specify your own value using the **Number of Starts** option. Increasing the number of random starts tends to improve the optimality of the resulting design.

Unless you specify a value for Number of Starts and click OK, the number of starts is controlled by Design Search Time. To see how many starts were used to construct a design, click Make Design. Then select Number of Starts. The value in the text box is the number of starts used to construct the specific design.

In certain special cases, the globally optimal design is known from theory. If the coordinate-exchange algorithm detects that it has found an optimal design, it stops searching and returns that design.

Tip: To reproduce a specific design, you need to specify the Number of Starts and the Random Seed originally used to produce the design. Obtain these values from the red triangle options after you click Make Design.

In examples of custom designs in the documentation, the random seed and number of starts are often provided so that you can reconstruct the exact design being discussed.

Design Search Time

Design Search Time is the amount of time allocated to finding an optimal design. Custom Design's coordinate-exchange algorithm consists of finding near-optimal designs based on random starting designs. See "Coordinate-Exchange Algorithm" on page 127. The Design Search Time determines how many designs are constructed based on random starting designs.

You can specify your own value using the **Design Search Time** option. Increasing the search time tends to improve the optimality of the resulting design.

Keep in mind that designs produced by rerunning the algorithm can differ. Even with the same random seed, the numbers of starting designs used to construct the final design might differ because of variations in computing speed and other factors.

Note: The number of starting designs is given by the value in the Number of Starts text box. However, this value is not updated until after you construct your design by clicking Make Design.

In certain special cases, the globally D-optimal design is known from theory. These cases include:

- Two-level fractional factorial designs or nonregular orthogonal arrays. These are globally D-optimal for all main effect and two-factor interaction models.
- Latin-square designs. These are D-optimal for main effect models assuming the right sample size and numbers of levels of the factors.
- Plackett-Burman designs. These are D-optimal for main effect models.

If the coordinate-exchange algorithm detects that it has found an optimal design, it stops searching and returns that design.

Set Delta for Power

This option specifies the difference in the mean response that you want to be able to detect for model effects. Power calculations appear in the Power Analysis outline within the Design Evaluation outline. Power is calculated for each model parameter based on detecting the specified difference of delta. For categorical effects, the power calculation is based on detecting a maximum change of delta between any two levels.

For example, suppose that you want to detect a change of 3 units in the mean response. All of your factors are continuous. Because your factors are expressed in coded units (the coded levels are -1 and 1), a change of 3 units in the response corresponds to parameter coefficient values of 1.5. When you specify 3 in the Set Delta for Power text box, the Anticipated Coefficients in the Power Analysis outline are set to 1.5. For each parameter, the probability of detecting the change of 3 units in the response appears in the Power column to the right of the parameter.

By default, delta is set to 2. The default coefficient for each continuous effect is set to 1. An *n*-level categorical factor is represented by *n–1* indicator variables. The default coefficients for the *n–1* terms (which represent the categorical factor) are alternating values of 1 and -1. The default coefficients for an interaction effect with more than one degree of freedom are also alternating values of 1 and -1.

Note: The order in which parameters appear in the Power Analysis report might not be identical to their order in the Parameter Estimates report obtained using Standard Least Squares. This difference can occur only when the model contains an interaction with more than one degree of freedom.

Given a specified value of delta, each coefficient in the Anticipated Coefficients list is set at delta/2 multiplied by the default coefficient. For a continuous factor, this assignment ensures that a difference of delta is detected with the calculated power. For a categorical factor, this assignment of coefficients ensures that a maximum difference of delta between any two levels is detected with the calculated power.

Technical Details

This section contains technical details for the following topics:

- "Designs with Randomization Restrictions" on page 113
- "Covariates with Hard-to-Change Levels" on page 122
- "Numbers of Whole Plots and Subplots" on page 122
- "Optimality Criteria" on page 123
- "Coordinate-Exchange Algorithm" on page 127

Designs with Randomization Restrictions

This section describes how the Custom Design platform handles various types of designs where random assignment of experimental units to factor level settings is restricted. Random block designs and various types of split-plot designs are included.

Random Block Designs

A random block design groups the runs of an experiment into blocks that are considered to be randomly chosen from a larger population. Runs within a block of runs are usually more homogeneous than runs in different blocks. In these instances, you are often better able to discern other effects if you account for the variation explained by the blocking variables.

Scenario for a Random Block Design

Goos (2002) presents an example involving a pastry dough mixing experiment. The purpose of the experiment is to understand how certain properties of the dough depend on three factors: feed flow rate, initial moisture content, and rotational screw speed. Since it was possible to only conduct four runs a day, the experiment required several days to run. It is likely that random day-to-day differences in environmental variables have some effect on all of the runs that are performed on a given day. To account for the day-to-day variation, the runs were grouped into blocks of size four so that this variation would not compromise the information about the three factors.

The blocking factor, Day, consists of each day's runs. The days on which the trials were conducted are representative of a large population of days with different environmental conditions. It follows that Day is a random blocking factor.

Setup for a Random Block Design

To create a random block design, use the Custom Design platform to enter responses and factors and define your model as usual. In the Design Generation outline, select the **Group runs into random blocks of size** option and enter the number of runs you want in each block. See "Design Structure Options" on page 99.

Note: To define a fixed blocking factor, enter a blocking factor in the Factors outline. To define a random blocking factor, do not enter a blocking factor in the Factors outline. Instead, select the **Group runs into random blocks of size** option under Design Generation.

Split-Plot Designs

Split-plot designs are used in situations where the settings of certain factors are held constant for groups of runs. In industry, these are usually factors that are difficult or expensive to change from run to run. Factors whose settings need to be held constant for groups of runs are classified as *hard-to-change* in JMP.

Because certain factors are hard-to-change, it is not practical to randomly allocate them to experimental units. Instead, they are allocated to groups of units. This imposes a restriction on randomization that must be considered in generating a design and in analyzing the results.

Scenario for a Split-Plot Design

Box et al. (2005) presents an experiment to study the corrosion resistance of steel bars. The bars are placed in a furnace for curing. Afterward, a coating is applied to increase resistance to corrosion. The two factors of interest are:

- Furnace Temp in degrees centigrade, with levels 360, 370, and 380
- Coating, with levels C1, C2, C3, and C4 depicting four different types of coating

Furnace Temp is a hard-to-change factor, due to the time it takes to reset the temperature in the furnace. For this reason, four bars are processed for each setting of furnace temperature. At a later stage, the four coatings are randomly assigned to the four bars.

The experimental units are the bars. Furnace Temp is a hard-to-change factor whose levels define whole plots. Within each whole plot, the Coating factor is randomly assigned to the experimental units to which the whole plot factor was applied.

Figure 4.29 Factors and Design Outlines for Split-Plot Design

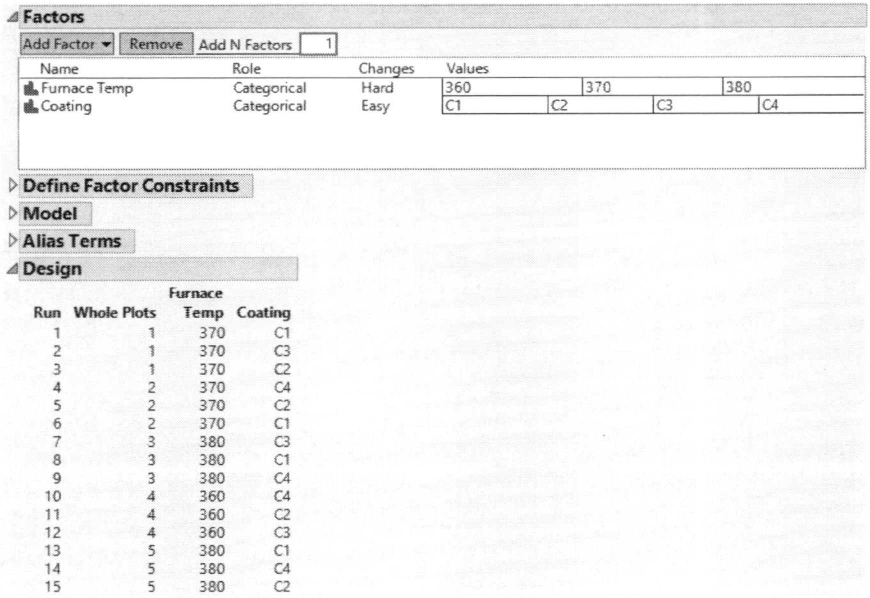

Factors

Add Factor | Remove | Add N Factors 1

Name	Role	Changes	Values			
Furnace Temp	Categorical	Hard	360	370	380	
Coating	Categorical	Easy	C1	C2	C3	C4

Define Factor Constraints

Model

Alias Terms

Design

Run	Whole Plots	Furnace Temp	Coating
1	1	370	C1
2	1	370	C3
3	1	370	C2
4	2	370	C4
5	2	370	C2
6	2	370	C1
7	3	380	C3
8	3	380	C1
9	3	380	C4
10	4	360	C4
11	4	360	C2
12	4	360	C3
13	5	380	C1
14	5	380	C4
15	5	380	C2

The Factors outline for the corrosion experiment has Changes set to Hard for Furnace Temp and Easy for Coating. The 15-run design consists of five whole plots, within which the settings of Temperature are held constant.

Setup for a Split-Plot Design

In general, several factors can be applied to a processing step where settings are hard-to-change. In the furnace example, you might consider a furnace location factor, as well as temperature. In the Factors outline, under the Changes column, you would specify a Changes value of Hard for such factors.

When a custom design involves only easy-to-change and hard-to-change factors, the runs of the hard-to-change factors are grouped using a new factor called Whole Plots. The values of Whole Plots designate blocks of runs with identical settings for the hard-to-change factors. The Model script that is saved to the design table treats Whole Plots as a random effect. For details, see "Changes" on page 86 and "Design Structure Options" on page 99.

For an example of creating a split-plot design and analyzing the experimental data, see "Split-Plot Experiment" on page 188 in the "Examples of Custom Designs" chapter.

Split-Split-Plot Designs

A split-split-plot design is used when there are two levels of factors that are hard-to-change. In industry, such designs often occur when batches of material or experimental units from one processing stage pass to a second processing stage. Factors are applied to batches of material at the first stage. Then those batches are divided for second-stage processing, where additional factors are studied. The first stage factors are considered *very-hard*-to-change, and the second-stage factors are considered *hard*-to-change. Additional factors can be applied to experimental units after the second processing stage. These factors are considered *easy*-to-change.

In a split-split-plot design, the batches are considered to be random blocks. Since the batches are divided for second-stage processing, the second-stage factors are nested within the first-stage factors.

Scenario for a Split-Split-Plot Design

Schoen (1999) presents an example of a split-split-plot design that relates to cheese quality. The factors are given in the Cheese Factors.jmp data table found in the Design Experiment folder. The experiment consists of three stages of processing:

- Milk is received from farmers and stored in a large tank.
- Milk from this tank is distributed to smaller tanks used for curd processing.
- The curds from each tank are transported to presses for processing individual cheeses.

The experiment consists of testing:

- Two factors that are applied when the milk is in the large storage tank.
- Five factors that are applied to the smaller curd processing tanks.
- Three factors that are applied to the individual cheeses from a curds processing tank.

Notice that the levels of factors applied to the curd processing tanks (subplots) are nested within the levels of factors applied to the milk storage tank (whole plots).

The Factors outline for the cheese experiment have Changes set as follows:

- Very Hard for the two storage tank factors
- Hard for the five curd processing tank factors
- Easy for the three factors that can be randomly assigned to cheeses

Figure 4.30 Factors and Design Generation Outline for Split-Split-Plot Design

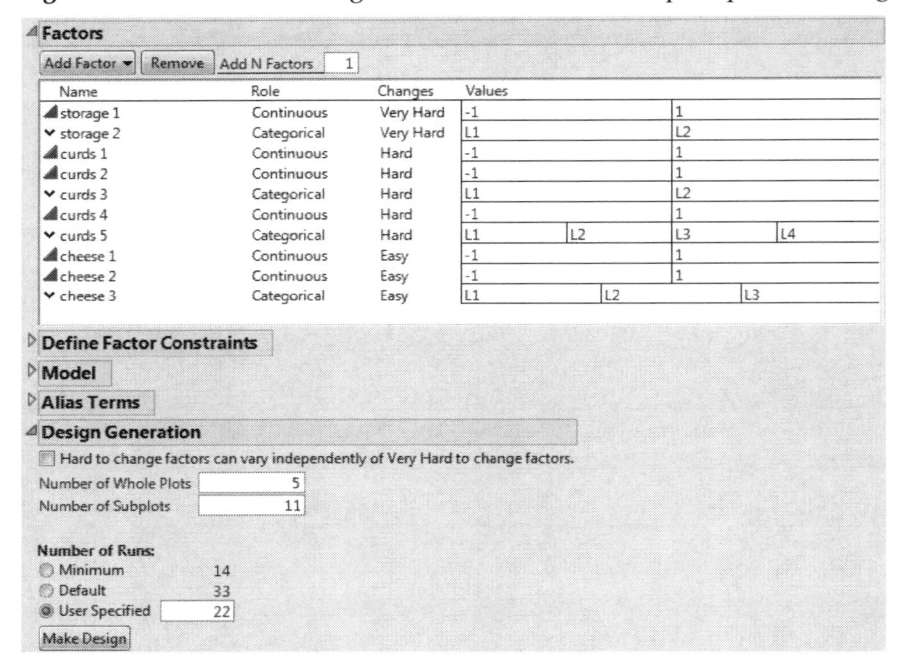

The default number of whole plots is 5 and the default number of subplots is 11. Click **Make Design** to see a 22-run design.

Figure 4.31 Split-Split-Plot Design for Cheese Scenario

Design

Run	Whole Plots	Subplots	storage 1	storage 2	curds 1	curds 2	curds 3	curds 4	curds 5	cheese 1	cheese 2	cheese 3
1	1	1	-1	L2	1	1	L1	-1	L2	1	-1	L3
2	1	1	-1	L2	1	1	L1	-1	L2	-1	1	L1
3	1	2	-1	L2	-1	-1	L2	-1	L3	1	1	L1
4	1	2	-1	L2	-1	-1	L2	-1	L3	-1	-1	L3
5	1	3	-1	L2	1	-1	L2	1	L1	1	1	L3
6	2	4	1	L1	-1	1	L1	-1	L1	1	-1	L2
7	2	4	1	L1	-1	1	L1	-1	L1	1	1	L3
8	2	4	1	L1	-1	1	L1	-1	L1	-1	1	L1
9	2	5	1	L1	1	1	L1	1	L3	-1	-1	L1
10	2	5	1	L1	1	1	L1	1	L3	1	1	L2
11	3	6	1	L2	1	-1	L2	-1	L2	1	-1	L1
12	3	6	1	L2	1	-1	L2	-1	L2	-1	1	L2
13	3	7	1	L2	-1	1	L1	1	L4	1	1	L2
14	3	7	1	L2	-1	1	L1	1	L4	-1	-1	L3
15	4	8	1	L1	-1	1	L2	1	L2	1	1	L3
16	4	8	1	L1	-1	1	L2	1	L2	-1	-1	L1
17	4	9	1	L1	1	-1	L1	-1	L4	1	-1	L1
18	4	9	1	L1	1	-1	L1	-1	L4	-1	1	L2
19	5	10	-1	L1	1	1	L2	-1	L4	-1	-1	L2
20	5	10	-1	L1	1	1	L2	-1	L4	1	1	L1
21	5	11	-1	L1	-1	-1	L1	1	L2	-1	1	L3
22	5	11	-1	L1	-1	-1	L1	1	L2	1	-1	L2

The five whole plots correspond to the storage factors, storage 1 and storage 2. The settings of the storage factors are constant within a whole plot. If consecutive whole plots have the same setting for a whole plot factor, the factor should be reset between the plots. For example, you should reset the level for storage 1 between runs 10 and 11 and between runs 14 and 15, and your should reset the level for storage 2 between runs 18 and 19. Resetting the factor between whole plots, even when the specified settings are the same, is required in order to capture whole plot variation.

The 11 subplots correspond to the curds factors. Within a subplot, the settings of the curds factors are constant. Each level of Subplots only appears within one level of Whole Plots, indicating that the levels of Subplots are nested within the levels of Whole Plots.

Levels of the cheese factors vary randomly from run to run.

Setup for a Split-Split-Plot Design

In a split-split-plot design, the Factors outline contains factors with Changes set to Very Hard and Hard. The design can also contain factors with Changes set to Easy. Two factors are created:

- A factor called Whole Plots represents the blocks of constant levels of the factors with Changes set to Very Hard.
- A factor called Subplots represents the blocks of constant levels of the factors with Changes set to Hard.
- The factor Subplots reflects the nesting of the levels of the factors with Changes set to Hard within the levels of the factors with Changes set to Very Hard.

- The levels of factors with Changes set to Easy are randomly assigned to units within subplots.
- The factors Whole Plots and Subplots are treated as random effects in the Model script that is saved to the design table.

For details, see the Changes description under "Factors Outline" on page 86 and "Design Structure Options" on page 99.

Two-Way Split-Plot Designs

A two-way split-plot (also known as strip plot or split block) design consists of two split-plot components. In industry, these designs arise when batches of material or experimental units from one processing stage pass to a second processing stage. But, after the first processing stage, it is possible to divide the batches into sub-batches. The second-stage processing factors are applied randomly to these sub-batches. For a specific second-stage experimental setting, all of the sub-batches assigned to that setting can be processed simultaneously. Additional factors can be applied to experimental units after the second processing stage.

In contrast to a split-split-plot design, the second-stage factors are *not nested* within the first-stage factors. After the first stage, the batches are subdivided and formed into new batches. Therefore, both the first- and second-stage factors are applied to whole batches.

Although factors at both stages might be equally hard-to-change, to distinguish these factors, JMP denotes the first stage factors as *very-hard*-to-change, and the second-stage factors as *hard*-to-change. Additional factors applied to experimental units after the second processing stage are considered *easy*-to-change.

Scenario for a Two-Way Split-Plot Design

Vivacqua and Bisgaard (2004) describe an experiment to improve the open circuit voltage in battery cells. Two stages of processing are of interest:

- First stage: A continuous assembly process
- Second stage: A curing process with a 5-day cycle time

The engineers want to study six two-level factors:

- Four factors, X1, X2, X3, and X4, that are applied to the assembly process
- Two factors, X5 and X6, that are applied to the curing process

A full factorial design with all factors at two levels would require $2^6 = 64$ runs, and would require a prohibitive 64*5 = 320 days. Also, it is not practical to vary assembly conditions for individual batteries. However, assembly conditions can be changed for large batches, such as batches of 2000 batteries.

Both the first- and second-stage factors are hard-to-change. In a sense, there are two split-plot designs. However, the batches of 2,000 batteries from the first-stage experiment can be divided into four sub-batches of 500 batteries each. These sub-batches can be randomly assigned to the four settings of the two second-stage factors. All of the batches assigned to a given set of curing conditions can be processed simultaneously. In other words, the first- and second-stage factors are *crossed*.

To distinguish between the first- and second-stage factors, you designate the Changes for the first-stage factors as Very Hard, and the Changes for the second-stage factors as Hard. See Figure 4.32. Also, under Design Generation, note the following option: **Hard to change factors can vary independently of Very Hard to change factors**. If this is not checked, the design is treated as a split-split-plot design, with nesting of factors at the two levels. Check this option to create a two-way split-plot design.

Figure 4.32 Factors and Design Generation Outline for Two-Way Split Plot Design

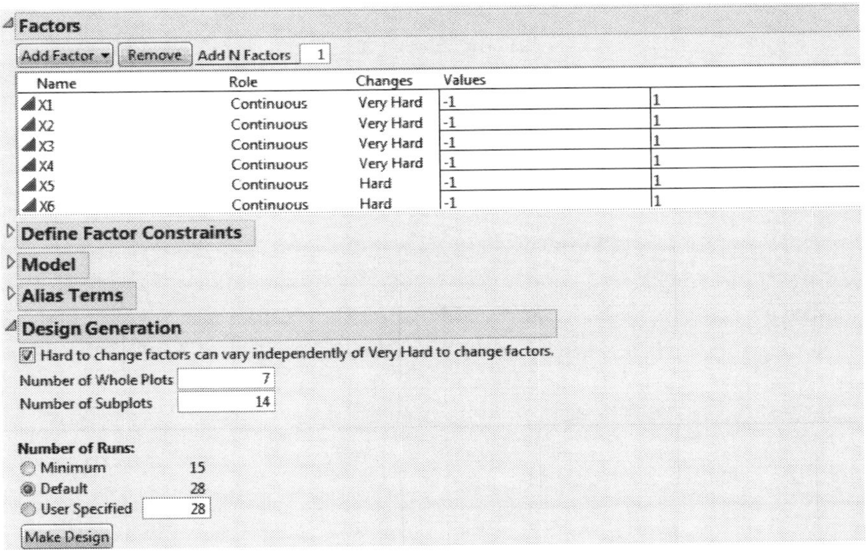

The default number of whole plots is 7; the default number of subplots is 14. Click **Make Design** to see the 28-run design.

Figure 4.33 Two-Way Split-Plot Design for Battery Cells

Design

Run	Whole Plots	Subplots	X1	X2	X3	X4	X5	X6
1	1	1	1	1	1	-1	1	-1
2	1	2	1	1	1	-1	-1	1
3	1	3	1	1	1	-1	-1	-1
4	1	4	1	1	1	-1	1	1
5	2	5	-1	-1	-1	-1	-1	1
6	2	6	-1	-1	-1	-1	1	-1
7	2	7	-1	-1	-1	-1	-1	-1
8	2	8	-1	-1	-1	-1	1	1
9	3	9	1	1	-1	-1	-1	1
10	3	10	1	1	-1	-1	1	-1
11	3	11	1	1	-1	-1	-1	-1
12	3	12	1	1	-1	-1	1	1
13	4	13	1	-1	-1	1	-1	-1
14	4	14	1	-1	-1	1	1	1
15	4	1	1	-1	-1	1	1	-1
16	4	2	1	-1	-1	1	-1	1
17	5	3	-1	1	-1	1	-1	-1
18	5	4	-1	1	-1	1	1	1
19	5	5	-1	1	-1	1	-1	1
20	5	6	-1	1	-1	1	1	-1
21	6	7	1	-1	1	1	-1	-1
22	6	8	1	-1	1	1	1	1
23	6	9	1	-1	1	1	-1	1
24	6	10	1	-1	1	1	1	-1
25	7	11	-1	-1	1	-1	-1	-1
26	7	12	-1	-1	1	-1	1	1
27	7	13	-1	-1	1	-1	-1	-1
28	7	14	-1	-1	1	-1	1	1

The seven whole plots correspond to the first-stage factors, X1, X2, X3, and X4. The settings of these factors are constant within a whole plot. The 14 subplots correspond to the second-stage factors, X5 and X6. For example, the sub-batches for runs 1 and 15 (from different whole plots) are subject to the same subplot treatment, where X5 is set at 1 and X6 at -1.

Setup for a Two-Way Split-Plot Design

A two-way split-plot design requires factors with Changes set to Very Hard and to Hard. As described in "Setup for a Split-Split-Plot Design" on page 118, factors called Whole Plots and Subplots are created. However, in a two-way split-plot design, Subplots does not nest the levels of factors with Changes set to Hard within the levels of factors with Changes set to Very Hard. Both Whole Plots and Subplots are treated as random effects in the Model script that is saved to the design table.

You need to ensure that the factor Subplots is *not nested* within the factor Whole Plots. Select the option **Hard to change factors can vary independently of Very Hard to change factor** in the Design Generation outline (Figure 4.32). For more details, see "Changes" on page 86 and "Design Structure Options" on page 99.

For an example of creating a split-plot design and analyzing the experimental data, see "Two-Way Split-Plot Experiment" on page 193 in the "Examples of Custom Designs" chapter.

Covariates with Hard-to-Change Levels

Suppose that you have measurements on batches of material that are available for use in testing experimental factors. Or suppose that you have measurements on individuals who might be selected to participate in testing experimental factors. The measurements on batches or individuals are known in advance of the experiment and are considered to be covariates.

The batches or individuals correspond to whole plots. You might want to use only some of these whole plots in your experiment. Because information about the whole plots in the form of covariates is available, the design should choose the whole plots in an optimal fashion.

In the Factors outline, the Custom Design platform enables you to designate covariates as hard-to-change. The model, as given by the terms that you include in the Model outline, can include interactions and powers constructed using covariates and experimental factors.

Note: When you set Changes for a Covariate factor to Hard, all other covariates are also set to Hard The remaining factors must be set to Easy. Because the algorithm requires a combination of row exchange and coordinate exchange, even moderately sized designs might take some time to generate.

Scenario for an Experiment with a Hard-to-Change Covariate

An experiment involving batches of polypropylene plates is discussed in Goos and Jones (2011, Chapter 9) and Jones and Goos (2015). Large batches of polypropylene plates are produced according to various formulations determined by several variables. Some plates are used immediately, and the remainder are stored for future experimental purposes. The compositions of these stored batches are known.

A customer has certain requirements regarding the plate formulation. Future experiments involve customizing the gas plasma treatment to the types of formulations required by the customer. The composition variables are considered hard-to-change covariates. Gas plasma treatment factors can be applied to sub-batches of plates with a given formulation.

The optimal design identifies the batches (defined by the covariates) to use, determines the number of plates from each batch to use, and provides settings for the gas plasma levels. Note that the optimal number of batches and plates from a given batch depend on the covariates.

An example is provided in "Examples of Custom Designs" chapter on page 129.

Numbers of Whole Plots and Subplots

JMP suggests default values for the Number of Whole Plots and Number of Subplots. These values are based on heuristics for creating a balanced design that allows for estimation of the effects specified in the Model outline.

If you enter missing values for Number of Whole Plots or Number of Subplots, JMP chooses values that maximize the D-efficiency of the design. The algorithm uses the values specified in the Split Plot Variance Ratio option. See "Advanced Options > Split Plot Variance Ratio" on page 105. The D-efficiency is given by the determinant of $X'V^{-1}X$, where $\mathbf{V}^{-1}$ is the inverse of the variance matrix of the responses. For further details, see Goos, 2002.

If you enter values for the Number of Whole Plots and Number of Subplots, Custom Design attempts to maximize the optimality of the resulting design. For details about split-plot designs, see Jones and Goos (2007). For details about designs with hard-to-change covariates, see Jones and Goos (2015).

Optimality Criteria

This section provides information about the following designs:

D-Optimality

By default, the Custom Design platform optimizes the D-optimality criterion except when a full quadratic model is created using the RSM button. In that case, an I-optimal design is constructed.

The D-optimality criterion minimizes the determinant of the covariance matrix of the model coefficient estimates. It follows that D-optimality focuses on precise estimates of the effects. This criterion is desirable in the following cases:

- screening designs
- experiments that focus on estimating effects or testing for significance
- designs where identifying the active factors is the experimental goal

The D-optimality criterion is dependent on the assumed model. This is a limitation because often the form of the true model is not known in advance. The runs of a D-optimal design optimize the precision of the coefficients of the assumed model. In the extreme, a D-optimal design might be saturated, with the same number of runs as parameters and no degrees of freedom for lack of fit.

Specifically, a D-optimal design maximizes *D,* where D is defined as follows:

$$D = \det[\mathbf{X}'\mathbf{X}]$$

and where **X** is the model matrix as defined in "Simulate Responses" on page 106.

D-optimal split-plot designs maximize D, where D is defined as follows:

$$D = \det[\mathbf{X}'\mathbf{V}^{-1}\mathbf{X}]$$

and $\mathbf{V}^{-1}$ is the block diagonal covariance matrix of the responses (Goos 2002).

Since a D-optimal design focuses on minimizing the standard errors of coefficients, it might not allow for checking that the model is correct. For example, a D-optimal design does not include center points for a first-order model. When there are potentially active terms that are not included in the assumed model, a better approach is to specify If Possible terms and to use a Bayesian D-optimal design.

Bayesian D-Optimality

Bayesian D-optimality is a modification of the D-optimality criterion. The Bayesian D-optimality criterion is useful when there are potentially active interactions or non-linear effects. See DuMouchel and Jones (1994) and Jones et al (2008).

Bayesian D-optimality estimates a specified set of model parameters precisely. These are the effects whose Estimability you designate as Necessary in the Model outline. But at the same time, Bayesian D-optimality has the ability to estimate other, typically higher-order effects, as allowed by the run size. These are the effects whose Estimability you designate as If Possible in the Model outline. To the extent possible given the run size restriction, a Bayesian D-optimal design allows for detecting inadequacy in a model that contains only the Necessary effects.

The Bayesian D-optimality criterion is most effective when the number of runs is larger than the number of Necessary terms, but smaller than the sum of the Necessary and If Possible terms. When this is the case, the number of runs is smaller than the number of parameters that you would like to estimate. Using prior information in the Bayesian setting allows for precise estimation of all of the Necessary terms while providing the ability to detect and estimate some If Possible terms.

To allow for a meaningful prior distribution to apply to the parameters of the model, responses and factors are scaled to have certain properties (DuMouchel and Jones, 1994, Section 2.2).

Consider the following notation:

- **X** is the model matrix as defined in "Simulate Responses" on page 106
- **K** is a diagonal matrix with values as follows:
  - $k = 0$ for Necessary terms
  - $k = 1$ for If Possible main effects, powers, and interactions involving a categorical factor with more than two levels

– $k = 4$ for all other If Possible terms

The prior distribution imposed on the vector of If Possible parameters is multivariate normal, with mean vector **0** and diagonal covariance matrix with diagonal entries $1/k^2$. Therefore, a value k^2 is the reciprocal of the prior variance of the corresponding parameter.

The values for *k* are empirically determined. If Possible main effects, powers, and interactions with more than one degree of freedom have a prior variance of 1. Other If Possible terms have a prior variance of 1/16. In the notation of DuMouchel and Jones, 1994, $k = 1/\tau$.

To control the weights for If Possible terms, select **Advanced Options > Prior Parameter Variance** from the red triangle menu. See "Advanced Options > Prior Parameter Variance" on page 106.

The posterior distribution for the parameters has the covariance matrix $(\mathbf{X'X} + \mathbf{K}^2)^{-1}$. The Bayesian D-optimal design is obtained by maximizing the determinant of the inverse of the posterior covariance matrix:

$$\det(\mathbf{X'X} + \mathbf{K}^2)$$

I-Optimality

I-optimal designs minimize the average variance of prediction over the design space. The I-optimality criterion is more appropriate than D-optimality if your primary experimental goal is not to estimate coefficients, but rather to do the following:

- predict a response
- determine optimum operating conditions
- determine regions in the design space where the response falls within an acceptable range

In these cases, precise prediction of the response takes precedence over precise estimation of the parameters.

The prediction variance relative to the unknown error variance at a point $\mathbf{x}_0$ in the design space can be calculated as follows:

$$\mathrm{var}(\hat{\mathbf{Y}}|\mathbf{x}_0) = f(\mathbf{x}_0)'(\mathbf{X'X})^{-1}f(\mathbf{x}_0)$$

where **X** is the model matrix as defined in "Simulate Responses" on page 106.

I-optimal designs minimize the integral *I* of the prediction variance over the entire design space, where *I* is given as follows:

$$I = \int_R f(\mathbf{x})'(\mathbf{X'X})^{-1}f(\mathbf{x})d\mathbf{x} = \mathrm{trace}[(\mathbf{X'X})^{-1}\mathbf{M}]$$

Here **M** is the moments matrix:

$$\mathbf{M} = \int_R f(\mathbf{x}) f(\mathbf{x})' d\mathbf{x}$$

See “Simulate Responses” on page 106. For further details, see Goos and Jones (2011).

The moments matrix does not depend on the design and can be computed in advance. The row vector $f(\mathbf{x})'$ consists of a 1 followed by the effects corresponding to the assumed model. For example, for a full quadratic model in two continuous factors, $f(\mathbf{x})'$ is defined as follows:

$$f(\mathbf{x})' = (1, x_1, x_2, x_1 x_2, x_1^2, x_2^2)$$

Bayesian I-Optimality

The Bayesian I-optimal design minimizes the average prediction variance over the design region for Necessary and If Possible terms.

Consider the following notation:

- **X** is the model matrix, defined in “Simulate Responses” on page 106
- **K** is a diagonal matrix with values as follows:
  - $k = 0$ for Necessary terms
  - $k = 1$ for If Possible main effects, powers, and interactions involving a categorical factor with more than two levels
  - $k = 4$ for all other If Possible terms

The prior distribution imposed on the vector of If Possible parameters is multivariate normal, with mean vector **0** and diagonal covariance matrix with diagonal entries $1/k^2$. (See “Bayesian D-Optimality” on page 124 for more details about the values k.)

The posterior variance of the predicted value at a point $\mathbf{x}_0$ is as follows:

$$\text{var}(\hat{\mathbf{Y}} | \mathbf{x}_0) = f(\mathbf{x}_0)'(\mathbf{X}'\mathbf{X} + \mathbf{K}^2)^{-1} f(\mathbf{x}_0)$$

The Bayesian I-optimal design minimizes the average prediction variance over the design region, as follows:

$$I_B = \text{trace}[(\mathbf{X}'\mathbf{X} + \mathbf{K}^2)^{-1}\mathbf{M}]$$

where **M** is the moments matrix. See “Simulate Responses” on page 106.

Alias Optimality

Alias optimality seeks to minimize the aliasing between effects that are in the assumed model and effects that are not in the model but are potentially active. Effects that are not in the model but that are of potential interest are called *alias effects*. For details about alias-optimal designs, see Jones and Nachtsheim (2011).

Specifically, let $\mathbf{X}_1$ be the model matrix corresponding to the terms in the assumed model, as defined in "Simulate Responses" on page 106. The design defines the model that corresponds to the alias effects. Denote the matrix of model terms for the alias effects by $\mathbf{X}_2$.

The *alias matrix* is the matrix **A**, defined as follows:

$$\mathbf{A} = (\mathbf{X}_1'\mathbf{X}_1)^{-1}\mathbf{X}_1'\mathbf{X}_2$$

The entries in the alias matrix represent the degree of bias associated with the estimates of model terms. See "The Alias Matrix" on page 690 in the "Technical Details" appendix for the derivation of the alias matrix.

The sum of squares of the entries in **A** provides a summary measure of bias. This sum of squares can be represented in terms of a trace as follows:

$$\text{trace}(\mathbf{A}'\mathbf{A})$$

Designs that reduce the trace criterion generally have lower D-efficiency than the D-optimal design. Consequently, alias optimality seeks to minimize the trace of $\mathbf{A}'\mathbf{A}$ subject to a lower bound on D-efficiency. For the definition of D-efficiency, see "Optimality Criteria" on page 123. The lower bound on D-efficiency is given by the D-efficiency weight, which you can specify under Advanced Options. See "Advanced Options > D Efficiency Weight" on page 106.

D-Efficiency

Let **X** denote the design, or model, matrix for a given assumed model with p parameters. For the definition of the model matrix, see "Simulate Responses" on page 106. Let $\mathbf{X}_D$ denote the model matrix for a D-optimal design for the assumed model. Then the D-efficiency of the design given by **X** is as follows:

$$\text{D-Efficiency} = \left[\frac{\det(\mathbf{X}'\mathbf{X})}{\det(\mathbf{X}_D'\mathbf{X}_D)}\right]^{1/p}$$

Coordinate-Exchange Algorithm

Custom Design constructs a design that seeks to optimize one of several optimality criteria. (See "Optimality Criteria" on page 123.) To optimize the criterion, Custom Design uses the *coordinate-exchange algorithm* (Meyer and Nachtsheim, 1995). The algorithm begins by randomly selecting values within the specified design region for each factor and each run to construct a starting design.

Suppose your study requires continuous factors, no factor constraints, and a main-effects model. An iteration consists of testing each value of the model matrix, as follows:

- The current value of each factor is replaced by its two most extreme values.
- The optimality criterion is computed for both of these replacements.
- If one of the values increases the optimality criterion, this value replaces the old value.

The process continues until no replacement occurs for an entire iteration.

Appropriate adjustments are made to the algorithm to account for polynomial terms, nominal factors, and factor constraints.

The design obtained using this process is optimal in a large class of neighboring designs. But it is only *locally* optimal. To improve the likelihood of finding a globally optimal design, the coordinate-exchange algorithm is repeated a large number of times. Goos and Jones (2011, p. 36) recommend using at least 1,000 random starts for all but the most trivial design situations. The number of starting designs is controlled by the Number of Starts option. See "Number of Starts" on page 110. Custom Design provides the design that maximizes the optimality criterion among all the constructed designs.

Chapter 5

Examples of Custom Designs

Perform Experiments That Meet Your Needs

Use the Custom Design platform as your primary tool for constructing a wide range of experimental designs. You can construct a variety of design types and fine tune them to your specific experimental needs and resource budget.

Custom Design provides more options and control than the Screening, Response Surface, Full Factorial, and Mixture Design platforms. The designs that you construct are created specifically to meet your goals. This eliminates the struggle to find a classical design that only comes close to meeting your goals.

The flexible special-purpose designs that you can construct using Custom Design include:

- Screening designs, including supersaturated screening designs
- Response surface designs, including those with categorical factors
- Mixture designs, including those with process factors, and mixture of mixture designs
- Designs that include covariates or that are robust to linear time trends
- Fixed and random block designs
- Split-plot, split-split-plot, and two-way split-plot (strip-plot) designs

In this chapter you construct most of these design types within the Custom Design platform. In many cases, you also analyze the experimental results. For help with using the Custom Design platform, see the "Custom Designs" chapter on page 63.

Figure 5.1 Fraction of Design Space Plot

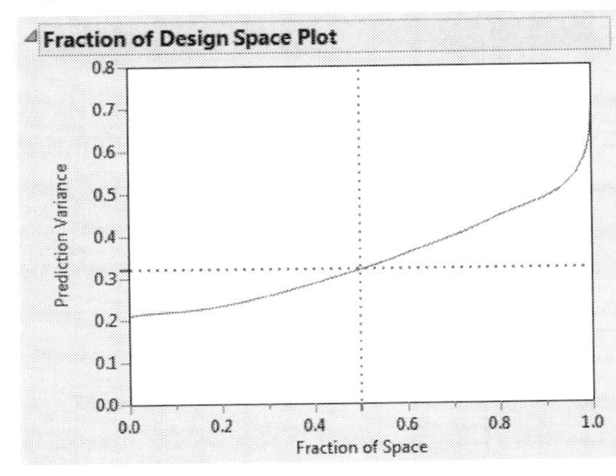

Screening Experiments

In the early stages of studying a process, you identify a list of factors that potentially affect your response or responses. You are interested in identifying the *active* factors, that is, the factors that actually do affect your response or responses. A *screening design* helps you determine which factors are likely to be active. Once the active factors are identified, you can construct more sophisticated designs, such as response surface designs, to model interactions and curvature.

Screening designs constructed using the Custom Design platform are often equivalent to the classical designs provided in the Screening Design platform. However, Custom Design can construct designs for cases where classical screening designs are not available.

The Custom Design platform constructs screening designs using either the D-optimality or Bayesian D-optimality criterion. The D-optimality criterion minimizes the determinant of the covariance matrix of the model coefficient estimates. It follows that D-optimality focuses on precise estimates of the effects. For details, see "Optimality Criteria" on page 123 in the "Custom Designs" chapter.

This section contains the following examples:

- "Design That Estimates Main Effects Only" on page 130
- "Design That Estimates All Two-Factor Interactions" on page 133
- "Design That Avoids Aliasing of Main Effects and Two-Factor Interactions" on page 135
- "Supersaturated Screening Designs" on page 139
- "Design for Fixed Blocks" on page 146

Design That Estimates Main Effects Only

Note: For details about main effects only screening designs, see the "Screening Designs" chapter on page 279.

In this example, you are interested in studying the main effects of six factors. You construct a screening design where all of the main effects are orthogonal. However, the main effects are aliased with two-factor interactions.

1. Select **DOE > Custom Design**.
2. In the Factors outline, type 6 next to **Add N Factors**.
3. Click **Add Factor > Continuous**.
4. Click **Continue**.

   The Model outline appears. It includes only main effects with Estimability designated as Necessary. This means that all main effects are estimable in the design that is generated.

Figure 5.2 Custom Design Window Showing Model Outline

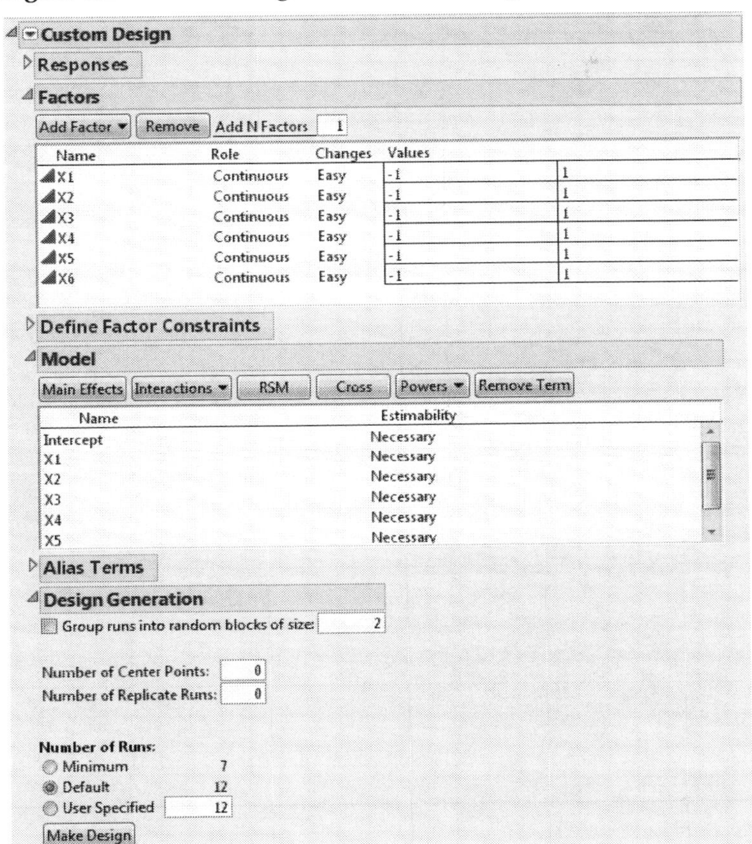

Keep the default of 12 runs.

Note: Setting the Random Seed in step 5 and Number of Starts in step 6 reproduces the exact results shown in this example. When constructing a design on your own, these steps are not necessary.

5. (Optional) From the Custom Design red triangle menu, select **Set Random Seed**, type 1839634787, and click **OK**.
6. (Optional) From the Custom Design red triangle menu, select **Number of Starts**, type 1, and click **OK**.
7. Click **Make Design**.

Figure 5.3 Design for Main Effects Only

Design

Run	X1	X2	X3	X4	X5	X6
1	-1	-1	1	1	-1	-1
2	1	1	1	1	-1	1
3	-1	1	-1	-1	-1	-1
4	1	1	1	-1	-1	-1
5	-1	-1	-1	-1	-1	1
6	1	-1	1	-1	1	-1
7	-1	-1	1	-1	1	1
8	1	-1	-1	1	1	-1
9	-1	1	1	1	1	1
10	1	1	-1	-1	1	1
11	-1	1	-1	1	1	-1
12	1	-1	-1	1	-1	1

8. Open the **Design Evaluation > Color Map on Correlations** outline.

Figure 5.4 Color Map on Correlations

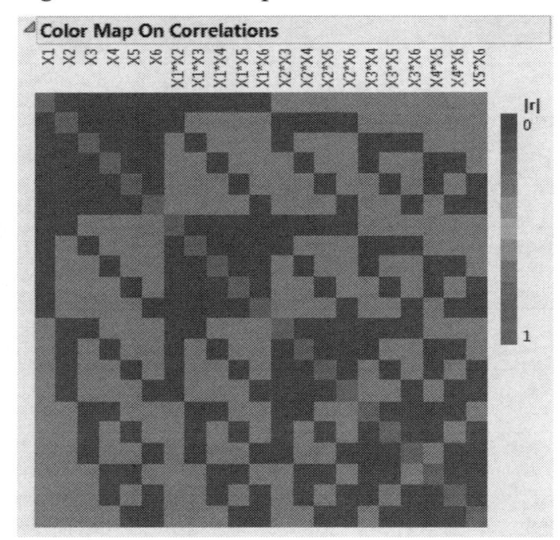

Note: The colors shown in Figure 5.4 are the JMP default colors.

Notice the following:

– The main effects are represented by the six terms in the upper left corner of the map.

– The deep blue color corresponding to the correlations of the six main effects with other main effects indicates correlations of 0. This means that all main effects are orthogonal and can be estimated independently of each other.

– The light blue color in the squares corresponding to some two-way interactions indicates that the corresponding effects are correlated. This means that these effects cannot be estimated independently of other effects. Place your cursor over these squares to see the exact correlation.

– Notice that no effects are completely confounded with each other. The only red squares, indicating absolute correlations of 1, are on the main diagonal.

9. Open the **Design Evaluation > Alias Matrix** outline.

Figure 5.5 Alias Matrix

Alias Matrix

Effect	X1*X2	X1*X3	X1*X4	X1*X5	X1*X6	X2*X3	X2*X4	X2*X5	X2*X6	X3*X4	X3*X5	X3*X6	X4*X5	X4*X6	X5*X6
Intercept	0	0	0	0	0	0	0	0	0	0	0	0	0	0	0
X1	0	0	0	0	0	0.333	-0.33	-0.33	0.333	-0.33	-0.33	-0.33	-0.33	0.333	-0.33
X2	0	0.333	-0.33	-0.33	0.333	0	0	0	0	0.333	-0.33	0.333	0.333	0.333	0.333
X3	0.333	0	-0.33	-0.33	-0.33	0	0.333	-0.33	0.333	0	0	0	-0.33	0.333	0.333
X4	-0.33	-0.33	0	-0.33	0.333	0.333	0	0.333	0.333	0	-0.33	0.333	0	0	-0.33
X5	-0.33	-0.33	-0.33	0	-0.33	-0.33	0.333	0	0.333	-0.33	0	0.333	0	-0.33	0
X6	0.333	-0.33	0.333	-0.33	0	0.333	0.333	0.333	0	0.333	0.333	0	-0.33	0	0

The Alias Matrix shows how the coefficients of the main effect terms in the model are biased by potentially active two-factor interaction effects. The column labels identify interactions. For example, in the X1 row, the column X2*X3 has a value of 0.333 and the column X2*X4 has a value of -0.33. This means that the expected value of the main effect of X1 is the sum of the main effect of X1 plus 0.333 times the effect of X2*X3, plus -0.33 times the effect of X2*X4, and so on, for the rest of the X1 row. In order for the estimate of the main effect of X1 to be meaningful, you must assume that these interactions are negligible in size compared to the effect of X1.

Tip: The Alias Matrix is a generalization of the confounding pattern in fractional factorial designs.

Design That Estimates All Two-Factor Interactions

The Alias Matrix in Figure 5.5 shows partial aliasing of effects. In other cases, main effects might be fully aliased, or *confounded*, with two-factor interactions. In both of these cases, strong two-factor interactions can confuse the results of main effects only experiments. To avoid this risk, create a design that resolves all two-factor interactions.

In this example, you create a resolution V screening design. Two-factor interactions are orthogonal, but they are confounded with three-factor interactions.

1. Select **DOE > Custom Design**.
2. Type 5 next to **Add N Factors**.
3. Click **Add Factor > Continuous**.
4. Click **Continue**.
5. In the Model outline, select **Interactions > 2nd**.

Figure 5.6 Model Outline Showing Interactions

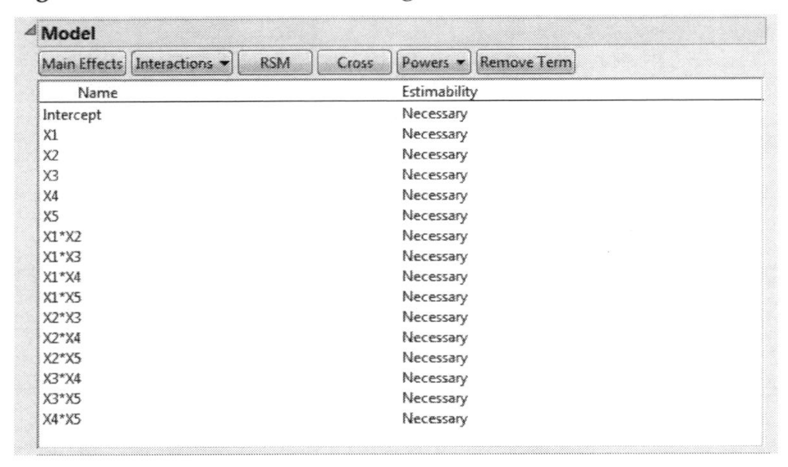

6. Click **Minimum** to accept 16 for the number of runs.

Note: Setting the Random Seed in step 7 and Number of Starts in step 8 reproduces the exact results shown in this example. In constructing a design on your own, these steps are not necessary.

7. (Optional) From the Custom Design red triangle menu, select **Set Random Seed**, type 819994207, and click **OK**.
8. (Optional) From the Custom Design red triangle menu, select **Number of Starts**, type 1, and click **OK**.
9. Click **Make Design**.

   Figure 5.7 shows the runs of the design. All main effects and two-factor interactions are estimable because their Estimability was designated as Necessary (by default) in the Model outline.

Figure 5.7 Design to Estimate All Two-Factor Interactions

Design

Run	X1	X2	X3	X4	X5
1	-1	-1	-1	-1	1
2	1	1	1	-1	-1
3	-1	1	1	-1	1
4	-1	1	1	1	-1
5	1	1	-1	-1	1
6	1	-1	1	1	-1
7	-1	-1	-1	1	-1
8	-1	-1	1	1	1
9	-1	1	-1	-1	-1
10	1	1	-1	1	-1
11	1	1	1	1	1
12	1	-1	1	-1	1
13	-1	-1	1	-1	-1
14	1	-1	-1	-1	-1
15	1	-1	-1	1	1
16	-1	1	-1	1	1

10. Open the **Design Evaluation > Color Map on Correlations** outline.

Figure 5.8 Color Map on Correlations

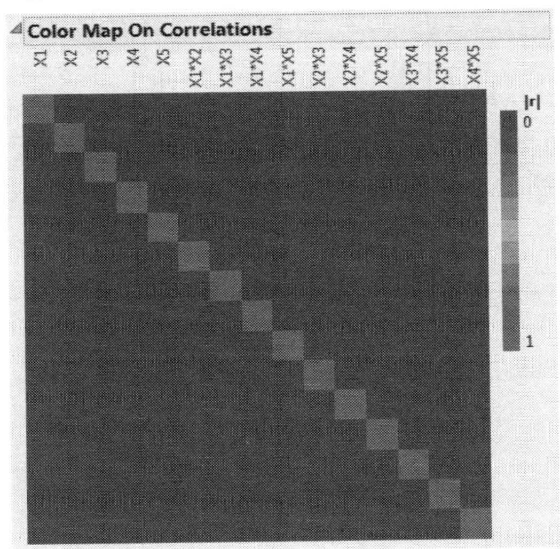

The Color Map indicates that the five main effects and the ten two-way interactions are all mutually orthogonal. (Figure 5.8 uses the JMP default colors.)

Design That Avoids Aliasing of Main Effects and Two-Factor Interactions

Suppose that your primary interest is in estimating the main effects of six continuous factors. However, you want to do this in a way that minimizes aliasing of main effects with potentially active two-factor interactions.

Your budget allows for only 16 runs. With six factors, there are 15 possible two-factor interactions. The minimum number of runs required to fit the constant, the six main effects, and the 15 two-factor interactions is 22.

In this example, you find a compromise between an 8-run main effects only design (see "Design That Estimates Main Effects Only" on page 130) and a 22-run design capable of fitting all the two-factor interactions. You use Alias Optimality as the optimality criterion to achieve your goal.

1. Select **DOE > Custom Design**.
2. Type 6 next to **Add N Factors**.
3. Click **Add Factor > Continuous**.
4. Click **Continue**.

   The model includes the main effect terms by default. The default estimability of these terms is Necessary. In the Alias Terms outline, notice that second-order interactions are added. By default, all two-way interactions not included in the assumed model are added to the Alias Terms list.
5. Select **Optimality Criterion > Make Alias Optimal Design** from the red triangle menu.

   The Make Alias Optimal Design selection tells JMP to generate a design that balances reduction in aliasing with D-efficiency. See "Alias Optimality" on page 127 in the "Custom Designs" chapter.
6. Click **User Specified** and change the number of runs to 16.

Figure 5.9 Factors, Model, Alias Terms, and Number of Runs

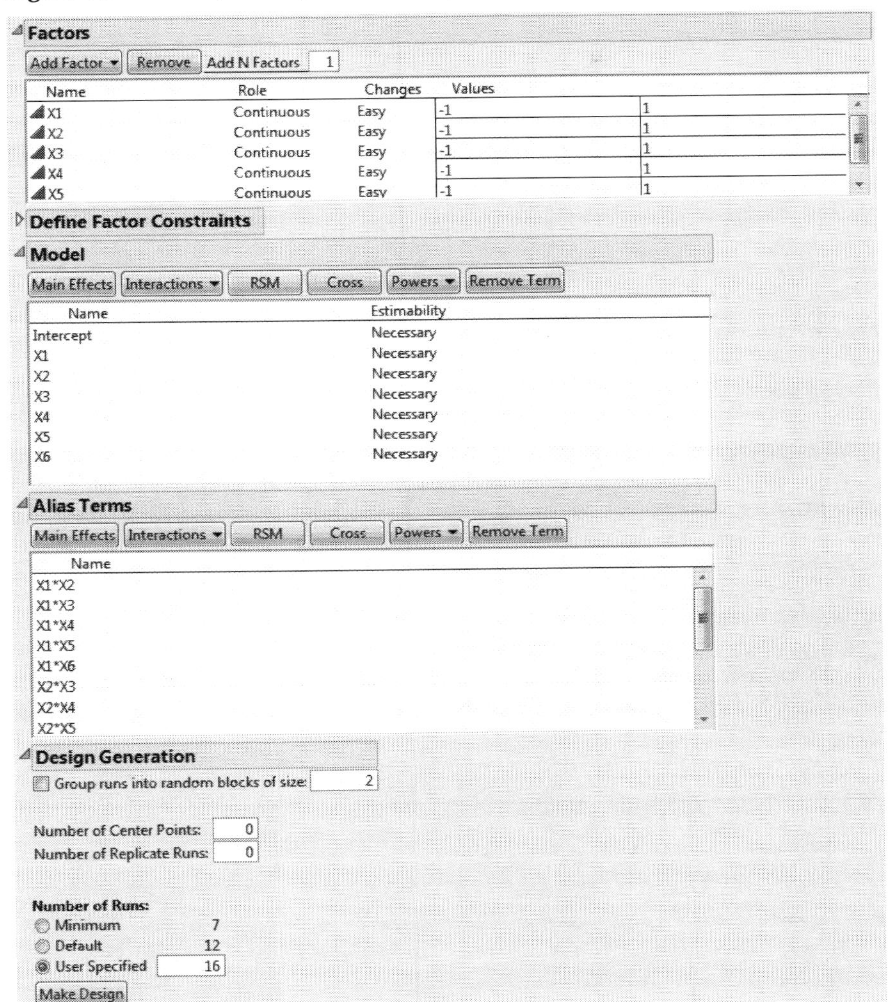

Note: Setting the Random Seed in step 7 and Number of Starts in step 8 reproduces the exact results shown in this example. In constructing a design on your own, these steps are not necessary.

7. (Optional) From the Custom Design red triangle menu, select **Set Random Seed**, type 1692819077, and click **OK**.

8. (Optional) From the Custom Design red triangle menu, select **Number of Starts**, type 161, and click **OK**.

9. Click **Make Design**.

10. Open the **Design Evaluation > Alias Matrix** outline.

Figure 5.10 Alias Matrix

Alias Matrix

Effect	X1*X2	X1*X3	X1*X4	X1*X5	X1*X6	X2*X3	X2*X4	X2*X5	X2*X6	X3*X4	X3*X5	X3*X6	X4*X5	X4*X6	X5*X6
Intercept	0	0	0	0	0	0	0	0	0	0	0	0	0	0	0
X1	0	0	0	0	0	0	0	0	0	0	0	0	0	0	0
X2	0	0	0	0	0	0	0	0	0	0	0	0	0	0	0
X3	0	0	0	0	0	0	0	0	0	0	0	0	0	0	0
X4	0	0	0	0	0	0	0	0	0	0	0	0	0	0	0
X5	0	0	0	0	0	0	0	0	0	0	0	0	0	0	0
X6	0	0	0	0	0	0	0	0	0	0	0	0	0	0	0

All rows contain only zeros, which means that the Intercept and main effect terms are not biased by any two-factor interactions.

11. Open the **Design Evaluation > Color Map on Correlations** outline.

Figure 5.11 Color Map on Correlations

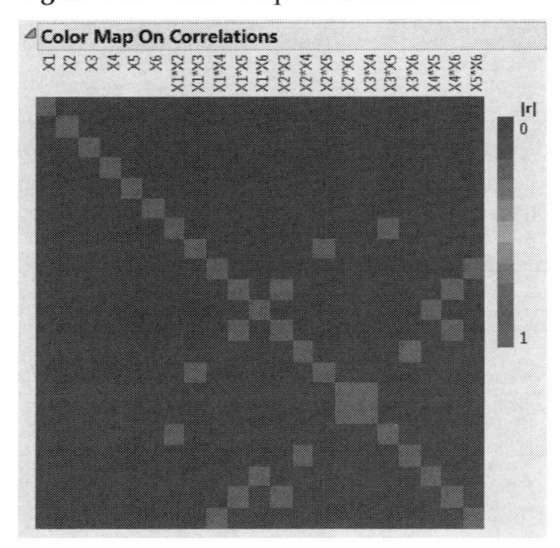

The Color Map on Correlations shows that main effects can be estimated independently of two-way interactions. However, some two-way interactions are fully aliased with other two-way interactions. Place your cursor over the off-diagonal red squares to see which two-way interactions are confounded.

It turns out that this particular design is a resolution IV orthogonal screening design. Main effects are not aliased with each other or with two-way interactions. But two-way interactions are fully aliased with other two-way interactions.

Supersaturated Screening Designs

It is common for brainstorming sessions to identify dozens of potentially active factors. Rather than reduce the list without the benefit of data, you can use a supersaturated design.

In a *saturated* design, the number of runs equals the number of model terms. In a *supersaturated* design, the number of model terms exceeds the number of runs (Lin, 1993). A supersaturated design can examine dozens of factors using fewer than half as many runs as factors. This makes it an attractive choice for factor screening when there are many factors and experimental runs are expensive.

Limitations of Supersaturated Designs

There are drawbacks to supersaturated designs:

- If the number of *active* factors is more than half the number of runs in the experiment, then it is likely that these factors will be impossible to identify. A general rule is that the number of runs should be at least four times larger than the number of active factors. In other words, if you expect that there might be as many as five active factors, you should plan on at least 20 runs.
- Analysis of supersaturated designs cannot yet be reduced to an automatic procedure. However, using forward stepwise regression is reasonable. In addition, the Screening platform (**Analyze** > **Specialized Modeling > Specialized DOE Models > Fit Two Level Screening**) offers a streamlined analysis.

Generate a Supersaturated Design

In this example, you want to construct a supersaturated design to study 12 factors in 8 runs. To create a supersaturated design, you set the Estimability of all model terms (except the intercept) to If Possible.

Note: This example is for illustration only. You should have at least 14 runs in any supersaturated design. If there are as many as four active factors, it is very difficult to interpret the results of an 8-run design. See "Limitations of Supersaturated Designs" on page 139.

1. Select **DOE > Custom Design.**
2. Type 12 next to **Add N Factors.**
3. Click **Add Factor > Continuous.**
4. Click **Continue.**
5. In the Model outline, select all terms except the Intercept.
6. Click Necessary next to any effect and change it to If Possible.

   Setting the effects to If Possible ensures that JMP uses the Bayesian D-optimality criterion to obtain the design.

Figure 5.12 Factors, Model, and Number of Runs

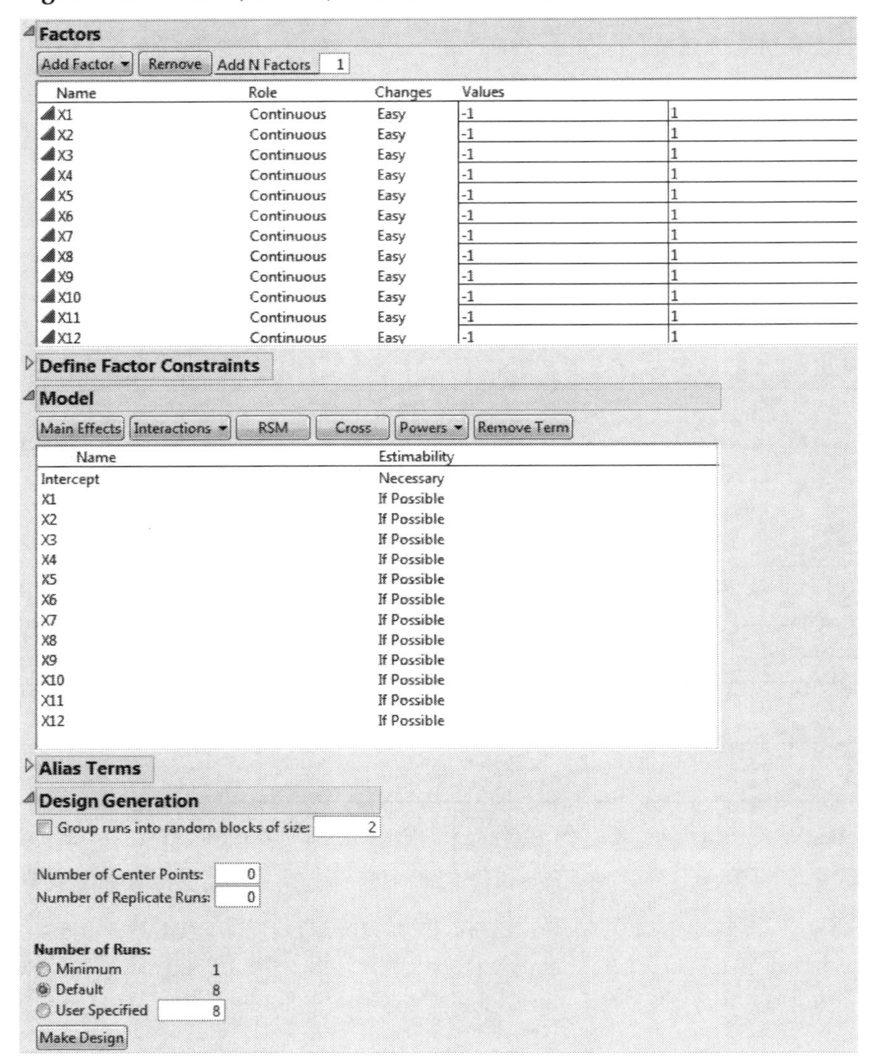

7. In the Alias Terms outline, select all effects and click **Remove Term**.

   This ensures that only the main effects appear in the Color Map on Correlations. This plot is constructed once the design is created.

8. Select **Simulate Responses** from the red triangle menu.

   This option generates random responses that appear in your design table. You will use these responses to see how to analyze experimental data.

   Keep the Number of Runs set to the Default of 8.

Note: Setting the Random Seed in step 9 and Number of Starts in step 10 reproduces the exact results shown in this example. In constructing a design on your own, these steps are not necessary.

9. (Optional) From the Custom Design red triangle menu, select **Set Random Seed**, type 1008705125, and click **OK**.
10. (Optional) From the Custom Design red triangle menu, select **Number of Starts**, type 100, and click **OK**.
11. Click **Make Design**.
12. Click **Make Table**.

    Do not close your Custom Design window. You return to it later in this example.

    The design table (Figure 5.13) and the Simulate Responses window (Figure 5.14) appear.

Figure 5.13 Design Table with Simulated Responses

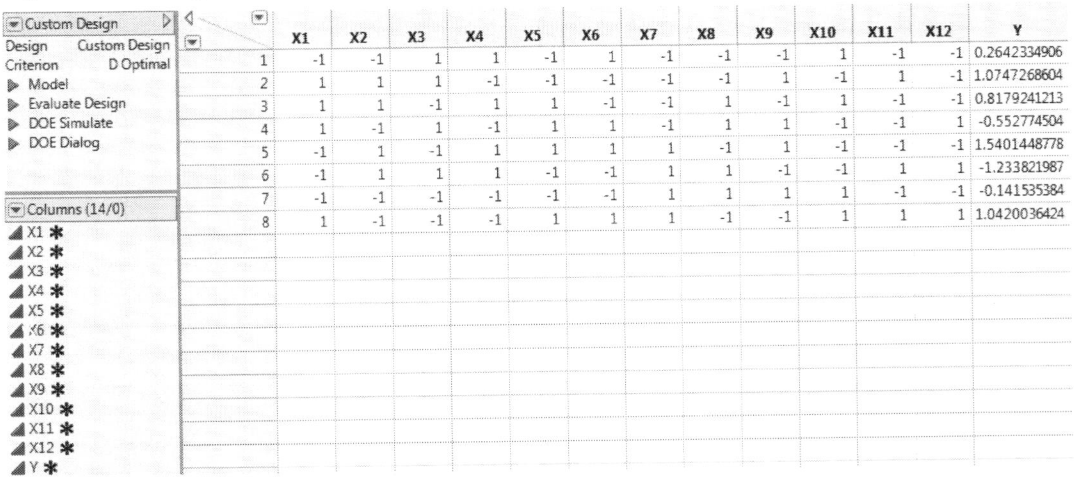

Custom Design
Design: Custom Design
Criterion: D Optimal
▶ Model
▶ Evaluate Design
▶ DOE Simulate
▶ DOE Dialog

Columns (14/0)
X1 * X2 * X3 * X4 * X5 * X6 * X7 * X8 * X9 * X10 * X11 * X12 * Y *

	X1	X2	X3	X4	X5	X6	X7	X8	X9	X10	X11	X12	Y
1	-1	-1	1	1	-1	1	-1	-1	-1	1	-1	-1	0.2642334906
2	1	1	1	-1	-1	-1	-1	-1	1	-1	1	-1	1.0747268604
3	1	1	-1	1	1	-1	-1	1	-1	1	-1	-1	0.8179241213
4	1	-1	1	-1	1	1	-1	1	1	-1	-1	1	-0.552774504
5	-1	1	-1	1	1	1	1	-1	1	-1	-1	-1	1.5401448778
6	-1	1	1	1	-1	-1	1	1	-1	-1	1	1	-1.233821987
7	-1	-1	-1	-1	-1	-1	1	1	1	1	-1	-1	-0.141535384
8	1	-1	-1	-1	1	1	1	-1	-1	1	1	1	1.0420036424

The response column, Y, contains simulated values. These are randomly generated using the model defined by the parameter values in the Simulate Responses window.

Figure 5.14 Simulate Responses Window

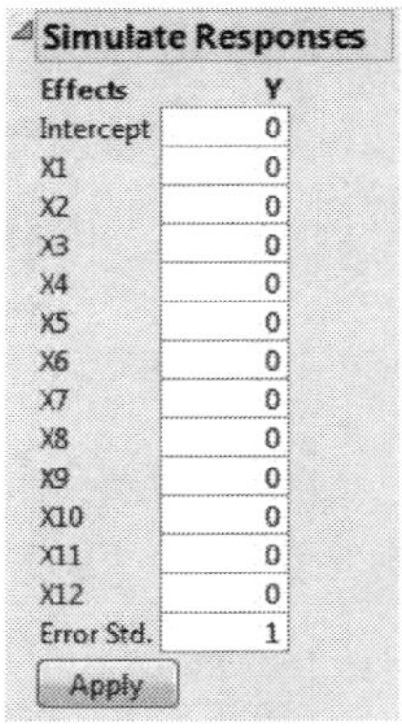

Simulate Responses

Effects	Y
Intercept	0
X1	0
X2	0
X3	0
X4	0
X5	0
X6	0
X7	0
X8	0
X9	0
X10	0
X11	0
X12	0
Error Std.	1

Apply

The Simulate Responses window shows coefficients of 0 for all terms, with an Error Std of 1. The values in the Y column currently reflect only random variation. Notice that the model coefficients are set to 0 because not all coefficients are estimable.

13. Change the values of the coefficients in the Simulate Responses window as shown in Figure 5.15.

Figure 5.15 Parameter Values for Simulated Responses

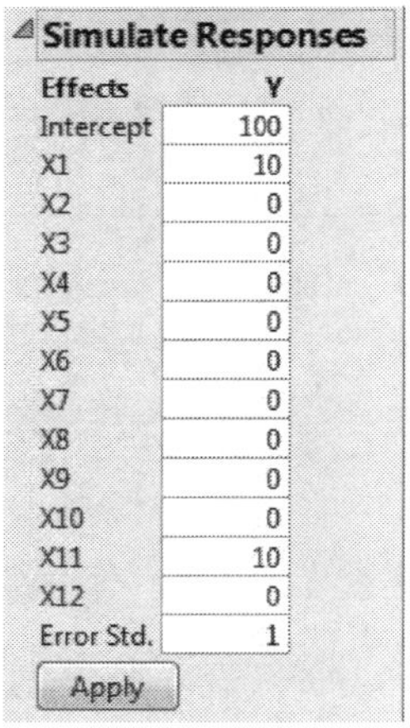

Simulate Responses

Effects	Y
Intercept	100
X1	10
X2	0
X3	0
X4	0
X5	0
X6	0
X7	0
X8	0
X9	0
X10	0
X11	10
X12	0
Error Std.	1

Apply

14. Click **Apply**.

The response values in the Y column change. See Figure 5.16.

Note: If you did not set the random seed and the number of starts, or if you click Apply more than once, your response values will not match those in Figure 5.16.

Figure 5.16 Response Column with X1 and X11 Active

Custom Design
Design Custom Design
Criterion D Optimal
▶ Screening
▶ Model
▶ DOE Dialog

Columns (13/0)
X1 ✱
X2 ✱

	X1	X2	X3	X4	X5	X6	X7	X8	X9	X10	X11	X12	Y
1	-1	-1	1	1	-1	1	-1	-1	-1	1	-1	-1	79.426379697
2	1	1	1	-1	-1	-1	-1	-1	1	-1	1	-1	122.73566401
3	1	1	-1	1	1	-1	-1	1	-1	1	-1	-1	99.902581852
4	1	-1	1	-1	1	1	-1	1	1	-1	-1	1	97.774291632
5	-1	1	-1	1	1	1	1	-1	1	-1	-1	-1	81.034856436
6	-1	1	1	1	-1	-1	1	1	-1	-1	1	1	100.99800916
7	-1	-1	-1	-1	-1	-1	1	1	1	1	-1	-1	79.332944607
8	1	-1	-1	-1	1	1	1	-1	-1	1	1	1	119.37709329

In your simulation, you specified X1 and X11 as active factors with large effects relative to the error variation. For this reason, your analysis of the data should identify these two factors as active.

Analyze a Supersaturated Design Using the Screening Platform

The Screening platform provides a way to identify active factors. The design table in Figure 5.16 contains three scripts. The Screening script analyzes your data using the Screening platform (located under the **Analyze > Specialized Modeling > Specialized DOE Models > Fit Two Level Screening** menu).

1. In the Table panel of the design table, click the green triangle next to the **Screening** script.

Figure 5.17 Screening Report for Supersaturated Design

Screening for Y

Contrasts

Term	Contrast	Lenth t-Ratio	Individual p-Value	Simultaneous p-Value
X11	13.0113	33.89	[illegible]	[illegible]
X1	9.3316	24.31	[illegible]	[illegible]
X12	-0.9575	-2.49	[illegible]	0.1803
X2	0.5483	1.43	0.1451	0.6070
X10	-0.2559	-0.67	0.5627	1.0000
X5	0.0669	0.17	0.8861	1.0000
X9	-0.0234	-0.06	0.9614	1.0000

Half Normal Plot

Lenth PSE=0.38392
P-Values derived from a simulation of 10000 Lenth t ratios.
Supersaturated main effects; bias makes p-values too small.
Make Model | Run Model

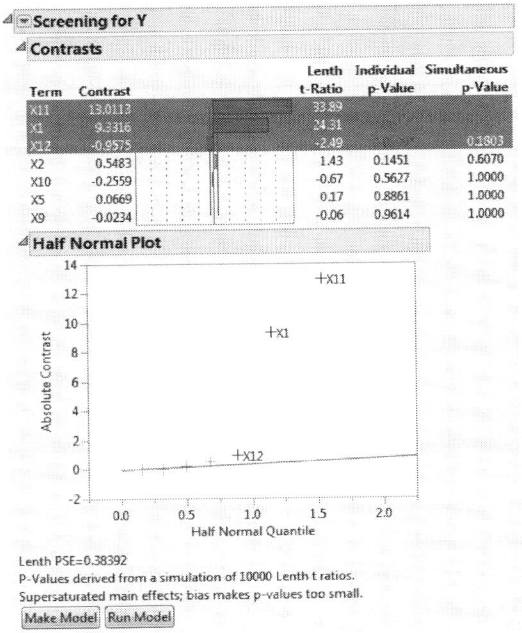

The factors X1 and X11 have large contrast and Lenth t-Ratio values. Also, their Simultaneous p-Values are small. In the Half Normal Plot, both X1 and X11 fall far from the line. The Contrasts and the Half Normal Plot reports indicate that X1 and X11 are active. Although X12 has an Individual p-Value less than 0.05, its effect is much smaller than that of X1 and X11.

Because the design is supersaturated, *p*-values might be smaller than they would be in a model where all effects are estimable. This is because effect estimates are biased by other potentially active main effects. In Figure 5.17, a note directly above the Make Model button warns you of this possibility.

You might also want to check whether the effects that appear active could be highly correlated with other effects. When this occurs, one effect can mask the true significance of another effect. The Color Map in Figure 5.19 displays absolute correlations between effects.

2. Click **Make Model.**

   The constructed model contains only the effects X1, X11, and X12.

3. Click **Run** in the Model Specification window.

Figure 5.18 Parameter Estimates for Model

Parameter Estimates

Term	Estimate	Std Error	t Ratio	Prob>|t|
Intercept	100.12645	0.319502	313.38	<.0001*
X11	11.347787	0.360447	31.48	<.0001*
X1	9.8209584	0.319502	30.74	<.0001*
X12	-1.1329	0.360447	-3.14	0.0347*

   Note that the parameter estimates for X11 and X1 are close to the theoretical values that you used to simulate the model. See Figure 5.15, where you specified a model with X1 = 10 and X11 = 10. The significance of the factor X12 is an example of a false positive.

4. In your Custom Design window, open the **Design Evaluation > Color Map on Correlations** outline.

Figure 5.19 Color Map on Correlations Outline

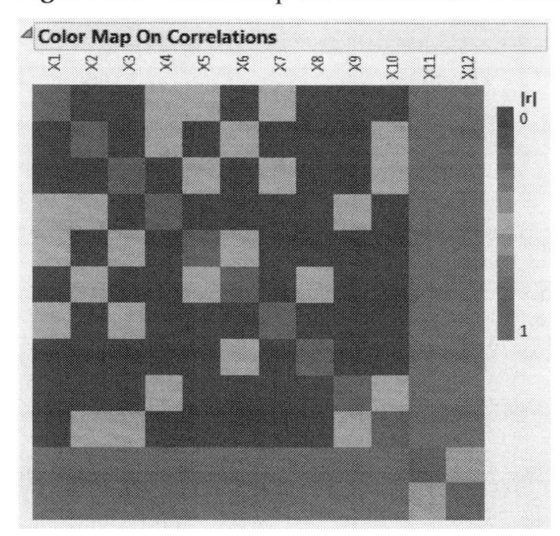

With your cursor, place your mouse pointer over cells to see the absolute correlations. Notice that X1 has correlations as high as 0.5 with other main effects (X4, X5, X7). (Figure 5.19 uses JMP default colors.)

Analyze a Supersaturated Design Using Stepwise Regression

Stepwise regression is another way to identify active factors. The design table in Figure 5.16 contains three scripts. The Model script analyzes your data using stepwise regression in the Fit Model platform.

1. In the Table panel of the design table, click the green triangle next to the **Model** script.
2. Change the **Personality** from **Standard Least Squares** to **Stepwise**.
3. Click **Run**.
4. In the Stepwise Fit for Y report, change the **Stopping Rule** to **Minimum AICc**.

   For designed experiments, BIC is typically a more lenient stopping rule than AICc as it tends to allow inactive effects into the model.
5. Click **Go**.

Figure 5.20 Stepwise Regression for Supersaturated Design

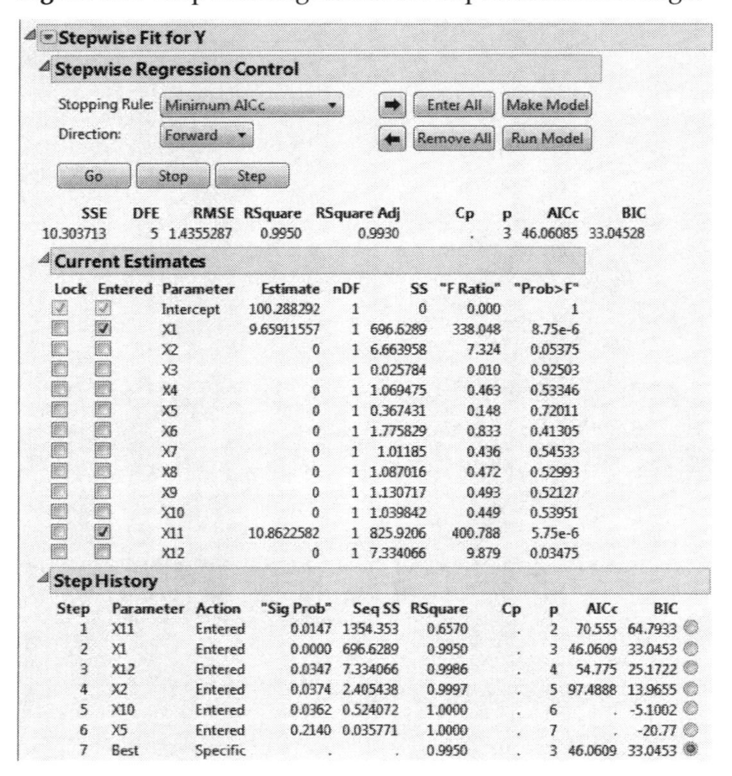

SSE	DFE	RMSE	RSquare	RSquare Adj	Cp	p	AICc	BIC
10.303713	5	1.4355287	0.9950	0.9930	.	3	46.06085	33.04528

Current Estimates

Lock	Entered	Parameter	Estimate	nDF	SS	"F Ratio"	"Prob>F"
☑	☑	Intercept	100.288292	1	0	0.000	1
☐	☑	X1	9.65911557	1	696.6289	338.048	8.75e-6
☐	☐	X2	0	1	6.663958	7.324	0.05375
☐	☐	X3	0	1	0.025784	0.010	0.92503
☐	☐	X4	0	1	1.069475	0.463	0.53346
☐	☐	X5	0	1	0.367431	0.148	0.72011
☐	☐	X6	0	1	1.775829	0.833	0.41305
☐	☐	X7	0	1	1.01185	0.436	0.54533
☐	☐	X8	0	1	1.087016	0.472	0.52993
☐	☐	X9	0	1	1.130717	0.493	0.52127
☐	☐	X10	0	1	1.039842	0.449	0.53951
☐	☑	X11	10.8622582	1	825.9206	400.788	5.75e-6
☐	☐	X12	0	1	7.334066	9.879	0.03475

Step History

Step	Parameter	Action	"Sig Prob"	Seq SS	RSquare	Cp	p	AICc	BIC
1	X11	Entered	0.0147	1354.353	0.6570	.	2	70.555	64.7933
2	X1	Entered	0.0000	696.6289	0.9950	.	3	46.0609	33.0453
3	X12	Entered	0.0347	7.334066	0.9986	.	4	54.775	25.1722
4	X2	Entered	0.0374	2.405438	0.9997	.	5	97.4888	13.9655
5	X10	Entered	0.0362	0.524072	1.0000	.	6	.	-5.1002
6	X5	Entered	0.2140	0.035771	1.0000	.	7	.	-20.77
7	Best	Specific	.	.	0.9950	.	3	46.0609	33.0453

Figure 5.20 shows that the selected model consists of the two active factors, X1 and X11. The step history appears in the bottom part of the report. Keep in mind that correlations between X1 and X11 and other factors could mask the effects of other active factors. See Figure 5.19.

Note: This example defines two large main effects and sets the rest to zero. Real-world situations can be less likely to have such clearly differentiated effects.

Design for Fixed Blocks

Traditional screening designs require block sizes to be a power of two. However, the Custom Design platform can create designs with fixed blocks of any size.

Suppose that you want to study three factors. You can run only three trials per day and you expect substantial day-to-day variation. Consequently, you need to block your design over multiple days. Also, in this study, you are interested in estimating all two-factor interactions. In this example, you construct a design with three runs per block.

1. Select **DOE > Custom Design**.
2. In the Factors outline, type 3 next to **Add N Factors**.
3. Click **Add Factor > Continuous**.
4. Click **Add Factor > Blocking > 3 runs per block**.

   The blocking factor X4 shows only one level under Values. This is because the run size is unknown at this point.

Figure 5.21 Factors Outline Showing One Block for X4

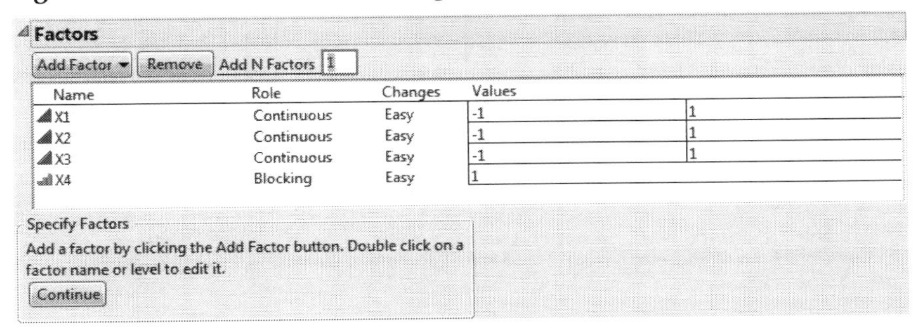

5. Click **Continue**.

Figure 5.22 Factors Outline Showing Three Blocks for X4

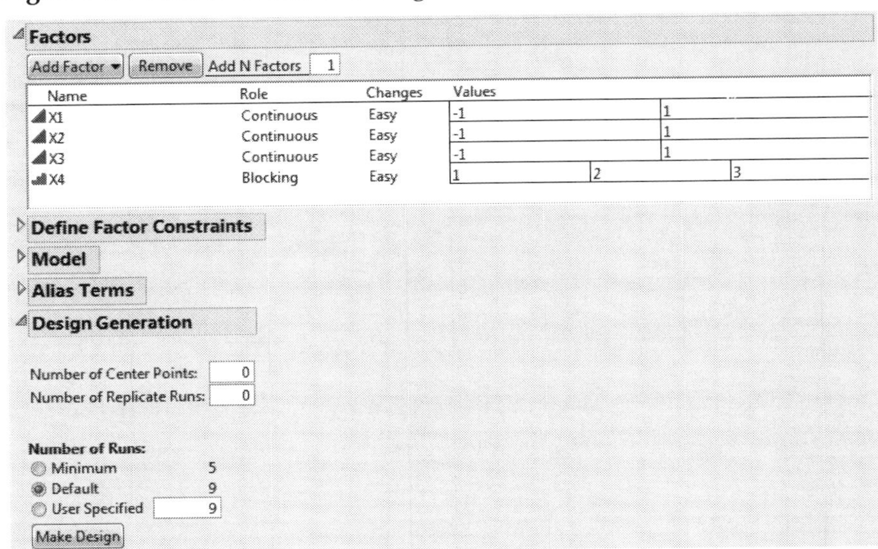

The Factors outline now shows an appropriate number of blocks, calculated as the Default run size divided by the number of runs per block. For this example, the default sample size of 9 requires three blocks. The Factors outline now shows that X4 has three values, indicating the three blocks.

Note: If you specify a different number of runs, the Factors outline updates to show the appropriate number of values for the blocking factor.

6. Select the three continuous factors, X1, X2, and X3, in the Factors outline.
7. In the Model outline, click **Interactions > 2nd**.

Figure 5.23 Factors Outline Showing Six Blocks for X4

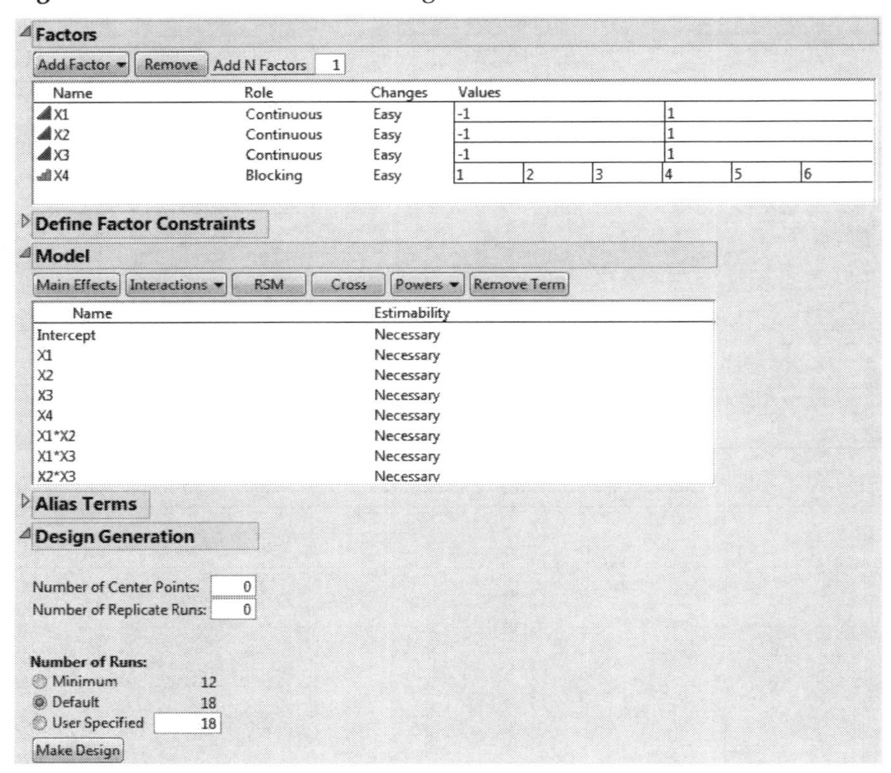

The Number of Runs panel now shows that 18 is the Default run size. The Factors outline now shows six values for X4, indicating six blocks.

Note: Setting the Random Seed in step 8 and Number of Starts in step 9 reproduces the exact results shown in this example. In constructing a design on your own, these steps are not necessary.

8. (Optional) From the Custom Design red triangle menu, select **Set Random Seed**, type 458027747, and click **OK**.
9. (Optional) From the Custom Design red triangle menu, select **Number of Starts**, type 10, and click **OK**.
10. Click **Make Design.**

Figure 5.24 Fixed Block Design

Design

Run	X1	X2	X3	X4
1	1	1	1	4
2	1	-1	-1	2
3	-1	1	-1	2
4	1	1	-1	3
5	-1	-1	-1	5
6	-1	-1	1	6
7	1	1	1	6
8	-1	-1	1	2
9	-1	1	-1	4
10	1	-1	1	1
11	1	-1	1	5
12	1	-1	-1	6
13	1	1	-1	5
14	-1	-1	-1	3
15	-1	1	1	1
16	1	-1	-1	4
17	1	1	-1	1
18	-1	1	1	3

In the design, look at the blocking factor, X4. The six blocks are represented. When you conduct your experiment, each day you will run three trials, where X4 = 1 on the first day, X4 = 2 on the second day, and so on. So you want the design table to randomize the trials within blocks. In the Output Options panel, the Randomize within Blocks option is already selected for Run Order.

11. Click **Make Table**.

Figure 5.25 Design Table for Fixed Block Design

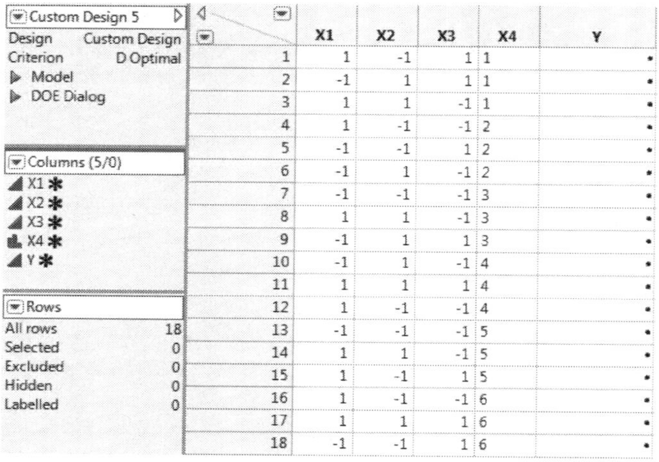

Custom Design 5
Design Custom Design
Criterion D Optimal
▶ Model
▶ DOE Dialog

Columns (5/0)
X1 *
X2 *
X3 *
X4 *
Y *

Rows
All rows 18
Selected 0
Excluded 0
Hidden 0
Labelled 0

	X1	X2	X3	X4	Y
1	1	-1	1	1	•
2	-1	1	1	1	•
3	1	1	-1	1	•
4	1	-1	-1	2	•
5	-1	-1	1	2	•
6	-1	1	-1	2	•
7	-1	-1	-1	3	•
8	1	1	-1	3	•
9	-1	1	1	3	•
10	-1	1	-1	4	•
11	1	1	1	4	•
12	1	-1	-1	4	•
13	-1	-1	-1	5	•
14	1	1	-1	5	•
15	1	-1	1	5	•
16	1	-1	-1	6	•
17	1	1	1	6	•
18	-1	-1	1	6	•

The rows in the design table are grouped by each day's runs. This design enables you to estimate the block effect, all main effects, and two-factor interactions.

Response Surface Experiments

Response surface experiments typically involve a small number (generally 2 to 8) of continuous factors that have been identified as active. The main goal of a response surface experiment is to develop a predictive model of the relationship between the factors and the response. Often, you use the predictive model to find better operating settings for your process. For this reason, your assumed model for a response surface experiment is usually quadratic.

Because a screening design is focused on identifying active effects, a measure of its quality is the size of the relative variance of the coefficients. You want these relative variances to be small. D-optimality addresses these relative variances.

In response surface experiments, the prediction variance over the range of the factors is more important than the variance of the coefficients. The prediction variance over the design space is addressed by I-optimality. An I-optimal design tends to place fewer runs at the extremes of the design space than does a D-optimal design. For details about D- and I-optimality, see "Optimality Criteria" on page 123 in the "Custom Designs" chapter.

By default, Custom Design uses the Recommended option for the Optimality Criterion. Custom Design uses the I-optimality criterion as the Recommended criterion whenever you add quadratic effects using the RSM button in the Model outline. Otherwise, Custom Design uses the D-optimality criterion as the Recommended criterion. See "Optimality Criteria" on page 123 in the "Custom Designs" chapter.

This section contains the following examples:

- "Response Surface Design" on page 151
- "Response Surface Design with Flexible Blocking" on page 157
- "Comparison of a D-Optimal and an I-Optimal Response Surface Design" on page 160

Response Surface Design

The following example contains these sections:

- "Construct a Response Surface Design" on page 151
- "Analyze the Experimental Results" on page 154

Construct a Response Surface Design

Construct a response surface design for three continuous factors that you have identified as active. You want to find process settings to maintain your response(Y) within specifications. The lower and upper specification limits for Y are 54 and 56, respectively, with a target of 55.

1. Select **DOE > Custom Design.**
2. In the Responses outline, click **Maximize** and select **Match Target.**
3. Type 54 as the **Lower Limit** and 56 as the **Upper Limit.**
4. Leave **Importance** blank.

   Because there is only one response, the Importance value is set to 1 by default.
5. Type 3 next to **Add N Factors.**
6. Click **Add Factor > Continuous.**

   This adds three continuous factors: X1, X2, and X3.
7. Click **Continue.**
8. In the Model outline, click the **RSM** button.

   This adds quadratic and interaction terms to the model. It also sets the value of the Recommended optimality criterion to I-optimality. You can verify this in the Design Diagnostics outline once you click Make Design.

   Leave the Default Number of Runs set to 16.

   Note: Setting the Random Seed in step 9 and Number of Starts in step 10 reproduces the exact results shown in this example. In constructing a design on your own, these steps are not necessary.
9. (Optional) From the Custom Design red triangle menu, select **Set Random Seed**, type 929281409, and click **OK.**

10. (Optional) From the Custom Design red triangle menu, select **Number of Starts**, type 40, and click **OK**.
11. Click **Make Design**.

    The Design outline shows the design.

Figure 5.26 RSM Design

Design

Run	X1	X2	X3
1	0	-1	0
2	1	-1	1
3	0	0	0
4	-1	1	1
5	-1	0	0
6	1	1	1
7	-1	1	-1
8	-1	-1	-1
9	0	0	1
10	1	0	0
11	0	1	0
12	0	0	0
13	1	1	-1
14	0	0	-1
15	-1	-1	1
16	1	-1	-1

In order to estimate quadratic effects, a response surface design uses three levels for each factor. Note that the design in Figure 5.26 is a face-centered Central Composite Design with two center points.

12. Open the **Design Evaluation > Design Diagnostics** outline.

Figure 5.27 Design Diagnostics Outline

Design Diagnostics

I Optimal Design	
D Efficiency	42.99905
G Efficiency	88.62743
A Efficiency	30.68134
Average Variance of Prediction	0.340517
Design Creation Time (seconds)	0.033333

The first line in the Design Diagnostics outline identifies the optimality criterion being used. This design is I-optimal.

13. Open the **Design Evaluation > Prediction Variance Profile** outline.

Figure 5.28 Prediction Variance Profile

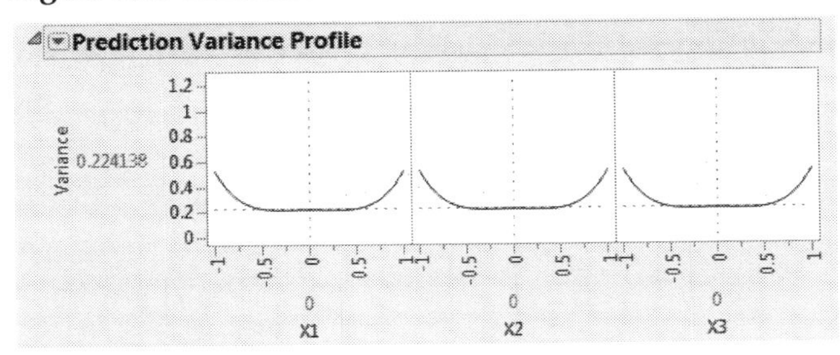

The vertical axis shows the relative prediction variance of the expected value of the response. The *relative prediction variance* is the prediction variance divided by the error variance. When the relative prediction variance is one, its absolute variance equals the error variance of the regression model.

The profiler shows values of the relative prediction variance over the design space. You can move the sliders to explore the prediction variance's behavior. The prediction variance is smallest in the center of the design space. It is fairly constant, with values only slightly larger than 0.2, for factor settings between -0.5 and 0.5. The prediction variance increases as the settings approach the design space boundaries.

14. Select **Optimization and Desirability > Maximize Desirability** from the red triangle menu next to Prediction Variance Profile outline.

Figure 5.29 Prediction Variance Profile with Relative Variance Maximized

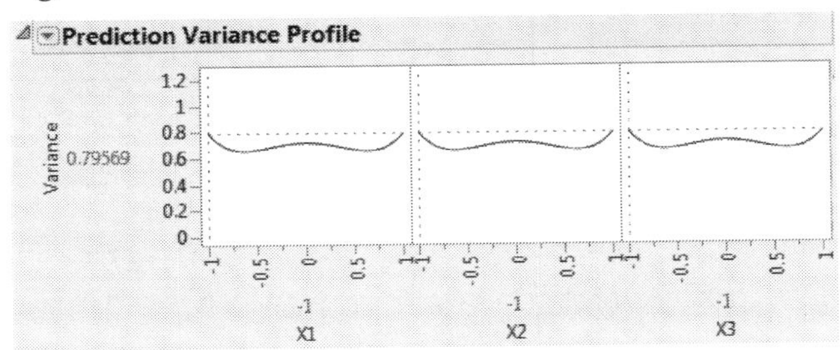

The profiler shows that the maximum value of the relative prediction variance is 0.79569.

15. Open the **Design Evaluation > Fraction of Design Space Plot** outline.

Figure 5.30 Fraction of Design Space Plot

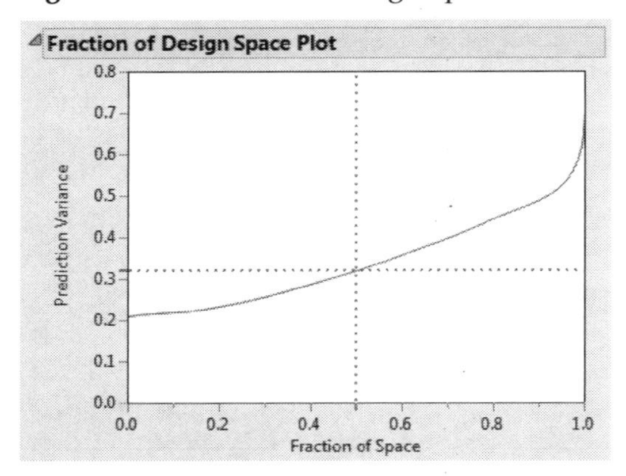

The blue curve in the plot shows the relative prediction variance as a function of the fraction of design space. The red dashed cross hairs indicate that, for 50% of the design space, the prediction variance is about 0.32 or less. Use the cross hair tool to draw other inferences. For example, when the Fraction of Space is 0.95, the Prediction Variance is about 0.52. This means that for 95% of the design space, the relative prediction variance is below 0.52.

Analyze the Experimental Results

The Custom RSM.jmp sample data table contains the results of the experiment. The Model script opens a Fit Model window showing all of the effects specified in the DOE window's Model outline. This script was saved to the data table by the Custom Design platform.

1. Select **Help > Sample Data Library** and open Design Experiment/Custom RSM.jmp.
2. In the Table panel, click the green triangle next to the **Model** script.
3. Click **Run**.

   The Effect Summary report shows the LogWorth and PValue for each effect in the model. The vertical blue line in the plot is set at the value 2. A LogWorth that exceeds 2 is significant at the 0.01 level.

Figure 5.31 Effect Summary Report

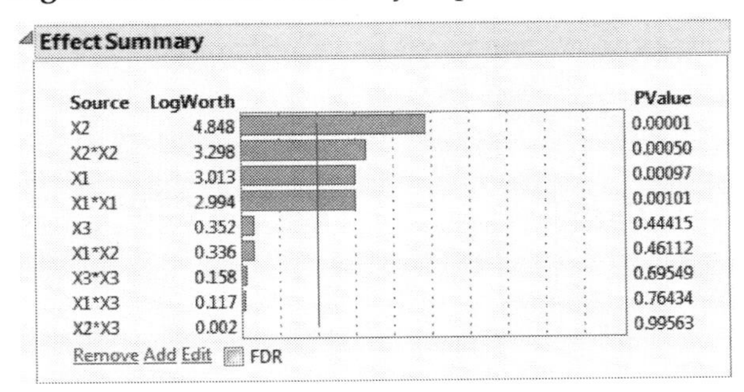

Effect Summary

Source	LogWorth	PValue
X2	4.848	0.00001
X2*X2	3.298	0.00050
X1	3.013	0.00097
X1*X1	2.994	0.00101
X3	0.352	0.44415
X1*X2	0.336	0.46112
X3*X3	0.158	0.69549
X1*X3	0.117	0.76434
X2*X3	0.002	0.99563

Remove Add Edit FDR

The report shows that X1, X2, X1*X1, and X2*X2 are significant at the 0.01 level. None of the other effects are significant at even the 0.10 level. Reduce the model by removing these insignificant effects.

4. In the Effect Summary report, select X3, X1*X2, X3*X3, X1*X3, and X2*X3.

Figure 5.32 Effect Summary Report with Insignificant Effects Selected

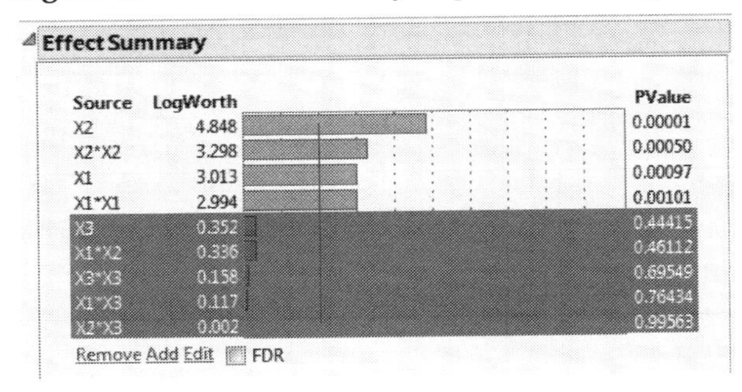

Effect Summary

Source	LogWorth	PValue
X2	4.848	0.00001
X2*X2	3.298	0.00050
X1	3.013	0.00097
X1*X1	2.994	0.00101
X3	0.352	0.44415
X1*X2	0.336	0.46112
X3*X3	0.158	0.69549
X1*X3	0.117	0.76434
X2*X3	0.002	0.99563

Remove Add Edit FDR

5. Click **Remove**.

   The Fit Least Squares report is updated to show a model containing only the significant effects: X1, X2, X1*X1, and X2*X2.

   Use the Prediction Profiler (at the bottom of the Fit Least Squares window) to explore how the predicted response (Y) changes as you vary the factors X1 and X2. Note the quadratic behavior of Y across the values of X1 and X2.

   Remember that you entered response limits for Y in the Responses outline of the Custom Design window. As a result, the Response Limit column property is attached to the Y column in the design table. The Desirability function for Y (in the top plot at right) is based on the information contained in the Response Limit column property. JMP uses this function to calculate Desirability as a function of the settings of X1 and X2. The traces of the Desirability function appear in the bottom row of plots.

6. In the Prediction Profiler report, select **Optimization and Desirability > Maximize Desirability** from the red triangle options.

Figure 5.33 Prediction Profiler with Desirability Maximized

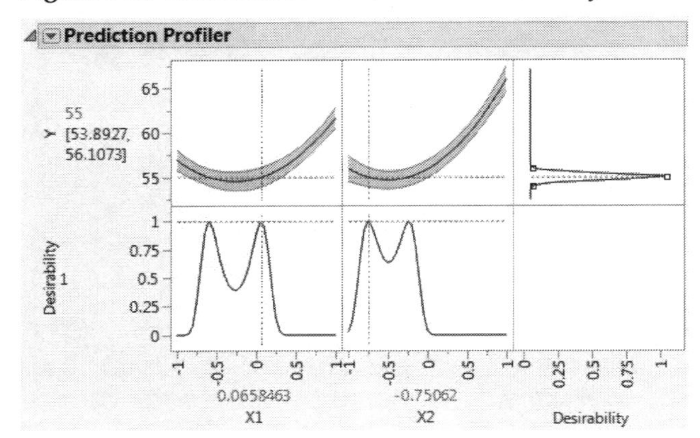

The predicted response achieves the target value of 55 at the process settings shown in red above X1 and X2. Figure 5.33 shows that a value of X1 near –0.65 also achieves a predicted value of 55 when X2 = -0.75062. In fact, your Prediction Profiler might show different settings as those that maximize desirability. This is because the predicted response is 55 for many settings of X1 and X2.

7. Select **Factor Profiling > Contour Profiler** from the red triangle next to Response Y.
8. In the Contour Profiler report, type 55 as the value for **Contour**.

Figure 5.34 Contour Profiler

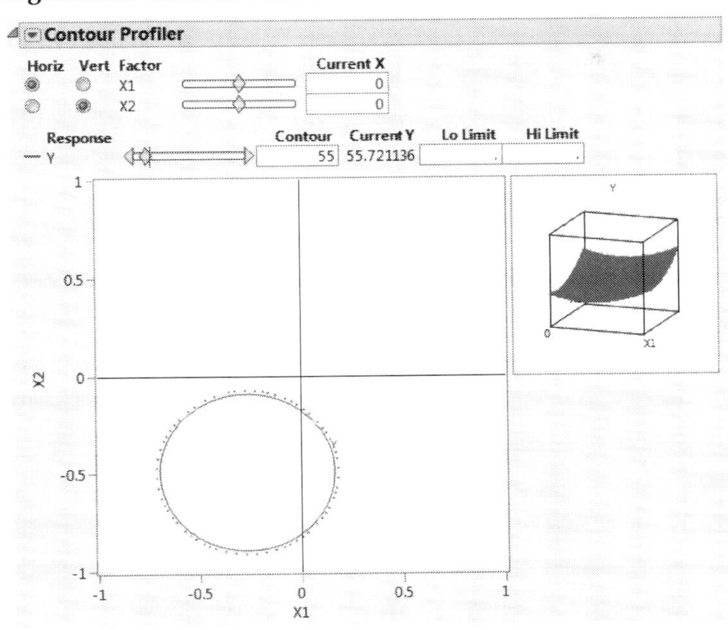

The settings of X1 and X2 that correspond to the red contour have predicted response values of 55. You might want to select from among these process settings based on cost efficiency.

Response Surface Design with Flexible Blocking

When optimizing a process, you might need to include qualitative factors in your experiment as well as continuous factors. You might need to block by qualitative factors such as batch or day, or include qualitative factors such as machine or delivery mechanism. But the Response Surface Design platform supports only continuous factors. To obtain a response surface design with a qualitative factor, you can replicate the design over each level of the factor. However, this is inefficient. The Custom Design platform constructs an optimal design with fewer runs.

In this example, you construct a response surface design that accommodates two continuous factors and a blocking factor with four runs per block. You can include categorical or discrete numeric factors in a similar fashion.

1. Select **DOE > Custom Design**.
2. Type 2 next to **Add N Factors**.
3. Click **Add Factor > Continuous**.

4. Click **Add Factor** > **Blocking** > **4 runs per block.**

   Notice that only one level appears under Values. This is because the number of blocks cannot be determined until the number of runs is determined.

Figure 5.35 Factors Outline with Two Continuous Factors and a Blocking Factor

Factors

Add Factor ▾ | Remove | Add N Factors 1

Name	Role	Changes	Values	
X1	Continuous	Easy	-1	1
X2	Continuous	Easy	-1	1
X3	Blocking	Easy	1	

5. Click **Continue.**

   The Default number of runs is 12. The Factors outline updates to show three levels for the Blocking factor, X3. Because you required X3 to have four runs per block, the 12 runs allow three blocks.

6. Click **RSM.**

   An informational JMP Alert window reminds you that the blocking factor cannot appear in interaction or quadratic terms. JMP adds only the appropriate RSM terms to the list.

7. Click **OK** to dismiss the message.

   Quadratic and interactions terms for X1 and X2 are added to the model. Because you added RSM terms, the Recommended optimality criterion changes from D-Optimal to I-Optimal. You can see this later in the Design Diagnostics outline.

Figure 5.36 Model Outline with Response Surface Effects

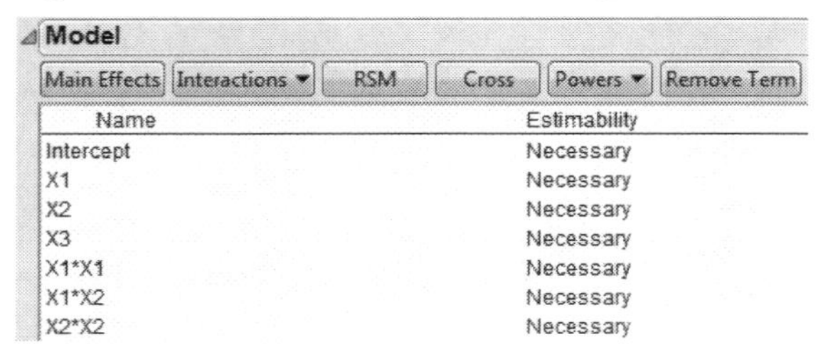
Model

Main Effects | Interactions ▾ | RSM | Cross | Powers ▾ | Remove Term

Name	Estimability
Intercept	Necessary
X1	Necessary
X2	Necessary
X3	Necessary
X1*X1	Necessary
X1*X2	Necessary
X2*X2	Necessary

Note: Setting the Random Seed in step 8 and Number of Starts in step 9 reproduces the exact results shown in this example. In constructing a design on your own, these steps are not necessary.

8. (Optional) From the Custom Design red triangle menu, select **Set Random Seed**, type 1415408414, and click **OK**.

9. (Optional) From the Custom Design red triangle menu, select **Number of Starts**, type 21, and click **OK**.

10. Click **Make Design**.
11. Open the **Design Evaluation > Design Diagnostics** outline.

Figure 5.37 Design Diagnostics Outline

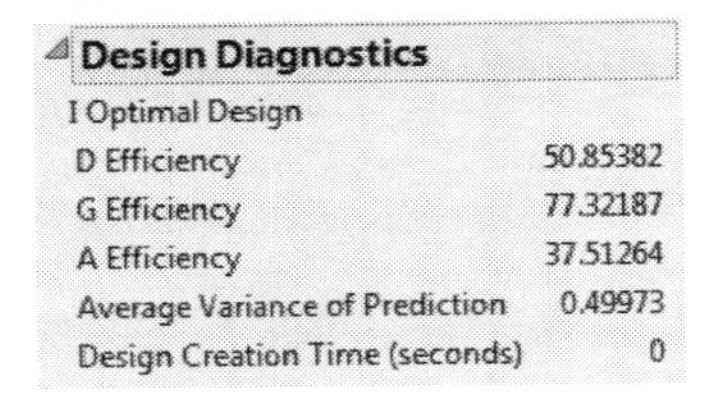

The first line in the Design Diagnostics outline identifies the optimality criterion being used. This design is I-optimal.

12. Click **Make Table**.

Figure 5.38 Design Table with Blocking Factor

Custom Design
Design Custom Design
Criterion I Optimal
Model
Evaluate Design
DOE Dialog

Columns (4/0)
X1 *
X2 *
X3 *
Y *

	X1	X2	X3	Y
1	0	0	1	•
2	0	-1	1	•
3	1	1	1	•
4	-1	0	1	•
5	0	-1	2	•
6	-1	1	2	•
7	0	0	2	•
8	1	0	2	•
9	0	1	3	•
10	0	0	3	•
11	-1	-1	3	•
12	1	-1	3	•

Because the default Run Order was Randomize within Blocks, the levels of the blocking factor (X3) are sorted.

13. In the Table panel of the design table, click the green triangle next to the **Model** script.

Figure 5.39 Fit Model Window

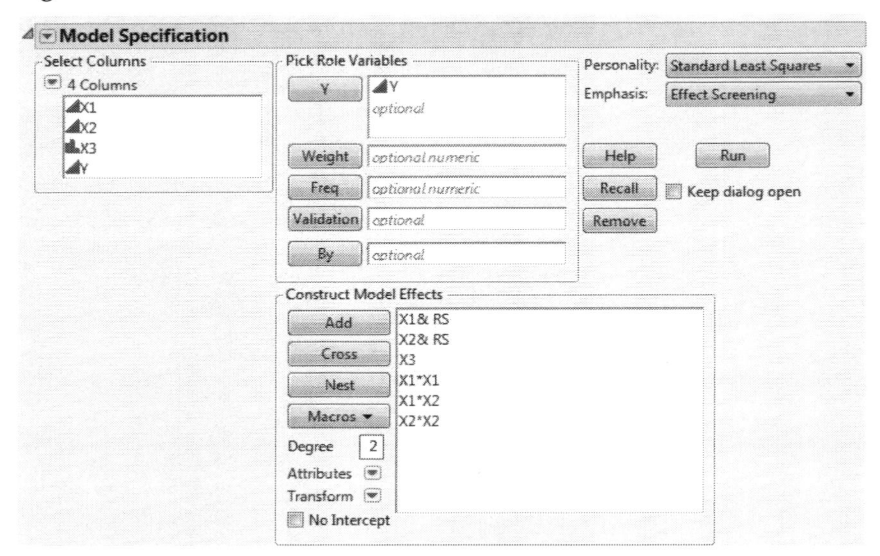

Notice the following:

– The blocking factor (X3) is entered as an effect.

– No interactions involving X3 are included.

– The other five effects define a response surface model for X1 and X2.

Comparison of a D-Optimal and an I-Optimal Response Surface Design

In this example, you explore the differences between I-optimality and D-optimality in the context of a two-factor response surface design.

I-Optimal Design

1. Select **DOE > Custom Design.**
2. Type 2 next to **Add N Factors.**
3. Click **Add Factor > Continuous.**
4. Click **Continue.**
5. Click **RSM.**

   Quadratic and interactions terms for X1 and X2 are added to the model. Because you added RSM terms, the Recommended optimality criterion changes from D-Optimal to I-Optimal. You can see this later in the Design Diagnostics outline.

Note: Setting the Random Seed in step 6 and Number of Starts in step 7 reproduces the exact results shown in this example. In constructing a design on your own, these steps are not necessary.

6. (Optional) From the Custom Design red triangle menu, select **Set Random Seed**, type 383570403, and click **OK**.
7. (Optional) From the Custom Design red triangle menu, select **Number of Starts**, type 8, and click **OK**.
8. Click **Make Design**.

Figure 5.40 I-Optimal Design

Design

Run	X1	X2
1	0	0
2	1	0
3	-1	-1
4	0	0
5	0	1
6	-1	0
7	0	0
8	-1	1
9	1	-1
10	0	0
11	0	-1
12	1	1

In this I-optimal design, runs 1, 4, 7, and 10 are at the center point (X1 = 0 and X2 = 0). I-optimal designs tend to place more runs in the center (and consequently fewer runs at the extremes) of the design space than do D-optimal designs. You can compare this design to the D-optimal design shown in Figure 5.42.

9. Open the **Design Evaluation** > **Prediction Variance Profile** outline.

Figure 5.41 Prediction Variance Profile for I-Optimal Model

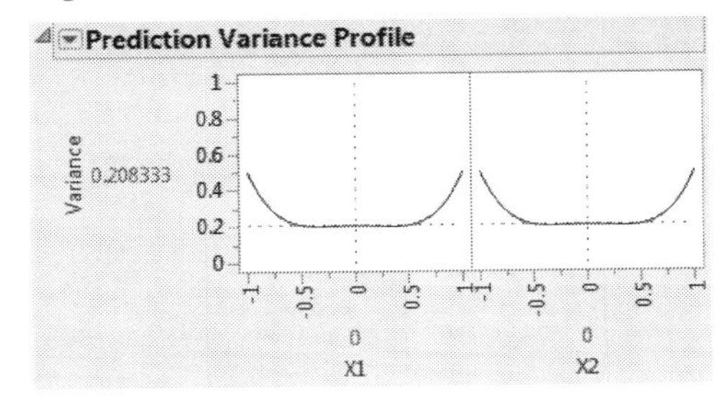

The relative prediction variance of the expected response is smallest in the center of the design space.

10. Open the **Fraction of Design Space Plot** outline.

The Fraction of Design Space Plot appears on the left in Figure 5.44. When the Fraction of Space is 0.95, the vertical coordinate of the blue curve is about 0.5. This means that for about 95% of the design space, the relative prediction variance is below 50% of the error variance.

This Custom Design window contains your I-optimal design. Keep this window open. In the next section, you generate a D-optimal design, and compare the two.

D-Optimal Design

To compare Prediction Variance Profile and Fraction of Design Space plots for the I- and D-optimal designs:

1. In the Custom Design window containing your I-optimal design, from the Custom Design red triangle menu, select **Save Script to Script Window**.

   A window appears, showing a script that reproduces your work.
2. In this new script window, select **Edit > Run Script**.

   A duplicate Custom Design window appears, but with the Design Evaluation outlines closed.
3. In this new Custom Design window, click **Back**.
4. From the red triangle next to Custom Design, select **Optimality Criterion > Make D-Optimal Design**.
5. Click **Make Design**.

   You current Custom Design window contains your D-optimal design.

Figure 5.42 D-Optimal Design

Design

Run	X1	X2
1	1	-1
2	-1	1
3	-1	-1
4	1	1
5	-1	-1
6	0	-1
7	0	0
8	0	1
9	1	-1
10	-1	0
11	1	0
12	-1	1

   In this D-optimal design, run 7 is the only run at the center point. D-optimal designs tend to place more runs at the extremes of the design space than do I-optimal designs. Recall that the I-optimal design had four center runs (Figure 5.40).
6. Open the **Design Evaluation** > **Prediction Variance Profile** outline.

Figure 5.43 Prediction Variance Profile for D-Optimal Model

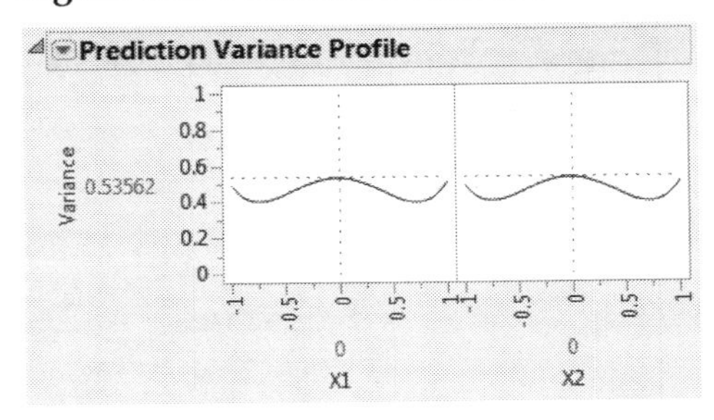

At the center of the design region, the relative prediction variance is 0.53562, as compared to 0.208333 for the I-optimal design (Figure 5.41). This means that the relative standard error is 0.732 for the D-optimal design and 0.456 for the I-optimal design. All else being equal, at the center of the design region, confidence intervals for the expected response based on the D-optimal design are about 60% wider than those based on the I-optimal design.

The Design outline shows that the D-optimal design has nine design points, one for every combination of X1 and X2 set to -1, 0, 1. The D-optimality criterion attempts to keep the relative prediction variance low at each of these design points. Explore the variance at the extremes of the design region by moving the sliders for X1 and X2 to -1 and 1. Note that the variance at these extreme points is usually smaller than the variance for the I-optimal design at these points.

7. Open the **Design Evaluation > Fraction of Design Space Plot** outline.

   The Fraction of Design Space Plot appears on the right in Figure 5.44.

Figure 5.44 Fraction of Design Space Plots (I-Optimal on left, D-Optimal on right)

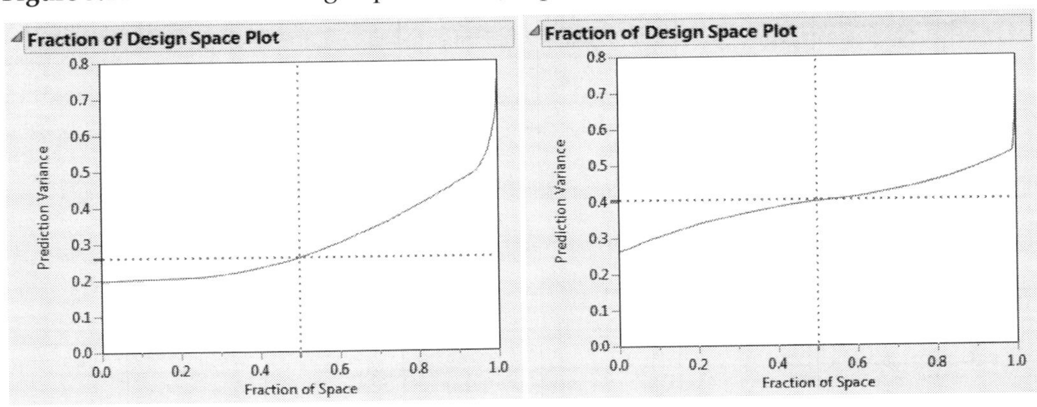

The red cross-hairs in each plot indicate the maximum prediction variance for 50% of the design space. For 50% of the design space, the prediction variance for the I-optimal design falls below about 0.27. For the D-optimal design, the prediction variance is about 0.4.

8. Right-click in the Fraction of Design Space Plot for your I-optimal design. Select **Edit > Copy Frame Contents**.
9. Right-click in the Fraction of Design Space Plot for your D-optimal design. Select **Edit > Paste Frame Contents**.

Figure 5.45 Fraction of Design Space Plots Superimposed

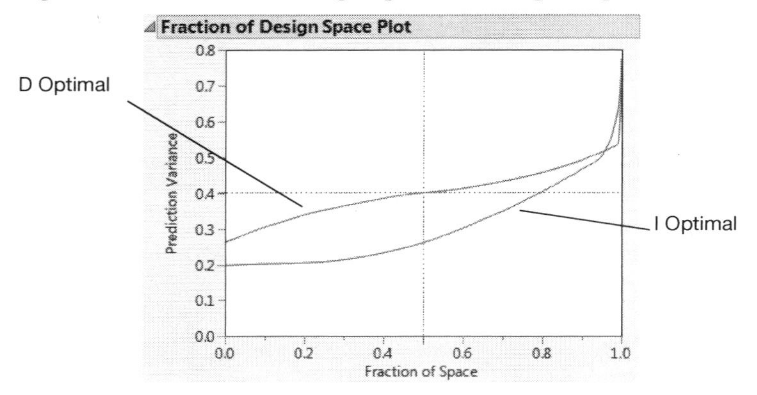

The variance curve for the I-optimal design is below the curve for the D-optimal design over at least 90% of the design space. This reflects the fact that I-optimality attempts to minimize prediction variance over all of the design space. In contrast, D-optimality focuses on reducing prediction variance at the design points.

Response Surface Design With Constraints and Categorical Factor

In this example, you create a design to optimize the yield of a chemical reaction. Your experimental factors include a categorical factor at three levels, where constraints involve two of the levels. In this example, you will use the Disallowed Combinations Filter to enter the constraints.

Your response is Yield. Your factors are the following:

- Time: The range of interest is 500 to 560 seconds.
- Temperature: The range of interest is 350 to 750 degrees Kelvin.
- Catalyst: Three catalysts A, B, and C, must be tested.

Your design must allow for constraints on two of the levels of Catalyst:

- When catalyst B is used, the temperature must be above 400.

- When catalyst C is used, the temperature must be below 650.

Define the Response and Factors

1. Select **DOE > Custom Design**.
2. In the Response outline, double-click Y and change it to Yield.

   Because your goal is to maximize Yield, leave the **Goal** set to Maximize.
3. In the Factors outline, type 2 next to **Add N Factors**.
4. Click **Add Factor > Continuous**.
5. Click **Add Factor > Categorical > 3 Level**.
6. Rename the factors Time, Temperature, and Catalyst.
7. Change the **Values** for Time to 500 and 560.
8. Change the **Values** for Temperature to 350 and 750.
9. Change the **Values** for Catalyst to A, B, and C.

Figure 5.46 Factor Settings

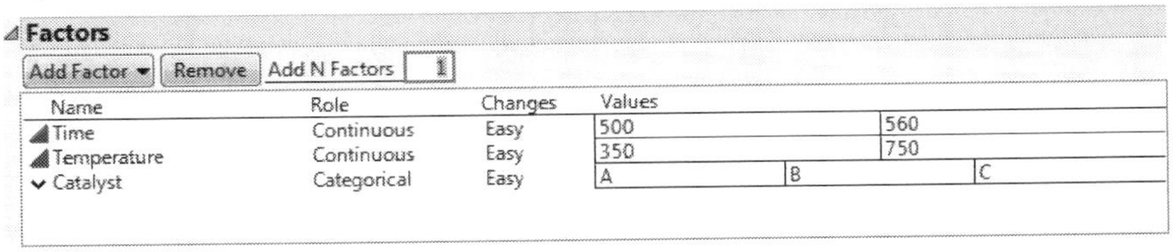

Factors

Add Factor | Remove | Add N Factors 1

Name	Role	Changes	Values		
Time	Continuous	Easy	500	560	
Temperature	Continuous	Easy	350	750	
Catalyst	Categorical	Easy	A	B	C

10. Click **Continue**.

Define the Constraints

1. Select **Use Disallowed Combinations Filter** in the Define Factor Constraints outline.
2. Select Temperature and Catalyst from the **Add Filter Factors** list.
3. Click **Add**.
4. Click 750 in the equation that appears above the Temperature slider and change it to 400. Press Enter (or click elsewhere).

   This disallows factor settings with Temperature values below 400.
5. Click the B block under Catalyst.

   Together with the constraint on Temperature, this disallows factor settings for which Catalyst is B and Temperature is below 400.
6. Click **OR**.

   This allows you to define your second constraint.

7. Select **Temperature** and **Catalyst** from the **Add Filter Factors** list. (Your earlier selection may have been retained.)
8. Click **Add**.
9. In the panel that appears beneath the word OR, click 350 in the equation that appears above the Temperature slider and change it to 650. Press Enter (or click elsewhere).

   This disallows factor settings with **Temperature** values above 650.
10. Click the C block under **Catalyst**.

    Together with the constraint on **Temperature**, this disallows factor settings for which **Catalyst** is C and **Temperature** is above 650.

Figure 5.47 Constraints Defined

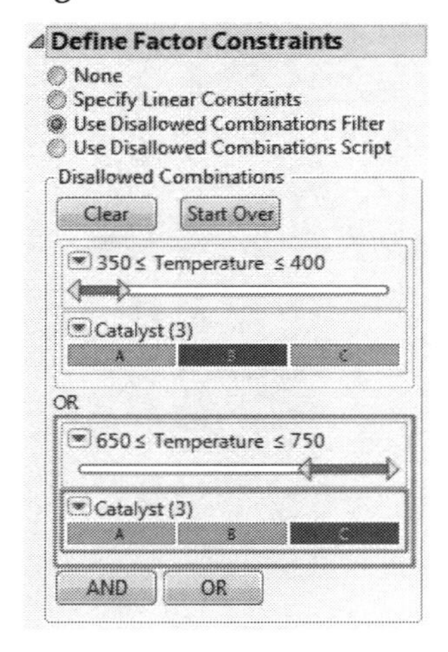

Add Response Surface Terms and Make Design

1. In the Model outline, select **RSM**.

   A JMP Alert informs you that only quadratic terms for continuous factors are being added to the model.
2. Click **OK** to dismiss the alert.

   JMP adds only the appropriate response surface terms to the model.

   Note: Setting the Random Seed in step 20 and Number of Starts in step 21 reproduces the exact results shown in this example. In constructing a design on your own, these steps are not necessary.

3. (Optional) From the Custom Design red triangle menu, select **Set Random Seed**, type 654321, and click **OK**.

4. (Optional) From the Custom Design red triangle menu, select **Number of Starts** and set it to 1000. Click **OK**.
5. Click **Make Design**.

Figure 5.48 Design Satisfying Constraints

Design

Run	Time	Temperature	Catalyst
1	500	750	B
2	500	410	B
3	560	750	A
4	521	750	A
5	542	350	A
6	500	648.7299	C
7	500	350	A
8	533	630	C
9	506	350	C
10	560	530	A
11	560	410	B
12	530	570	B
13	530	550	A
14	530	570	B
15	557	645.1294	C
16	500	570	A
17	560	750	B
18	560	350	C

Because you added RSM terms to the Model outline, this is an I-optimal design. It satisfies your constraints:

- When catalyst B is used, the temperature must be above 500.
- When catalyst C is used, the temperature must be below 600.

View the Design

Recall that the design region consists of settings of Time between 500 and 560 and Temperature between 350 and 750. To obtain a geometric view of the design points within the design region, do the following:

1. Click **Make Table**.
2. Select **Graph > Graph Builder**.
3. Select Time and Temperature and drag them to the center of the template.
4. Deselect the **Smoother** icon. This is the second icon from the left above the template.
5. Select Catalyst and drag it to the **Group X** zone at the top of the template.

Figure 5.49 Design Points for Three Levels of Catalyst

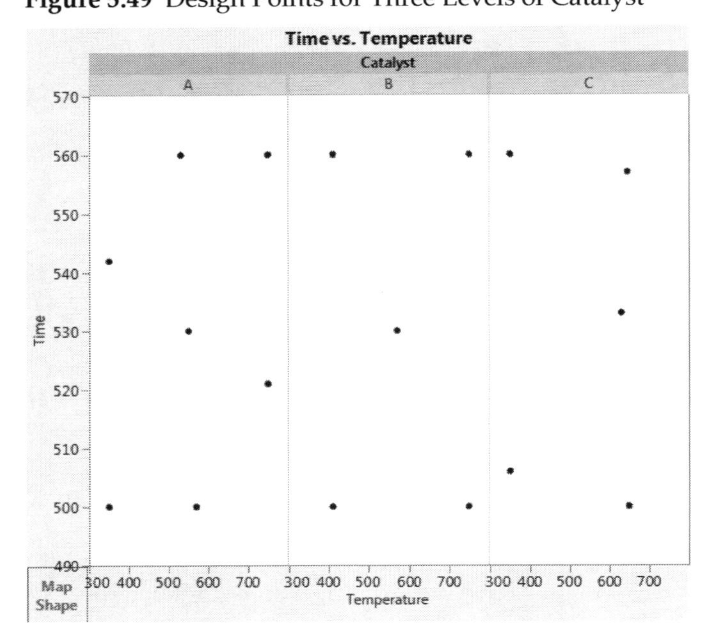

6. Double-click the **Temperature** axis.
7. Under Reference Lines, do the following:
   – Next to **Value**, enter 350.
   – Click **Add**.
   – Next to **Value**, enter 750.
   – Click **Add**.
8. Click **OK**.

Figure 5.50 Design Bounds on Temperature for Three Levels of Catalyst

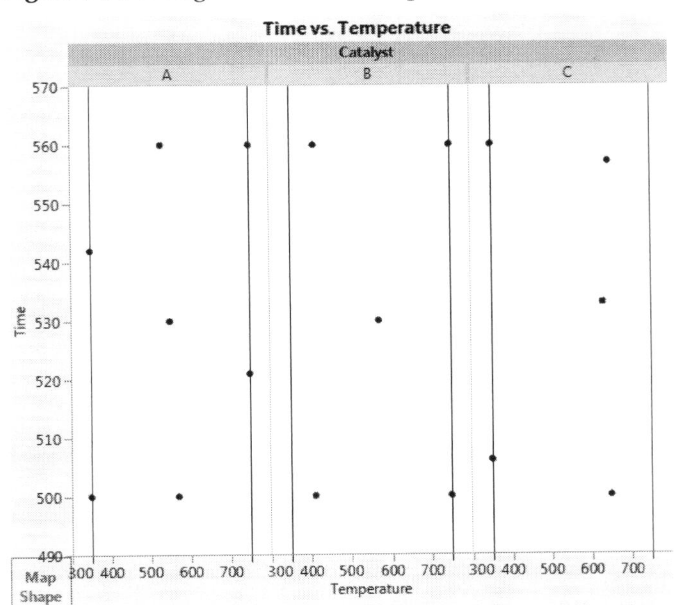

9. Double-click the Time axis.
10. Under Reference Lines, do the following:
    – Next to **Value**, enter 500.
    – Click **Add**.
    – Next to **Value**, enter 560.
    – Click **Add**.
11. Click **OK**.
12. Click **Done**.

Note: The plot in Figure 5.51 is vertically resized.

Figure 5.51 Design Regions for Three Levels of Catalyst

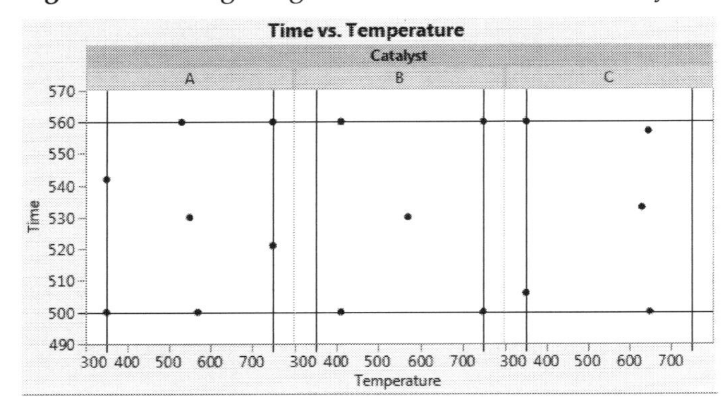

Six of the settings for Catalyst A fall on the edges of the design region. All Temperature setting for Catalyst B are above the 400 degree constraint and all Temperature settings for Catalyst C are below the 650 degree constraint.

Mixture Experiments

Both the Custom Design and Mixture Design platforms construct designs for situations where all of your factors are ingredients in a mixture. However, mixture experiments can involve non-mixture process variables, or *process factors*. The Custom Design platform can construct a design to accommodate both mixture ingredients and process factors. The Custom Design platform also allows the mixture components to sum to any positive number. See the Mixture description in "Factor Types" on page 87 in the "Custom Designs" chapter.

This section contains the following examples:

- "Mixture Design with Nonmixture Factors" on page 170
- "Mixture of Mixtures Design" on page 174

Mixture Design with Nonmixture Factors

In the following example from Atkinson and Donev (1992), you create a design for an experiment involving both mixture factors and process factors. The design is an 18-run design that is balanced with respect to the levels of a categorical factor. The design enables you to fit a full response surface. You use Design Evaluation plots and results to examine the relative prediction variance of the design.

The response and factors involved in the design are as follows:

- The response is Damping, which measures the electromagnetic damping of an acrylonitrile powder.

- The three mixture ingredients are:
  - CuSO4 (copper sulphate), ranging from 0.2 to 0.8
  - Na2S2O3 (sodium thiosulphate), ranging from 0.2 to 0.8
  - Glyoxal (glyoxal), ranging from 0 to 0.6
- The nonmixture environmental factor of interest is **Wavelength** (the wavelength of an electromagnetic wave) at three levels denoted L1, L2, and L3.

  Wavelength is a continuous variable. However, the researchers were interested only in predictions at three specific wavelengths. For this reason, you treat **Wavelength** as a categorical factor with three levels.

This example contains the following steps:

- "Create the Design" on page 171
- "Evaluate the Design" on page 173

Create the Design

1. Select **DOE > Custom Design**.
2. Double-click Y under Response Name and type Damping.
3. Click **Maximize** under Goal and change it to **None**.

   The goal is set to None because the authors of the study do not mention how much damping is desirable.
4. From the Custom Design red triangle menu, select **Load Factors**.
5. Select **Help > Sample Data Library** and open Design Experiment/Donev Mixture Factors.jmp.

   This loads the three mixture ingredients and the categorical environmental factor. Note that the bounds on the values of the three mixture factors are also loaded.

Figure 5.52 Responses Outline and Factors Outline

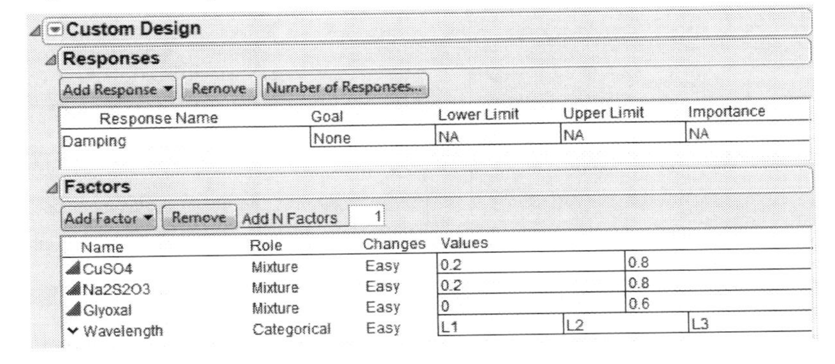

6. In the Model outline, click **Interactions > 2nd**.

An informational JMP Alert window reminds you that JMP removes the main effect terms for non-mixture factors that interact with all the mixture factors. This means that the main effect of **Wavelength** is removed, but all two-way interactions of mixture factors with **Wavelength** are added.

7. Click **OK** to dismiss the message.

   The effects in the Model outline define a response surface model in the mixture ingredients along with the additive effect of the wavelength. See Scheffé (1958).

Figure 5.53 Model and Design Generation Outlines

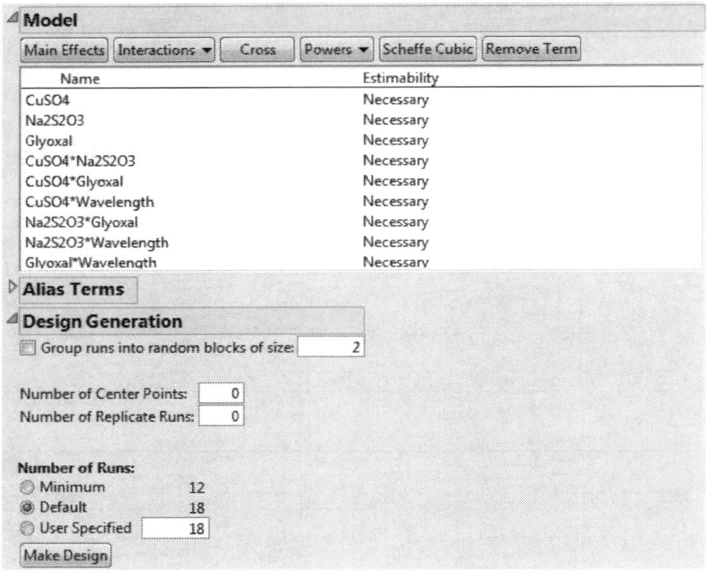

Model

Main Effects | Interactions | Cross | Powers | Scheffe Cubic | Remove Term

Name	Estimability
CuSO4	Necessary
Na2S2O3	Necessary
Glyoxal	Necessary
CuSO4*Na2S2O3	Necessary
CuSO4*Glyoxal	Necessary
CuSO4*Wavelength	Necessary
Na2S2O3*Glyoxal	Necessary
Na2S2O3*Wavelength	Necessary
Glyoxal*Wavelength	Necessary

Alias Terms

Design Generation

Group runs into random blocks of size: 2

Number of Center Points: 0
Number of Replicate Runs: 0

Number of Runs:
Minimum 12
Default 18
User Specified 18

Make Design

8. Leave the default number of runs at 18.

   The choice of 18 runs allows six runs for each of the three levels of the wavelength factor.

   Note: Setting the Random Seed in step 9 and Number of Starts in step 10 reproduces the exact results shown in this example. In constructing a design on your own, these steps are not necessary.

9. (Optional) From the Custom Design red triangle menu, select **Set Random Seed**, type 858576648, and click **OK**.
10. (Optional) From the Custom Design red triangle menu, select **Number of Starts**, type 10, and click **OK**.
11. Click **Make Design**.

Figure 5.54 Design Outline Showing 18-Run Design

Design

Run	CuSO4	Na2S2O3	Glyoxal	Wavelength
1	0.5	0.5	0	L3
2	0.5	0.2	0.3	L2
3	0.8	0.2	0	L3
4	0.2	0.2	0.6	L2
5	0.5	0.5	0	L2
6	0.2	0.2	0.6	L1
7	0.5	0.2	0.3	L1
8	0.8	0.2	0	L1
9	0.8	0.2	0	L2
10	0.2	0.2	0.6	L3
11	0.5	0.2	0.3	L3
12	0.2	0.8	0	L1
13	0.2	0.5	0.3	L2
14	0.2	0.5	0.3	L1
15	0.2	0.8	0	L2
16	0.2	0.8	0	L3
17	0.2	0.5	0.3	L3
18	0.5	0.5	0	L1

You can check that there are six runs for each level of **Wavelength**.

Evaluate the Design

1. Open the **Design Evaluation** > **Prediction Variance Profile** outline.

Figure 5.55 Prediction Variance Profile for 18-Run Design

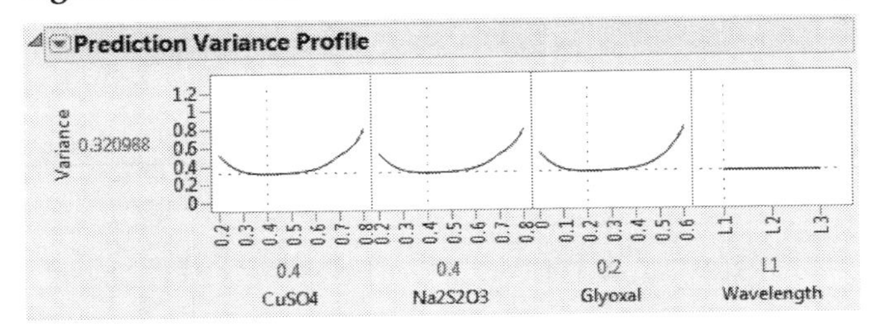

Move the slider for **Wavelength** to verify that the relative prediction variance profiles for the mixture factors do not change across the levels of **Wavelength**. Move the slider for any one of the mixture factors. The sliders for the other two mixture factors adjust to make the mixture ingredients sum to one. Notice that the smallest relative prediction variances occur near the center settings for the mixture factors.

2. From the Prediction Variance Profile red triangle menu, select **Maximize Desirability**.

   Notice that the maximum relative prediction variance over the design space is 0.8 times the error variance.

3. Open the **Fraction of Design Space Plot** outline.

Figure 5.56 Fraction of Design Space Plot for 18-Run Design

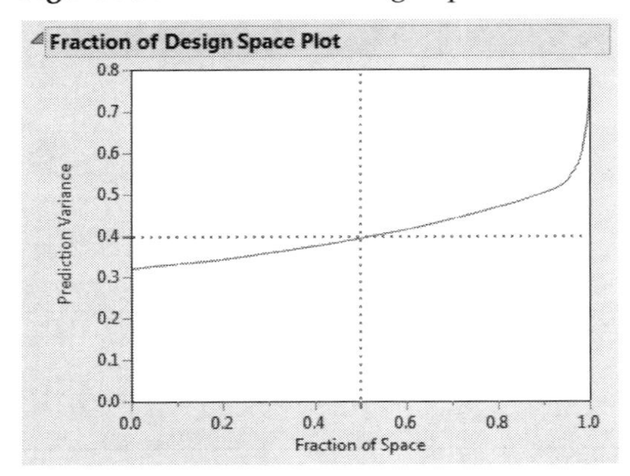

Over the entire design space, the relative prediction variance is below 0.8. The minimum relative prediction variance is about 0.32. As seen in Figure 5.55, the minimum occurs near the center settings for the mixture factors.

4. Open the **Design Diagnostics** outline.

Figure 5.57 Design Diagnostics Outline for 18-Run Design

Design Diagnostics	
D Optimal Design	
D Efficiency	3.627407
G Efficiency	91.28709
A Efficiency	0.328715
Average Variance of Prediction	0.410395
Design Creation Time (seconds)	0.1

The design is optimal relative to the D-optimality criterion, even though its D-efficiency is very low (3.6%). Because mixture designs are far from orthogonal due to the mixture constraint, they typically have very low D-efficiencies. The Average (relative) Variance of Prediction is 0.410395. This is consistent with the Fraction of Design Space plot in Figure 5.56.

Mixture of Mixtures Design

In this example, construct a design for a mixture of mixtures situation.

Consider the ingredients that go into a cake. Dry ingredients include flour, sugar, and cocoa. Wet ingredients include milk, melted butter, and eggs. The wet and dry components of the cake are two mixtures that are first mixed separately and then blended together. Table 5.1 lists the factors and the ranges over which you vary them as part of your experiment.

Table 5.1 Dry and Wet Components and Experimental Ranges

Mixture	Ingredient	Lower and Upper Levels
Dry	Cocoa	0.1 - 0.2
	Sugar	0 - 0.15
	Flour	0.2 - 0.3
Wet	Butter	0.1 - 0.2
	Milk	0.25 - 0.35
	Eggs	0.05 - 0.20

The dry components (the mixture of cocoa, sugar, and flour) comprise 45% of the combined mixture. The wet components (butter, milk, and eggs) comprise 55%.

The goal of your experiment is to optimize a Taste rating. Taste is rated on a scale of 1 to 10, with 10 representing the best taste.

You construct a 10-run design to fit a main effects model. Because of the constraint on the proportions of dry and wet ingredients, you need to include only five factors in the Model outline to avoid singularity. The choice of which factor not to include is arbitrary.

This example contains the following steps:

- "Create the Design" on page 175
- "Analyze the Experimental Results" on page 177

Create the Design

1. Select **DOE > Custom Design**.
2. Double-click Y under Response Name and type Taste.

   Note that the default goal is Maximize. Because you want to maximize the Taste rating, do not change the goal.
3. Click under Lower Limit and type 0.

   The least desirable rating is 0.
4. Click under Upper Limit and type 10.

   The most desirable rating is 10.
5. Leave the area under Importance blank.

   Because there is only one response, that response is given Importance 1 by default.
6. From the Custom Design red triangle menu, select **Load Factors**.

7. Open Cake Factors.jmp from the Design Experiment sample data folder.

Figure 5.58 Completed Responses and Factors Outlines

Responses

Add Response | Remove | Number of Responses...

Response Name	Goal	Lower Limit	Upper Limit	Importance
Taste	Maximize	1	10	.
optional item				

Factors

Add Factor | Remove | Add N Factors 1

Name	Role	Changes	Values	
Cocoa	Mixture	Easy	0.1	0.2
Sugar	Mixture	Easy	0	0.15
Flour	Mixture	Easy	0.2	0.3
Butter	Mixture	Easy	0.1	0.2
Milk	Mixture	Easy	0.25	0.35
Eggs	Mixture	Easy	0.05	0.2

Note that the factors are all mixture factors. The Values that define the range of settings for the experiment vary from factor to factor.

8. In the Define Factor Constraints outline, select **Specify Linear Constraints**.
9. In the Linear Constraints panel, click **Add** twice.
10. Enter the constraints shown in Figure 5.59.

Figure 5.59 Define Factor Constraints

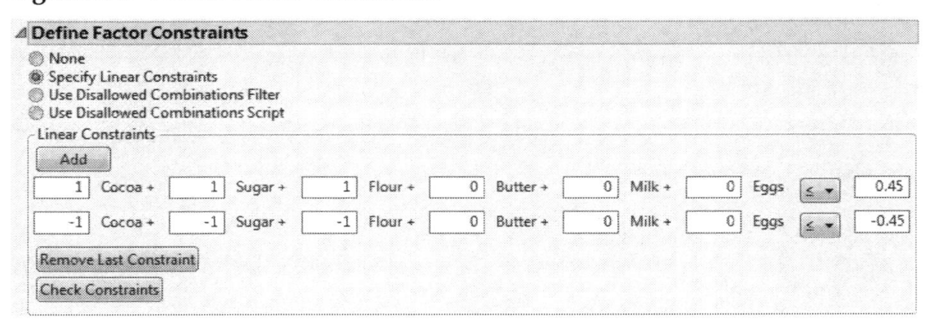

Note that the second constraint is equivalent to specifying that the sum of Cocoa, Sugar, and Flour is greater than or equal to 0.45. The two constraints together imply that Cocoa, Sugar, and Flour comprise exactly 45% of the mixture, ensuring that the wet factors constitute the remaining 55%.

11. In the Model outline, select any effect and click **Remove Term**.

    Because of the equality constraint, a model containing all six effects would be singular.

12. Type 10 next to **User Specified**.

Your experiment requires baking 10 cakes.

Note: Setting the Random Seed in step 13 and Number of Starts in step 14 reproduces the exact results shown in this example. In constructing a design on your own, these steps are not necessary.

13. (Optional) From the Custom Design red triangle menu, select **Set Random Seed**, type 1992991263, and click **OK**.
14. (Optional) From the Custom Design red triangle menu, select **Number of Starts**, type 40, and click **OK**.
15. Click **Make Design**.

    A JMP Alert informs you that your factor constraints include an equality constraint. This was what you intended, because the sum of the dry ingredient proportions is constrained to 45%.
16. Click **OK** to dismiss the JMP Alert.
17. Click **Make Table**.

Figure 5.60 Mixture of Mixtures Design

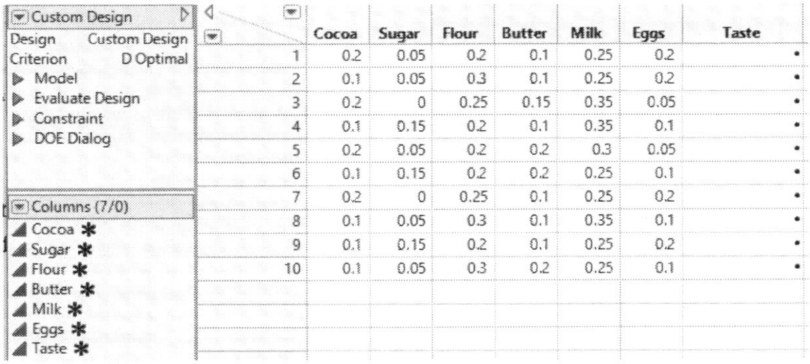

Custom Design
Design Custom Design
Criterion D Optimal
Model
Evaluate Design
Constraint
DOE Dialog

Columns (7/0)
Cocoa *
Sugar *
Flour *
Butter *
Milk *
Eggs *
Taste *

	Cocoa	Sugar	Flour	Butter	Milk	Eggs	Taste
1	0.2	0.05	0.2	0.1	0.25	0.2	•
2	0.1	0.05	0.3	0.1	0.25	0.2	•
3	0.2	0	0.25	0.15	0.35	0.05	•
4	0.1	0.15	0.2	0.1	0.35	0.1	•
5	0.2	0.05	0.2	0.2	0.3	0.05	•
6	0.1	0.15	0.2	0.2	0.25	0.1	•
7	0.2	0	0.25	0.1	0.25	0.2	•
8	0.1	0.05	0.3	0.1	0.35	0.1	•
9	0.1	0.15	0.2	0.1	0.25	0.2	•
10	0.1	0.05	0.3	0.2	0.25	0.1	•

The settings for the dry ingredients, Cocoa, Sugar, and Flour, sum to 45% of the mixture and the settings for the wet ingredients, Butter, Milk, and Eggs, sum to 55% of the mixture. The settings also conform to the upper and lower limits given in the Factors outline.

Analyze the Experimental Results

The Cake Data.jmp sample data table shows the results of the experiment. (Note that the runs in Cake Data.jmp differ from those shown in Figure 5.60.) The design table contains a Model script that opens a Fit Model window showing the five main effects specified in the DOE window's Model outline. Notice that the main effect of Egg is not included in the Model outline for this design. This script was saved to the data table when it was created by Custom Design.

1. Open the Cake Data.jmp sample data table, located in the Design Experiment folder.
2. In the Table panel of the design table, click the green triangle next to the **Model** script.

   The main effect due to Egg is not included because it was excluded from the Model outline in the Custom Design window. All five effects are designated as Response Surface and Mixture effects.
3. Click **Run**.

   A JMP Alert appears, notifying you that the Profiler cannot be shown because of the additional constraint.
4. Click **OK** to dismiss the JMP Alert.

The Parameter Estimates report indicates that Sugar, Flour, and Butter are significant at the 0.05 level.

Figure 5.61 Parameter Estimates Report

Parameter Estimates

Term	Estimate	Std Error	t Ratio	Prob>|t|
(Cocoa-0.1)/0.35	7.4556695	3.264525	2.28	0.0712
Sugar/0.35	14.141998	2.69585	5.25	0.0033*
(Flour-0.2)/0.35	12.957954	3.603964	3.60	0.0156*
(Butter-0.1)/0.35	11.457133	4.038675	2.84	0.0364*
(Milk-0.25)/0.35	6.0783774	3.853774	1.58	0.1756

Experiments with Covariates

Sometimes measurements on the experimental units that are intended for an experiment are available. These measurements might affect the experimental response. It is useful to include these variables, called *covariates*, as design factors. Although you cannot directly control these values, you can ensure that the levels of the other design factors are chosen to yield the most precise estimates of all the effects.

The Custom Design platform constructs a design that selects covariate values in an optimal fashion. Covariate values are selected from an existing data table that provides covariate information about the potential experimental units. You can specify a number of runs that is smaller than the number of experimental units listed in your data table. You can also specify covariates that are hard-to-change.

Note: The number of rows in the covariate data table where covariate factors have nonmissing values must be greater than or equal to the specified Number of Runs.

Design with Fixed Covariates

In this example, you are interested in modeling the Shrinkage of parts produced by an injection molding process. The Thermoplastic.jmp sample data table in the Design Experiment folder lists 25 batches of raw (thermoplastic) material for potential use in your study. For each batch, material was removed to obtain measurements of Specific Gravity and Tensile Strength. A third covariate, Supplier, is also available.

You want to study the effects of three controllable factors, Temperature (mold temperature), Speed (screw speed), and Time (hold time), on Shrinkage. But you also want to study the effects of the covariates: Specific Gravity, Tensile Strength, and Supplier. Your resources allow for 12 runs.

Create the Design

1. Select **Help > Sample Data Library** and open Design Experiment/Thermoplastic.jmp.
2. Select **DOE > Custom Design**.
3. Double-click Y under Response Name and type Shrinkage.
4. Click **Maximize** under Goal and change it to **Minimize**.
5. Click **Add Factor** and select **Covariate**.
6. Select Specific Gravity, Tensile Strength, and Supplier from the list and click **OK**.

   These are covariates and cannot be controlled.
7. Type 3 next to **Add N Factors**.
8. Click **Add Factor > Continuous**.
9. Rename the three continuous factors Temperature, Speed, and Time.

   These are factors that can be controlled.

Figure 5.62 Responses and Factors Outlines

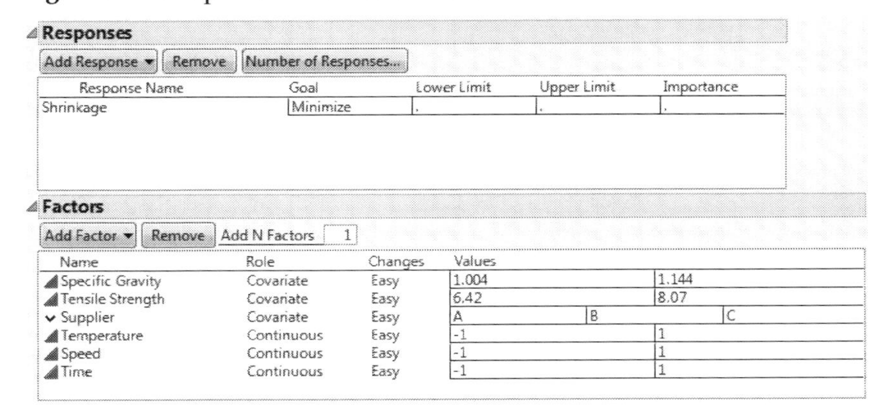

10. Click **Continue.**

    The Number of Runs shows the number of rows with covariate values available. You have 25 batches with measured covariates.

11. Type 12 next to **Number of Runs.**

Note: Setting the Random Seed in step 12 and Number of Starts in step 13 reproduces the exact results shown in this example. In constructing a design on your own, these steps are not necessary.

12. (Optional) From the Custom Design red triangle menu, select **Set Random Seed**, type 84951, and click **OK**.
13. (Optional) From the Custom Design red triangle menu, select **Number of Starts**, type 40, and click **OK**.
14. Click **Make Design.**

Figure 5.63 Twelve-Run Optimal Design

Design

Run	Specific Gravity	Tensile Strength	Supplier	Temperature	Speed	Time
1	1.107	8.07	B	1	1	1
2	1.144	6.82	B	1	-1	-1
3	1.004	6.52	C	1	1	-1
4	1.015	7.46	A	-1	1	-1
5	1.129	6.97	A	1	-1	-1
6	1.139	6.94	C	-1	1	-1
7	1.004	7.12	C	1	-1	1
8	1.067	6.7	B	-1	1	1
9	1.047	6.42	A	-1	-1	1
10	1.03	7.54	B	-1	-1	-1
11	1.094	6.48	A	1	1	1
12	1.113	6.89	C	-1	-1	1

This design is D-optimal, given the potential covariate values. It selects the best sets of covariate values and the best settings for the three controllable factors.

Evaluate the Design

1. Open the **Design Evaluation > Color Map On Correlations** outline.

Figure 5.64 Color Map on Correlations

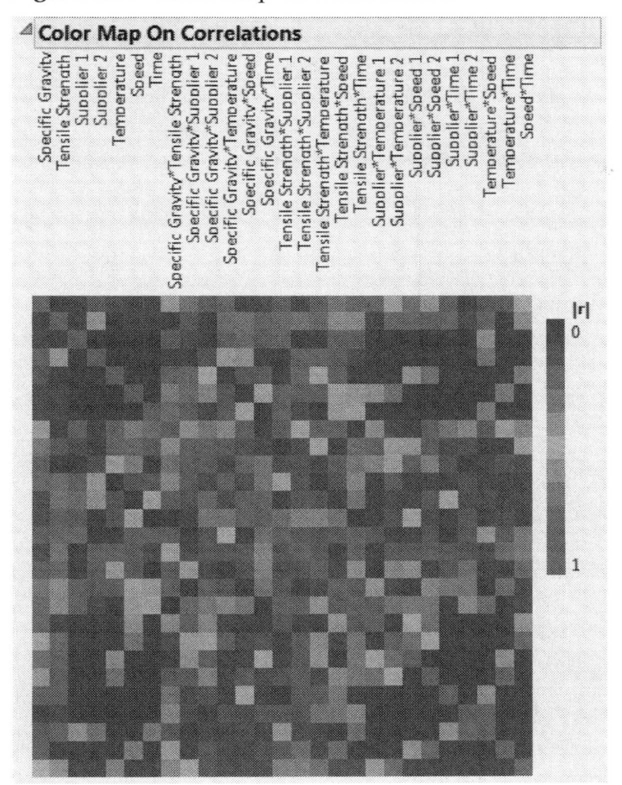

The seven terms corresponding to main effects appear in the upper left corner of the color map. Notice that these seven terms are close to orthogonal. The largest absolute correlation is between Tensile Strength and Supplier 2. This absolute correlation of about 0.43 is a consequence of the available covariate values. (Figure 5.64 uses JMP default colors.)

Design with Hard-to-Change Covariates

In this example, you construct a design for developing a running shoe for serious runners that has good wear (Wear) properties. Your experimental factors include the following:

- sole thickness (Thickness)
- amount of gel cushioning (Gel)
- outsole material (Outsole)
- midsole material (Midsole)

To obtain generalizable conclusions, you need to test your shoes on a broad base of serious runners. To accommodate your experimental budget, each runner must test several experimental combinations.

Your company has collected data on 100 suitable runners willing to participate in your study. The concomitant variables (*covariates*) measured on these runners are average daily miles run (Miles), weight (Weight), and the foot's strike point (Strike Point).

Create your design as follows:

1. Select **Help > Sample Data Library** and open Design Experiment/Runners Covariates.jmp.
2. Select **DOE > Custom Design**.
3. Double-click Y under Response Name and type Wear.
4. Click **Maximize** under Goal and change it to **Minimize**.
5. Click **Add Factor** and select **Covariate**.
6. Select Miles, Weight, and Strike Point from the list and click **OK**.

   These are the hard-to-change covariates associated with the runners.
7. For one of the factors Miles, Weight, and Strike Point, under Changes, click **Easy** and change it to **Hard**.

   Note that Changes for all three covariates turn to Hard.

To add the remaining factors manually, follow step 8 through step 16. Or, to load factors from a saved table, select **Load Factors** from the red triangle menu next to Custom Design. Open the Runners Factors.jmp sample data table, located in the Design Experiment folder. If you select **Load Factors**, skip step 8 through step 16.

8. Type 2 next to **Add N Factors**.
9. Click **Add Factor > Continuous**.
10. Rename the two factors Thickness and Gel.
11. Change the **Values** for Thickness to 5 and 20.
12. Change the **Values** for Gel to 1 and 10.
13. Type 2 next to **Add N Factors**.
14. Click **Add Factor > Categorical > 3 Level**.
15. Rename the two factors Outsole and Midsole.

    Keep the default **Values** for these factors.

Figure 5.65 Responses and Factors Outlines

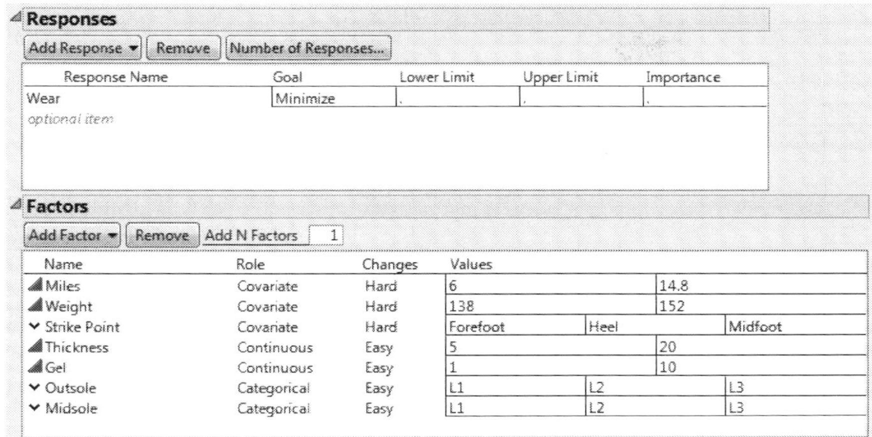

16. Click **Continue**.
17. Select **Interactions > 2nd**.

    The specified model fits all two-factor interactions, including covariate by experimental factor interactions.
18. Set the Number of Whole Plots, or runners, to 32 (if it is not already set to that number).
19. Type 64 next to User Specified under Number of Runs (if it is not already set to that number).

Note: Setting the Random Seed in step 20 and Number of Starts in step 21 reproduces the exact results shown in this example. In constructing a design on your own, these steps are not necessary.

20. (Optional) From the Custom Design red triangle menu, select **Set Random Seed**, type 12345, and click **OK**.
21. (Optional) From the Custom Design red triangle menu, select **Number of Starts** and set it to 1(if it is not already set to that number). Click **OK**.
22. Click **Make Design**.

Figure 5.66 First 40 Runs of Design for Hard-to-Change Covariates

Design

Run	Whole Plots	Miles	Weight	Strike Point	Thickness	Gel	Outsole	Midsole
1	1	7.6	144	Midfoot	5	1	L1	L3
2	1	7.6	144	Midfoot	20	10	L3	L2
3	2	7.6	145	Midfoot	20	1	L3	L1
4	2	7.6	145	Midfoot	5	10	L2	L2
5	3	12.4	140	Midfoot	5	10	L1	L2
6	3	12.4	140	Midfoot	20	1	L2	L3
7	4	11.2	148	Heel	5	1	L3	L2
8	4	11.2	148	Heel	20	1	L2	L3
9	5	11.2	138	Heel	5	1	L3	L1
10	5	11.2	138	Heel	20	10	L1	L1
11	6	6	150	Heel	20	1	L1	L3
12	6	6	150	Heel	5	10	L3	L1
13	7	6.4	146	Heel	20	10	L3	L3
14	7	6.4	146	Heel	5	10	L1	L2
15	8	14	139	Midfoot	20	1	L3	L2
16	8	14	139	Midfoot	5	1	L1	L1
17	9	13.2	140	Midfoot	20	10	L1	L3
18	9	13.2	140	Midfoot	5	10	L3	L3
19	10	10.8	143	Forefoot	20	1	L2	L2
20	10	10.8	143	Forefoot	5	10	L1	L3
21	11	12.8	139	Heel	20	10	L3	L2
22	11	12.8	139	Heel	5	1	L2	L3
23	12	14.8	141	Midfoot	5	1	L2	L2
24	12	14.8	141	Midfoot	20	10	L2	L1
25	13	8	148	Forefoot	5	10	L3	L3
26	13	8	148	Forefoot	5	1	L1	L2
27	14	11.2	141	Heel	5	1	L1	L3
28	14	11.2	141	Heel	5	10	L2	L1
29	15	13.2	146	Midfoot	5	10	L2	L3
30	15	13.2	146	Midfoot	20	1	L1	L1
31	16	9.2	152	Midfoot	20	10	L3	L3
32	16	9.2	152	Midfoot	20	1	L2	L2
33	17	6.4	146	Heel	20	10	L2	L1
34	17	6.4	146	Heel	5	1	L3	L2
35	18	11.2	144	Forefoot	20	10	L3	L3
36	18	11.2	144	Forefoot	5	1	L3	L1
37	19	6.4	147	Forefoot	20	1	L1	L1
38	19	6.4	147	Forefoot	5	1	L2	L3
39	20	6.4	148	Heel	5	1	L1	L1
40	20	6.4	148	Heel	20	1	L2	L2

Of the 100 runners, 32 are selected based on their covariate values. The rows corresponding to the selected runners are selected in the RunnersCovariates.jmp sample data table. Settings of the experimental factors Thickness, Gel, Insole, and Outsole are determined so that the design is optimal for the model described in the Model Outline.

23. With the RunnersCovariates.jmp sample data table as the active table, select **Analyze > Distribution**.
24. Select all three columns as **Y, Columns**.
25. Check **Histograms Only**.
26. Click **OK**.

Figure 5.67 Histograms for 100 Runners with Selected Runner Data Shaded

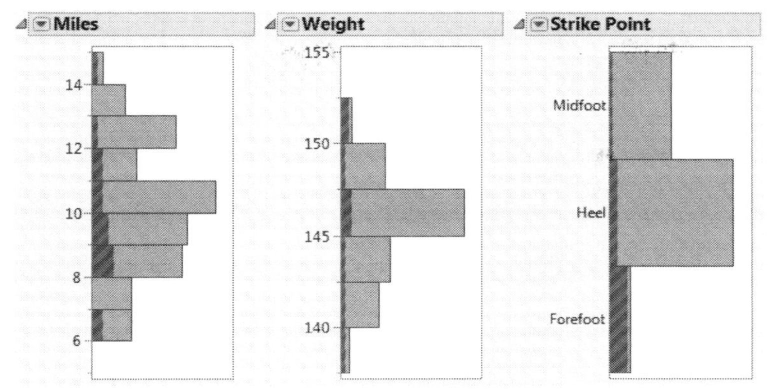

The histograms indicate that all of the runners with **Miles** of 14.0 or higher were selected. Runners at the extremes of the Weight distribution were selected. Almost all of the runners with a Strike Point of Forefoot were selected. Notice that the design is somewhat balanced in terms of Strike Point.

Design with a Linear Time Trend

Often, experiments conducted in a time sequence experience a linear drift in the response. If you randomize the order of the runs, then the drift's effect does not generally bias the estimated factor effects. However, by accounting for the drift, you can reduce the variance of those effects.

Suppose that there is reason to suspect a strong linear trend in the response over time independent of the factor changes. Then you can construct a design that includes a linear covariate to account for the trend. The resulting design is optimal, given this trend covariate.

In this example, you design an experiment for 7 factors. You construct a 16-run design that is robust to linear trend.

1. Select **File > New > Data Table**.
2. Right-click Column1 and select **Column Info**.
3. Change the column name to Run Order.
4. From the list of **Initialize Data** options, select **Sequence Data**.
5. Type 16 next to **To**.
6. Click **OK**.

   Consecutive integers from 1 to 16 have been entered in the data table.
7. Select **DOE > Custom Design**.
8. Click **Add Factor > Covariate**.

9. Select Run Order and click **OK**.
10. Type 7 next to **Add N Factors**.
11. Click **Add Factor > Continuous**.

Figure 5.68 Responses and Factors Outlines

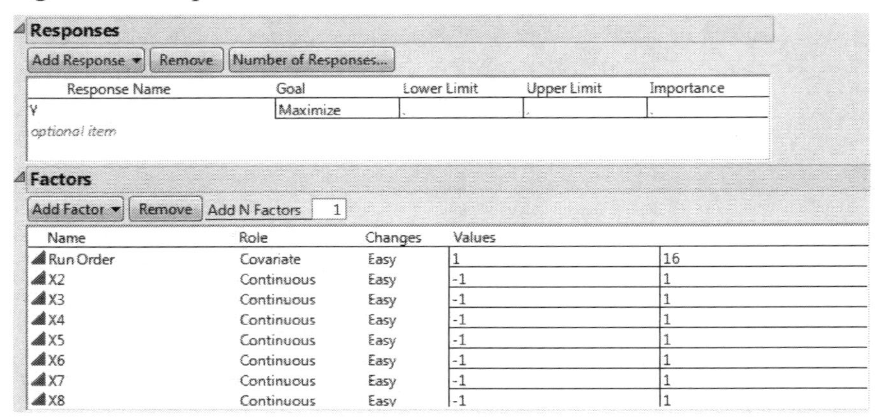

12. Click **Continue**.
13. Open the **Alias Terms** outline.
14. Select all of the effects in the list and click **Remove Term**.

    This omits the interaction effects from the correlation color map, leaving only the main effects.

    Note: Setting the Random Seed in step 15 and Number of Starts in step 16 reproduces the exact results shown in this example. In constructing a design on your own, these steps are not necessary.

15. (Optional) From the Custom Design red triangle menu, select **Set Random Seed**, type 1084680980, and click **OK**.
16. (Optional) From the Custom Design red triangle menu, select **Number of Starts**, type 30000, and click **OK**.
17. Click **Make Design**.

    A progress bar displays elapsed time and D Efficiency.

Figure 5.69 Design Outline

Design

Run	Run Order	X2	X3	X4	X5	X6	X7	X8
1	1	1	1	-1	-1	-1	-1	-1
2	2	1	-1	1	1	-1	-1	-1
3	3	1	-1	-1	1	1	1	1
4	4	-1	-1	1	-1	1	-1	1
5	5	-1	1	1	1	-1	1	-1
6	6	-1	1	-1	1	1	-1	1
7	7	-1	1	1	-1	-1	1	1
8	8	-1	1	-1	1	1	1	-1
9	9	1	1	1	-1	1	1	1
10	10	-1	-1	-1	-1	1	-1	-1
11	11	1	-1	1	-1	1	1	-1
12	12	1	-1	-1	1	-1	1	1
13	13	-1	-1	-1	-1	-1	1	-1
14	14	-1	-1	1	1	-1	-1	1
15	15	1	1	-1	-1	-1	-1	1
16	16	1	1	1	1	1	-1	-1

18. Open the **Design Evaluation > Color Map On Correlations** outline.

Figure 5.70 Color Map Showing Absolute Correlations with Run Order

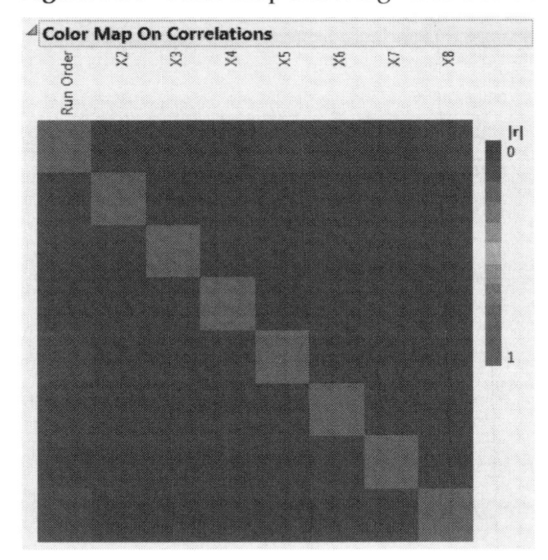

The Color Map shows the following:

- The seven continuous factors, X2 through X8, are orthogonal to each other.
- Run Order, the linear time trend variable, has extremely low absolute correlation with X2 through X8.

19. Open the **Design Evaluation > Estimation Efficiency** outline.

Figure 5.71 Estimation Efficiency Outline

Estimation Efficiency

Parameter	Fractional Increase in CI Length	Relative Std Error of Parameters
Intercept	0	0.25
Run Order	0.634	0.408
X2	4e-4	0.25
X3	4e-4	0.25
X4	0	0.25
X5	0.001	0.25
X6	4e-4	0.25
X7	0	0.25
X8	0.001	0.25

The small absolute correlations of Run Order with X2 through X8 result in very small increases in confidence interval lengths, relative to an ideal orthogonal design. The increases in the lengths of confidence intervals for X2 through X8 are all less than 0.1%.

In this example, the run order factor is nearly orthogonal to the factor effects. In some cases, your design might have more substantial correlations between the run order factor and other factors. Even in such a situation, including the run order as a factor accounts for any linear trend effect. Including the run order also allows for more precise estimation of the other factor effects.

Experiments with Randomization Restrictions

The Custom Design platform constructs split-plot, split-split-plot, and two-way split-plot (strip-plot) designs that are D-optimal or I-optimal. For details about constructing these designs, see Goos (2002). In this section, you construct examples of two of these designs: a split-plot design that involves mixture factors, and a two-way split-plot design.

Split-Plot Experiment

Split-plot designs originated in agriculture, but are commonplace in manufacturing and engineering studies. In a split-plot experiment, hard-to-change factors are reset only between one whole plot and the next whole plot. The whole plot is divided into subplots, and the levels of the easy-to-change factors are randomly assigned to each subplot.

The example in this section is adapted from Kowalski, Cornell, and Vining (2002). You are interested in the effects of five factors on the thickness of vinyl that is used to make automobile seat covers. The response and factors in the experiment are described below:

- The response is the thickness of the vinyl that is produced. You want to maximize thickness. A lower limit for thickness values is 10.
- The whole plot factors are the rate of extrusion (extrusion rate) and the temperature (temperature) of drying. These are process variables and are hard to change.

- The subplot factors are three plasticizers whose proportions (m1, m2, and m3) sum to one. These factors are mixture components.

Your experimental budget allows for running 7 settings of these whole plot factors. For each whole plot, you can conduct 4 runs of the subplot factors. This gives you a total of 28 runs.

Create the Design

1. Select **DOE > Custom Design**.
2. Double-click Y under Response Name and type thickness.

   Keep the default goal set to Maximize.
3. Enter a Lower Limit of 10.

To add factors manually, follow step 4 through step 11. Or, to load factors from a saved table, select **Load Factors** from the red triangle menu next to Custom Design. Open the Vinyl Factors.jmp sample data table, located in the Design Experiment folder. If you select **Load Factors**, skip step 4 through step 11.

4. Type 2 next to **Add N Factors**.
5. Click **Add Factor > Continuous**.
6. Rename these factors extrusion rate and temperature.

   Keep the default Values of –1 and 1 for these two factors.
7. Click **Easy** and select **Hard** for both extrusion rate and temperature.

   This defines extrusion rate and temperature to be whole plot factors.
8. Type 3 next to **Add N Factors**.
9. Click **Add Factor > Mixture**.
10. Rename the three mixture factors m1, m2, and m3.

    Keep the default Values of 0 and 1 for those three factors.

Figure 5.72 Responses and Factors Outlines

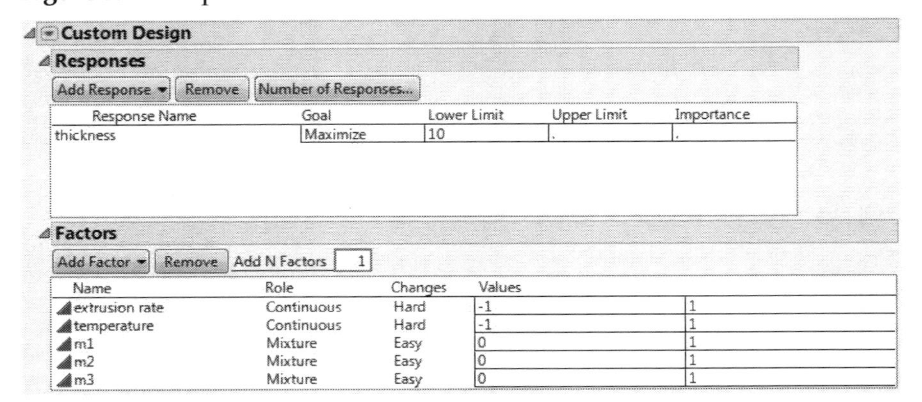

11. Click **Continue**.
12. Click **Interactions > 2nd**.
13. Click **OK** to dismiss the informative message.

    Note that 3 is the default value for the Number of Whole Plots.
14. Type 7 next to **Number of Whole Plots**.
15. Type 28 next to **User Specified**.

Note: Setting the Random Seed in step 16 and Number of Starts in step 17 reproduces the exact results shown in this example. In constructing a design on your own, these steps are not necessary.

16. (Optional) From the Custom Design red triangle menu, select **Set Random Seed**, type 123686, and click **OK**.
17. (Optional) From the Custom Design red triangle menu, select **Number of Starts**, type 10, and click **OK**.
18. Click **Make Design**.

Figure 5.73 Design Outline

Design

Run	Whole Plots	extrusion rate	temperature	m1	m2	m3
1	1	-1	-1	1	0	0
2	1	-1	-1	0	1	0
3	1	-1	-1	0	0	1
4	1	-1	-1	0.5	0.5	0
5	2	1	1	0	0	1
6	2	1	1	0	0.5	0.5
7	2	1	1	1	0	0
8	2	1	1	0	1	0
9	3	1	-1	0	0.6	0.4
10	3	1	-1	0	0	1
11	3	1	-1	0.5	0.5	0
12	3	1	-1	1	0	0
13	4	-1	1	0	1	0
14	4	-1	1	1	0	0
15	4	-1	1	0	0.5	0.5
16	4	-1	1	0	0	1
17	5	-1	-1	0	0.4	0.6
18	5	-1	-1	0	1	0
19	5	-1	-1	1	0	0
20	5	-1	-1	0.5	0	0.5
21	6	1	1	0	0	1
22	6	1	1	0	1	0
23	6	1	1	0.5	0	0.5
24	6	1	1	0.565335	0.434665	0
25	7	1	-1	0.5	0	0.5
26	7	1	-1	1	0	0
27	7	1	-1	0	1	0
28	7	1	-1	0	0	1

Note that the whole plot factors, extrusion rate and temperature, are reset seven times in accordance with the levels of the factor Whole Plots. Within each level of Whole Plots, the settings for the mixture ingredients, m1, m2, and m3, are assigned at random.

Analyze the Results

The Vinyl Data.jmp sample data table contains experimental results using a design created in a previous version of JMP.

1. Select **Help > Sample Data Library** and open Design Experiment/Vinyl Data.jmp.

   This sample data table contains 28 runs and response values. The design settings in the table that you created using the Custom Design platform might differ from those used in the Vinyl Data.jmp design.

2. In the Table panel, click the green triangle next to the **Model** script.

Figure 5.74 Fit Model Window

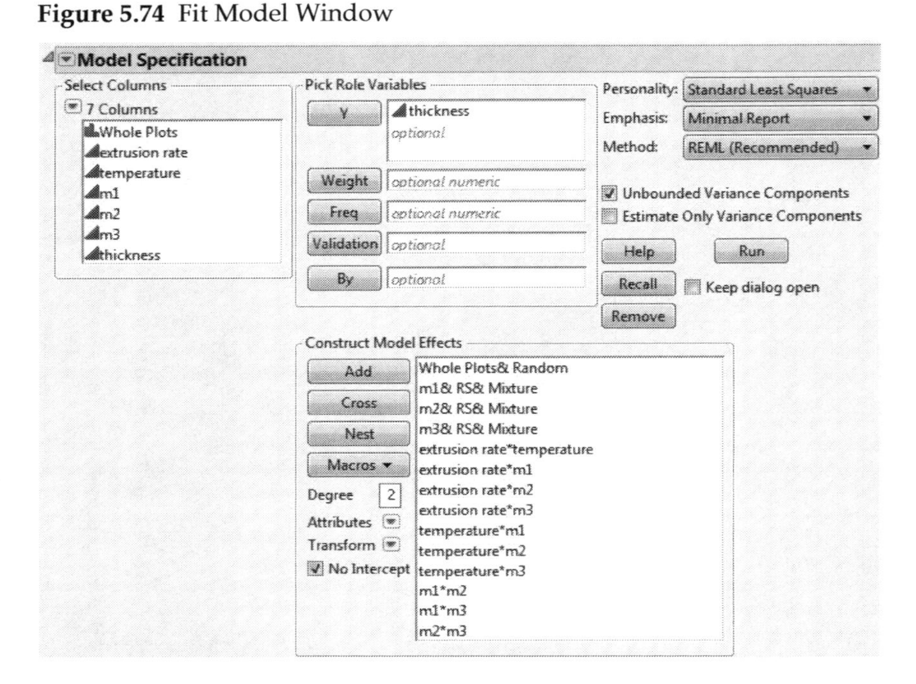

Notice the following in the Fit Model window:

– The factor Whole Plots has the Attribute called Random Effects (**&Random**). This specifies that the levels of Whole Plots are random realizations. They have an associated error term.

– The analysis method is **REML (Recommended)**. This method is specified precisely because the model contains a random effect. For more information about REML models, see the Standard Least Squares chapter in the *Fitting Linear Models* book.

Tip: In the Fit Model window, JMP Pro users can change the Personality to Mixed Model.

3. Click **Run**.

Figure 5.75 Split-Plot Analysis Results

Parameter Estimates

Term	Estimate	Std Error	DFDen	t Ratio	Prob>|t|
m1	9.4464782	0.989604	7.99	9.55	<.0001*
m2	5.378571	1.120267	10.35	4.80	0.0007*
m3	5.5546291	1.030955	8.786	5.39	0.0005*
extrusion rate*temperature	-1.866921	0.740748	2.858	-2.52	0.0903
extrusion rate*m1	-0.647611	0.958111	7.261	-0.68	0.5200
extrusion rate*m2	0.2944425	0.997579	8.162	0.30	0.7752
extrusion rate*m3	2.6611621	0.979087	7.733	2.72	0.0272*
temperature*m1	-1.8151	0.961061	7.328	-1.89	0.0990
temperature*m2	1.2044413	0.993319	8.065	1.21	0.2596
temperature*m3	-1.498647	0.97904	7.731	-1.53	0.1657
m1*m2	-6.458845	5.83806	13.36	-1.11	0.2881
m1*m3	8.5532871	5.607183	13.15	1.53	0.1508
m2*m3	10.575333	5.226599	13.04	2.02	0.0640

Random Effect Predictions

REML Variance Component Estimates

Random Effect	Var Ratio	Var Component	Std Error	95% Lower	95% Upper	Pct of Total
Whole Plots	0.6350069	2.476748	2.9711536	-3.346606	8.3001021	38.838
Residual		3.9003483	1.5898101	2.0073487	10.607283	61.162
Total		6.3770963	3.1515558	2.9308168	22.945461	100.000

-2 LogLikelihood = 82.032495402
Note: Total is the sum of the positive variance components.
Total including negative estimates = 6.3770963

The Parameter Estimates report shows that the three mixture ingredients, as well as the extrusion rate*m3 interaction, are significant at the 0.05 level.

The REML Variance Component Estimates report indicates that the variance component associated with Whole Plots is 2.476748. This is 38.838% of the total variation. It follows that the error term associated with whole plot replication is smaller than the residual (or within-plot) error term.

Two-Way Split-Plot Experiment

A two-way split-plot (also known as strip-plot or split-block) design consists of two split-plot components. In industry, these designs arise when batches of material or experimental units from one processing stage pass to a second processing stage. To use a two-way split-plot design, you must be able to reorder the units between stages.

After the first processing stage, you must be able to divide the batches into sub-batches. The second-stage processing factors are applied randomly to these sub-batches. For a specific second-stage experimental setting, all of the sub-batches assigned to that setting can be processed simultaneously. Additional factors can be applied to experimental units after the second processing stage.

In contrast to a split-split-plot design, the second-stage factors are *not nested* within the first-stage factors. After the first stage, the batches are subdivided and formed into new batches. Therefore, both the first- and second-stage factors are applied to whole batches.

Although factors at both stages might be equally hard-to-change, in order to distinguish these factors, JMP denotes the first stage factors as *very-hard*-to-change and the second-stage factors as *hard*-to-change. Additional factors applied to experimental units after the second processing stage are considered *easy*-to-change.

Scenario for a Two-Way Split-Plot Design

This example is based on an experiment to improve the open circuit voltage (OCV) in battery cells (Vivacqua and Bisgaard, 2004). You need to minimize the OCV in order to keep the cells from discharging on their own.

Battery cells move through two stages of processing:

- First stage: A continuous assembly process where batteries are processed in batches of 2000.
- Second stage: A curing process with a 5-day cycle time in a chamber that can accommodate 4000 batteries.

You want to study six two-level continuous factors:

- Four factors (A1, A2, A3, and A4) are applied to the assembly process. You can run 16 trials for the first-stage factors.
- Two factors (C5 and C6) are applied to the curing process. Because curing requires a 5-day cycle time, you can run only 6 cycles (30 days) for the second-stage factors. Using six curing cycles gives you partial replication of the curing settings, enabling you to test for curing effects.

Both the first- and second-stage factors are hard-to-change, suggesting two split-plots. However, the batches of 2,000 batteries from the first-stage experiment can be divided into sub-batches of 500 batteries each. Eight of these sub-batches can be randomly selected and processed simultaneously in the curing chamber.

The experiment has 48 experimental units. Note that the first- and second-stage factors are *crossed*.

Create the Design

To design a two-way split-plot experiment:

1. Select **DOE > Custom Design**.
2. Double-click Y under Response Name and type OCV.
3. Under Goal, click **Maximize** and select **Minimize**.
4. To add factors manually, follow step 5 through step 10. Or, to load factors from a saved table:
   a. Select **Load Factors** from the Custom Design red triangle menu.

b. Open the **Battery Factors.jmp** sample data table, located in the Design Experiment folder.

c. Proceed to step 11.

5. Type 6 next to **Add N Factors**.
6. Click **Add Factor > Continuous**.
7. Rename the factors A1, A2, A3, A4, C1, and C2.

   Keep the default Values of -1 and 1 for these factors.

8. For each of the factors A1, A2, A3, and A4, under Changes, click **Easy** and change it to **Very Hard**.

   To distinguish between the first- and second-stage factors, you designate the Changes for the first-stage factors as Very Hard, and the Changes for the second-stage factors as Hard.

9. For each of the factors C1 and C2, under Changes, click **Easy** and change it to **Hard**.

Figure 5.76 Responses and Factors Outlines

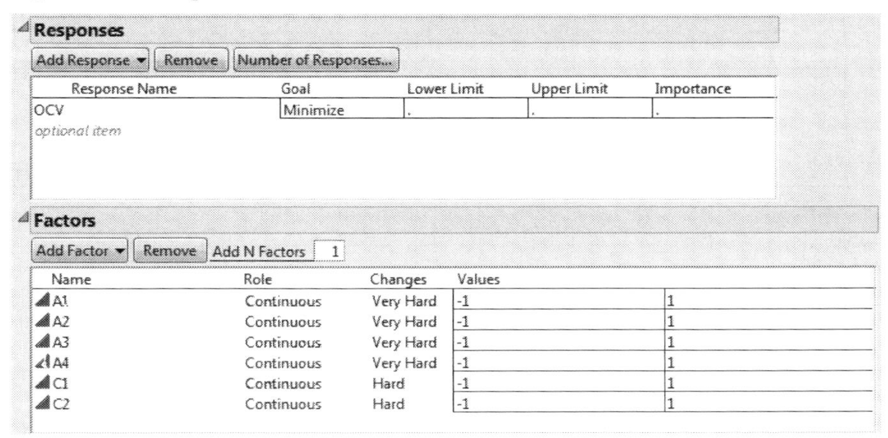

10. Click **Continue**.
11. Select **Interactions> 2nd** in the Model outline.
12. In the Design Generation outline, select the option **Hard to change factors can vary independently of Very Hard to change factors**.

    See Figure 5.77. Checking this option creates a two-way split-plot design. If this option is not checked, the design is treated as a split-split-plot design, with nesting of factors at the two levels.

13. Type 16 as the **Number of Whole Plots**.

    This is the number of trials that you can run for the first-stage factors.

14. Type 6 as the **Number of Subplots**.

This is the number of trials that you can run for the second-stage factors.

15. Under Number of Runs, type 48 next to **User Specified**.

    This is the total number of experimental units.

Figure 5.77 Design Generation Outline

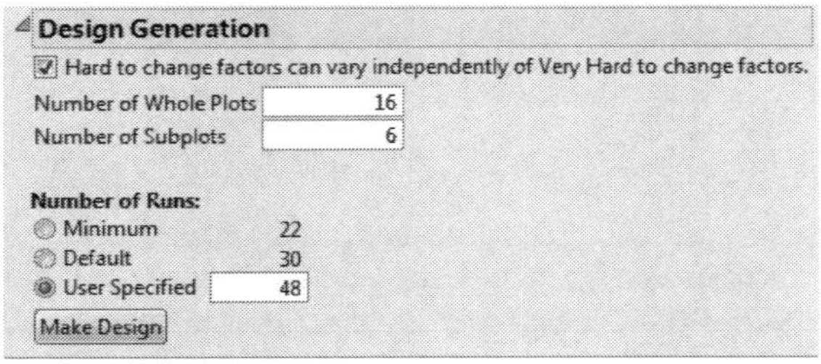

Note: Setting the Random Seed in step 16 and Number of Starts in step 17 reproduces the exact results shown in this example. In constructing a design on your own, these steps are not necessary.

---

16. (Optional) From the Custom Design red triangle menu, select **Set Random Seed**, type 1866762673, and click **OK**.
17. (Optional) From the Custom Design red triangle menu, select **Number of Starts**, type 21, and click **OK**.
18. Click **Make Design**.
19. Click **Make Table**.

Figure 5.78 Partial View of Design Table

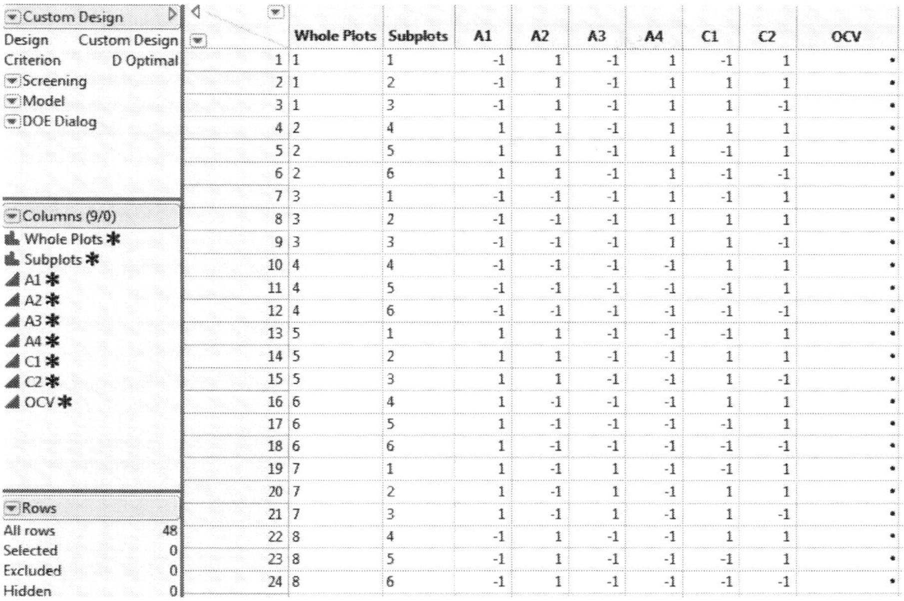

Custom Design
Design Custom Design
Criterion D Optimal
Screening
Model
DOE Dialog

Columns (9/0)
Whole Plots *
Subplots *
A1 *
A2 *
A3 *
A4 *
C1 *
C2 *
OCV *

Rows
All rows 48
Selected 0
Excluded 0
Hidden 0

	Whole Plots	Subplots	A1	A2	A3	A4	C1	C2	OCV
1	1	1	-1	1	-1	1	-1	1	•
2	1	2	-1	1	-1	1	1	1	•
3	1	3	-1	1	-1	1	1	-1	•
4	2	4	1	1	-1	1	1	1	•
5	2	5	1	1	-1	1	-1	1	•
6	2	6	1	1	-1	1	-1	-1	•
7	3	1	-1	-1	-1	1	-1	1	•
8	3	2	-1	-1	-1	1	1	1	•
9	3	3	-1	-1	-1	1	1	-1	•
10	4	4	-1	-1	-1	-1	1	1	•
11	4	5	-1	-1	-1	-1	-1	1	•
12	4	6	-1	-1	-1	-1	-1	-1	•
13	5	1	1	1	-1	-1	-1	1	•
14	5	2	1	1	-1	-1	1	1	•
15	5	3	1	1	-1	-1	1	-1	•
16	6	4	1	-1	-1	-1	1	1	•
17	6	5	1	-1	-1	-1	-1	1	•
18	6	6	1	-1	-1	-1	-1	-1	•
19	7	1	1	-1	1	-1	-1	1	•
20	7	2	1	-1	1	-1	1	1	•
21	7	3	1	-1	1	-1	1	-1	•
22	8	4	-1	1	-1	-1	1	1	•
23	8	5	-1	1	-1	-1	-1	1	•
24	8	6	-1	1	-1	-1	-1	-1	•

The design table shows 16 levels for Whole Plots. For each level of Whole Plots, the settings of the four assembly factors are constant. From each level of Whole Plots, three batches of 500 batteries (Subplots) are randomly assigned to settings of the curing factors. Two sets of curing conditions are replicated (C1 = -1, C2 = 1 and C1 = 1, C2 = 1). To see this, select columns C1 and C2, right-click in the header area, and select **Sort > Ascending**.

Analyze the Results

The Battery Data.jmp sample data table contains experimental results for the design that you generated.

1. Select **Help > Sample Data Library** and open Design Experiment/Battery Data.jmp.
2. In the Table panel, click the green triangle next to the **Model** script.

   Notice the following in the Fit Model window:

   – The factor Whole Plots has the Attribute called Random Effects (**&Random**). This specifies that the levels of Whole Plots are random realizations. They have an associated error term.

   – The factor Subplots also has the Random Effects Attribute (**&Random**).

   – The analysis Method is **REML (Recommended)**. This method is specified precisely because the model contains random effects. For more information about REML models, see the Standard Least Squares chapter in the *Fitting Linear Models* book.

Tip: In the Fit Model window, JMP Pro users can change the Personality to Mixed Model.

3. Check the option to **Keep dialog open**.
4. Click **Run**.

Figure 5.79 Report for Full Model

Parameter Estimates

Term	Estimate	Std Error	DFDen	t Ratio	Prob>|t|
Intercept	38.183196	1.673404	2.075	22.82	0.0016*
A1	-4.657571	1.093074	2.916	-4.26	0.0251*
A2	0.147742	1.093074	2.916	0.14	0.9013
A3	0.9486795	1.093074	2.916	0.87	0.4510
A4	4.0716925	1.190175	3.214	3.42	0.0376*
C1	-1.827487	1.606207	0.868	-1.14	0.4804
C2	-15.73288	1.35446	0.719	-11.62	0.1065
A1*A2	0.1289012	1.080526	2.784	0.12	0.9131
A1*A3	-1.114015	1.080526	2.784	-1.03	0.3837
A1*A4	-1.92398	1.176342	3.04	-1.64	0.1993
A1*C1	-3.271232	0.480531	20.65	-6.81	<.0001*
A1*C2	3.737258	0.478635	20.09	7.81	<.0001*
A2*A3	0.3768179	1.080526	2.784	0.35	0.7520
A2*A4	1.4693535	1.176342	3.04	1.25	0.2992
A2*C1	1.9072313	0.480531	20.65	3.97	0.0007*
A2*C2	-0.161179	0.478635	20.09	-0.34	0.7398
A3*A4	0.0764369	1.176342	3.04	0.06	0.9522
A3*C1	0.1761619	0.480531	20.65	0.37	0.7176
A3*C2	-0.445242	0.478635	20.09	-0.93	0.3633
A4*C1	0.5306856	0.511896	20.84	1.04	0.3118
A4*C2	1.3506119	0.502894	20.12	2.69	0.0142*
C1*C2	-0.425326	1.606207	0.868	-0.26	0.8401

Random Effect Predictions

REML Variance Component Estimates

Random Effect	Var Ratio	Var Component	Std Error	95% Lower	95% Upper	Pct of Total
Whole Plots	1.6095523	15.451509	15.876576	-15.66601	46.569027	45.968
Subplots	0.8919389	8.562507	16.314188	-23.41271	40.537728	25.473
Residual		9.5998804	3.0299487	5.6239301	19.985723	28.559
Total		33.613897	16.824615	15.325965	123.59174	100.000

-2 LogLikelihood = 227.810084

Note: Total is the sum of the positive variance components.

Total including negative estimates = 33.613897

The Parameter Estimates report indicates that four two-way interactions, A1*C1, A1*C2, A2*C1, and A4*C2, and two main effects, A1 and A4, are significant at the 0.05 level.

5. In the Table panel of Battery Data.jmp, click the green triangle next to the **Reduced Model 1** script.

   The script opens a Fit Model window where insignificant interactions have been removed. The remaining effects are all main effects and the four two-way interactions A1*C1, A1*C2, A2*C1, and A4*C2. You are reducing the model in a conservative fashion.

6. Click **Run**.

Figure 5.80 Report for Preliminary Reduced Model

Parameter Estimates

Term	Estimate	Std Error	DFDen	t Ratio	Prob>|t|
Intercept	38.249531	1.373127	7.723	27.86	<.0001*
A1	-4.723906	1.175023	8.5	-4.02	0.0034*
A2	0.0497917	1.165418	8.228	0.04	0.9669
A3	0.7560417	1.165418	8.228	0.65	0.5342
A4	4.2328205	1.207654	9.16	3.50	0.0065*
C1	-2.195626	0.866042	1.083	-2.54	0.2234
C2	-15.79922	0.840903	1.008	-18.79	0.0331*
A1*C1	-3.283225	0.452505	25.17	-7.26	<.0001*
A1*C2	3.8035938	0.4498	24.42	8.46	<.0001*
A2*C1	1.8797279	0.452505	25.17	4.15	0.0003*
A4*C2	1.3877006	0.482556	24.95	2.88	0.0081*

Notice that the main effect C2 is now significant at the 0.05 level (Prob>|t| = 0.0331)

7. In the Fit Model window, remove A3.

   The main effect A3 is the only main effect that is not significant and not involved in a two-way interaction.

8. Click **Run**.

Figure 5.81 Report for Reduced Model

Parameter Estimates

Term	Estimate	Std Error	DFDen	t Ratio	Prob>|t|
Intercept	38.249531	1.362184	6.639	28.08	<.0001*
A1	-4.723906	1.137313	9.131	-4.15	0.0024*
A2	0.0497917	1.127389	8.818	0.04	0.9658
A4	4.2131272	1.173862	9.921	3.59	0.0050*
C1	-2.166553	0.899138	1.037	-2.41	0.2432
C2	-15.79922	0.874257	0.963	-18.07	0.0389*
A1*C1	-3.282201	0.452234	25.24	-7.26	<.0001*
A1*C2	3.8035938	0.449735	24.43	8.46	<.0001*
A2*C1	1.8843473	0.452234	25.24	4.17	0.0003*
A4*C2	1.3851686	0.4823	24.91	2.87	0.0082*

Random Effect Predictions

REML Variance Component Estimates

Random Effect	Var Ratio	Var Component	Std Error	95% Lower	95% Upper	Pct of Total
Whole Plots	2.0231563	17.459497	9.7154555	-1.582446	36.50144	60.025
Subplots	0.3473621	2.9976761	5.8732929	-8.513766	14.509119	10.306
Residual		8.6298312	2.4692519	5.2816799	16.588508	29.669
Total		29.087005	9.1941655	17.028461	60.633509	100.000

-2 LogLikelihood = 254.17861349

Note: Total is the sum of the positive variance components.

Total including negative estimates = 29.087005

The REML Variance Component Estimates report shows that the variance component associated with Whole Plots is about six times as large as the variance component for Subplots. This suggests that the assembly process is more variable than the curing process. Also, the within (Residual) error is larger than that for Subplots. Efforts to reduce variation should focus on the assembly process and on battery-to-battery differences.

9. From the red triangle next to Response OCV, select **Factor Profiling > Profiler**.

10. From the red triangle next to Prediction Profiler, select **Optimization and Desirability > Maximize Desirability**.

Figure 5.82 Prediction Profiler with Settings That Minimize OCV

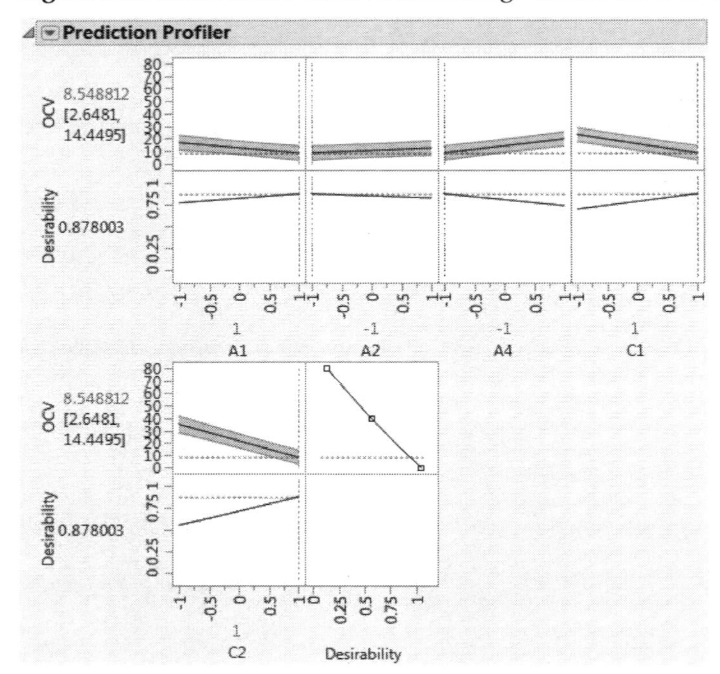

The profiler shows the five factors identified as active and settings that minimize OCV.

Chapter 6

Augment Designs

Experimental design is best treated as a sequential process. Ambiguities that result from a single design can be resolved by conducting further experimental runs.

For an existing design table, the Augment Design platform constructs additional runs in a way that optimizes the overall design process. You can add runs to accomplish the following objectives:

- Replicate the design a specified number of times.
- Add center points.
- Create a foldover design.
- Add axial points together with center points to transform a screening design to a response surface design.
- Add space filling points to a design.
- Add runs to the design using a model that can have more terms than the original model.

This chapter provides an overview of the Augment Design platform. It also presents a case study of design augmentation.

Example of Augment Design

This example demonstrates how to use the Augment Design platform to resolve ambiguities left by a screening design. In this study, a chemical engineer investigates the effects of five factors on the percent reaction of a chemical process.

1. Select **Help > Sample Data Library** and open Design Experiment/Reactor 8 Runs.jmp.
2. Select **DOE > Augment Design**.
3. Select Percent Reacted and click **Y, Response**.
4. Select all other variables except Pattern and click **X, Factor**.
5. Click **OK**.

Figure 6.1 Augment Design Dialog for the Reactor Example

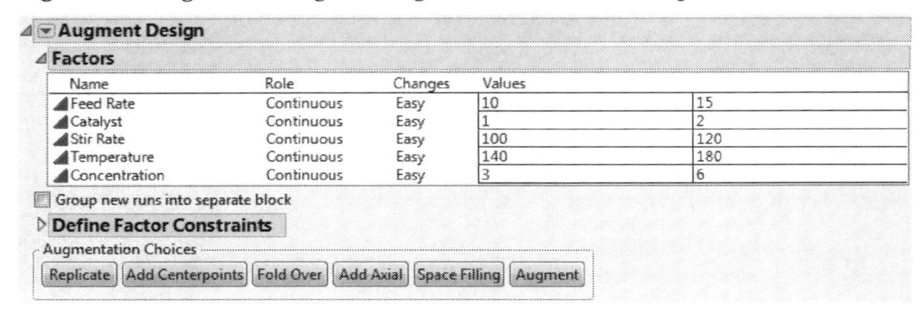

Note: You can check **Group new runs into separate block** to add a blocking factor to any design. However, the purpose of this example is to estimate all two-factor interactions in 16 runs, which cannot be done when there is the additional blocking factor in the model.

6. Click **Augment**.

   The model shown in Figure 6.2 is defined using the Model script in the data table.

Figure 6.2 Initial Augmented Model

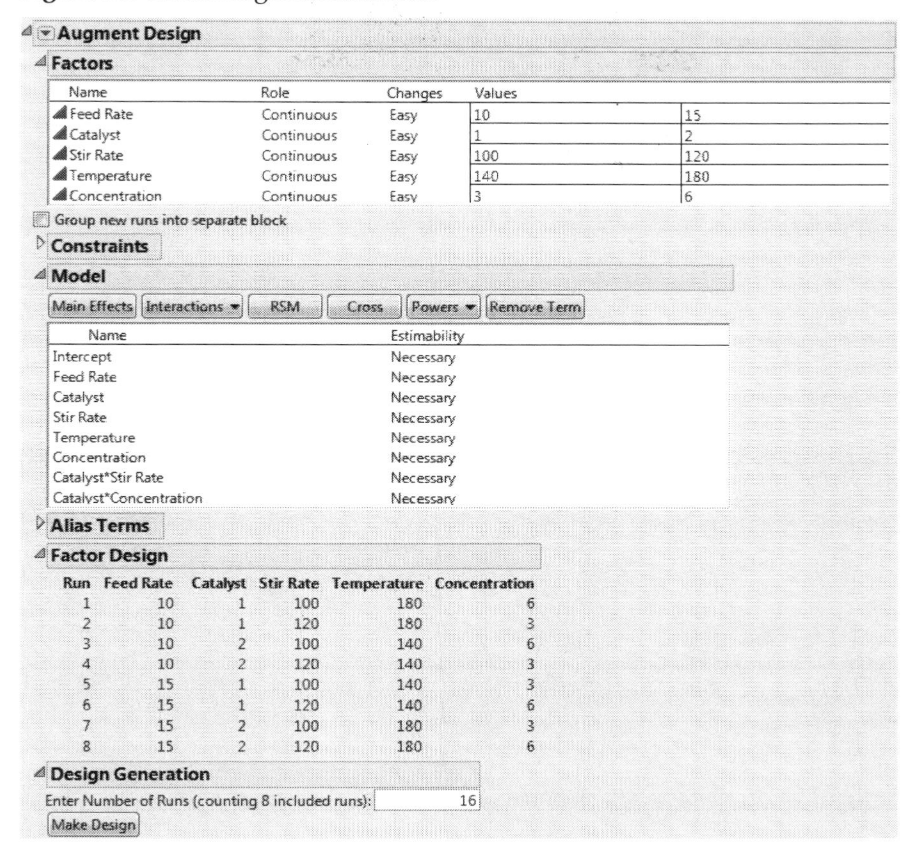

Run	Feed Rate	Catalyst	Stir Rate	Temperature	Concentration
1	10	1	100	180	6
2	10	1	120	180	3
3	10	2	100	140	6
4	10	2	120	140	3
5	15	1	100	140	3
6	15	1	120	140	6
7	15	2	100	180	3
8	15	2	120	180	6

7. In the Model outline, select **2nd** from the Interactions menu.

   This adds all the two-factor interactions to the model. The Minimum number of runs given for the specified model is 16, as shown in the Design Generation text edit box.

Figure 6.3 Augmented Model with All Two-Factor Interactions

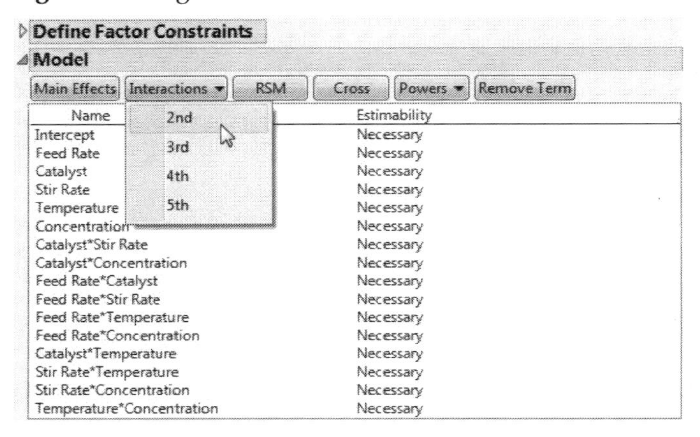

Note: Setting the Random Seed in step 8 and Number of Starts in step 9 reproduces the exact results shown in Figure 6.4. When constructing a design on your own, these steps are not necessary.

8. (Optional) From the Augment Design red triangle menu, select **Set Random Seed**, type 282322901, and click **OK**.
9. (Optional) From the Augment Design red triangle menu, select **Number of Starts**, type 800, and click **OK**.
10. Click **Make Design**.

    JMP computes settings for a *D*-optimally augmented design.

Figure 6.4 D-Optimally Augmented Design

Design

Run	Feed Rate	Catalyst	Stir Rate	Temperature	Concentration
1	15	2	100	180	3
2	15	1	120	140	6
3	15	1	100	140	3
4	10	1	120	180	3
5	10	1	100	180	6
6	10	2	120	140	3
7	10	2	100	140	6
8	15	2	120	180	6
9	15	2	120	140	3
10	10	2	100	180	3
11	10	2	120	180	3
12	10	2	120	140	6
13	15	1	120	180	3
14	10	1	120	180	6
15	15	2	100	180	6
16	10	1	120	140	3

11. Click **Make Table** to generate a design table containing the original design and results and the *D*-Optimally augmented factor settings.

Analyze the Augmented Design

Suppose you have already run the experiment on the augmented data and recorded results in the Percent Reacted column of the data table.

1. Select **Help > Sample Data Library** and open Design Experiment/Reactor Augment Data.jmp.

   You want to maximize Percent Reacted. However, the column's Response Limits column property in this sample data table is set to **Minimize**.

2. Click the asterisk next to the Percent Reacted column name in the Columns panel of the data table and select **Response Limits**, as shown on the left in Figure 6.5.
3. In the Column Info dialog that appears, change the response limit to **Maximize**, as shown on the right in Figure 6.5.

Figure 6.5 Change the Response Limits Column Property for the Percent Reacted Column

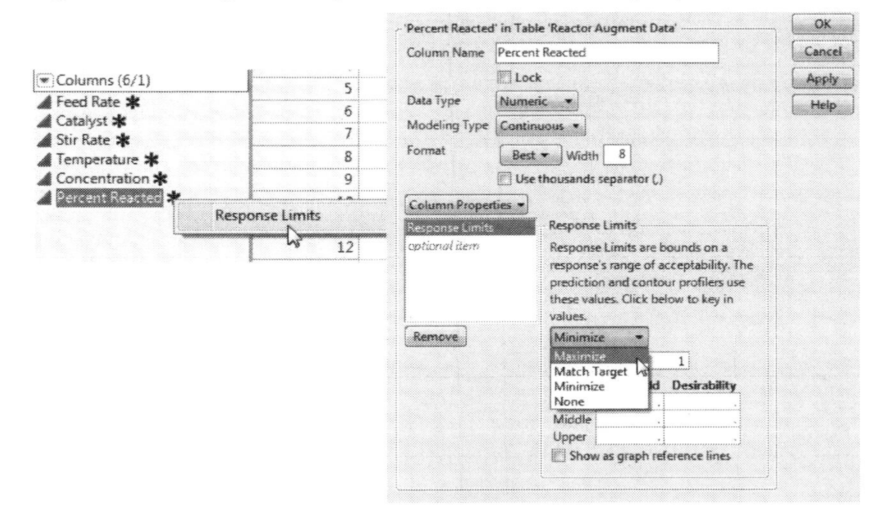

   You are now ready to run the analysis.

4. Click the green triangle next to the **Model** script.

Figure 6.6 Completed Augmented Experiment (Reactor Augment Data.jmp)

Reactor Augment Data
Design Augmented Design
Model
Click the green triangle to run, or right-click for more choices.
Columns (6/0)
Feed Rate
Catalyst
Stir Rate
Temperature
Concentration
Percent Reacted
Rows
All rows 16
Selected 0
Excluded 0

	Feed Rate	Catalyst	Stir Rate	Temperature	Concentration	Percent Reacted
1	10	1	100	180	6	44
2	10	1	120	180	3	66
3	10	2	100	140	6	70
4	10	2	120	140	3	54
5	15	1	100	140	3	53
6	15	1	120	140	6	55
7	15	2	100	180	3	93
8	15	2	120	180	6	82
9	15	2	120	180	3	98
10	15	2	120	140	6	65
11	10	2	100	140	3	63
12	10	1	120	180	6	49
13	15	1	100	140	6	63
14	15	1	100	180	3	61
15	10	1	120	140	6	59
16	10	2	100	180	3	94

The **Model** script opens the Fit Model window with all main effects and two-factor interactions as effects.

5. Change the fitting personality on the Fit Model dialog from **Standard Least Squares** to **Stepwise**.

Figure 6.7 Fit Model Dialog for Stepwise Regression on Generated Model

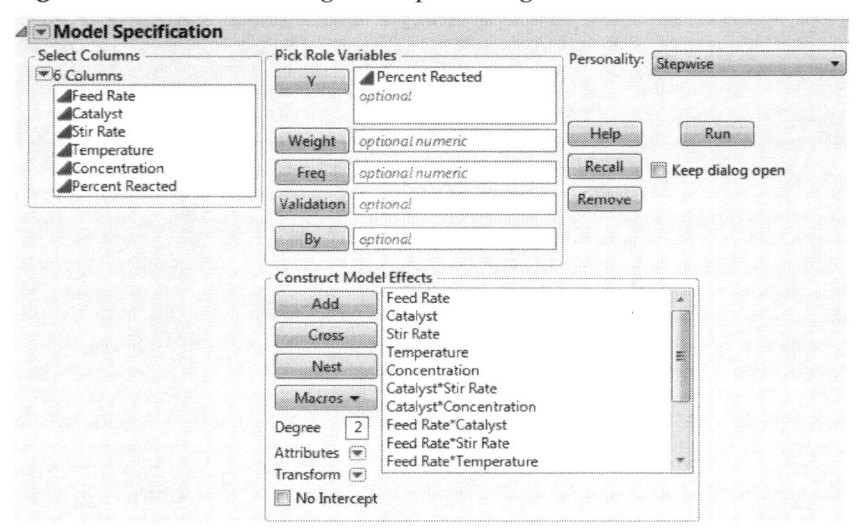

6. Click **Run.**

The stepwise regression control panel appears. Click the check boxes for all the main effect terms.

Note: Select **P-value Threshold** from the **Stopping Rule** menu, **Mixed** from the **Direction** menu, and make sure **Prob to Enter** is 0.050 and **Prob to Leave** is 0.100. These are not the default values. Follow the dialog shown in Figure 6.8.

Figure 6.8 Initial Stepwise Model

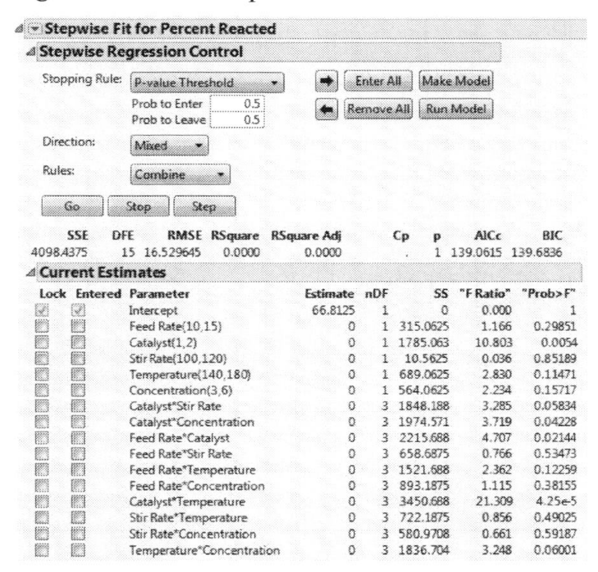

SSE	DFE	RMSE	RSquare	RSquare Adj	Cp	p	AICc	BIC
4098.4375	15	16.529645	0.0000	0.0000	.	1	139.0615	139.6836

Current Estimates

Lock	Entered	Parameter	Estimate	nDF	SS	"F Ratio"	"Prob>F"
☑	☑	Intercept	66.8125	1	0	0.000	1
☐	☐	Feed Rate{10,15}	0	1	315.0625	1.166	0.29851
☐	☐	Catalyst{1,2}	0	1	1785.063	10.803	0.0054
☐	☐	Stir Rate{100,120}	0	1	10.5625	0.036	0.85189
☐	☐	Temperature{140,180}	0	1	689.0625	2.830	0.11471
☐	☐	Concentration{3,6}	0	1	564.0625	2.234	0.15717
☐	☐	Catalyst*Stir Rate	0	3	1848.188	3.285	0.05834
☐	☐	Catalyst*Concentration	0	3	1974.571	3.719	0.04228
☐	☐	Feed Rate*Catalyst	0	3	2215.688	4.707	0.02144
☐	☐	Feed Rate*Stir Rate	0	3	658.6875	0.766	0.53473
☐	☐	Feed Rate*Temperature	0	3	1521.688	2.362	0.12259
☐	☐	Feed Rate*Concentration	0	3	893.1875	1.115	0.38155
☐	☐	Catalyst*Temperature	0	3	3450.688	21.309	4.25e-5
☐	☐	Stir Rate*Temperature	0	3	722.1875	0.856	0.49025
☐	☐	Stir Rate*Concentration	0	3	580.9708	0.661	0.59187
☐	☐	Temperature*Concentration	0	3	1836.704	3.248	0.06001

7. Click **Go**.

   This starts the stepwise regression. The process continues until all terms that meet the **Prob to Enter** and **Prob to Leave** criteria in the Stepwise Regression Control panel are entered into the model.

   Figure 6.9 shows the result of this example analysis. Note that Feed Rate is out of the model while the Catalyst*Temperature, Stir Rate*Temperature, and the Temperature*Concentration interactions have entered the model.

Figure 6.9 Completed Stepwise Model

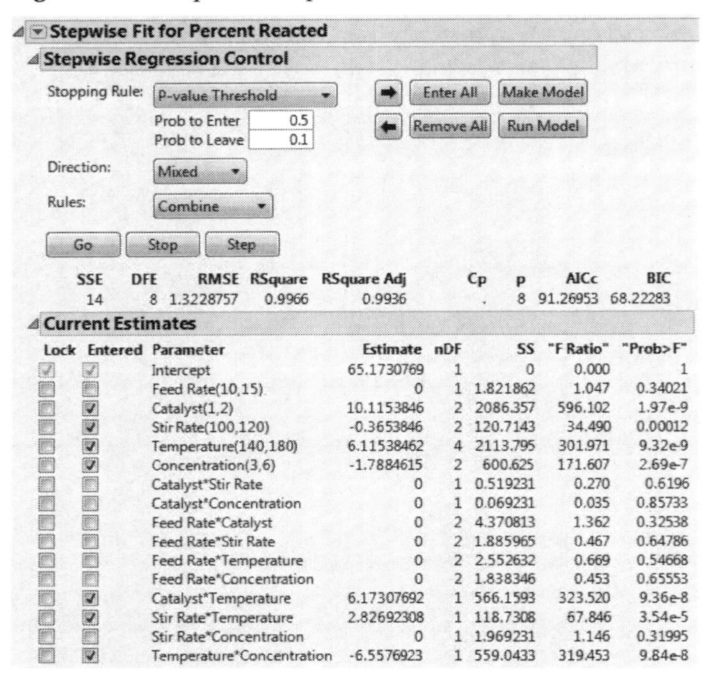

8. Click **Make Model** on the Stepwise control panel to generate the reduced model.

Figure 6.10 New Prediction Model Dialog

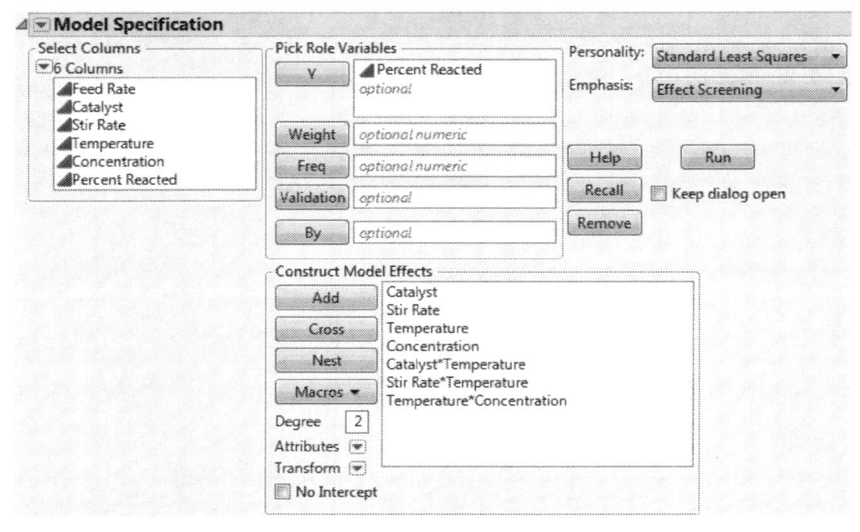

9. Click **Run.**

   The Actual by Predicted Plot indicates that the overall model is significant (P < 0.0001). Both the Actual by Predicted Plot and the Lack of Fit Test show no evidence of model misspecification.

Figure 6.11 Prediction Model Analysis of Variance and Lack of Fit Tests

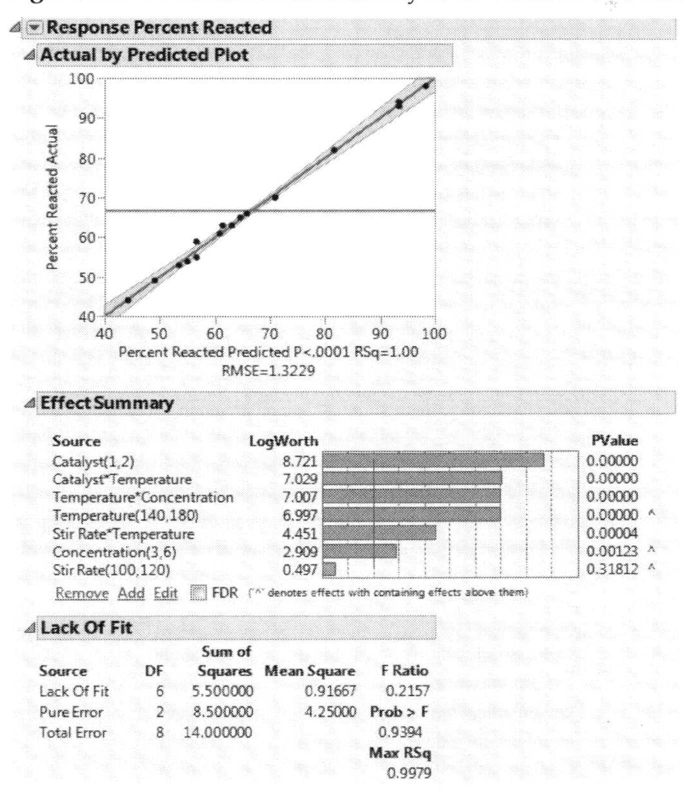

Source	LogWorth	PValue
Catalyst(1,2)	8.721	0.00000
Catalyst*Temperature	7.029	0.00000
Temperature*Concentration	7.007	0.00000
Temperature(140,180)	6.997	0.00000 ^
Stir Rate*Temperature	4.451	0.00004
Concentration(3,6)	2.909	0.00123 ^
Stir Rate(100,120)	0.497	0.31812 ^

Source	DF	Sum of Squares	Mean Square	F Ratio
Lack Of Fit	6	5.500000	0.91667	0.2157
Pure Error	2	8.500000	4.25000	Prob > F
Total Error	8	14.000000		0.9394
				Max RSq
				0.9979

The Effect Summary report shows that Catalyst is the most significant effect. However, note that the three two-factor interactions are also significant.

10. Choose **Optimality and Desirability > Maximize Desirability** from the menu on the Prediction Profiler title bar.

    The prediction profile plot in Figure 6.12 shows that maximum occurs at the high levels of Catalyst, Stir Rate, and Temperature and the low level of Concentration. At these extreme settings, the estimate of Percent Reacted increases from 65.17 to 98.38.

Figure 6.12 Maximum Percent Reacted

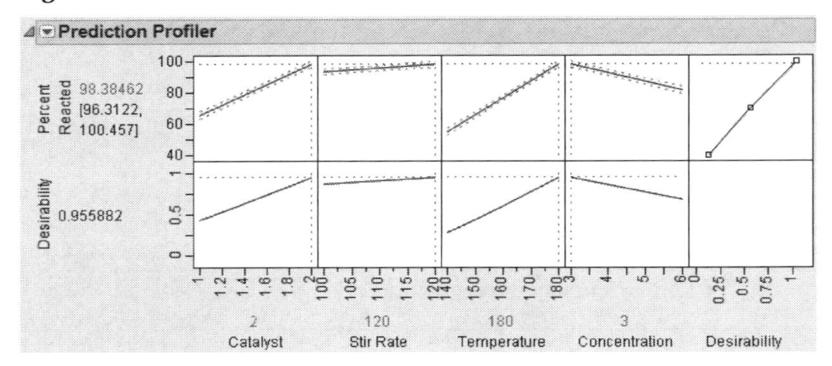

This example illustrates how an iterative approach to DOE can reduce costs and provide valuable information.

Augment Design Launch Window

To launch the Augment Design platform, open the data table that contains the design that you would like to augment and select **DOE** > **Augment Design**. The example in Figure 6.13 uses the Reactor 8 Runs.jmp sample data table, which is located in the Design Experiment folder.

Figure 6.13 Augment Design Launch Window

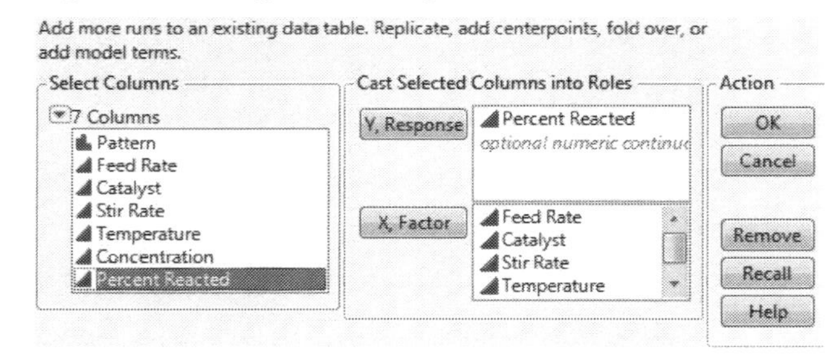

The launch window contains the following buttons:

Y, Response Enter the response column or columns. Entering a response is required. Responses must be numeric.

X, Factor Enter the factor columns. Factors can be of any data type or modeling type.

Augment Design Window

The initial Augment Design window consists of the Factors and Define Factor Constraints outlines, and the Augmentation Choices panel.

Figure 6.14 Initial Augment Design Window Using Reactor 8 Runs.jmp

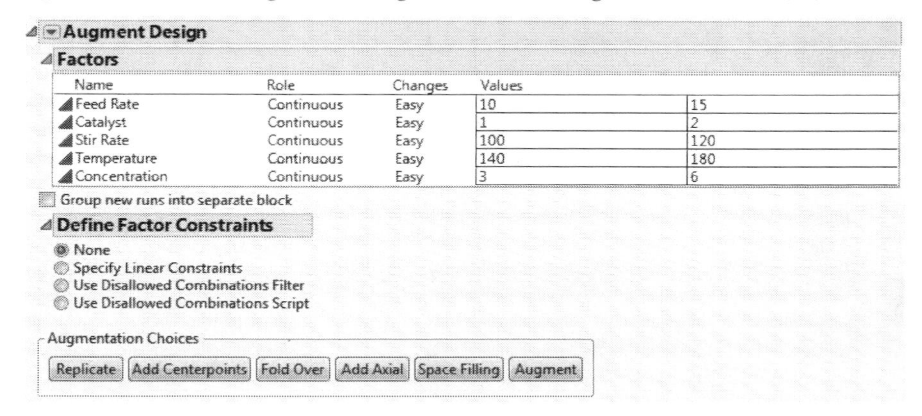

Factors

When the Augment Design window opens, the Factors outline shows the following:

Name All factors listed as X, Factor, in the Augment Design launch window except for factors with the Random Block design role column property.

Role If the factor has a Design Role column property specified in the data table, that role is shown in the Role column. If the factor does not have a Design Role column property and is constant, then Constant appears in the Role column. Otherwise, the factor's modeling type appears in the Role column.

Changes If the factor has a Factor Changes column property specified in the data table, that value is shown in the Changes column. If the factor does not have a Factor Changes column property, then Changes is specified as Easy.

Note: If a factor has a Factor Changes column property that is set to Hard or Very Hard, then the corresponding whole plot factor must be included in the X, Factor list in the Augment Design launch window.

Values For continuous factors, shows the minimum and maximum values. For categorical factors, shows the levels.

Tip: Factors that have a role of Categorical or Constant appear in the Name column with a down arrow icon. Click on the down arrow to add levels. If the factor is Constant and has a categorical modeling type, multiple levels can be added. If the factor is Constant and has a continuous modeling type, only one level can be added.

Define Factor Constraints

If you augment a design using the Space Filling or Augment options, you can define restrictions on the design space for the added runs.

Use Define Factor Constraints to restrict the design space. Unless you have loaded a constraint or included one as part of a script, the **None** option is selected. To specify constraints, select one of the other options:

Specify Linear Constraints Specifies inequality constraints on linear combinations of factors. Only available for factors with a Role of Continuous or Mixture. See "Specify Linear Constraints".

Note: When you save a script for a design that involves a linear constraint, the script expresses the linear constraint as a *less than or equal to* inequality ($\leq$).

Use Disallowed Combinations Filter Defines sets of constraints based on restricting values of individual factors. You can define both AND and OR constraints. See "Use Disallowed Combinations Filter".

Use Disallowed Combinations Script Defines disallowed combinations and other constraints as Boolean JSL expressions in a script editor box. See "Use Disallowed Combinations Script".

Specify Linear Constraints

In cases where it is impossible to vary continuous factors independently over the design space, you can specify linear inequality constraints. Linear inequalities describe factor level settings that are allowed.

Click **Add** to enter one or more linear inequality constraints.

Add Adds a template for a linear expression involving all the continuous factors in your design. Enter coefficient values for the factors and select the direction of the inequality to reflect your linear constraint. Specify the constraining value in the box to the right of the inequality. To add more constraints, click **Add** again.

Note: The Add option is disabled if you have already constrained the design region by specifying a Sphere Radius.

Remove Last Constraint Removes the last constraint.

Check Constraints Checks the constraints for consistency. This option removes redundant constraints and conducts feasibility checks. A JMP alert appears if there is a problem. If constraints are equivalent to bounds on the factors, a JMP alert indicates that the bounds in the Factors outline have been updated.

Use Disallowed Combinations Filter

This option uses an adaptation of the Data Filter to facilitate specifying disallowed combinations. For detailed information about using the Data Filter, see the JMP Reports chapter in the *Using JMP* book.

Select factors from the Add Filter Factors list and click **Add**. Then specify the disallowed combinations by using the slider (for continuous factors) or by selecting levels (for categorical factors).

The red triangle options for the Add Filter Factors menu are those found in the Select Columns panel of many platform launch windows. See the Get Started chapter in the *Using JMP* book for additional details about the column selection menu.

When you click Add, the Disallowed Combinations control panel shows the selected factors and provides options for further control. Factors are represented as follows, based on their modeling types:

Continuous Factors For a continuous factor, a double-arrow slider that spans the range of factor settings appears. An expression that describes the range using an inequality appears above the slider. You can specify disallowed settings by dragging the slider arrows or by clicking on the inequality bounds in the expression and entering your desired constraints. In the slider, a solid blue highlight represents the disallowed values.

Categorical Factor For a categorical factor, the possible levels are displayed either as labeled blocks or, when the number of levels is large, as list entries. Select a level to disallow it. To select multiple levels, hold the Control key. The block or list entries are highlighted to indicate the levels that have been disallowed. When you add a categorical factor to the Disallowed Combinations panel, the number of levels of the categorical factor is given in parentheses following the factor name.

Disallowed Combinations Options

The control panel has the following controls:

Clear Clears all disallowed factor level settings that you have specified. This does not clear the selected factors.

Start Over Removes all selected factors and returns you to the initial list of factors.

AND Opens the Add Filter Factors list. Selected factors become an AND group. Any combination of factor levels specified within an AND group is disallowed.

To add a factor to an AND group later on, click the group's outline to see a highlighted rectangle. Select AND and add the factor.

To remove a single factor, select **Delete** from its red triangle menu.

OR Opens the Add Filter Factors list. Selected factors become a separate AND group. For AND groups separated by OR, a combination is disallowed if it is specified in at least one AND group.

Red Triangle Options for Factors

A factor can appear in several OR groups. An occurrence of the factor in a specific OR group is referred to as an *instance* of the factor.

Delete Removes the selected instance of the factor from the Disallowed Combinations panel.

Clear Selection Clears any selection for that instance of the factor.

Invert Selection Deselects the selected values and selects the values not previously selected for that instance of the factor.

Display Options Available only for categorical factors. Changes the appearance of the display. Options include:

- **Blocks Display** shows each level as a block.
- **List Display** shows each level as a member of a list.
- **Single Category Display** shows each level.
- **Check Box Display** adds a check box next to each value.

Find Available only for categorical factors. Provides a text box beneath the factor name where you can enter a search string for levels of the factor. Press the Enter key or click outside the text box to perform the search. Once **Find** is selected, the following Find options appear in the red triangle menu:

- **Clear Find** clears the results of the Find operation and returns the panel to its original state.
- **Match Case** uses the case of the search string to return the correct results.
- **Contains** searches for values that include the search string.
- **Does not contain** searches for values that do not include the search string.
- **Starts with** searches for values that start with the search string.
- **Ends with** searches for values that end with the search string.

Use Disallowed Combinations Script

Use this option to disallow particular combinations of factor levels using a JSL script. This option can be used with continuous factors or mixed continuous and categorical factors.

This option opens a script window where you insert a script that identifies the combinations that you want to disallow. The script must evaluate as a Boolean expression. When the expression evaluates as true, the specified combination is disallowed.

When forming the expression for a categorical factor, use the ordinal value of the level instead of the name of the level. If a factor's levels are high, medium, and low, specified in that order in the Factors outline, their associated ordinal values are 1, 2, and 3. For example, suppose that you have two continuous factors, X1 and X2, and a categorical factor X3 with three levels: L1, L2, and L3, in order. You want to disallow levels where the following holds:

$$e^{X_1} + 2X_2 < 0 \text{ and } X_3 = L2$$

Enter the expression `(Exp(X1) + 2*X2 < 0) & (X3 == 2)` into the script window.

Figure 6.15 Expression in Script Editor

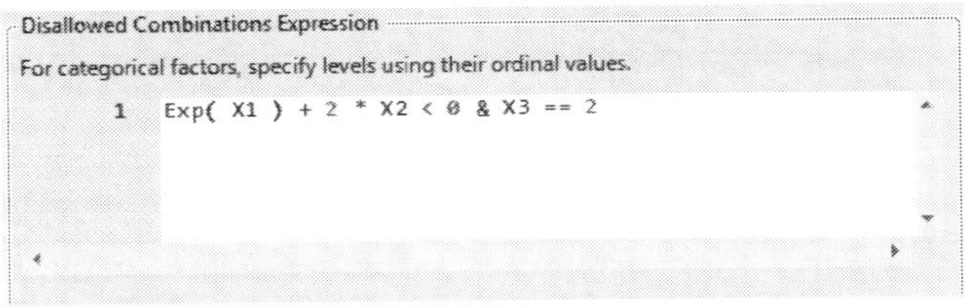

```
Disallowed Combinations Expression
For categorical factors, specify levels using their ordinal values.
1   Exp( X1 ) + 2 * X2 < 0 & X3 == 2
```

(In the figure, unnecessary parentheses were removed by parsing.) Notice that functions can be entered as part of the Boolean expression.

Augmentation Choices

The Augment Design platform requires an existing design data table. It gives the following five choices:

Replicate replicates the design a specified number of times. See "Replicate a Design" on page 216.

Add Centerpoints Adds center points. Specify how many additional runs you want to add as center points to the design. A center point is a run whose setting for each continuous factor is midway between the high and low settings. See "Center Points, Replicate Runs, and Testing" on page 60 in the "Starting Out with DOE" chapter.

If a design contains both continuous and other types of factors, center points might not be balanced relative to the levels of the other factors. Augment Design chooses the center points to maximize the D-, I-, or alias efficiency of the design.

See "Add Center Points" on page 219.

Fold Over creates a foldover design. See "Creating a Foldover Design" on page 219.

Add Axial adds axial points together with center points to transform a screening design to a response surface design. See "Adding Axial Points" on page 220.

Space Filling Adds additional runs to any design consisting of continuous factors. Additional runs are constructed using the fast flexible filling methodology. See "Space Filling" on page 221.

Augment adds runs to the design (augment) using a model, which can have more terms than the original model. See "Augment" on page 222.

Replicate a Design

Replication provides a direct check on the assumption that the error variance is constant. It also reduces the variability of the regression coefficients in the presence of large process or measurement variability.

To replicate the design a specified number of times:

1. Open a data table that contains a design that you want to augment. This example uses Reactor 8 Runs.jmp from the Design Experiment sample data folder installed with JMP.
2. Select **DOE > Augment Design** to see the initial dialog for specifying factors and responses.
3. Select Percent Reacted and click **Y, Response**.
4. Select all other variables except Pattern and click **X, Factor** to identify the factors that you want to use for the augmented design.
5. Click **OK** to see the Augment Design panel shown in Figure 6.16.
6. If you want the original runs and the resulting augmented runs to be identified by a blocking factor, select **Group New Runs into Separate Block** on the Augment Design panel.

Figure 6.16 Choose an Augmentation Type

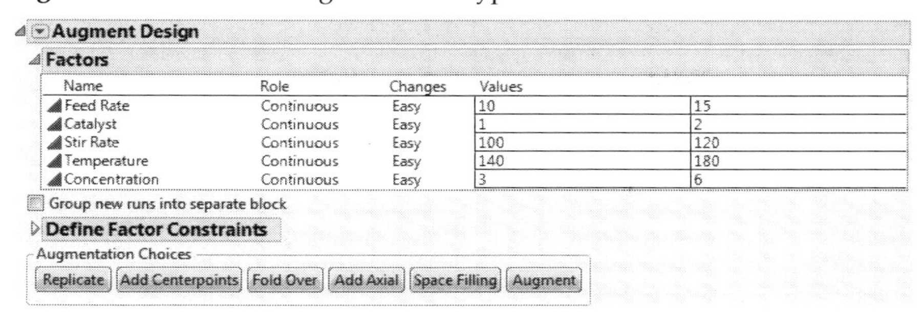

7. Click the **Replicate** button to see the dialog shown on the left in Figure 6.17. Enter the number of times you want JMP to perform each run and then click **OK**.

   Note: Entering 2 specifies that you want each run to appear twice in the resulting design. This is the same as one replicate (Figure 6.17).

8. View the design, shown on the right in Figure 6.17.

Figure 6.17 Reactor Data Design Augmented with Two Replicates

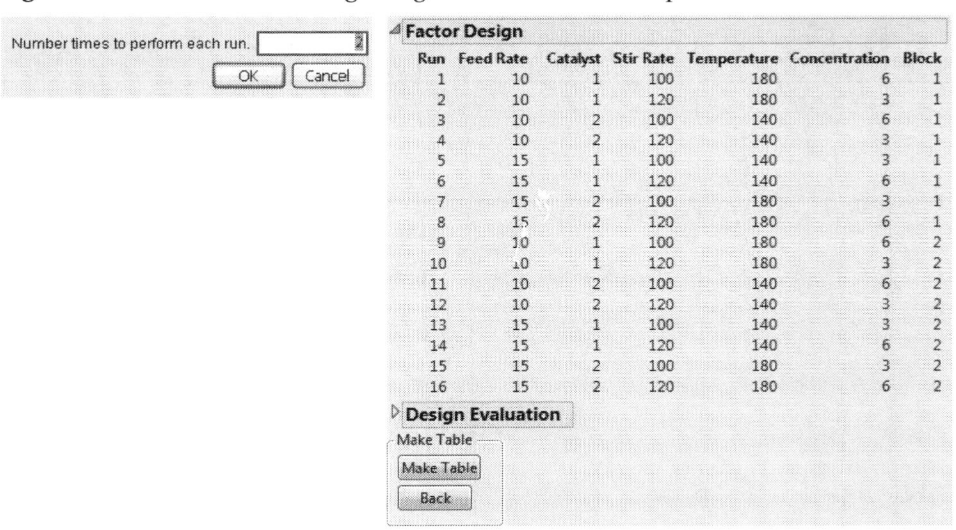

Run	Feed Rate	Catalyst	Stir Rate	Temperature	Concentration	Block
1	10	1	100	180	6	1
2	10	1	120	180	3	1
3	10	2	100	140	6	1
4	10	2	120	140	3	1
5	15	1	100	140	3	1
6	15	1	120	140	6	1
7	15	2	100	180	3	1
8	15	2	120	180	6	1
9	10	1	100	180	6	2
10	10	1	120	180	3	2
11	10	2	100	140	6	2
12	10	2	120	140	3	2
13	15	1	100	140	3	2
14	15	1	120	140	6	2
15	15	2	100	180	3	2
16	15	2	120	180	6	2

9. In the Design Evaluation section, click the disclosure icons next to Prediction Variance Profile and Prediction Variance Surface to see the profile and surface plots shown in Figure 6.18.

Figure 6.18 Prediction Profiler and Surface Plot

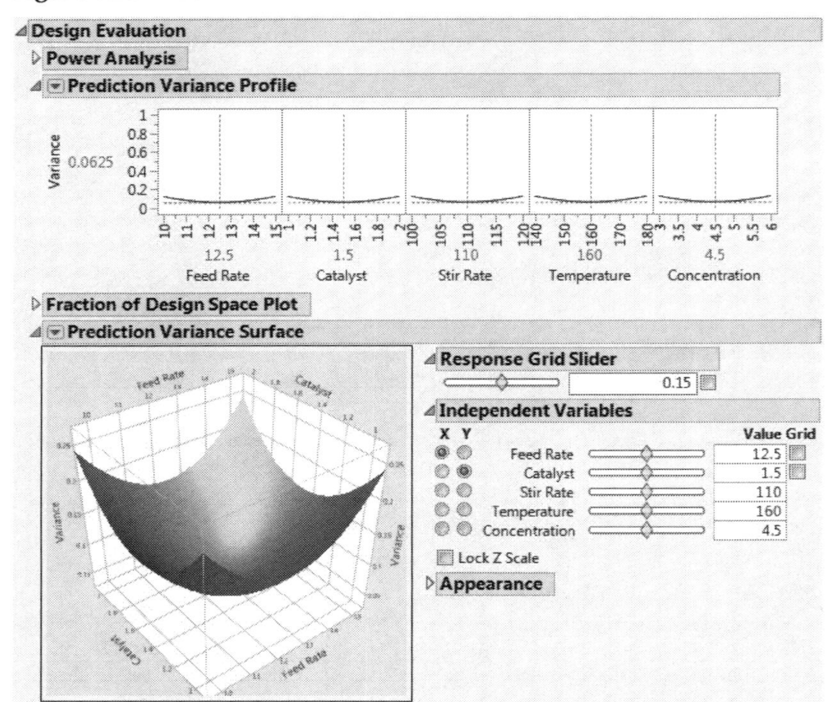

10. Click **Make Table** to produce the design table shown in Figure 6.19.

Figure 6.19 The Replicated Design

Augmented Design
Design Augmented Design
- Screening
- Model
- Original Data Table
- DOE Dialog

Columns (6/0)
- Feed Rate *
- Catalyst *
- Stir Rate *
- Temperature *
- Concentration *
- Percent Reacted *

Rows
All rows 16
Selected 0
Excluded 0
Hidden 0
Labelled 0

	Feed Rate	Catalyst	Stir Rate	Temperature	Concentration	Percent Reacted
1	15	2	100	180	3	93
2	15	1	120	140	6	55
3	15	1	100	140	3	53
4	10	1	120	180	3	66
5	10	1	100	180	6	44
6	10	2	120	140	3	54
7	10	2	100	140	6	70
8	15	2	120	180	6	82
9	15	2	100	180	3	•
10	15	1	120	140	6	•
11	15	1	100	140	3	•
12	10	1	120	180	3	•
13	10	1	100	180	6	•
14	10	2	120	140	3	•
15	10	2	100	140	6	•
16	15	2	120	180	6	•

Add Center Points

Adding center points is useful to check for curvature and reduce the prediction error in the center of the factor region. Center points are usually replicated points that allow for an independent estimate of pure error, which can be used in a lack-of-fit test.

To add center points:

1. Open a data table that contains a design that you want to augment. This example uses Reactor 8 Runs.jmp found in the Design Experiment sample data folder installed with JMP.
2. Select **DOE > Augment Design**.
3. In the initial Augment Design dialog, identify the response and factors that you want to use for the augmented design (see Figure 5) and click **OK**.
4. If you want the original runs and the resulting augmented runs to be identified by a blocking factor, check the box beside **Group new runs into separate block.** (Figure 6.16 shows the check box location directly under the Factors panel.)
5. Click the **Add Centerpoints** button and enter the number of center points that you want to add. For this example, add two center points, and click **OK**.
6. Click **Make Table** to see the data table in Figure 6.20.

The table shows two center points appended to the end of the design.

Figure 6.20 Design with Two Center Points Added

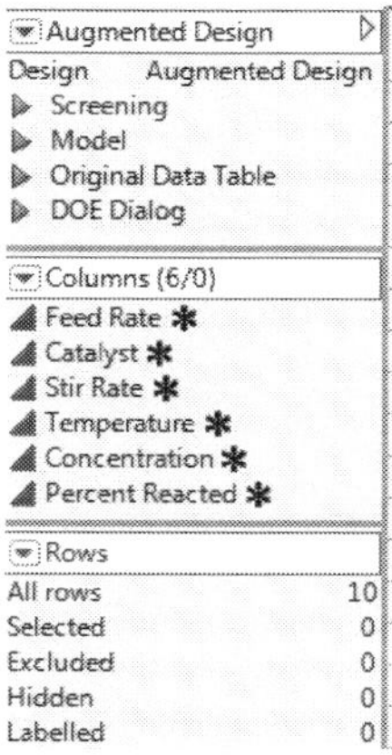

Augmented Design
Design Augmented Design
- Screening
- Model
- Original Data Table
- DOE Dialog

Columns (6/0)
- Feed Rate *
- Catalyst *
- Stir Rate *
- Temperature *
- Concentration *
- Percent Reacted *

Rows

All rows	10
Selected	0
Excluded	0
Hidden	0
Labelled	0

	Feed Rate	Catalyst	Stir Rate	Temperature	Concentration	Percent Reacted
1	15	2	100	180	3	93
2	15	1	120	140	6	55
3	15	1	100	140	3	53
4	10	1	120	180	3	66
5	10	1	100	180	6	44
6	10	2	120	140	3	54
7	10	2	100	140	6	70
8	15	2	120	180	6	82
9	12.5	1.5	110	160	4.5	•
10	12.5	1.5	110	160	4.5	•

Creating a Foldover Design

A foldover design removes the confounding of two-factor interactions and main effects. This is especially useful as a follow-up to saturated or near-saturated fractional factorial or Plackett-Burman designs.

To create a foldover design:

1. Open a data table that contains a design that you want to augment. This example uses Reactor 8 Runs.jmp, found in the Design Experiment sample data folder installed with JMP.
2. Select **DOE > Augment Design**.
3. In the initial Augment Design dialog, identify the response and factors that you want to use for the augmented design (see Figure 5) and click **OK**.
4. Check the box to the left of **Group new runs into separate block.** (Figure 6.16 shows the check box location directly under the Factors panel.) This identifies the original runs and the resulting augmented runs with a blocking factor.
5. Click the **Fold Over** button. A dialog appears that lists all the design factors.
6. Select the factors to fold. The default, if you select no factors, is to fold on all design factors. If you choose a subset of factors to fold over, the remaining factors are replicates of the original runs. The example in Figure 6.21 folds on all five factors and includes a blocking factor.
7. Click **Make Table**. The design data table that results lists the original set of runs as block 1 and the new (foldover) runs are block 2.

Figure 6.21 Listing of a Foldover Design on All Factors

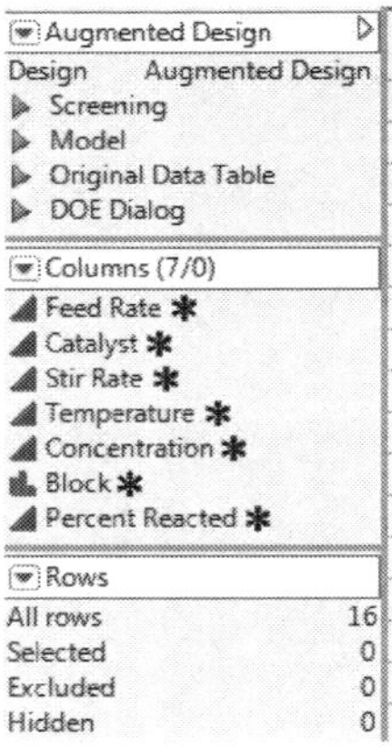

Augmented Design
Design Augmented Design
Screening
Model
Original Data Table
DOE Dialog

Columns (7/0)
Feed Rate *
Catalyst *
Stir Rate *
Temperature *
Concentration *
Block *
Percent Reacted *

Rows
All rows 16
Selected 0
Excluded 0
Hidden 0

	Feed Rate	Catalyst	Stir Rate	Temperature	Concentration	Block	Percent Reacted
1	15	2	100	180	3	1	93
2	15	1	120	140	6	1	55
3	15	1	100	140	3	1	53
4	10	1	120	180	3	1	66
5	10	1	100	180	6	1	44
6	10	2	120	140	3	1	54
7	10	2	100	140	6	1	70
8	15	2	120	180	6	1	82
9	10	1	120	140	6	2	•
10	10	2	100	180	3	2	•
11	10	2	120	180	6	2	•
12	15	2	100	140	6	2	•
13	15	2	120	140	3	2	•
14	15	1	100	180	6	2	•
15	15	1	120	180	3	2	•
16	10	1	100	140	3	2	•

Adding Axial Points

You can add axial points together with center points, which transforms a screening design to a response surface design. Follow these steps:

1. Open a data table that contains a design that you want to augment. This example uses Reactor 8 Runs.jmp, from the Design Experiment sample data folder installed with JMP.
2. Select **DOE > Augment Design**.

3. In the initial Augment Design dialog, identify the response and factors that you want to use for the augmented design (see Figure 5) and click **OK**.
4. If you want the original runs and the resulting augmented runs to be identified by a blocking factor, check the box beside **Group New Runs into Separate Block** (Figure 6.16).
5. Click **Add Axial.**
6. Enter the axial values in units of the factors scaled from –1 to +1, and then enter the number of center points that you want. When you click **OK**, the augmented design includes the number of center points specified and constructs two axial points for each variable in the original design.

Figure 6.22 Entering Axial Values

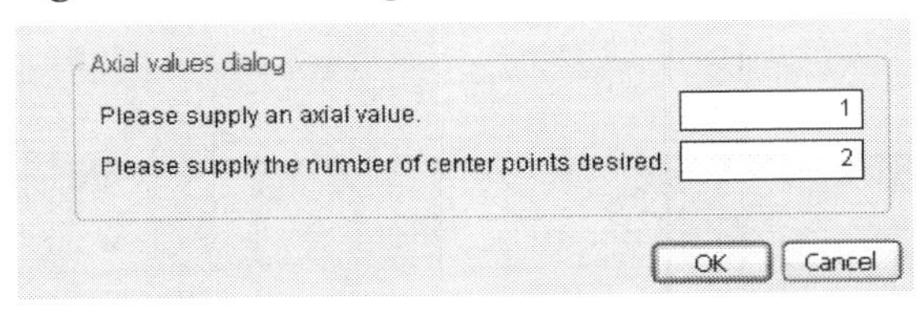

7. Click **Make Table**. The design table appears. Figure 6.23 shows a table augmented with two center points and two axial points for five variables.

Figure 6.23 Design Augmented with Two Center and Ten Axial Points

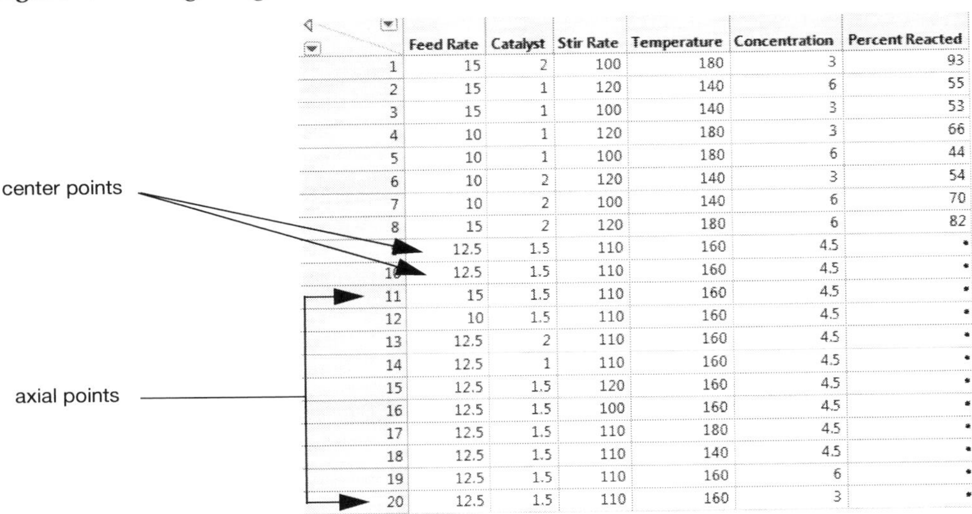

	Feed Rate	Catalyst	Stir Rate	Temperature	Concentration	Percent Reacted
1	15	2	100	180	3	93
2	15	1	120	140	6	55
3	15	1	100	140	3	53
4	10	1	120	180	3	66
5	10	1	100	180	6	44
6	10	2	120	140	3	54
7	10	2	100	140	6	70
8	15	2	120	180	6	82
[illegible]	12.5	1.5	110	160	4.5	•
10	12.5	1.5	110	160	4.5	•
11	15	1.5	110	160	4.5	•
12	10	1.5	110	160	4.5	•
13	12.5	2	110	160	4.5	•
14	12.5	1	110	160	4.5	•
15	12.5	1.5	120	160	4.5	•
16	12.5	1.5	100	160	4.5	•
17	12.5	1.5	110	180	4.5	•
18	12.5	1.5	110	140	4.5	•
19	12.5	1.5	110	160	6	•
20	12.5	1.5	110	160	3	•

Space Filling

The Space Filling augmentation choice adds points to a design consisting of continuous factors using the fast flexible filling method with the MaxPro criterion. The Space Filling choice accommodates constraints on the design space. You can specify linear constraints or disallowed combinations.

The algorithm that is used to augment designs begins by generating a large number of random points within the specified design region. These points are then clustered using a Fast Ward algorithm into a number of clusters that equals the Number of Additional Runs that you specify.

The final design points are obtained by optimizing the MaxPro (*maximum projection*) criterion over the existing and additional runs. For p factors and n equal to the number of existing and additional runs, the MaxPro criterion strives to find points in the clusters that minimize the following criterion:

$$C_{MaxPro} = \sum_{i}^{n-1} \sum_{j=i+1}^{n} \left[1 / \prod_{k=1}^{p} (x_{ik} - x_{jk})^2 \right]$$

The MaxPro criterion maximizes the product of the distances between design points in a way that involves all factors. This supports the goal of providing good space-filling properties on projections of factors. See Joseph et al. (2015).

Augment

A powerful use of the Augment Design platform is to add runs using a model that can have more terms than the original model. For example, you can achieve the objectives of response surface methodology by changing a linear model to a full quadratic model and adding the necessary number of runs.

D-optimal augmentation is a powerful tool for sequential design. Using this feature you can add terms to the original model and find optimal new test runs with respect to this expanded model. You can also group the two sets of experimental runs into separate blocks, which optimally blocks the second set with respect to the first.

When you select to Augment a design, Model and Alias Terms outlines appear. Use these outlines to add effects to the Model and Alias Terms lists.

Model Outline

The Model outline lists the effects that are in the Model script in the design table containing the design that you want to augment. If there is no Model script in the design table, the Model outline shows only the main effects.

Add or remove effects to specify your model for the augmented design. For details on how to add and remove effects, see “Model” on page 432 in the “Evaluate Designs” chapter.

Alias Outline

The Alias Terms outline contains all two-factor interactions that are not in the Model outline. When you generate your augmented design, a Design Evaluation outline is provided. The effects in Alias Terms list control the calculations in the Alias Matrix and Color Map on Correlations outlines under Design Evaluation. See "Alias Matrix" on page 446 in the "Evaluate Designs" chapter and "Color Map on Correlations" on page 448 in the "Evaluate Designs" chapter.

Add or remove effects to compare your designs for effects that may be active. For details on how to add and remove effects, see "Model" on page 432 in the "Evaluate Designs" chapter.

Example of Using Augment

This example illustrates how to add new runs and model terms to the original model:

1. Select Help > Sample Data Library and open Design Experiment/Reactor Augment Data.jmp.
2. Select **DOE > Augment Design**.
3. Select Percent Reacted and click **Y, Response**.
4. Select Feed Rate, Catalyst, Stir Rate, Temperature, and Concentration, and click **X, Factor**.
5. Click **OK**.

   Notice the check **Group New Runs into Separate Block**. This creates a blocking factor that places the original runs and the augmented runs in separate blocks. You will not use this option in this example.
6. Click **Augment**.

   The original 16 runs are shown in the Factor Design panel. You will add 6 new runs with the goal of estimating all 5 quadratic terms.
7. In the Model outline, click the **Powers** button and select **2nd**.

   This adds the five quadratic terms to the model, shown in Figure 6.24. The augmented design will enable you to estimate these quadratic terms.
8. In the Design Generation outline, enter 22 in the box to the right of **Enter Number of Runs (counting 16 included runs)**.

Figure 6.24 Model, Factor Design, and Design Generation Outlines

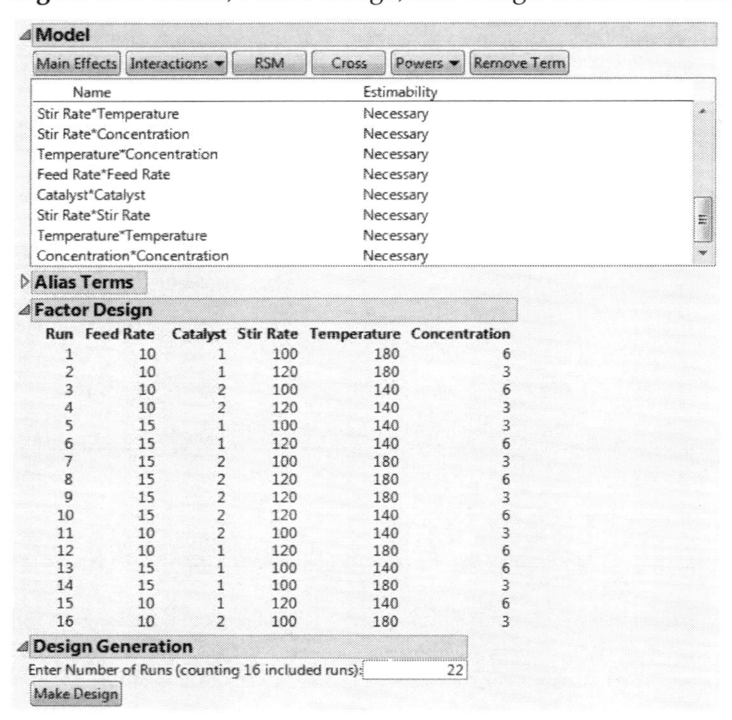

Run	Feed Rate	Catalyst	Stir Rate	Temperature	Concentration
1	10	1	100	180	6
2	10	1	120	180	3
3	10	2	100	140	6
4	10	2	120	140	3
5	15	1	100	140	3
6	15	1	120	140	6
7	15	2	100	180	3
8	15	2	120	180	6
9	15	2	120	180	3
10	15	2	120	140	6
11	10	2	100	140	3
12	10	1	120	180	6
13	15	1	100	140	6
14	15	1	100	180	3
15	10	1	120	140	6
16	10	2	100	180	3

9. Click **Make Design**.

   The six new runs appear in the Design panel. Note that you can explore various properties of the augmented design in the Design Evaluation outline.

Figure 6.25 24 Total Runs

Design

Run	Feed Rate	Catalyst	Stir Rate	Temperature	Concentration
1	10	1	100	180	6
2	10	1	120	180	3
3	10	2	100	140	6
4	10	2	120	140	3
5	15	1	100	140	3
6	15	1	120	140	6
7	15	2	100	180	3
8	15	2	120	180	6
9	15	2	120	180	3
10	15	2	120	140	6
11	10	2	100	140	3
12	10	1	120	180	6
13	15	1	100	140	6
14	15	1	100	180	3
15	10	1	120	140	6
16	10	2	100	180	3
17	12.5	2	110	160	3
18	10	1.5	100	160	3
19	10	1.5	110	140	4.5
20	12.5	1.5	100	180	6
21	15	2	100	160	4.5
22	12.5	1	100	140	4.5

10. Click **Make Table**.

    This creates the augmented design table (Figure 6.26) with the additional runs.

Figure 6.26 The Augmented Design Table with New Runs

Augmented Design
Design Augmented Design
Screening
Model
Original Data Table
DOE Dialog

Columns (6/0)
Feed Rate
Catalyst
Stir Rate
Temperature
Concentration
Percent Reacted

Rows
All rows 22
Selected 0

	Feed Rate	Catalyst	Stir Rate	Temperature	Concentration	Percent Reacted
1	10	1	100	180	6	44
2	10	1	120	180	3	66
3	10	2	100	140	6	70
4	10	2	120	140	3	54
5	15	1	100	140	3	53
6	15	1	120	140	6	55
7	15	2	100	180	3	93
8	15	2	120	180	6	82
9	15	2	120	180	3	98
10	15	2	120	140	6	65
11	10	2	100	140	3	63
12	10	1	120	180	6	49
13	15	1	100	140	6	63
14	15	1	100	180	3	61
15	10	1	120	140	6	59
16	10	2	100	180	3	94
17	12.5	2	110	160	3	•
18	10	1.5	100	160	3	•
19	10	1.5	110	140	4.5	•
20	12.5	1.5	100	180	6	•
21	15	2	100	160	4.5	•
22	12.5	1	100	140	4.5	•

11. Click on the green triangle next to the **Model** script.

    Note that all five quadratic effects are listed as model effects in the Model Specification window.

12. Enter fictional **Percent Reacted** data values for the six new runs, rows 17 to 22, and click **Run**.

    Note that you can test for all effects, including the five new quadratic effects.

Augment Design Options

This section describes the options available under the Augment Design red triangle menu.

Save Responses Saves the information in the Responses panel to a new data table. You can then quickly load the responses and their associated information into most DOE windows. This option is helpful if you anticipate re-using the responses.

Load Responses Loads responses that you saved using the Save Responses option.

Save Factors Saves the information in the Factors panel to a new data table. Each factor's column contains its levels. Other information is stored as column properties. You can then quickly load the factors and their associated information into most DOE windows.

> **Note:** It is possible to create a factors table by entering data into an empty table, but remember to assign each column an appropriate Design Role. Do this by right-clicking on the column name in the data grid and selecting **Column Properties > Design Role**. In the Design Role area, select the appropriate role.

Load Factors Loads factors that you saved using the Save Factors option.

Save Constraints (Unavailable for some platforms) Saves factor constraints that you defined in the Define Factor Constraints or Linear Constraints outline into a data table, with a column for each constraint. You can then quickly load the constraints into most DOE windows.

In the constraint table, the first rows contain the coefficients for each factor. The last row contains the inequality bound. Each constraint's column contains a column property called ConstraintState that identifies the constraint as a "less than" or a "greater than" constraint. See "ConstraintState" on page 686 in the "Column Properties" appendix.

Load Constraints (Unavailable for some platforms) Loads factor constraints that you saved using the Save Constraints option.

Set Random Seed Sets the random seed that JMP uses to control certain actions that have a random component. These actions include the following:

- simulating responses using the Simulate Responses option
- randomizing Run Order for design construction

– selecting a starting design for designs based on random starts

To reproduce a design or simulated responses, enter the random seed that generated them. For designs using random starts, set the seed before clicking Make Design. To control simulated responses or run order, set the seed before clicking Make Table.

Note: The random seed associated with a design is included in the DOE Dialog script that is saved to the design data table.

Simulate Responses Adds response values and a column containing a simulation formula to the design table. Select this option before you click Make Table.

When you click Make Table, the following occur:

– A set of simulated response values is added to each response column.

– For each response, a new a column that contains a simulation model formula is added to the design table. The formula and values are based on the model that is specified in the design window.

– A Model window appears where you can set the values of coefficients for model effects and specify one of three distributions: Normal, Binomial, or Poisson.

– A script called **DOE Simulate** is saved to the design table. This script re-opens the Model window, enabling you to re-simulate values or to make changes to the simulated response distribution.

Make selections in the Model window to control the distribution of simulated response values. When you click Apply, a formula for the simulated response values is saved in a new column called <Y> Simulated, where Y is the name of the response. Clicking Apply again updates the formula and values in <Y> Simulated.

For additional details, see "Simulate Responses" on page 106 in the "Custom Designs" chapter.

Note: JMP PRO You can use Simulate Responses to conduct simulation analyses using the JMP Pro Simulate feature. For information about Simulate and some DOE examples, see the Simulate chapter in the *Basic Analysis* book.

Save X Matrix Saves scripts called Moments Matrix and Model Matrix to the design data table. These scripts contain the moments and design matrices. See "Save X Matrix" on page 108 in the "Custom Designs" chapter.

Caution: For a design with nominal factors, the matrix in the Model Matrix script saved by the Save X Matrix option is *not* the coding matrix used in fitting the linear model. You can obtain the coding matrix used for fitting the model by selecting the option Save Columns > Save Coding Table in the Fit Model report that you obtain when you run the Model script.

Optimality Criterion Changes the design optimality criterion. The default criterion, **Recommended**, specifies D-optimality for all design types, unless you added quadratic effects using the RSM button in the Model outline. For more information about the D-, I-, and alias-optimal designs, see "Optimality Criteria" on page 123 in the "Custom Designs" appendix.

Note: You can set a preference to always use a given optimality criterion. Select File > Preferences > Platforms > DOE. Check Optimality Criterion and select your preferred criterion.

Number of Starts Enables you to specify the number of random starts used in constructing the design. See "Number of Starts" on page 110 in the "Custom Designs" chapter.

Design Search Time Maximum number of seconds spent searching for a design. The default search time is based on the complexity of the design. See "Design Search Time" on page 111 in the "Custom Designs" chapter and "Number of Starts" on page 110 in the "Custom Designs" chapter.

If the iterations of the algorithm require more than a few seconds, a Computing Design progress window appears. If you click **Cancel** in the progress window, the calculation stops and gives the best design found at that point. The progress window also displays D-efficiency for D-optimal designs that do not include factors with Changes set to Hard or Very Hard or with Estimability set to If Possible.

Note: You can set a preference for Design Search Time. Select File > Preferences > Platforms > DOE. Check Design Search Time and enter the maximum number of seconds. In certain situations where more time is required, JMP extends the search time.

Sphere Radius Constrains the continuous factors in a design to a hypersphere. Specify the radius and click **OK**. Design points are chosen so that their distance from 0 equals the Sphere Radius. Select this option before you click Make Design.

Note: Sphere Radius constraints cannot be combined with constraints added using the Specify Linear Constraints option. Also, the option is not available when hard-to-change factors are included (split-plot designs).

Advanced Options > Mixture Sum Set the sum of the mixture factors to any positive value. Use this option to keep a component of a mixture constant throughout an experiment.

Advanced Options > Split Plot Variance Ratio Specify the ratio of the variance of the random whole plot and the subplot variance (if present) to the error variance. Before setting this value, you must define a hard-to-change factor for your split-plot design, or hard and very-hard-to-change factors for your split-split-plot design. Then you can enter one or two positive numbers for the variance ratios, depending on whether you have specified a split-plot or a split-split-plot design.

Advanced Options > Prior Parameter Variance (Available only when the Model outline is available) Specify the weights that are used for factors whose Estimability is set to If Possible. The option updates to show the default weights when you click Make Design. Enter a positive number for each of the terms for which you want to specify a weight. The value that you enter is the square root of the reciprocal of the prior variance. A larger value represents a smaller variance and therefore more prior information that the effect is not active.

Bayesian D- or I-optimality is used in constructing designs with If Possible factors. The default values used in the algorithm are 0 for Necessary terms, 4 for interactions involving If Possible terms, and 1 for If Possible terms. For more details, see "The Alias Matrix" on page 690 in the "Technical Details" appendix and "Optimality Criteria" on page 123 in the "Custom Designs" chapter.

Advanced Options > D Efficiency Weight Specify the relative importance of D-efficiency to alias optimality in constructing the design. Select this option to balance reducing the variance of the coefficients with obtaining a desirable alias structure. Values should be between 0 and 1. Larger values give more weight to D-Efficiency. The default value is 0.5. This option has an effect only when you select Make Alias Optimal Design as your Optimality Criterion.

For the definition of D-efficiency, see "Optimality Criteria" on page 123 in the "Custom Designs" chapter. For details about alias optimality, see "Alias Optimality" on page 127 in the "Custom Designs" chapter.

Advanced Options > Set Delta for Power Specify the difference in the mean response that you want to detect for model effects. See "Set Delta for Power" on page 112 in the "Custom Designs" chapter.

Save Script to Script Window Creates the script for the design that you specified in the Custom Design window and places it in an open script window.

Chapter 7

Definitive Screening Designs

Definitive screening designs work for factor screening if you have any combination of continuous or two-level categorical factors. These designs are particularly useful if you suspect active two-factor interactions, or you suspect that a plot of a continuous factor's effect on the response might exhibit strong curvature.

Definitive screening designs are small designs. For six or more factors, there are only about twice as many runs as factors. Yet, they often conclusively identify which of several factors affect the response. In particular, they can detect and identify any factors causing strong nonlinear effects on the response.

Here are areas where definitive screening designs are superior to standard screening designs:

- They help identify the causes of nonlinear effects by fielding each continuous factor at three levels. In standard screening designs, continuous factors have only two levels. You can add center points to screening designs, but these points establish only if curvature exists. They do not allow you to identify the factors responsible for quadratic effects.
- They avoid confounding between any effects up through the second order. For continuous factors, definitive screening designs have main effects that are orthogonal to each other and orthogonal to two-factor interactions and quadratic effects. Two-factor interactions are not completely confounded with each other. Confounding occurs in many standard screening designs with a similar number of runs.
- They avoid the need for costly additional experimentation to resolve ambiguity from the initial results of standard screening designs.

Figure 7.1 Plot of Response against Factor Values Showing Curvature

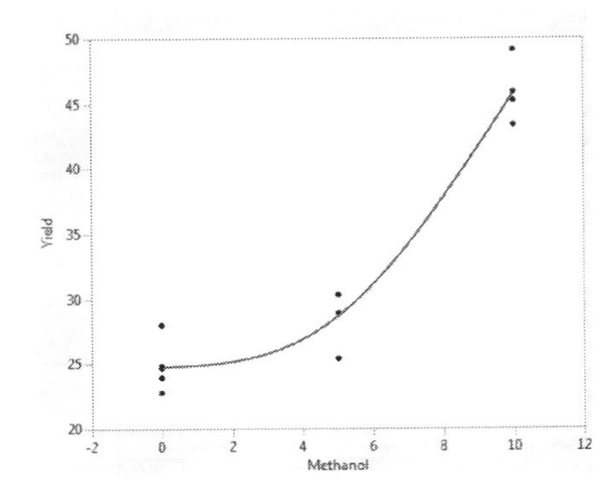

Overview of Definitive Screening Design

Investigators use screening designs when they want to identify the factors that have the most substantial effects on a response. A screening design enables you to study a large number of factors in a fairly small experiment.

Many standard screening designs focus on estimating main effects. Definitive screening designs offer advantages over standard screening designs. They avoid confounding of effects and can identify factors having a nonlinear effect on the response. For details about the advantages and construction of definitive screening designs, see Jones and Nachtsheim (2011).

For designs containing only continuous factors, compare these properties of definitive screening designs versus standard screening designs:

Note: When quadratic effects are mentioned, the standard screening designs are assumed to have center points.

- Main effects are orthogonal to two-factor interactions.
  - Definitive Screening Designs: Always
  - Standard Screening Designs: Only for Resolution IV or higher
- No two-factor interaction is completely confounded with any other two-factor interaction.
  - Definitive Screening Designs: Always
  - Standard Screening Designs: Only for Resolution V or higher
- All quadratic effects are estimable in models containing only main and quadratic effects.
  - Definitive Screening Designs: Always
  - Standard Screening Designs: Never

These properties are described more fully in the remainder of this section.

Standard Screening Designs

Standard screening designs, such as fractional factorial or Plackett-Burman designs, attempt to study many factors with a relatively small allocation of resources. However, standard screening designs have several undesirable features:

- They can alias some main effects with two-factor interactions. In Plackett-Burman designs, for example, main effects are correlated with several two-factor interactions. If one or more two-factor interaction effects are substantial, then the experimenter must perform additional runs to resolve the ambiguities.
- They can also confound some two-factor interactions with each other. Consequently, if a two-factor interaction effect is substantial, then the experimenter must perform additional runs to resolve the remaining ambiguities.

- Continuous factors are usually set at only two levels (low and high). However, engineers and scientists often prefer designs where continuous factors are set at three levels (low, middle, and high). This is because two levels are not sufficient to detect nonlinearity, which is common in physical systems. You can use a traditional screening design with added center points to detect nonlinearity, but such a design does not identify the responsible factors.

Definitive Screening Designs

Using definitive screening designs, you can do the following:

- Avoid model ambiguity, enabling you to identify important factors more quickly and efficiently.
- Identify the cause of nonlinear effects while avoiding confounding any terms up to second order. So not only can you *detect* nonlinearity, as you might with center points in a traditional screening design, but you can *identify* the responsible factors.

Definitive screening designs offer the following advantages:

- Definitive screening designs require only a small number of runs. For six or more factors, the number of required runs is usually only a few more than twice the number of factors. For more detail on the number of runs, see "Conference Matrices and the Number of Runs" on page 261.
- Main effects are orthogonal to two-factor interactions. This means that estimates of main effects are not biased by the presence of active two-factor interactions, whether these interactions are included in the model or not. Note that resolution III screening designs confound some main and interaction effects. Also, Plackett-Burman designs produce biased main effect estimates if there are active two-factor interactions.
- No two-factor interaction is completely confounded with any other two-factor interaction. However, a two-factor interaction might be correlated with other two-factor interactions. Note that resolution IV screening designs completely confound some two-factor interaction effects.
- All quadratic effects are estimable in models comprised only of main effects and quadratic terms. This enables you to identify the factors that account for nonlinearity. Note that traditional screening designs with added center points do *not* allow estimation of all quadratic effects in models consisting of main and quadratic effects.
- Quadratic effects are orthogonal to main effects and not completely confounded with two-factor interactions. A quadratic effect might be correlated with interaction effects.
- For 6 through at least 30 factors, it is possible to estimate the parameters of any full quadratic model involving three or fewer factors with high precision.
- For 18 factors or more, they can fit full quadratic models in any 4 factors. For 24 factors or more, they can fit full quadratic models in any 5 factors.

Definitive Screening Design Platform

The Definitive Screening Design platform enables you to construct definitive screening designs for continuous factors and for two-level categorical factors. It also enables you to construct blocked designs. You can add extra non-center runs that enhance the ability of the design to reliably detect effects when many effects are active.

To view the absolute values of the correlations among effects, use the Color Map on Correlations provided as part of the Design Evaluation outline in the Definitive Screening Design window. You can compare the aliasing structure of definitive screening designs to that of other designs by comparing their color maps on correlations. For details, see "Color Map on Correlations" on page 448 in the "Evaluate Designs" chapter.

For details, see "Structure of Definitive Screening Designs" on page 260. For information about definitive screening designs with blocks, see "Blocking in Definitive Screening Designs" on page 251. For suggestions on how to analyze data obtained using definitive screening designs, see "Analysis of Experimental Data" on page 262.

Fit Definitive Screening Platform

After you run a Definitive Screening Design (DSD), analyze your results using the Fit Definitive Screening platform. Standard model selection methods applied to DSDs can fail to identify active effects. To identify active main effects and second-order effects, the Fit Definitive Screening platform uses an algorithm called *Effective Model Selection for DSDs*. This algorithm leverages the special structure of DSDs. See Chapter 8, "The Fit Definitive Screening Platform".

If you create your DSD in JMP, the design table contains a script called Fit Definitive Screening that automatically runs an analysis using the Effective Model Selection for DSDs methodology.

Examples of Definitive Screening Designs

This section contains the following examples:

- "Definitive Screening Design" on page 235
- "Comparison with a Fractional Factorial Design" on page 237
- "Definitive Screening Design with Blocking" on page 239
- "Comparison of a Definitive Screening Design with a Plackett-Burman Design" on page 242

Definitive Screening Design

Suppose that you need to determine which of six factors have an effect on the yield of an extraction process.

Create the Design

The factors and their settings are given in the data table Extraction Factors.jmp. You create a definitive screening design to investigate.

1. Select **DOE > Definitive Screening > Definitive Screening Design.**
2. Double-click Y under Response Name and type Yield.
3. From the red triangle menu, select **Load Factors.**
4. Open the Extraction Factors.jmp sample data table, located in the Design Experiment folder.

   The factor names and ranges are added to the Factors outline.

Figure 7.2 Responses and Factors for Extraction Design

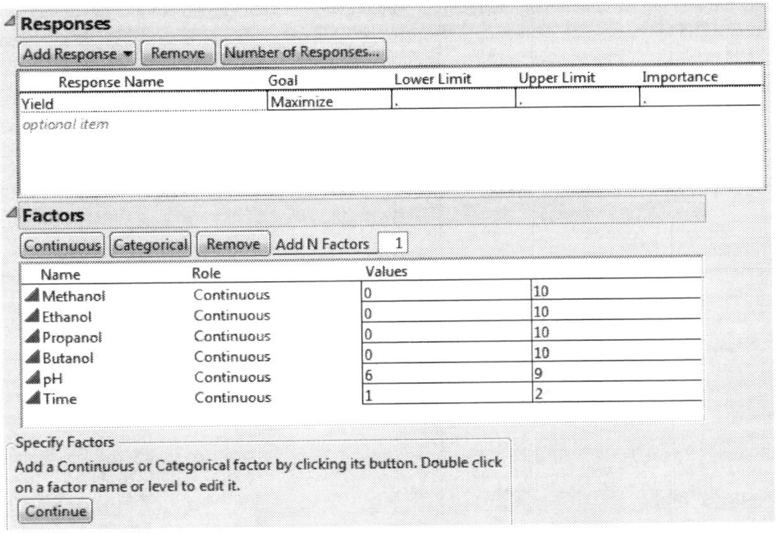

5. Click **Continue.**

   The Design Options outline opens. Here you can specify a blocking structure. There is no need to block in this example, so you accept the default selection of **No Blocks Required.**

   You can also choose to add Extra Runs, which greatly enhance your ability to detect second-order effects. A minimum of four Extra Runs is highly recommended and is the default.
6. Click **Make Design.**

The Definitive Screening Design window updates to show a Design outline and a Design Evaluation outline.

Figure 7.3 Design Outline

Design

Run	Methanol	Ethanol	Propanol	Butanol	pH	Time
1	5	10	10	10	9	2
2	5	0	0	0	6	1
3	10	5	10	10	6	2
4	0	5	0	0	9	1
5	10	0	5	10	9	1
6	0	10	5	0	6	2
7	10	0	0	5	9	2
8	0	10	10	5	6	1
9	10	10	0	0	7.5	2
10	0	0	10	10	7.5	1
11	10	0	10	0	6	1.5
12	0	10	0	10	9	1.5
13	10	10	0	10	6	1
14	0	0	10	0	9	2
15	10	10	10	0	9	1
16	0	0	0	10	6	2
17	5	5	5	5	7.5	1.5

7. Open the **Design Evaluation > Color Map on Correlations** outline.

   The Color Map on Correlations assigns a color intensity scale to the absolute values of correlations among all main effects and two-factor interactions.

Figure 7.4 Color Map on Correlations for Extraction Design

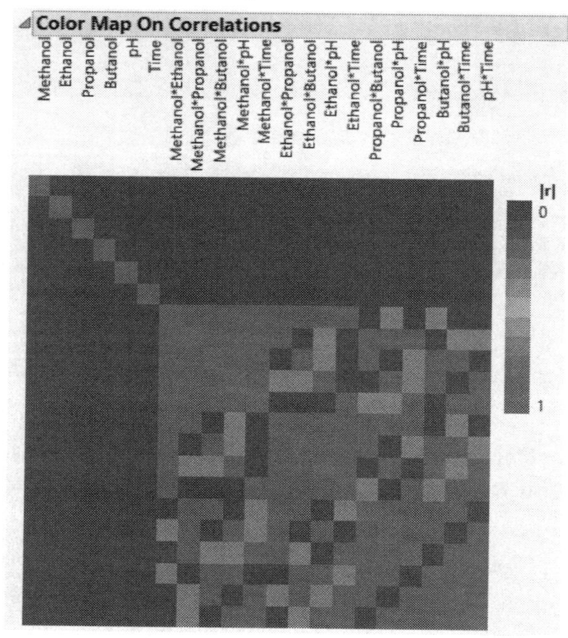

Note the following:

- The solid deep blue area shows that there is no correlation between main effects or between main effects and two-factor interactions.
- The lighter blue and gray areas indicate that the absolute correlations between two-factor interactions are small.
- The solid red squares indicate absolute correlations of 1. These all appear on the diagonal, reflecting the expected correlation of an effect with itself.

In the Output Options panel, note that the Run Order is set to **Randomize**.

8. Click **Make Table** to obtain the data table shown in Figure 7.5.

Note: The runs in your design might appear in a different order than the order shown in Figure 7.5.

Figure 7.5 Definitive Screening Design for Extraction Process

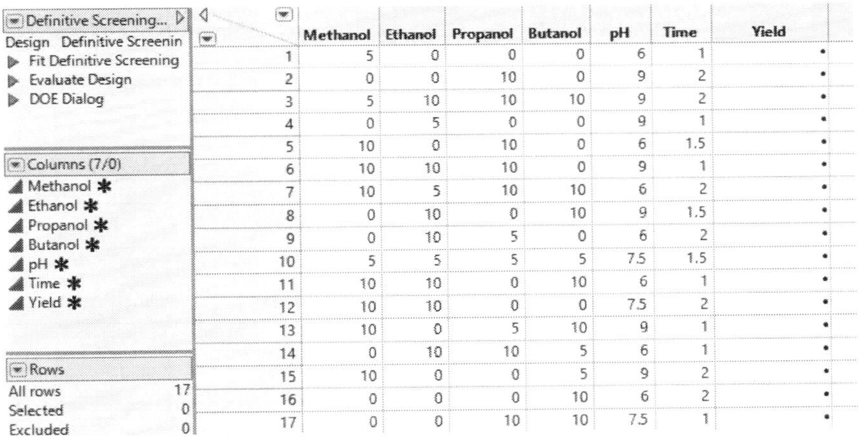

Definitive Screening...
Design Definitive Screenin
▶ Fit Definitive Screening
▶ Evaluate Design
▶ DOE Dialog

Columns (7/0)
Methanol *
Ethanol *
Propanol *
Butanol *
pH *
Time *
Yield *

Rows
All rows 17
Selected 0
Excluded 0

	Methanol	Ethanol	Propanol	Butanol	pH	Time	Yield
1	5	0	0	0	6	1	•
2	0	0	10	0	9	2	•
3	5	10	10	10	9	2	•
4	0	5	0	0	9	1	•
5	10	0	10	0	6	1.5	•
6	10	10	10	0	9	1	•
7	10	5	10	10	6	2	•
8	0	10	0	10	9	1.5	•
9	0	10	5	0	6	2	•
10	5	5	5	5	7.5	1.5	•
11	10	10	0	10	6	1	•
12	10	10	0	0	7.5	2	•
13	10	0	5	10	9	1	•
14	0	10	10	5	6	1	•
15	10	0	0	5	9	2	•
16	0	0	0	10	6	2	•
17	0	0	10	10	7.5	1	•

Comparison with a Fractional Factorial Design

Suppose that you had chosen a traditional screening design instead of the definitive screening design in "Definitive Screening Design" on page 235. This example compares the two designs in terms of confounding.

1. Select **DOE > Classical > Screening Design**.
2. Double-click Y under Response Name and type Yield.
3. Select **Load Factors** From the red triangle next to Definitive Screening Design.
4. Open the Extraction Factors.jmp sample data table, located in the Design Experiment folder.

   The factor names and ranges are added to the Factors outline.

5. Click **Continue.**
6. Select **Choose from a list of fractional factorial designs.**
7. Click **Continue.**

   Potential designs appear in the Design List.

Figure 7.6 Screening Design List for Six Continuous Factors

Design List

Choose a design by clicking on its row in the list.

Number Of Runs	Block Size	Design Type	Resolution - what is estimable
8		Fractional Factorial	3 - Main Effects Only
8	4	Fractional Factorial	3 - Main Effects Only
12		Plackett-Burman	3 - Main Effects Only
16		Fractional Factorial	4 - Some 2-factor interactions
16	8	Fractional Factorial	4 - Some 2-factor interactions
16	4	Fractional Factorial	4 - Some 2-factor interactions
16	2	Fractional Factorial	4 - Some 2-factor interactions
32		Fractional Factorial	5+ - All 2-factor interactions
32	16	Fractional Factorial	5+ - All 2-factor interactions
32	8	Fractional Factorial	4 - Some 2-factor interactions
32	4	Fractional Factorial	4 - Some 2-factor interactions
32	2	Fractional Factorial	4 - Some 2-factor interactions
64		Full Factorial	>6 - Full Resolution
64	32	Full Factorial	5+ - All 2-factor interactions
64	16	Full Factorial	5+ - All 2-factor interactions
64	8	Full Factorial	5+ - All 2-factor interactions
64	4	Full Factorial	4 - Some 2-factor interactions
64	2	Full Factorial	4 - Some 2-factor interactions

8. Select the sixteen-run fractional factorial design with no blocks, shown highlighted in Figure 7.6.
9. Click **Continue.**
10. Open the **Display and Modify Design > Aliasing of Effects** outline.

Figure 7.7 Aliasing of Effects for Fractional Factorial Design

Aliasing of Effects

Effects	Aliases
Methanol*Ethanol	= pH*Time
Methanol*Propanol	= Butanol*Time
Methanol*Butanol	= Propanol*Time
Methanol*pH	= Ethanol*Time
Methanol*Time	= Ethanol*pH = Propanol*Butanol
Ethanol*Propanol	= Butanol*pH
Ethanol*Butanol	= Propanol*pH

The Aliasing of Effects outline for the 16-run fractional factorial design shows that every two-factor interaction is confounded with at least one other two-factor interaction. In this fractional factorial design, the Ethanol*Time interaction is confounded with Methanol*pH. To determine which interaction is active, you need to run additional trials. If the factors had been entered in a different order, the Ethanol*Time interaction might have been aliased with two other two-factor interactions.

In the section "Definitive Screening Design" on page 235, you constructed a 17-run definitive screening design. The Color Map on Correlations for this DSD (Figure 7.4) shows that no two-factor interactions are confounded with any other two-factor interactions. For the fractional factorial design, there are seven instances of confounded two-factor interactions. If you suspect that there are active two-factor effects, the DSD is the better choice.

You can conduct a more thorough comparison of the two designs using the Compare Designs platform (DOE > Design Diagnostics > Compare Designs). See Chapter 16, "Compare Designs".

Definitive Screening Design with Blocking

Suppose that, due to raw material constraints, the extraction experiment requires that you run it using material from two separate lots. You can generate a definitive screening design with a blocking variable to account for the potential lot variation.

Create the Design

The extraction factors and their settings are given in the data table Extraction Factors.jmp. Generate a definitive screening design with a block as follows.

1. Select **DOE > Definitive Screening > Definitive Screening Design**.
2. Double-click Y under Response Name and type Yield.
3. From the red triangle menu, select **Load Factors**.
4. Open the Extraction Factors.jmp sample data table, located in the Design Experiment folder.

   The factor names and ranges are added to the Factors outline.
5. Click **Continue**.

   The Design Options outline opens. Here you can specify a blocking structure.
6. Select **Add Blocks with Center Runs to Estimate Quadratic Effects**.

   Leave **Number of Blocks** set at 2.

   You are recreating the design for the Extraction2 Data.jmp sample data table, which was created without Extra Runs. Although four Extra Runs are strongly recommended, you will not add Extra Runs in this example.
7. Next to **Number of Extra Runs**, select 0.
8. Click **Make Design**.

The Definitive Screening Design window updates to show a Design outline and a Design Evaluation outline.

Check that Block has been added to the Factors outline and to the Design.

9. In the Factors outline, Double-click Block and type Lot.

   In the Output Options panel, note that the Run Order is set to Randomize within Blocks.

10. Click **Make Table.**

Note: The runs in your design might appear in a different order than the order shown in Figure 7.8.

Figure 7.8 Definitive Screening Design with Block for Extraction Process

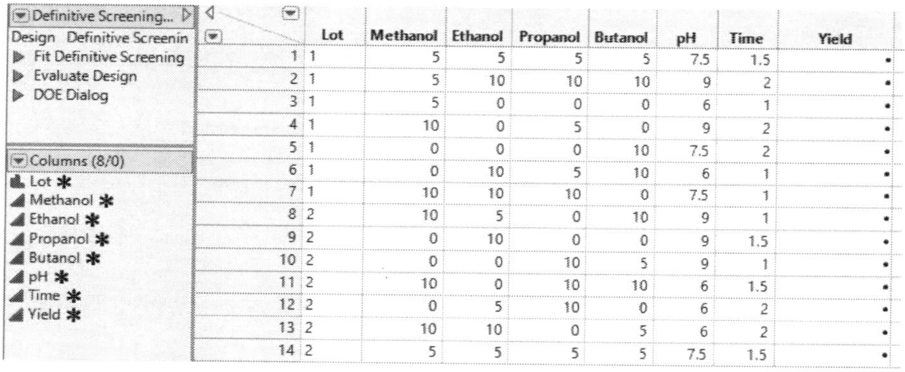

Definitive Screening...
Design Definitive Screenin
▶ Fit Definitive Screening
▶ Evaluate Design
▶ DOE Dialog

Columns (8/0)
Lot *
Methanol *
Ethanol *
Propanol *
Butanol *
pH *
Time *
Yield *

	Lot	Methanol	Ethanol	Propanol	Butanol	pH	Time	Yield
1	1	5	5	5	5	7.5	1.5	•
2	1	5	10	10	10	9	2	•
3	1	5	0	0	0	6	1	•
4	1	10	0	5	0	9	2	•
5	1	0	0	0	10	7.5	2	•
6	1	0	10	5	10	6	1	•
7	1	10	10	10	0	7.5	1	•
8	2	10	5	0	10	9	1	•
9	2	0	10	0	0	9	1.5	•
10	2	0	0	10	5	9	1	•
11	2	10	0	10	10	6	1.5	•
12	2	0	5	10	0	6	2	•
13	2	10	10	0	5	6	2	•
14	2	5	5	5	5	7.5	1.5	•

Notice that run 1 is a center point run in Lot 1 and run 14 is a center point run in Lot 2.

Analyze the Experimental Data

At this point, you conduct your experiment and record your data in the Yield column of the design table (Figure 7.8). The Extraction2 Data.jmp sample data table contains your experimental results. The runs in the Extraction2 Data.jmp sample data table are in a different order than those in Figure 7.8.

To explore all second-order effects, one option is to use All Possible Models regression. Another option is to use forward stepwise regression. However, these standard methods often fail to identify active effects. For this reason, you use the Fit Definitive Screening platform.

1. Select **Help > Sample Data Library** and open Design Experiment/Extraction2 Data.jmp.
2. In the Table panel of the design table, click the green triangle next to the **Fit** Definitive Screening script.

Figure 7.9 Fit Definitive Screening Report

Fit Definitive Screening for Yield

Stage 1 - Main Effect Estimates

Term	Estimate	Std Error	t Ratio	Prob>|t|
Methanol	9.831	0.4106	23.941	0.0002*
Ethanol	2.865	0.4106	6.9769	0.0060*
Time	4.476	0.4106	10.9	0.0017*

Statistic	Value
RMSE	1.2986
DF	3

Stage 2 - Even Order Effect Estimates

Term	Estimate	Std Error	t Ratio	Prob>|t|
Intercept	29.749	0.6758	44.017	0.0005*
Lot[1]	12.335	0.736	16.759	0.0035*
Ethanol*Time	6.9911	0.8436	8.2869	0.0143*
Methanol*Methanol	-0.686	1.6491	-0.416	0.7179
Ethanol*Ethanol	4.6532	0.9939	4.6817	0.0427*
Time*Time	4.0582	0.9939	4.083	0.0551

Statistic	Value
RMSE	1.0409
DF	2

Combined Model Parameter Estimates

Term	Estimate	Std Error	t Ratio	Prob>|t|
Intercept	29.749	0.7805	38.113	<.0001*
Lot[1]	12.335	0.85	14.511	<.0001*
Methanol	9.831	0.3802	25.861	<.0001*
Ethanol	2.865	0.3802	7.5365	0.0007*
Time	4.476	0.3802	11.774	<.0001*
Ethanol*Time	6.9911	0.9743	7.1755	0.0008*
Methanol*Methanol	-0.686	1.9045	-0.36	0.7335
Ethanol*Ethanol	4.6532	1.1479	4.0538	0.0098*
Time*Time	4.0582	1.1479	3.5354	0.0166*

Statistic	Value
RMSE	1.2021
DF	5

Make Model | Run Model

The effects identified by Fit Definitive Screening as potentially active are listed in the Combined Model Parameter Estimates report.

3. Click the **Run Model** button at the bottom of the Combined Model Parameter Estimates report.

   This fits a standard least squares model to the effects identified as potentially active.

Figure 7.10 Standard Least Squares Report

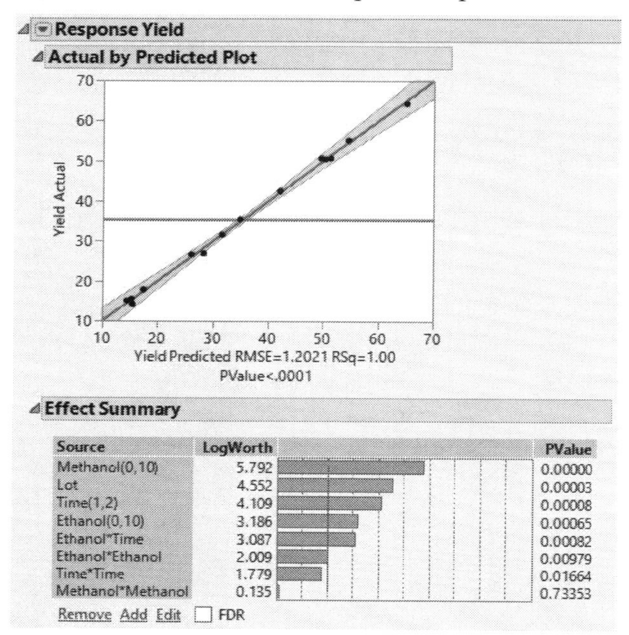

Source	LogWorth	PValue
Methanol(0,10)	5.792	0.00000
Lot	4.552	0.00003
Time(1,2)	4.109	0.00008
Ethanol(0,10)	3.186	0.00065
Ethanol*Time	3.087	0.00082
Ethanol*Ethanol	2.009	0.00979
Time*Time	1.779	0.01664
Methanol*Methanol	0.135	0.73353

The Actual by Predicted Plot shows no evidence of lack of fit. The Effect Summary report shows that Methanol*Methanol is not significant. You decide to remove this effect from the model.

4. Select Methanol*Methanol in the Effect Summary report and click **Remove**.

Figure 7.11 Final Set of Active Effects

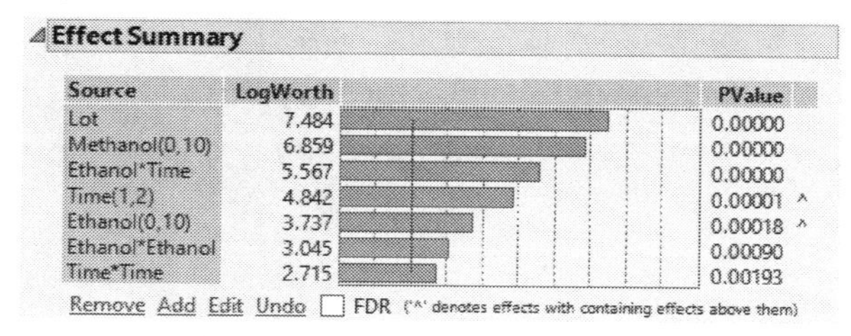

Source	LogWorth	PValue
Lot	7.484	0.00000
Methanol(0,10)	6.859	0.00000
Ethanol*Time	5.567	0.00000
Time(1,2)	4.842	0.00001 ^
Ethanol(0,10)	3.737	0.00018 ^
Ethanol*Ethanol	3.045	0.00090
Time*Time	2.715	0.00193

The remaining effects are all significant. You conclude that these are the active effects.

Comparison of a Definitive Screening Design with a Plackett-Burman Design

Plackett-Burman designs are an alternative to fractional factorials for screening. However, Plackett-Burman designs have complex aliasing of the main effects by two-factor interactions.

This example shows how to compare a definitive screening with a Plackett-Burman design using the Evaluate Design platform. For an extensive example using the Compare Designs platform, see "Designs of Same Run Size" on page 454 in the "Compare Designs" chapter.

The Definitive Screening Design

1. Select **DOE > Definitive Screening > Definitive Screening Design.**
2. Type **4** in the **Add N Factors** box and click **Continuous**.
3. Type **2** in the **Add N Factors** box and click **Categorical.**

   Your window should appear as shown in Figure 7.12.

Figure 7.12 Definitive Screening Dialog with 4 Continuous and 2 Categorical Factors

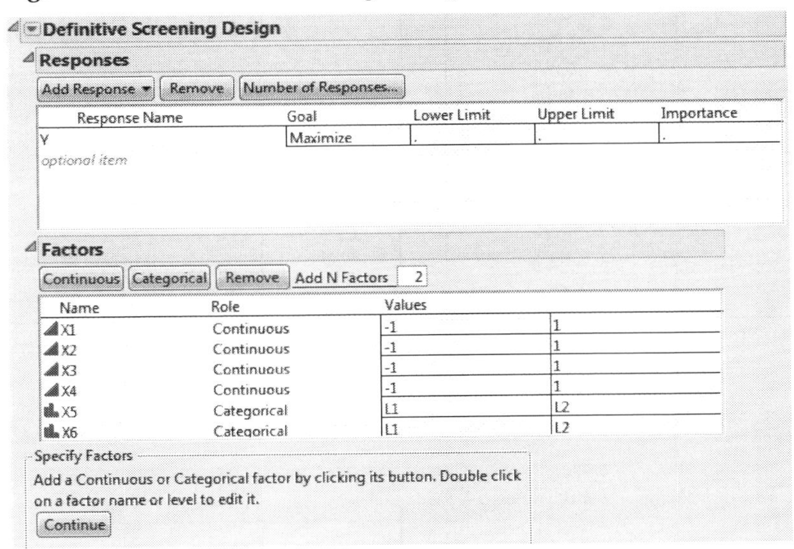

4. Click **Continue.**

   This example does not require a block. Under the Design Options Outline, check that the No Blocks Required option is selected.

   In order to compare designs of approximately equal sizes. do not add Extra Runs.
5. Next to **Number of Extra Runs**, select 0.
6. Click **Make Design**.

   The design that is generated has 14 runs.
7. Open the **Design Evaluation > Color Map On Correlations** outline.

Figure 7.13 Color Map for Definitive Screening Design

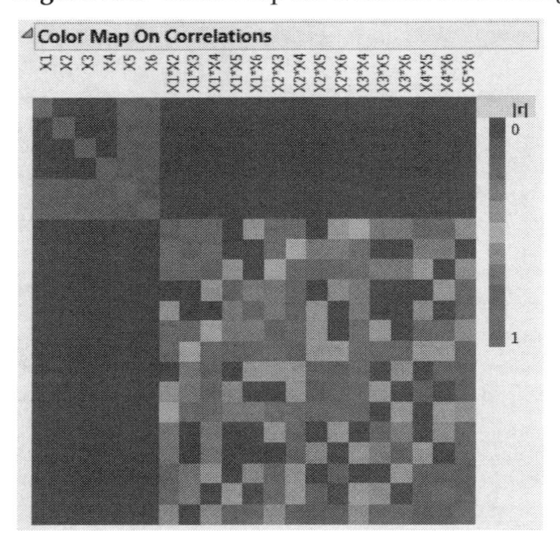

Notice that the categorical main effects have small correlations with each other and with the continuous factors' main effects. These correlations lead to a small reduction in the precision of the estimates.

8. Do not close your Definitive Design Screening window until you compare the color map with that of the Plackett-Burman design, below.

The Plackett-Burman Design

Now create a Plackett-Burman design using the same factor structure.

1. Select **DOE > Classical > Screening Design.**
2. Type **4** in the **Add N Factors** box and click **Continuous**.
3. Type **2** in the **Add N Factors** box and click **Categorical > 2 Level.**
4. Click **Continue**.
5. Select **Choose from a list of fractional factorial designs** and click **Continue**.

   Potential designs appear in the Design List, shown in Figure 7.14.

Figure 7.14 Plackett-Burman Design

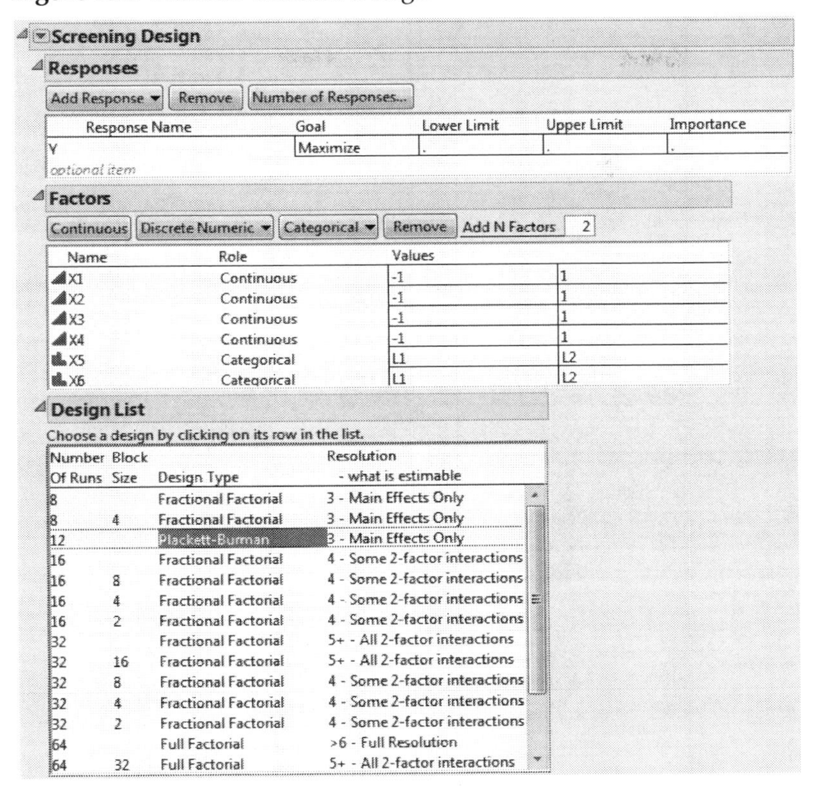

6. Select the 12 run Plackett-Burman design. See Figure 7.14.
7. Click **Continue.**
8. Open the **Design Evaluation > Color Map On Correlations** outline.

   Compare the color map for the 12-run Plackett-Burman design to the color map for the 14-run definitive screening design.

Figure 7.15 Plackett-Burman Correlations (left) and Definitive Screening Correlations (right)

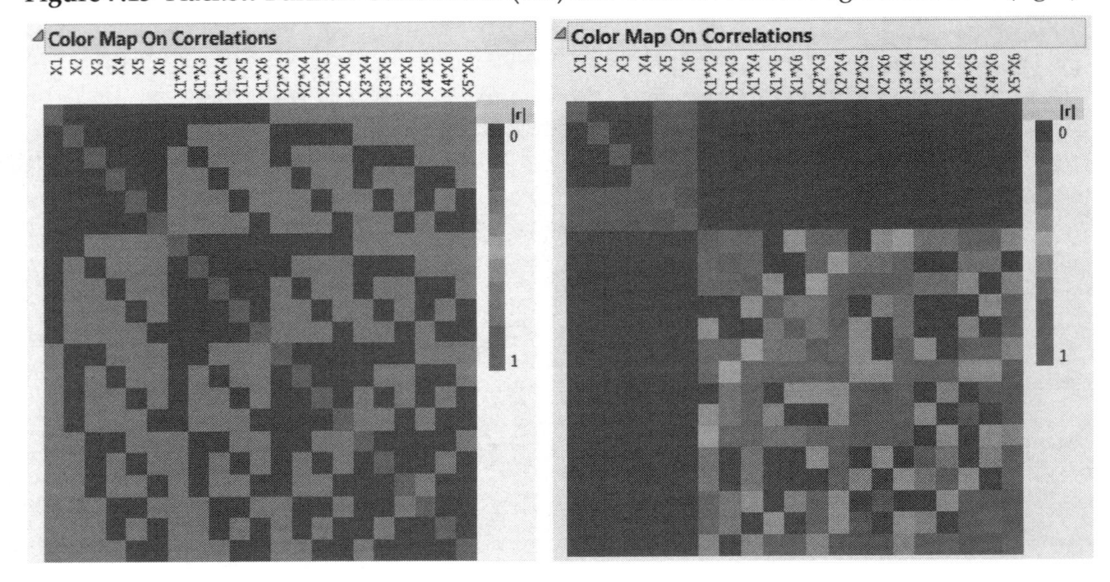

Figure 7.15 shows both color maps, but shows only the portion of the Plackett-Burman color map that involves main effects and two-way interactions. (To construct the color map for the Plackett-Burman design without the three-way interactions, construct the design. Then obtain the color map using Evaluate Design.)

In the color map for the Plackett-Burman design on the left, you see that most two-factor interactions are correlated with main effects. This means that any non-negligible two-factor interaction will bias several main effects. This can lead to a failure to identify an active main effect or the false conclusion that an inactive main effect is active.

Contrast this with the color map for the definitive screening design on the right. With only two additional runs, the definitive screening design trades off a small increase in the variance of the main effects for complete independence of main effects and two-factor interactions.

Definitive Screening Design Window

The definitive screening design window updates as you work through the design steps. For more information, see "The DOE Workflow: Describe, Specify, Design" on page 54 in the "Starting Out with DOE" chapter. The outlines, separated by buttons that update the outlines, follow the flow in Figure 7.16.

Figure 7.16 Definitive Screening Design Flow

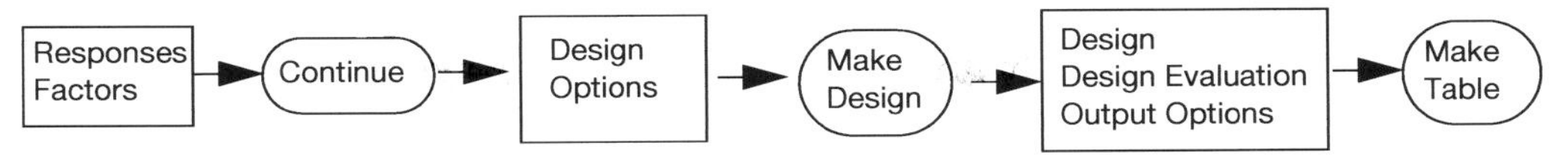

Responses

Use the Responses outline to specify one or more responses.

Tip: When you have completed the Responses outline, consider selecting **Save Responses** from the red triangle menu. This option saves the response names, goals, limits, and importance values in a data table that you can later reload in DOE platforms.

Figure 7.17 Responses Outline

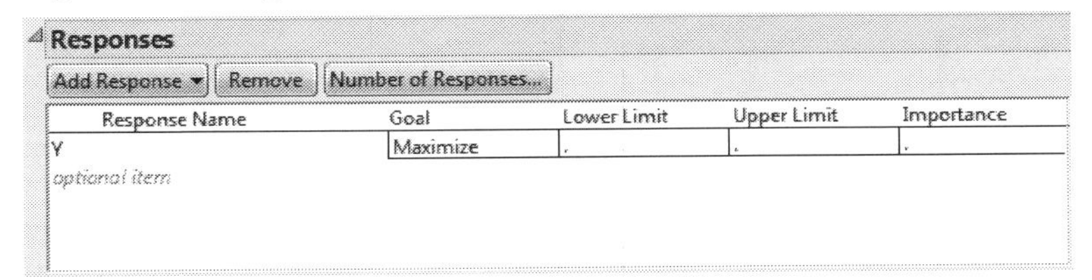

Add Response Enters a single response with a goal type of Maximize, Match Target, Minimize, or None. If you select Match Target, enter limits for your target value. If you select Maximize or Minimize, entering limits is not required but can be useful if you intend to use desirability functions.

Remove Removes the selected responses.

Number of Responses Enters additional responses so that the number that you enter is the total number of responses. If you have entered a response other than the default Y, the Goal for each of the additional responses is the Goal associated with the last response entered. Otherwise, the Goal defaults to Match Target. Click the Goal type in the table to change it.

The Responses outline contains the following columns:

Response Name The name of the response. When added, a response is given a default name of Y, Y2, and so on. To change this name, double-click it and enter the desired name.

Goal, Lower Limit, Upper Limit The Goal tells JMP whether you want to maximize your response, minimize your response, match a target, or that you have no response goal. JMP assigns a Response Limits column property, based on these specifications, to each response column in the design table. It uses this information to define a desirability function for each response. The Profiler and Contour Profiler use these desirability functions to find optimal factor settings. For further details, see the Profiler chapter in the *Profilers* book and "Response Limits" on page 653 in the "Column Properties" appendix.

– A Goal of Maximize indicates that the best value is the largest possible. If there are natural lower or upper bounds, you can specify these as the Lower Limit or Upper Limit.

– A Goal of Minimize indicates that the best value is the smallest possible. If there are natural lower or upper bounds, you can specify these as the Lower Limit or Upper Limit.

– A Goal of Match Target indicates that the best value is a specific target value. The default target value is assumed to be midway between the Lower Limit and Upper Limit.

– A Goal of None indicates that there is no goal in terms of optimization. No desirability function is constructed.

Note: If your target response is not midway between the Lower Limit and the Upper Limit, you can change the target after you generate your design table. In the data table, open the Column Info window for the response column (**Cols** > **Column Info**) and enter the desired target value.

Importance When you have several responses, the Importance values that you specify are used to compute an overall desirability function. These values are treated as weights for the responses. If there is only one response, then specifying the Importance is unnecessary because it is set to 1 by default.

Editing the Responses Outline

In the Responses outline, note the following:

- Double-click a response to edit the response name.
- Click the goal to change it.
- Click on a limit or importance weight to change it.
- For multiple responses, you might want to enter values for the importance weights.

Response Limits Column Property

The Goal, Lower Limit, Upper Limit, and Importance that you specify when you enter a response are used in finding optimal factor settings. For each response, the information is saved in the generated design data table as a Response Limits column property. JMP uses this information to define the desirability function. The desirability function is used in the Prediction Profiler to find optimal factor settings. For further details about the Response Limits column property and examples of its use, see "Response Limits" on page 653 in the "Column Properties" appendix.

If you do not specify a Lower Limit and Upper Limit, JMP uses the range of the observed data for the response to define the limits for the desirability function. Specifying the Lower Limit and Upper Limit gives you control over the specification of the desirability function. For more details about the construction of the desirability function, see the Profiler chapter in the *Profilers* book.

Factors

Add factors in the Factors outline.

Figure 7.18 Factors Outline

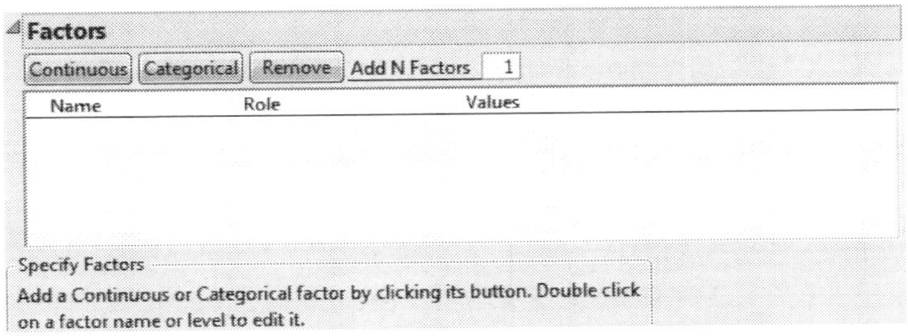

The factors outline contains the following buttons.

Continuous Enters the number of continuous factors specified in **Add N Factors**.

Categorical Enters the number of nominal factors specified in **Add N Factors.**

Remove Removes the selected factors.

Add N Factors Adds multiple factors of a given type. Enter the number of factors to add and click Continuous or Categorical. Repeat **Add N Factors** to add multiple factors of different types.

Tip: When you have completed your Factors panel, select Save Factors from the red triangle menu. This saves the factor names and values in a data table that you can later reload. See "Definitive Screening Design Options" on page 255.

The Factors outline contains the following columns:

Name The name of the factor. When added, a factor is given a default name of X1, X2, and so on. To change this name, double-click it and enter the desired name.

Role Specifies the Design Role of the factor. The Design Role column property for the factor is saved to the data table. This property ensures that the factor type is modeled appropriately.

Values The experimental settings for the factors. To insert Values, click on the default values and type the desired values.

Editing the Factors Outline

In the Factors outline, note the following:

- To edit a factor name, double-click the factor name.
- To edit a value, click the value in the **Values** column.

Factor Types

Continuous Numeric data types only. A continuous factor is a factor that you can conceptually set to any value between the lower and upper limits you supply, given the limitations of your measurement system.

Categorical Either numeric or character data types with two levels. For a categorical factor, the value ordering is the order of the values, as entered from left to right. This ordering is saved in the Value Ordering column property after the design data table is created.

Factor Column Properties

For each factor, various column properties are saved to the data table.

Design Role Each factor is assigned the Design Role column property. The Role that you specify in defining the factor determines the value of its Design Role column property. When you add a block under Design Options, that factor is assigned the Blocking value. The Design Role property reflects how the factor is intended to be used in modeling the experimental data. Design Role values are used in the Augment Design platform.

Factor Changes Each factor is assigned the Factor Changes column property with a setting of Easy. In definitive screening designs, it is assumed that factor levels can be changed for each experimental run. Factor Changes values are used in the Evaluate Design and Augment Design platforms.

Coding If the Design Role is Continuous, the Coding column property for the factor is saved. This property transforms the factor values so that the low and high values correspond to –1 and +1, respectively. The estimates and tests in the Fit Least Squares report are based on the transformed values.

Value Ordering If the Design Role is Categorical or Blocking, the Value Ordering column property for the factor is saved. This property determines the order in which levels of the factor appear.

RunsPerBlock Indicates the number of runs in each block. When a Block is selected in the Design Options outline and you then click Make Design, a factor with the default name Block is added to the Factors list. The RunsPerBlock column property is saved for that factor.

Design Options

The Design Options outline enables you to specify the blocking structure, the number of blocks, and the number of extra runs. Block effects are orthogonal to the main effects. Block sizes need not be equal.

The outline contains the following options:

No Blocks Required Indicates that the design will not contain a blocking factor. This is the default selection.

Add Blocks with Center Runs to Estimate Quadratic Effects Adds the number of blocks specified in the Number of Blocks text box. Constructs a design where block effects are orthogonal to main effects and where the model consisting of all main and quadratic effects is estimable. For details, see "Add Blocks with Center Runs to Estimate Quadratic Effects" on page 252

Add Blocks without Extra Center Runs Adds the number of blocks specified in the Number of Blocks text box. Adds only as many center runs as required by the design structure. Constructs a design where block effects are orthogonal to main effects, but the model consisting of all main effects and quadratic effects might *not* be estimable. For details, see "Add Blocks without Extra Center Runs" on page 252.

> **Note:** Use the **Add Blocks without Extra Center Runs** option only if you can assume that not all quadratic effects are important.

Number of Blocks Indicates the number of blocks to add. The number of blocks that you can add ranges from two to the number of factors.

Number of Extra Runs Adds non-center runs that enable you to conduct *effective model selection*. See "Extra Runs" on page 261 and "Effective Model Selection for DSDs" on page 266 in the "The Fit Definitive Screening Platform" chapter.

> **Tip:** Adding runs to your design with the Extra Runs option enhances your ability to detect effects in the presence of many active effects. The recommended number of Extra Runs is four, which dramatically improves the power of the design to identify active second-order effects.

Make Design Generates the design, presents it in the Design outline, and provides evaluation information in the Design Evaluation outline. The Output Options panel also appears, enabling you to create the design table.

Blocking in Definitive Screening Designs

This section describes the two blocking options:

- "Add Blocks with Center Runs to Estimate Quadratic Effects" on page 252

- "Add Blocks without Extra Center Runs" on page 252

If a design contains a categorical factor, a *center run* is a run where all continuous factors are set at their middle values. If all factors are categorical, the two blocking options are available. Both options produce designs whose blocks are orthogonal to main effects.

Add Blocks with Center Runs to Estimate Quadratic Effects

Note: For details about the construction and properties of blocked designs that estimate quadratic effects, see Jones and Nachtsheim (2016). The paper also contains information about treating the blocks as random effects.

The **Add Blocks with Center Runs to Estimate Quadratic Effects** option constructs a design with these properties:

- Block effects are orthogonal to main effects.
- The model consisting of all main and quadratic effects is estimable.

If a design contains only continuous factors, a blocked design for k factors having these properties can be constructed as follows:

- Remove the center run from the DSD design for k factors.
- Assign conference matrix foldover pairs to the same block.
- Add one center run to each block.

When some factors are categorical, the **Add Blocks with Center Runs to Estimate Quadratic Effects** option adds pairs of center runs within certain blocks. This structure ensures orthogonality and the ability to estimate all main and quadratic effects.

Because the only requirement on block size is that a block contains a foldover pair, the number of blocks can range from 2 to k, if k is even and from 2 to $k+1$, if k is odd. See "Conference Matrices and the Number of Runs" on page 261. JMP attempts to construct blocks of equal size.

Add Blocks without Extra Center Runs

The **Add Blocks without Extra Center Runs** option constructs a design that has a single center run when all factors are continuous and two center runs when some factors are categorical. The resulting design has these properties:

- Block effects are orthogonal to main effects.
- Each block effect is confounded with a linear combination of quadratic effects. This implies that the model consisting of all main and quadratic effects is *not* estimable.

For this reason, use this option only if you can assume that some quadratic effects are negligible.

A blocked design for *k* factors without extra center runs is constructed as follows:

- Assign conference matrix foldover pairs to the same block.
- If all factors are continuous, assign the single center run to a single block.
- If there are categorical factors, the unblocked definitive screening design requires the addition of two center runs to the foldover pairs defined by the conference matrix. See "Conference Matrices and the Number of Runs" on page 261. To construct the blocked design without extra center runs, these two center runs are added to a single block.

Because the only requirement on block size is that a block contains a foldover pair, the number of blocks can range from 2 to *k*, if *k* is even and from 2 to *k+1*, if *k* is odd. See "Conference Matrices and the Number of Runs" on page 261. JMP attempts to construct blocks of equal size.

Design

The Design outline shows the runs for the definitive screening design. The runs are given in a standard order. To change the run order for your design table, you can select Run Order options in the Output Options panel before generating the table.

Note: Definitive screening designs for four or fewer factors are based on a five-factor design. See "Definitive Screening Designs for Four or Fewer Factors" on page 262.

Design Evaluation

The Design Evaluation outline provides a number of ways to evaluate the properties of the generated design. Open the Design Evaluation outline to see the following options:

Power Analysis Enables you to explore your ability to detect effects of given sizes.

Prediction Variance Profile Shows the prediction variance over the range of factor settings.

Fraction of Design Space Plot Shows how much of the model prediction variance lies below (or above) a given value.

Prediction Variance Surface Shows a surface plot of the prediction variance for any two continuous factors.

Estimation Efficiency For each parameter, gives the fractional increase in the length of a confidence interval compared to that of an ideal (orthogonal) design, which might not exist. Also gives the relative standard error of the parameters.

Alias Matrix Gives coefficients that indicate the degree by which the model parameters are biased by effects that are potentially active, but not in the model.

Color Map on Correlations Shows the absolute correlation between effects on a plot using an intensity scale.

Design Diagnostics Indicates the optimality criterion used to construct the design. Also gives efficiency measures for your design.

For more details about the Design Evaluation panel, see "Design Evaluation" on page 433 in the "Evaluate Designs" chapter.

Output Options

Use the Output Options panel to perform the following tasks:

- specify the order for the runs in the design data table
- construct the design table
- return to a previous point in the Definitive Screening Design window

Figure 7.19 Output Options Panel

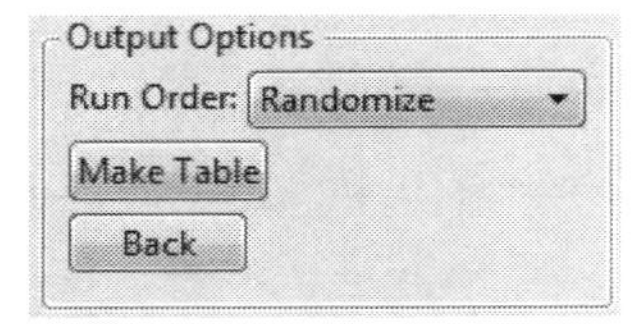

The Output Options panel contains these options:

Run Order

The **Run Order** options determine the order of the runs in the design table. Choices include the following:

Keep the Same Rows in the design table are in the same order as in the Design outline.

Sort Left to Right Columns in the design table are sorted from left to right.

Randomize Rows in the design table are in random order.

Sort Right to Left Columns in the design table are sorted from right to left.

Randomize within Blocks Rows in the design table are in random order within the blocks.

Make Table

Click **Make Table** to construct the Definitive Screening Design data table.

In the Definitive Screening Design table, the Table panel (in the upper left) contains the following scripts. To run a script, click the green triangle next to the script name.

Fit Definitive Screening Runs the **DOE** > **Definitive Screening** > **Fit Definitive Screening** platform. See Chapter 8, "The Fit Definitive Screening Platform".

Evaluate Design Runs the **DOE** > **Design Diagnostics** > **Evaluate Design** platform. See Chapter 15, "Evaluate Designs".

DOE Dialog Re-creates the Definitive Screening Design window that you used to generate the design table.

Figure 7.20 Definitive Screening Design Table Showing Scripts

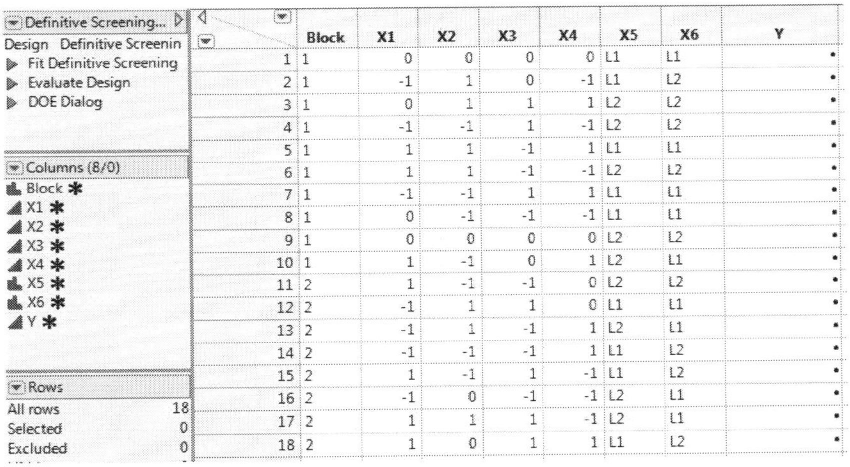

Definitive Screening...
Design Definitive Screenin
Fit Definitive Screening
Evaluate Design
DOE Dialog

Columns (8/0)
Block *
X1 *
X2 *
X3 *
X4 *
X5 *
X6 *
Y *

Rows
All rows 18
Selected 0
Excluded 0

	Block	X1	X2	X3	X4	X5	X6	Y
1	1	0	0	0	0	L1	L1	•
2	1	-1	1	0	-1	L1	L2	•
3	1	0	1	1	1	L2	L2	•
4	1	-1	-1	1	-1	L2	L2	•
5	1	1	1	-1	1	L1	L1	•
6	1	1	1	-1	-1	L2	L2	•
7	1	-1	-1	1	1	L1	L1	•
8	1	0	-1	-1	-1	L1	L1	•
9	1	0	0	0	0	L2	L2	•
10	1	1	-1	0	1	L2	L1	•
11	2	1	-1	-1	0	L2	L2	•
12	2	-1	1	1	0	L1	L1	•
13	2	-1	1	-1	1	L2	L1	•
14	2	-1	-1	-1	1	L1	L2	•
15	2	1	-1	1	-1	L1	L2	•
16	2	-1	0	-1	-1	L2	L1	•
17	2	1	1	1	-1	L2	L1	•
18	2	1	0	1	1	L1	L2	•

Back

The Back button takes you back to where you were before clicking Make Design. You can make changes to the previous outlines and regenerate the design.

Definitive Screening Design Options

The red triangle menu in the Definitive Screening Design platform contains these options:

Save Responses Saves the information in the Responses panel to a new data table. You can then quickly load the responses and their associated information into most DOE windows. This option is helpful if you anticipate re-using the responses.

Load Responses Loads responses that you saved using the Save Responses option.

Save Factors Saves the information in the Factors panel to a new data table. Each factor's column contains its levels. Other information is stored as column properties. You can then quickly load the factors and their associated information into most DOE windows.

Note: It is possible to create a factors table by entering data into an empty table, but remember to assign each column an appropriate Design Role. Do this by right-clicking on the column name in the data grid and selecting **Column Properties > Design Role**. In the Design Role area, select the appropriate role.

Load Factors Loads factors that you saved using the Save Factors option.

Save Constraints (Unavailable for some platforms) Saves factor constraints that you defined in the Define Factor Constraints or Linear Constraints outline into a data table, with a column for each constraint. You can then quickly load the constraints into most DOE windows.

In the constraint table, the first rows contain the coefficients for each factor. The last row contains the inequality bound. Each constraint's column contains a column property called ConstraintState that identifies the constraint as a "less than" or a "greater than" constraint. See "ConstraintState" on page 686 in the "Column Properties" appendix.

Load Constraints (Unavailable for some platforms) Loads factor constraints that you saved using the Save Constraints option.

Set Random Seed Sets the random seed that JMP uses to control certain actions that have a random component. These actions include the following:

– simulating responses using the Simulate Responses option
– randomizing Run Order for design construction
– selecting a starting design for designs based on random starts

To reproduce a design or simulated responses, enter the random seed that generated them. For designs using random starts, set the seed before clicking Make Design. To control simulated responses or run order, set the seed before clicking Make Table.

Note: The random seed associated with a design is included in the DOE Dialog script that is saved to the design data table.

Simulate Responses Adds response values and a column containing a simulation formula to the design table. Select this option before you click Make Table.

When you click Make Table, the following occur:

– A set of simulated response values is added to each response column.
– A Model window opens where you can set the values of coefficients for model effects and specify one of three distributions: Normal, Binomial, or Poisson.

– A script called **DOE Simulate** is saved to the design table. This script re-opens the Model window, enabling you to re-simulate values or to make changes to the simulated response distribution.

Make selections in the Model window to control the distribution of simulated response values. When you click Apply, a formula for the simulated response values is saved in a new column called <Y> **Simulated**, where Y is the name of the response.

For additional details, see "Simulate Responses" on page 257.

Note: JMP PRO You can use Simulate Responses to conduct simulation analyses using the JMP Pro Simulate feature. For information about Simulate and some DOE examples, see the Simulate chapter in the *Basic Analysis* book.

Advanced Options > Set Delta for Power Specify the difference in the mean response that you want to detect for model effects. See "Evaluate Design Options" on page 451 in the "Evaluate Designs" chapter.

Save Script to Script Window Creates the script for the design that you specified in the Definitive Screening Design window and saves it in an open script window.

Simulate Responses

When you click Make Table to create your design table, the Simulate Responses option does the following for each response:

- It adds random response values to the response column in your design table.
- It adds a new a column containing a simulation model formula to the design table. The formula and values are based on a main effects model.

A Model window opens where you can add and remove effects to define a model, specify parameter values, and select a response distribution for simulation. When you click Apply in the Model window, each column containing a simulation model formula is updated.

Control Window

Figure 7.21 shows the Model window for a design with two continuous factors (X1 and X2) and one two-level categorical factor (X3).

Figure 7.21 Simulate Responses Control Window

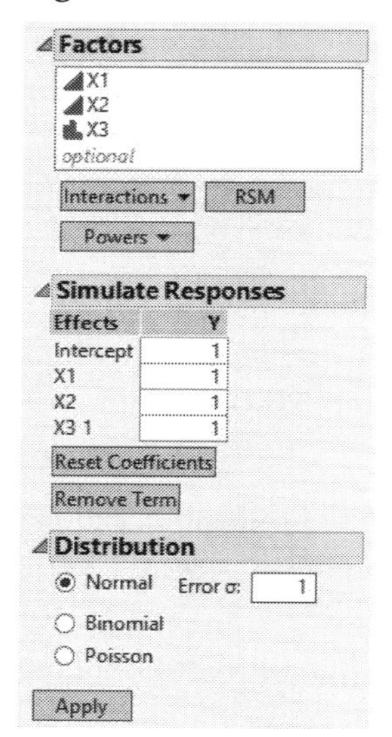

The window has three outlines:

- Factors
- Simulate Responses
- Distribution

The initial Simulate Responses outline shows terms for a main effects model with values of 1 for all coefficients. The Distribution outline shows a Normal distribution with error standard deviation equal to 1. If you have set Anticipated Coefficients as part of Power Analysis under Design Evaluation in the DOE window, then the initial values in the Simulate Responses outline are the values that you specified as Anticipated Coefficients and Anticipated RMSE (Error Std) in the Power Analysis outline.

Factors

Add terms to the simulation model using the Factors outline.

Interactions Select factors in the list. Then select the interaction order from the Interactions menu. Those interactions are added to the list of Effects in the Simulate Responses outline.

RSM Adds all possible response surface terms to the list of Effects in the Simulate Responses outline.

Powers Select factors in the list. Then select the order from the Powers menu. Those powers are added to the list of Effects in the Simulate Responses outline.

Simulate Responses

To specify a model for simulated values, do the following:

1. For each term in the list of Effects, enter coefficients for the linear model used to simulate the response values. These define a linear function, $L(\mathbf{x}, \beta) = \mathbf{x}'\beta$. See the Simulate Responses outline in Figure 7.21:
   - The vector **x** consists of the terms that define the effects listed under Effects.
   - The vector β is the vector of model coefficients that you specify under Y.
2. Under Distribution, select a response distribution.
3. Click **Apply**. A <Y> Simulated column containing simulated values and their formula is added to the design table, where Y is the name of the response column.

Reset coefficients Sets all coefficients to 0.

Remove Term Remove terms from the list of Effects. Select the effects to remove and click Remove Term. Note that you cannot remove main effects.

Distribution

Choose from one of the following distributions in the Simulate Responses window:

Normal Simulates values from a normal distribution. Enter a value for Error σ, the standard deviation of the normal error distribution. If you have designated factors to have Changes of Hard in the Factors outline, you can enter a value for Whole Plots σ, the whole plot error. If you have designated factors to have Changes of Hard and Very Hard, you can enter values for both the subplot and whole plot errors. When you click Apply, random values and a formula containing a random response vector based on the model are entered in the column <Y> Simulated.

Binomial Simulates values from a binomial distribution. Enter a value for N, the number of trials. Random integer values are generated according to a binomial distribution based on N trials with probability of success $1/(1 + exp(-L(\mathbf{x}, \beta))$. When you click Apply, random values and their formula are entered in the column <Y> Simulated. A column called N Trials that contains the value N is also added to the data table.

Poisson Simulates random integer values according to a Poisson distribution with parameter $exp((L(\mathbf{x}, \beta))$. When you click Apply, random values and their formula are entered in the column <Y> Simulated.

Note: You can set a preference to simulate responses every time you click Make Table. To do so, select **File > Preferences > Platforms > DOE**. Select **Simulate Responses**.

Technical Details

This section contains technical details for the following topics:

- "Structure of Definitive Screening Designs" on page 260
- "Analysis of Experimental Data" on page 262

Structure of Definitive Screening Designs

Figure 7.22 shows an example of a definitive design with eight continuous factors. and four Extra Runs that correspond to fake factors. Notice the following:

- Each pair of rows is a foldover pair; each even-numbered row is -1 times the previous row. The foldover aspect of the design removes the confounding of two-factor interactions and main effects.
- Each factor is set at its center value for three runs; this, together with the design's construction, makes all quadratic effects estimable.
- Rows 17 through 20 are the Extra Runs that correspond to the fake factors.
- Adding the center run in the last row enables you to fit a model that includes an intercept and all main and quadratic effects.

This structure is typical of definitive screening designs for continuous factors.

Figure 7.22 Definitive Screening Design for Eight Continuous Factors

Design

Run	X1	X2	X3	X4	X5	X6	X7	X8
1	0	1	1	1	1	1	1	1
2	0	-1	-1	-1	-1	-1	-1	-1
3	1	0	-1	-1	-1	-1	1	1
4	-1	0	1	1	1	1	-1	-1
5	1	-1	0	-1	1	1	-1	-1
6	-1	1	0	1	-1	-1	1	1
7	1	-1	-1	0	1	1	1	1
8	-1	1	1	0	-1	-1	-1	-1
9	1	-1	1	1	0	-1	-1	1
10	-1	1	-1	-1	0	1	1	-1
11	1	-1	1	1	-1	0	1	-1
12	-1	1	-1	-1	1	0	-1	1
13	1	1	-1	1	-1	1	0	-1
14	-1	-1	1	-1	1	-1	0	1
15	1	1	-1	1	1	-1	-1	0
16	-1	-1	1	-1	-1	1	1	0
17	1	1	1	-1	-1	1	-1	1
18	-1	-1	-1	1	1	-1	1	-1
19	1	1	1	-1	1	-1	1	-1
20	-1	-1	-1	1	-1	1	-1	1
21	0	0	0	0	0	0	0	0

Conference Matrices and the Number of Runs

Definitive screening designs in JMP are constructed using *conference matrices* (Xiao et al., 2012). A conference matrix is an $m \times m$ matrix $\mathbf{C}$ where m is even. The matrix $\mathbf{C}$ has *0*s on the diagonal, off-diagonal entries equal to *1* or *–1*, and satisfies $\mathbf{C'C} = (m-1)\mathbf{I}_{mxm}$.

Note: For certain even values of m, it is not known if a conference matrix exists.

Suppose that the number of factors, k, is five or larger. For the case of $k \leq 4$ factors, see "Definitive Screening Designs for Four or Fewer Factors" on page 262.

Consider the case of k continuous factors and suppose that a conference matrix is available.

- When k is even, the $k \times k$ conference matrix is used to define k runs of the design. Its negative, –**C**, defines the foldover runs. A center point is added to the design to ensure that a model containing an intercept, main effects, and quadratic effects is estimable. So, for k even, the minimum number of runs in the definitive screening design is $2k + 1$.
- When k is odd, a $(k+1) \times (k+1)$ conference matrix is used, with its last column deleted. A center point is added. Thus, for k odd, the minimum number of runs in the screening design is $2k + 3$.

A similar procedure is used when some factors are categorical and a conference matrix is available. See Jones and Nachtsheim (2013).

- Instead of a single center point, two additional runs are required. These two runs are center runs where all continuous factors are set at their middle values.
- When there are k factors and k is even, the number of runs in the design is $2k + 2$.
- When k is odd, the number of runs is $2k + 4$.

For those values of m for which a conference matrix is not available, a definitive screening design can be constructed using the next largest conference matrix. As a result, the required number of runs might exceed $2k + 3$, in the continuous case, and $2k + 4$, in the categorical case.

Extra Runs

Extra runs are constructed using fictitious, or *fake*, factors. Adding k_1 fake factors to a design results in 2^{k1} additional runs.

Denote the number of factors in your experimental study by k. Extra runs are constructed by creating a design for $k + k_1$ factors, as described in "Conference Matrices and the Number of Runs" on page 261, and then dropping the last k_1 columns. If $k_1 = 2$, four extra runs are added. If $k_1 = 3$, eight extra runs are added. As few as four extra runs can be highly beneficial in model selection.

For information about how extra runs are used, see "Effective Model Selection for DSDs" on page 266 in the "The Fit Definitive Screening Platform" chapter.

Definitive Screening Designs for Four or Fewer Factors

Definitive screening designs for four or fewer factors are constructed using the five-factor definitive screening design as a base. This is because designs for $k \leq 4$ factors constructed strictly according to the conference matrix approach have undesirable properties. In particular, it is difficult to separate second-order effects.

If you specify $k \leq 4$ factors, a definitive screening design for five factors is constructed and unnecessary columns are dropped. For this reason, the number of runs for an unblocked design with $k \leq 4$ factors is 13 if all factors are continuous or 14 if some factors are categorical.

Analysis of Experimental Data

In general, you want to fit a model that permits the possibility that two-way interactions are active. You also might want to include pure quadratic terms in your model. You might want to postulate a full second-order model, or you might want to specify an a priori model containing only certain second-order terms.

Two-Way Interactions

In fitting such a model, you need to be mindful of two facts:

- two-way interaction effects and quadratic effects are often correlated
- two-way interaction and quadratic effects cannot *all* be estimated simultaneously

Figure 7.23 shows a Color Map on Correlations for the design with eight continuous factors shown in Figure 7.22. The color map is for a full quadratic design. The eight pure quadratic effects are listed to the far right. You can construct this plot by using DOE > Design Diagnostics > Evaluate Design and entering the appropriate terms into the Alias Terms list. See "Alias Terms" on page 433 in the "Evaluate Designs" chapter for details.

Figure 7.23 Color Map on Correlations for Full Quadratic Model

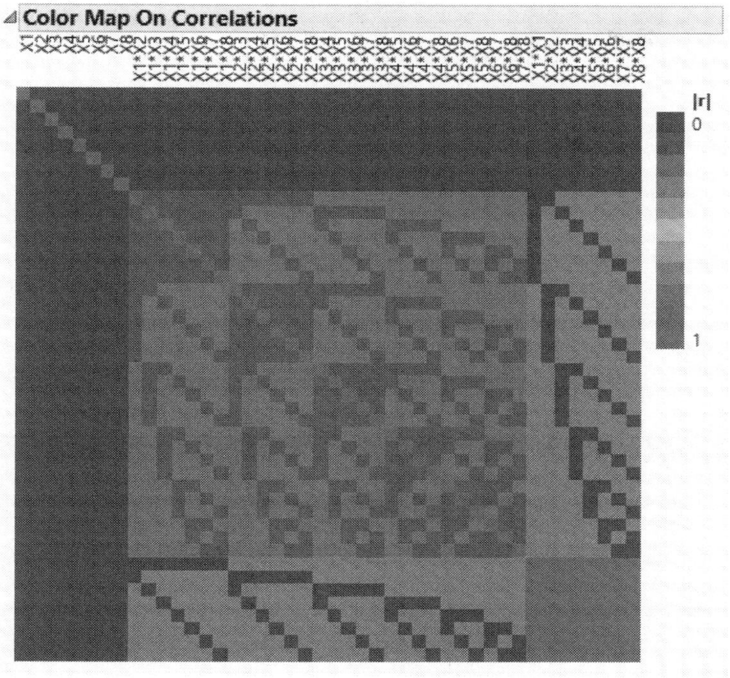

Use your cursor to hold your mouse pointer over the cells of the color map in order to see the absolute correlations between effects. You see that main effects are uncorrelated with all two-way interaction and pure quadratic effects. You also see that none of the effects are completely confounded with other effects because the only red cells are on the main diagonal. But note that some of the absolute correlations between two-factor interactions are substantial, with some at 0.75. Note also that absolute correlations between two-factor interactions and pure quadratic effects are either 0 or 0.3118.

If only main and pure quadratic effects are active, you can fit a saturated model that contains main effects and quadratic effects. This model will result in effect estimates that are unbiased, assuming no active three-way or higher order effects.

Because of the correlations involving second-order effects, you must be careful in fitting a model with two-way interactions. Analysis methodologies include the following, where the first is preferred:

- The method of Efficient Model Selection performs well, especially if many effects are likely to be active. See "Effective Model Selection for DSDs" on page 266 in the "The Fit Definitive Screening Platform" chapter.
- Forward stepwise or all possible subsets regression performs adequately if the following conditions hold:

- The number of active effects is no more than half the number of runs.
- There are at most two active two-way interactions or at most one active quadratic effect.

See "Forward Stepwise Regression or All Possible Subsets Regression" on page 264.

Forward Stepwise Regression or All Possible Subsets Regression

This method consists of first specifying a full response surface model. Then do one of the following:

- Use forward stepwise regression with the Stopping Rule set to Minimum AICc and the Rules set to Combine to ensure model heredity.
- Use All Possible Models regression, where you select the option that imposes the heredity restriction and use the AICc criterion for model selection.

You cannot fit the full response surface model because the number of runs is less than the number of parameters. So your analysis depends on the assumption of effect sparsity, where you assume that the number of active effects is less than the number of runs. This approach has some limitations:

- If the number of active effects exceeds half the number of runs, both stepwise and all possible models regression have difficulty finding the correct model.
- The power of tests to detect moderate quadratic effects is low. A quadratic effect must exceed three times the error standard deviation for the power to exceed 0.9.
- Because of effect confounding, several models might be equivalent. Additional runs will be necessary to resolve the confounding.

Chapter 8

The Fit Definitive Screening Platform

Analyze Data from Definitive Screening Experiments

Use the Fit Definitive Screening platform to analyze definitive screening designs (DSDs) using a methodology called *Effective Model Selection for DSDs*. This methodology takes advantage of the unique structure of definitive screening designs.

Standard model selection methods applied to DSDs can fail to identify active effects. To identify active main effects and second-order effects, the Effective Model Selection for DSDs algorithm leverages the structure of DSDs and assumes strong effect heredity.

Figure 8.1 Fit Definitive Screening Results

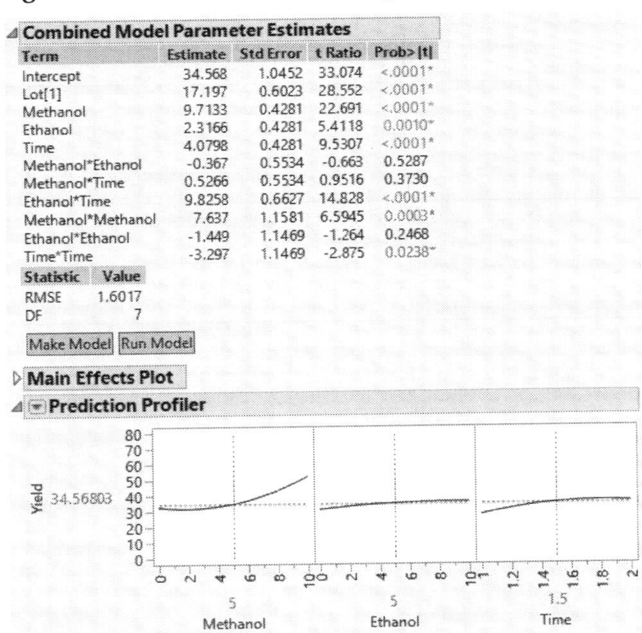

Combined Model Parameter Estimates

Term	Estimate	Std Error	t Ratio	Prob>|t|
Intercept	34.568	1.0452	33.074	<.0001*
Lot[1]	17.197	0.6023	28.552	<.0001*
Methanol	9.7133	0.4281	22.691	<.0001*
Ethanol	2.3166	0.4281	5.4118	0.0010*
Time	4.0798	0.4281	9.5307	<.0001*
Methanol*Ethanol	-0.367	0.5534	-0.663	0.5287
Methanol*Time	0.5266	0.5534	0.9516	0.3730
Ethanol*Time	9.8258	0.6627	14.828	<.0001*
Methanol*Methanol	7.637	1.1581	6.5945	0.0003*
Ethanol*Ethanol	-1.449	1.1469	-1.264	0.2468
Time*Time	-3.297	1.1469	-2.875	0.0238*

Statistic	Value
RMSE	1.6017
DF	7

Overview of the Fit Definitive Screening Platform

The Fit Definitive Screening platform analyzes definitive screening designs (DSDs) using a methodology that takes advantage of their special structure. The methodology is called *Effective Model Selection for DSDs*. If you created your design in JMP, the design table contains a script called Fit Definitive Screening that automatically runs an analysis using the Effective Model Selection for DSDs methodology.

Identification of Active Effects in DSDs

DSDs are three-level designs that are valuable for identifying main effects and second-order effects in a single experiment. A minimum run-size DSD is capable of correctly identifying active terms with high probability if the number of active effects is less than about half the number of runs and if the effects sizes exceed twice the standard deviation.

However, by augmenting a minimum run-size DSD with four or more properly selected runs, you can identify substantially more effects with high probability. These runs are called *Extra Runs*, and correspond to fictitious inactive factors, called *fake factors*. For information about Extra Runs, see "Structure of Definitive Screening Designs" on page 260 in the "Definitive Screening Designs" chapter.

Extra Runs substantially increase the design's ability to detect second-order effects. For this reason, Jones and Nachtsheim (2016) strongly encourage the inclusion of at least four Extra Runs.

Effective Model Selection for DSDs

When standard model selection methods are applied to DSDs, they can fail to identify active effects. See Errore et al. (2016). Also, standard selection methods do not leverage the structure of DSDs. The Fit Definitive Design platform uses the Effective Model Selection for DSDs approach, which takes full advantage of the structure of the DSD.

Jones and Nachtsheim (2016) report on simulation studies using Effect Model Selection for DSDs as well as standard approaches. Denote by c the sum of the number of factors and the number of fake factors in a DSD. In many situations, if the number of active main effects exceeds three, then up to $c/2$ active second-order effects can be reliably identified. Assuming strong effect heredity, if there are three or fewer active main effects, then all active second-order effects can be reliably identified. *Reliable identification* means that the ratio of the absolute value of the coefficient to the error standard deviation exceeds three and that the power to detect the effect exceeds 0.80.

The Fit Definitive Screening platform assumes *strong effect heredity*. Strong effect heredity means that the A*B interaction can only be considered for inclusion in the model if both A and B have been included. Strong effect heredity requires that all lower-order components of a model effect be included in the model. In identifying active second-order effects, the algorithm uses strong effect heredity and the results cited earlier about how many active second-order effects can be reliably identified.

In a DSD, main effects and second-order effects are orthogonal to each other. The Effective Model Selection for DSDs approach takes advantage of this fact. The linear space of the response is separated into the subspace spanned by the main effects and the orthogonal complement of this subspace. Miller and Sitter (2005) refer to the linear subspace spanned by the main effects as the *odd space*, because it contains all the information about odd effects: main effects, 3-factor effects, 5-factors effects, and so on. They refer to its orthogonal complement as the *even space*, because it contains all the information about even effects: the intercept, 2-factor effects, 4-factor effects, and so on.

Fit Definitive Screening follows this thinking. The subspace spanned by the main effects is the odd space. Its orthogonal complement, the even space, contains the second-order effects and the block variable, if one exists. For details of the algorithm, see "The Effective Model Selection for DSDs Algorithm" on page 275 and Jones and Nachtsheim (2016).

Example of the Fit Definitive Screening Platform

The design in the data table Extraction 3 Data.jmp is a definitive screening design for six factors in two blocks. The **Add Blocks with Center Runs to Estimate Quadratic Effects** option was selected, and 4 Extra Runs were added. The resulting design has 18 runs.

Fit the Model

1. Select **Help > Sample Data Library** and open Design Experiment/Extraction 3 Data.jmp.
2. Select **DOE > Definitive Screening > Fit Definitive Screening**.
3. Select Yield and click **Y**.
4. Select Lot through Time and click **X**.
5. Click OK.

The fit performs a two-stage analysis. For details about the algorithm, see "Technical Details" on page 275.

Examine Results

Stage 1: Main Effect Estimates

Stage 1 determines which main effects are likely to be active.

Figure 8.2 Stage 1 Report for Main Effects

Stage 1 - Main Effect Estimates

Term	Estimate	Std Error	t Ratio	Prob>|t|
Methanol	9.7133	0.3674	26.438	<.0001*
Ethanol	2.3166	0.3674	6.3055	0.0015*
Time	4.0798	0.3674	11.104	0.0001*

Statistic	Value
RMSE	1.3747
DF	5

Note: The fake factors do not appear in the design or as factors in the analysis.

A two-degree-of-freedom error sum of squares is computed from the four runs corresponding to the two fake factors. Because the fake factors are, by construction, inactive, this estimate of error variance is unbiased. For each main effect, the main effects response Y_{ME} is tested against this estimate. In this example, three factors, Methanol, Ethanol, and Time, have *p*-values smaller than the threshold value and are retained as active. For details about the threshold values, see “Stage 1 Methodology” on page 276.

The variability from the three inactive factors, Propanol, Butanol, and pH, is pooled with the fake factor sum of squares to produce the five-degree-of-freedom RMSE statistic shown in Figure 8.2.

Stage 2: Even Order Effect Estimates

Stage 2 uses guided subset selection to arrive at a list of second-order effects that are likely to be active.

Figure 8.3 Stage 2 Report for Even-Order Effects

Stage 2 - Even Order Effect Estimates

Term	Estimate	Std Error	t Ratio	Prob>|t|
Intercept	34.568	1.3459	25.683	0.0015*
Lot[1]	17.197	0.7757	22.171	0.0020*
Methanol*Ethanol	-0.367	0.7127	-0.515	0.6581
Methanol*Time	0.5266	0.7127	0.7389	0.5369
Ethanol*Time	9.8258	0.8534	11.514	0.0075*
Methanol*Methanol	7.637	1.4914	5.1208	0.0361*
Ethanol*Ethanol	-1.449	1.477	-0.981	0.4299
Time*Time	-3.297	1.477	-2.232	0.1552

Statistic	Value
RMSE	2.0626
DF	2

Because three main effects are identified as active in Stage 1, the guided subset selection procedure for active second-order effects can continue until all second-order effects are included. Because all six second-order effects are reported in Stage 2, it follows that the Stage 2 RMSE remained larger than the Stage 1 RMSE. See “Stage 2 Methodology” on page 277.

The two-degree-of-freedom RMSE given in the Stage 2 report is the error estimate obtained from the final subset of all six second-order effects.

Combined Results

The effects selected for the model are listed in the Combined Model Parameter Estimates report.

Figure 8.4 Combined Model Parameter Estimates Report

Combined Model Parameter Estimates

Term	Estimate	Std Error	t Ratio	Prob>|t|
Intercept	34.568	1.0452	33.074	<.0001*
Lot[1]	17.197	0.6023	28.552	<.0001*
Methanol	9.7133	0.4281	22.691	<.0001*
Ethanol	2.3166	0.4281	5.4118	0.0010*
Time	4.0798	0.4281	9.5307	<.0001*
Methanol*Ethanol	-0.367	0.5534	-0.663	0.5287
Methanol*Time	0.5266	0.5534	0.9516	0.3730
Ethanol*Time	9.8258	0.6627	14.828	<.0001*
Methanol*Methanol	7.637	1.1581	6.5945	0.0003*
Ethanol*Ethanol	-1.449	1.1469	-1.264	0.2468
Time*Time	-3.297	1.1469	-2.875	0.0238*

Statistic	Value
RMSE	1.6017
DF	7

The RMSE and degrees of freedom given at the bottom of the report are the usual standard least squares quantities. Use these effects as potential factors for your final model.

Reduce the Model

The Make Model button enters the model for the listed terms in a Fit Model specification window. To run the model directly using standard least squares, click the Run Model button.

1. Click **Run Model**.

   The Actual by Predicted Plot shows no lack of fit. The Effect Summary report suggests that you can reduce the model further.

Figure 8.5 Actual by Predicted Plot and Effect Summary Report

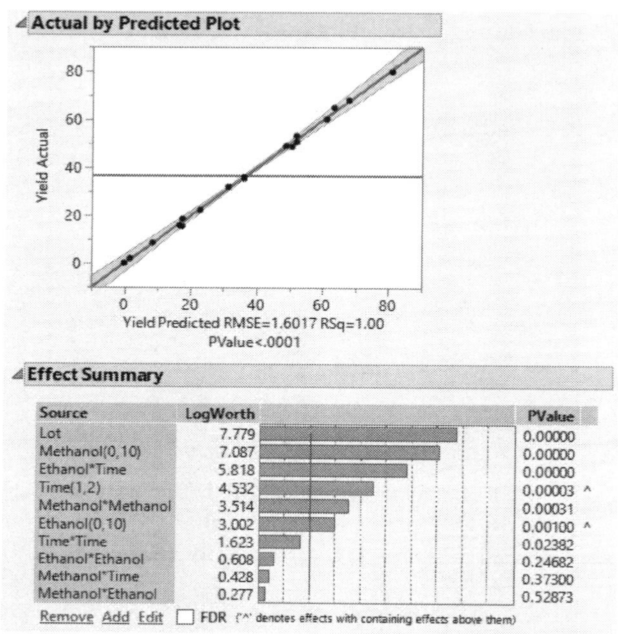

Source	LogWorth	PValue
Lot	7.779	0.00000
Methanol(0,10)	7.087	0.00000
Ethanol*Time	5.818	0.00000
Time(1,2)	4.532	0.00003 ^
Methanol*Methanol	3.514	0.00031
Ethanol(0,10)	3.002	0.00100 ^
Time*Time	1.623	0.02382
Ethanol*Ethanol	0.608	0.24682
Methanol*Time	0.428	0.37300
Methanol*Ethanol	0.277	0.52873

Remove Add Edit ☐ FDR ('^' denotes effects with containing effects above them)

2. Select **Methanol*Ethanol** in the Effect Summary report and click **Remove**.

   Methanol*Time has *p*-value 0.33750. Remove it next.

3. Select Methanol*Time in the Effect Summary report and click **Remove**.

   Ethanol*Ethanol has *p*-value 0.15885. Remove it next.

4. Select Ethanol*Ethanol in the Effect Summary report and click **Remove**.

Figure 8.6 Effect Summary Report Showing Effects in Final Model

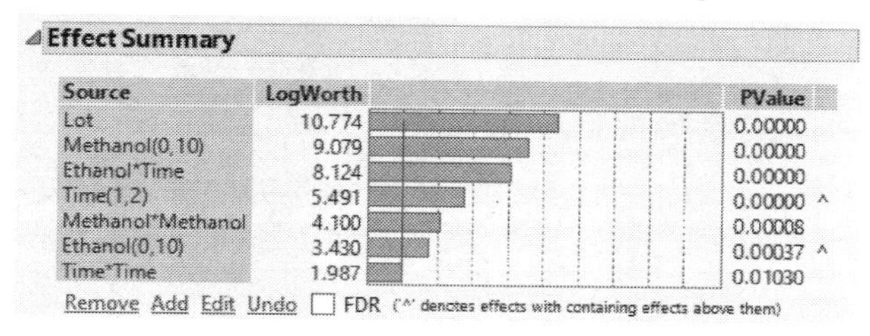

Source	LogWorth	PValue
Lot	10.774	0.00000
Methanol(0,10)	9.079	0.00000
Ethanol*Time	8.124	0.00000
Time(1,2)	5.491	0.00000 ^
Methanol*Methanol	4.100	0.00008
Ethanol(0,10)	3.430	0.00037 ^
Time*Time	1.987	0.01030

Remove Add Edit Undo ☐ FDR ('^' denotes effects with containing effects above them)

The remaining effects are significant. You conclude that these are the active effects.

Launch the Fit Definitive Screening Platform

To launch the Fit Definitive Screening platform, select **DOE > Definitive Screening > Fit Definitive Screening.** The launch window in Figure 8.7 uses Extraction3 Data.jmp.

Note: If you created your design in JMP, the design table contains a script called Fit Definitive Screening. Run this script to run the analysis directly.

Figure 8.7 Fit Definitive Screening Launch Window

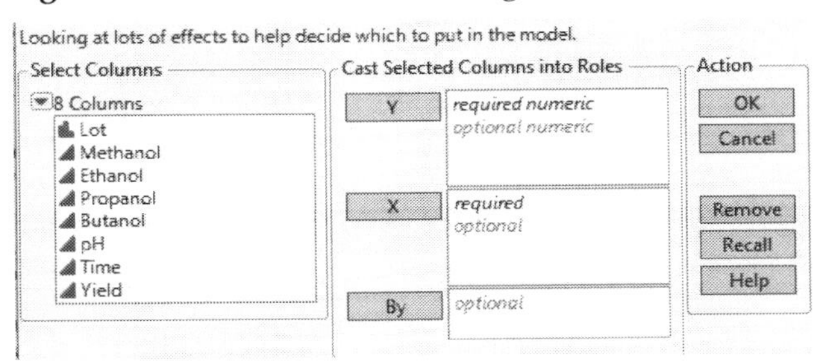

Y One or more numeric response variables.

X Continuous or two-level categorical factors. Because the platform uses the unique features of a DSD in performing the analysis, these factors must define a DSD or a fold-over design.

By A column whose levels define separate analyses. For each level of the specified column, the corresponding rows are analyzed. The results appear in separate reports. If more than one By variable is assigned, a separate analysis is produced for each possible combination of the levels of the By variables.

Fit Definitive Screening Report

The Fit Definitive Screening report provides these outlines:

- "Stage 1 - Main Effect Estimates" on page 272
- "Stage 2 - Even Order Effect Estimates" on page 273
- "Combined Model Parameter Estimates" on page 273
- "Main Effects Plot" on page 274
- "Prediction Profiler" on page 274

Stage 1 - Main Effect Estimates

The Main Effect Estimates report lists main effects that are identified as active. Main effects with *p*-values less than the threshold *p*-value are considered active. For additional details, see "Stage 1 - Main Effect Estimates" on page 272.

- If fake factors or center point replicates are available, an estimator of error variance that is independent of the model is constructed. The main effects are tested against this estimate.
- If no fake factors or center point replicates are available, subsets of main effects are tested sequentially against an estimate of error variance constructed from the inactive main effects. For this procedure to be viable, at least one of the main effects must be inactive.

In either case, variability from the inactive main effects is pooled into the error variance used to test the main effects.

Figure 8.8 Stage 1 Report

Stage 1 - Main Effect Estimates

Term	Estimate	Std Error	t Ratio	Prob>|t|
Methanol	9.7133	0.3674	26.438	<.0001*
Ethanol	2.3166	0.3674	6.3055	0.0015*
Time	4.0798	0.3674	11.104	0.0001*

Statistic	Value
RMSE	1.3747
DF	5

Term Main effects identified as active. These effects have *p*-values less than the threshold when tested as described in "Stage 1 Methodology" on page 276.

Estimate Parameter estimate for a regression fit of Y on the main effects.

Std Error The standard error of the estimate, computed using the Stage 1 RMSE.

t Ratio The Estimate divided by its Std Error.

Prob>|t| The *p*-value computed using the t Ratio and the degrees of freedom for error (DF).

RMSE The square root of the mean square error that results from the Stage 1 analysis.

- If fake factors or centerpoint replicates are available, the mean square error is the estimate of variance from fake factors and centerpoints pooled with the variance estimate constructed from the main effects that are not identified as active.
- If no fake factors or centerpoint replicates are available, the mean square error is the estimate of variance constructed from the main effects that are not identified as active.

DF The degrees of freedom associated with the error estimate used to construct RMSE.

- If fake factors or centerpoint replicates are available, DF is the sum of the number of fake factors, centerpoint replicates, and main effects not identified as active.
- If no fake factors or centerpoint replicates are available, DF is the number of main effects that are not identified as active.

Stage 2 - Even Order Effect Estimates

The Even Order Effect Estimates report lists second-order effects that are identified as active. Active second-order effects are identified using the guided variable selection procedure described in "Stage 2 Methodology" on page 277. The block effect (if one is included) is also listed, whether it is significant or not.

Figure 8.9 Stage 2 Report

Stage 2 - Even Order Effect Estimates

Term	Estimate	Std Error	t Ratio	Prob>|t|
Intercept	34.568	1.3459	25.683	0.0015*
Lot[1]	17.197	0.7757	22.171	0.0020*
Methanol*Ethanol	-0.367	0.7127	-0.515	0.6581
Methanol*Time	0.5266	0.7127	0.7389	0.5369
Ethanol*Time	9.8258	0.8534	11.514	0.0075*
Methanol*Methanol	7.637	1.4914	5.1208	0.0361*
Ethanol*Ethanol	-1.449	1.477	-0.981	0.4299
Time*Time	-3.297	1.477	-2.232	0.1552

Statistic	Value
RMSE	2.0626
DF	2

Term The block factor and second-order effects identified as active.

Estimate Parameter estimates for a regression fit of Y on the Stage 2 second-order effects defined by Y_{2nd}. See "Decomposition of Response" on page 275.

Std Error The standard error of the estimate, computed using the Stage 2 RMSE.

t Ratio The Estimate divided by its Std Error.

Prob>|t| The *p*-value computed using the t Ratio and the degrees of freedom for error (DF).

RMSE The square root of the mean square error that results from the Stage 2 analysis. RMSE is estimated as the residual variance from Y_{2nd} after fitting the second order effects identified as active. See "Decomposition of Response" on page 275.

DF The degrees of freedom associated with the error estimate used to construct RMSE.

Combined Model Parameter Estimates

The Combined Model Parameter Estimates report lists the terms in the final model and their usual standard least squares estimates, standard errors, t ratios, *p*-values, RMSE, and model degrees of freedom.

Below the report are buttons that construct or run the combined model.

Figure 8.10 Combined Model Parameter Estimates Report

Combined Model Parameter Estimates

Term	Estimate	Std Error	t Ratio	Prob>|t|
Intercept	34.568	1.0452	33.074	<.0001*
Lot[1]	17.197	0.6023	28.552	<.0001*
Methanol	9.7133	0.4281	22.691	<.0001*
Ethanol	2.3166	0.4281	5.4118	0.0010*
Time	4.0798	0.4281	9.5307	<.0001*
Methanol*Ethanol	-0.367	0.5534	-0.663	0.5287
Methanol*Time	0.5266	0.5534	0.9516	0.3730
Ethanol*Time	9.8258	0.6627	14.828	<.0001*
Methanol*Methanol	7.637	1.1581	6.5945	0.0003*
Ethanol*Ethanol	-1.449	1.1469	-1.264	0.2468
Time*Time	-3.297	1.1469	-2.875	0.0238*

Statistic	Value
RMSE	1.6017
DF	7

Make Model | Run Model

Make Model Creates a model for the Fit Model window containing the model terms in the Combined Model Parameter Estimates report and the response specified for the Fit Definitive Screening analysis. The Standard Least Squares personality is specified.

Run Model Runs a standard least squares fit for the model terms in the Combined Model Parameter Estimates report and the response specified for the Fit Definitive Screening analysis.

Main Effects Plot

Shows a plot of the response against each of the factors entered as X in the Fit Definitive Screening launch window. Notice that the block factor is not shown.

Prediction Profiler

Shows a prediction profiler for the main effects identified as active in the Stage 1 analysis. You can view a prediction profiler for the combined model terms in the report that you obtain by clicking the Run Model button. For details about the prediction profiler, see the Profiler chapter in the *Profilers* book.

Fit Definitive Screening Platform Options

See the JMP Reports chapter in the *Using JMP* book for more information about the following options:

Local Data Filter Shows or hides the local data filter that enables you to filter the data used in a specific report.

Redo Contains options that enable you to repeat or relaunch the analysis. In platforms that support the feature, the Automatic Recalc option immediately reflects the changes that you make to the data table in the corresponding report window.

Save Script Contains options that enable you to save a script that reproduces the report to several destinations.

Save By-Group Script Contains options that enable you to save a script that reproduces the platform report for all levels of a By variable to several destinations. Available only when a By variable is specified in the launch window.

Technical Details

The Effective Model Selection for DSDs Algorithm

This section provides a summary of the algorithm used in the Fit Definitive Screening platform. For further details, see Jones and Nachtsheim (2016).

Decomposition of Response

The Effective Model Selection algorithm expresses the response, Y, in terms of two responses Y_{ME} and Y_{2nd}, so that $Y = Y_{ME} + Y_{2nd}$.

– Y_{ME} is the predicted value obtained from a regression of Y on the main effects and fake factors.

  There is no need to include the block factor in Y_{ME} because of the fold-over structure of the design. The block factor is included in Y_{2nd}.

– Y_{2nd} is given by $Y_{2nd} = Y - Y_{ME}$.

Note: In a DSD, the columns Y_{ME} and Y_{2nd} are orthogonal.

The analysis proceeds in two stages:

- Stage 1: The response Y is used to identify main effects. Stage 1 identifies the main effects that are considered active.
- Stage 2: The response Y_{2nd} is used to identify second-order effects. Stage 2 considers all second-order terms in the active main effects from Stage 1 and determines a subset of these containing effects considered to be active.

Note: If there is a blocking factor, it is included in the Stage 2 list of effects even if it is not significant.

Stage 1 Methodology

The Stage 1 methodology depends on whether the design contains fake factors or centerpoint replicates.

Case 1: Fake Factors or Centerpoint Replicates Available

1. Using the fake factors or center point replicates, an estimator of error variance that is independent of the model is constructed. Assuming that there are no active third or higher odd order effects, this estimate is unbiased.
2. Using Y_{ME}, main effects are tested against this estimate. Main effects with *p*-values less than a threshold *p*-value are considered active. The threshold values are the following:
   - For one error degree of freedom, the threshold value is 0.20.
   - For two error degrees of freedom, the threshold value is 0.10.
   - For more than two error degrees of freedom, the threshold value is 0.05.
3. If no main effect has a *p*-value less than the threshold value, conclude that there are no active main effects and no active two-factor effects. The procedure terminates.
4. If active main effects are found, then variability from the inactive main effects is pooled into the error variance constructed in (1).

Case 2: No Fake Factors or Centerpoint Replicates Available

In this case, there is no model-independent estimator of error variance available. Subsets of main effects are tested sequentially against an estimate of error variance constructed from the inactive main effects. Suppose that there are m main effects.

1. The absolute values of the estimated effects, using Y_{ME} as the response, are ordered from largest to smallest.
2. For each $1 \leq i < m$, the effect with the ith largest absolute value is tested against the adjusted residual sum of squares for the model containing that effect and all effects with larger absolute values.
3. The effects in the model with the smallest *p*-value are considered to be the active effects.
4. If active main effects are found, then variability from the inactive main effects is used to construct an estimate of error variance, using Y_{ME} as the response.

Note: For the Fit Definitive Screening procedure to work properly in Case 2, at least one of the main effects must be active and at least one must be inactive. If no main effects are active, or if all main effects are active, the procedure will identify a set of main effects, but the procedure for arriving at that subset is compromised.

Stage 2 Methodology

In Stage 2, only second-order effects involving the factors whose main effects are active are considered. Stage 2 uses a guided subset selection procedure. The goal is to continue to add second-order effects to the model as long as the RMSE from Stage 2 exceeds the RMSE from Stage 1. When the Stage 2 RMSE is less than or equal to the Stage 1 RMSE, this indicates that there are no additional second-order effects to add to the model.

The same threshold values are used as in Stage 1:

- For one error degree of freedom, the threshold value is 0.20.
- For two error degrees of freedom, the threshold value is 0.10.
- For more than two error degrees of freedom, the threshold value is 0.05.

1. The variability for Y_{2nd} is tested against the error estimate from Stage 1 to determine if there is additional variability due to second-order effects.
   - If the *p*-value for this test exceeds the threshold value the procedure terminates and no active second-order effects are identified.
2. If the *p*-value for this test is less than or equal to the threshold value, then subsets of size *k*, $k = 1,2,3,...$ are successively tested, starting with $k = 1$.
3. For each *k*, the residual sum of squares for each subset of that size is tested against the error estimate from Stage 1. The subset with the smallest RMSE is identified.
4. The procedure continues until a *k* is found whose RMSE is smaller than the Stage 1 RMSE.
5. The effects in the subset preceding the one that corresponds to the terminal value of *k* are considered to be the active two-factor effects.

Chapter 9

Screening Designs

Screening designs are among the most popular designs for industrial experimentation. Typically used in the initial stages of experimentation, they examine many factors in order to identify those factors that have the greatest effect on the response. The factors that are identified are then studied using more sensitive designs. Because screening designs generally require fewer experimental runs than other designs, they are a relatively inexpensive and efficient way to begin improving a process.

If a standard screening design exists for your experimental situation, you can choose from several standard screening designs. The list includes blocked designs when applicable. Your factors can be two-level continuous factors, three-level categorical factors, or continuous factors that can assume only discrete values (*discrete numeric* factors).

However, there are situations where standard screening designs are not available. In these cases, the Screening Design platform constructs a *main effects screening design*. A main effects screening design is either orthogonal or near orthogonal. It focuses on estimating main effects in the presence of negligible interactions.

Note that JMP also provides two compelling alternatives to screening designs:

- Definitive screening designs are particularly useful if you suspect active two-factor interactions or if you suspect that a plot of a continuous factor's effect on the response might exhibit strong curvature. See Chapter 7, "Definitive Screening Designs".
- Custom designs are highly flexible and often more cost-effective than a design obtained using alternative methods. See Chapter 4, "Custom Designs".

Figure 9.1 Results from a Fractional Factorial Design

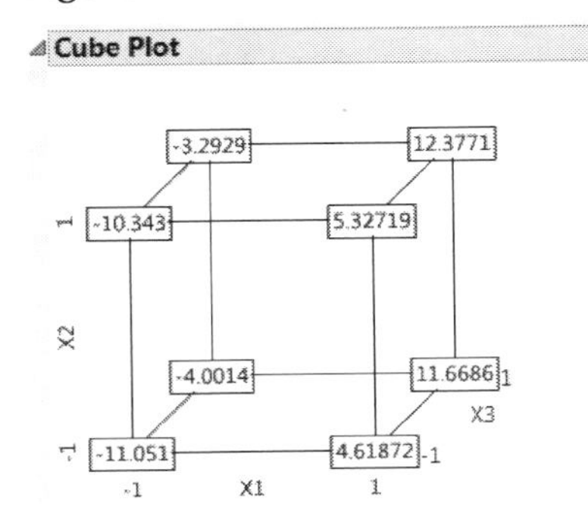

Overview of Screening Designs

Screening experiments tend to be small and are aimed at identifying the factors that affect a response. Because identification is the goal (rather than sophisticated modeling), continuous factors in a screening design are typically set at only two levels. However, a screening situation might also involve discrete numeric or categorical factors, in which case classical screening designs might not fit your situation. The Screening Design platform can handle all three types of factors: two-level continuous factors, categorical factors, and discrete numeric factors.

There are two types of designs:

- Classical designs: For situations where standard screening designs exist, you can choose from a list that includes fractional factorial designs, Plackett-Burman, Cotter, and mixed-level designs.
- Main effects screening designs: Whether a standard design is available, you can ask JMP to construct a main effects screening design. These designs are orthogonal or near orthogonal and focus on estimating main effects in the presence of negligible interactions. See "Main Effects Screening Designs" on page 304.

Underlying Principles

The emphasis on studying main effects early on in the experimentation process is supported by the empirical principle of effect hierarchy. This principle maintains that lower order effects are more likely to be important than higher order effects. For this reason, screening designs focus on identifying active main effects. In cases where higher order interactions *are* of interest, screening designs assume that two-factor interactions are more important than three-factor interactions, and so on. See "Effect Hierarchy" on page 59 in the "Starting Out with DOE" chapter and Wu and Hamada, 2009.

The efficiency of screening designs also depends on the principle of effect sparsity. Effect sparsity asserts that most of the variation in the response is explained by a relatively small number of effects. See "Effect Sparsity" on page 59 in the "Starting Out with DOE" chapter.

To appreciate the importance of effect sparsity, consider an example where you have seven two-level factors. Contrast a full factorial design to a screening design:

- A full factorial design consists of all combinations of the levels of the factors. The number of runs is the product of the numbers of levels for each factor. In this example, a full factorial design has $2^7 = 128$ runs.
- In contrast, a screening design requires only a fraction of the runs in the full factorial design. The main effects of the seven factors can be studied in an eight-run screening design.

Analysis of Screening Design Results

Screening designs are often used to test a large number of factors or interactions. When there are degrees of freedom for error, allowing construction of an error estimate, the experimental results can be analyzed using the usual regression techniques (Analyze > Fit Model).

However, sometimes there are no degrees of freedom for error. In this case, assuming effect sparsity, the Screening platform (Analyze > Specialized Modeling > Specialized DOE Models > Fit Two Level Screening) provides a way to analyze the results of a two-level design. The Screening platform accepts multiple responses and multiple factors. It automatically shows significant effects with plots and statistics. For details, see "The Fit Two Level Screening Platform" chapter on page 323. For an examples in the current chapter, see "Modify Generating Rules in a Fractional Factorial Design" on page 311 and "Plackett-Burman Design" on page 317.

Examples of Screening Designs

This section contains the following examples:

Compare a Fractional Factorial Design and a Main Effects Screening Design

In this example, suppose an engineer wants to investigate a process that uses an electron beam welding machine to join two parts. The engineer fits the two parts into a welding fixture that holds them snugly together. A voltage applied to a beam generator creates a stream of electrons that heats the two parts, causing them to fuse. The ideal depth of the fused region is 0.17 inches. The engineer wants to study the welding process to determine the best settings for the beam generator to produce the desired depth in the fused region.

For this study, the engineer wants to explore the following seven factors:

- Operator is the technician operating the welding machine. Two technicians typically operate the machine.
- Speed (in rpm) is the speed at which the part rotates under the beam.
- Current (in amps) is a current that affects the intensity of the beam.
- Mode is the welding method used.
- Wall Size (in mm) is the thickness of the part wall.
- Geometry indicates whether the joint is a single-bevel joint or a double-bevel joint.
- Material is the type of material being welded.

Notice that three of these factors are continuous: Speed, Current, and Wall Size. Four are categorical: Operator, Mode, Geometry, and Material. Each of these categorical factors has two levels.

After each processing run, the engineer cuts the part in half. This reveals an area where the two parts have fused. The length of this fused area, measured in inches, is the depth of penetration of the weld. The depth of penetration is the response for the study.

The goals of the study are the following:

- Find which factors affect the depth of the weld.
- Quantify those effects.
- Find specific factor settings that predict a weld depth of 0.17 inches with a tolerance of ±0.05 inches.

Your experimental budget allows you at most 12 runs. Construct and compare two designs for your experimental situation. The first is a classical fractional factorial design using eight runs. The second is a main effects screening design using 12 runs.

Constructing a Standard Screening Design

In this section, construct a standard screening design for this experimental situation.

Specify the Response

1. Select **DOE > Classical > Screening Design**.
2. In the Responses panel, double-click Y under Response Name and type Depth.

   Note that the default Goal is Maximize. Your goal is to find factor settings that enable you to obtain a target depth of 0.17 inches with limits of 0.12 and 0.22.
3. Click on the default Goal of Maximize and change it to **Match Target**.
4. Click under **Lower Limit** and type 0.12.
5. Click under **Upper Limit** and type 0.22.
6. Leave the area under **Importance** blank.

   Because there is only one response, that response is given Importance 1 by default.

The completed Responses outline appears in Figure 9.2. Now, specify the factors.

Specify Factors

You can enter the factors manually or automatically:

- To enter the factors manually, see "Specify Factors Manually" on page 283.
- To enter the factors automatically, use the Weld Factors.jmp data table:

1. Select **Help > Sample Data Library** and open Design Experiment/Weld Factors.jmp.
2. Select **Load Factors** from the Screening Design red triangle menu. Proceed to "Choose a Design" on page 284.

Specify Factors Manually

1. Type **3** in the **Add N Factors** box and click **Continuous**.
2. Double-click X1 and type Speed.
3. Use the Tab key to move through the rest of the values and factors. Make the following changes:
   a. Change the Speed values to 3 and 5.
   b. Change X2 to Current, with values of 150 and 165.
   c. Change X3 to Wall Size, with values of 20 and 30.
4. Type **4** in the **Add N Factors** box and select **Categorical > 2 Level**.
5. Double-click X4 and type Operator.
6. Use the Tab key to move through the rest of the values and factors. Make the following changes:
   a. Change the Operator values to John and Mary.
   b. Change X5 to Mode, with values of Conductance and Keyhole.
   c. Change X6 to Geometry, with values of Double and Single.
   d. Change X7 to Material, with values of Aluminum and Magnesium.

Your Responses and Factors outlines should appear as shown in Figure 9.2.

Figure 9.2 Responses and Factors Outlines for Weld Experiment

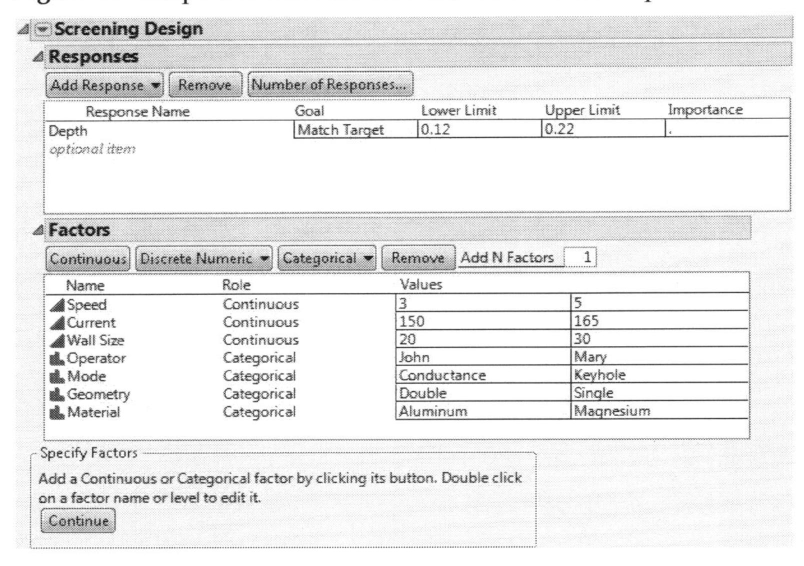

Choose a Design

1. Click **Continue**.

   Because the combination of factors and levels that you have specified can be accommodated by a standard fractional factorial design, the Choose Screening Type panel appears. You can either select a standard design from a list or construct a main effects design.

   Note: Setting the Random Seed in the next step reproduces the results shown in this example. When you are constructing a design on your own, this step is not necessary.

2. (Optional) From the Screening Design red triangle menu, select **Set Random Seed**, type 12345, and click **OK**.
3. Accept the default selection to **Choose from a list of fractional factorial designs** and click **Continue**.
4. Select the first Fractional Factorial design. See Figure 9.3.

Figure 9.3 Design List for Two Continuous Factors and One Categorical Factor

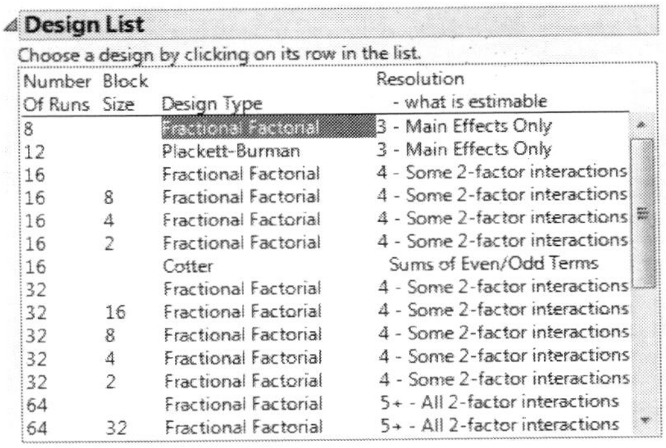

Design List

Choose a design by clicking on its row in the list.

Number Of Runs	Block Size	Design Type	Resolution - what is estimable
8		Fractional Factorial	3 - Main Effects Only
12		Plackett-Burman	3 - Main Effects Only
16		Fractional Factorial	4 - Some 2-factor interactions
16	8	Fractional Factorial	4 - Some 2-factor interactions
16	4	Fractional Factorial	4 - Some 2-factor interactions
16	2	Fractional Factorial	4 - Some 2-factor interactions
16		Cotter	Sums of Even/Odd Terms
32		Fractional Factorial	4 - Some 2-factor interactions
32	16	Fractional Factorial	4 - Some 2-factor interactions
32	8	Fractional Factorial	4 - Some 2-factor interactions
32	4	Fractional Factorial	4 - Some 2-factor interactions
32	2	Fractional Factorial	4 - Some 2-factor interactions
64		Fractional Factorial	5+ - All 2-factor interactions
64	32	Fractional Factorial	5+ - All 2-factor interactions

This specifies an eight-run Resolution 3 fractional factorial design. For information about resolution, see "Resolution as a Measure of Confounding" on page 301.

5. Click **Continue**.

   In the Output Options outline, note that **Run Order** is set to **Randomize**. This means that the design runs will appear in random order. This is the order you should use to conduct your experimental runs.

Figure 9.4 Completed Screening Design Window

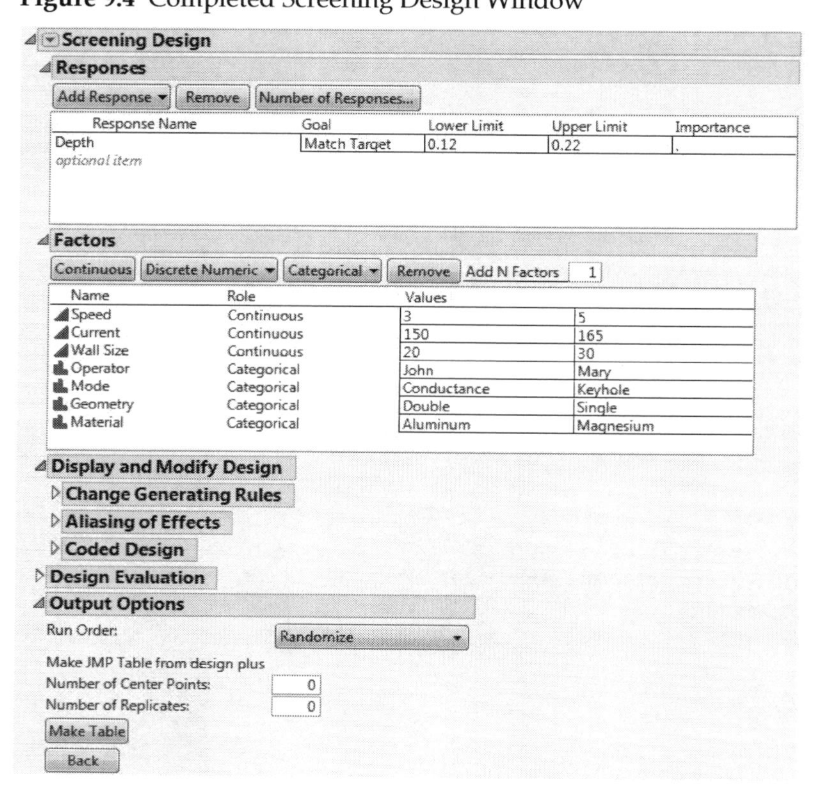

6. Open the Aliasing of Effects outline under Display and Modify Design.

Figure 9.5 Aliasing for an Eight-Run Fractional Factorial Design

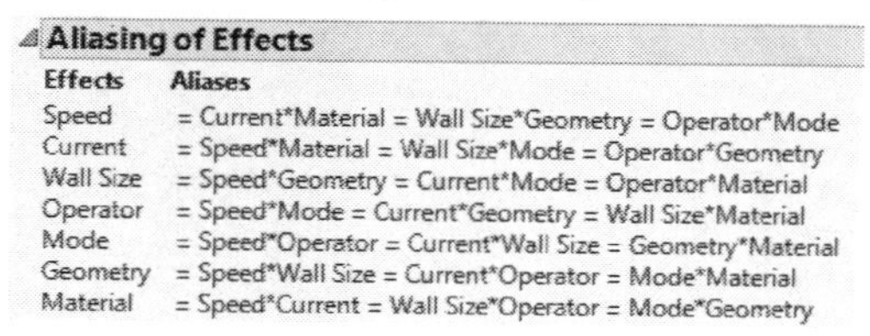

Aliasing of Effects

Effects	Aliases
Speed	= Current*Material = Wall Size*Geometry = Operator*Mode
Current	= Speed*Material = Wall Size*Mode = Operator*Geometry
Wall Size	= Speed*Geometry = Current*Mode = Operator*Material
Operator	= Speed*Mode = Current*Geometry = Wall Size*Material
Mode	= Speed*Operator = Current*Wall Size = Geometry*Material
Geometry	= Speed*Wall Size = Current*Operator = Mode*Material
Material	= Speed*Current = Wall Size*Operator = Mode*Geometry

Recall that you selected a Resolution 3 design (see Figure 9.3). In a Resolution 3 design, some main effects are confounded with two-way interactions. The Aliasing of Effects outline indicates that, for this Resolution 3 design, every main effect is completely confounded with three two-way interactions. If you suspect that two-way interactions are active, this is a poor design. For a description of confounding, see "Two-Level Regular Fractional Factorial" on page 298.

7. Click **Make Table**.

Figure 9.6 The Design Data Table

	Pattern	Speed	Current	Wall Size	Operator	Mode	Geometry	Material	Depth
1	+++++++	5	165	30	Mary	Keyhole	Single	Magnesium	•
2	-+-+-+-	3	165	20	Mary	Conductance	Single	Aluminum	•
3	+-+--+-	5	150	30	John	Conductance	Single	Aluminum	•
4	----+++	3	150	20	John	Keyhole	Single	Magnesium	•
5	-++-+--	3	165	30	John	Keyhole	Double	Aluminum	•
6	++----+	5	165	20	John	Conductance	Double	Magnesium	•
7	+--++--	5	150	20	Mary	Keyhole	Double	Aluminum	•
8	--++--+	3	150	30	Mary	Conductance	Double	Magnesium	•

Notice the following:

– The table uses the names for the responses, factors, and levels that you specified.

– The **Pattern** column shows the assignment of high and low settings for the design runs.

– This fractional factorial design is a Resolution 3 design. It enables you to study the main effects of seven factors in eight runs.

Constructing a Main Effects Screening Design

Main effects screening designs are orthogonal or near orthogonal designs. In this section, construct a main effects screening design for your seven factors.

1. Open your Screening Design window. If you have closed it, then run the **DOE Dialog** script in the Design Data table.
2. Click **Back**.
3. Click **Continue**.

Note: Setting the Random Seed and the Number of Starts in the next two steps reproduces the exact results shown in this example. When constructing a design on your own, these steps are not necessary.

4. (Optional) From the Screening Design red triangle menu, select **Set Random Seed**, type 12345, and click **OK**.
5. (Optional) From the Screening Design red triangle menu, select **Number of Starts**, type 50, and click **OK**.
6. In the Choose Screening Type panel, select the **Construct a main effects screening design** option.
7. Click **Continue**.

   Under Number of Runs, the selected option is **Default** with the number of runs set to 12. Keep this setting.
8. Click **Make Design**.

Figure 9.7 Main Effects Screening Design

Design

Run	Speed	Current	Wall Size	Operator	Mode	Geometry	Material
1	3	165	20	John	Conductance	Single	Aluminum
2	5	150	30	Mary	Conductance	Double	Aluminum
3	3	150	30	John	Keyhole	Double	Aluminum
4	5	165	20	Mary	Keyhole	Double	Magnesium
5	3	165	20	Mary	Conductance	Double	Aluminum
6	5	150	20	John	Conductance	Double	Magnesium
7	3	150	30	Mary	Conductance	Single	Magnesium
8	5	150	20	John	Keyhole	Single	Aluminum
9	3	165	30	John	Keyhole	Double	Magnesium
10	5	165	30	John	Conductance	Single	Magnesium
11	3	150	20	Mary	Keyhole	Single	Magnesium
12	5	165	30	Mary	Keyhole	Single	Aluminum

9. Open the Design Evaluation outline and then open the Color Map on Correlations outline.

Figure 9.8 Color Map on Correlations for 12-Run Main Effects Screening Design

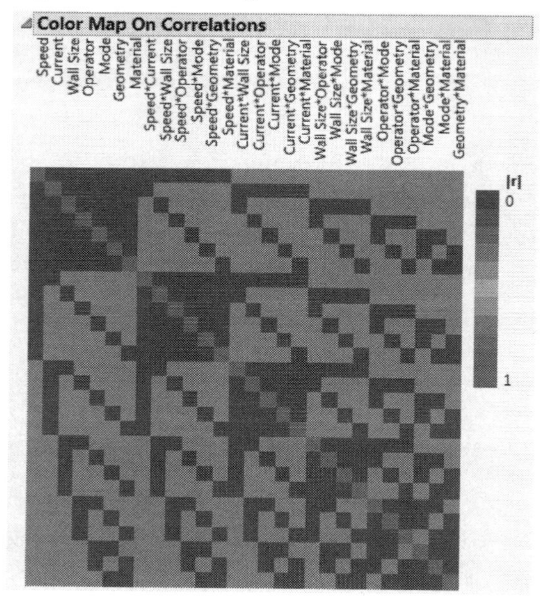

The Color Map on Correlations shows that the main effects are uncorrelated with each other. This is indicated by the solid blue off-diagonal cells in the upper left corner of the color map. Each main effect is partially aliased with some two-way interactions, indicated by the gray cells. You can see, by hovering your cursor above one of the gray cells, that the absolute correlations are 0.333.

In this case, the 12-run main effects screening design is a Plackett-Burman design, which you could have obtained in the Design List. However, in many design situations, the partial aliasing that occurs in a main effects design is preferable to the complete confounding that occurs in a fractional factorial design that you adapt to your experimental situation.

The next section shows an example of a situation where no standard design exists. In this case, JMP constructs a main effects screening design.

Main Effects Screening Design where No Standard Design Exists

Main effects screening designs are orthogonal or near orthogonal designs for the main effects. You can use them in place of standard designs and in situations where standard designs do not exist. Main effects screening designs are excellent for estimating main effects when interactions are negligible.

In the following experimental situation, no standard design exists. You need a design to study 13 factors: 2 are categorical, one with 4 levels and one with 6 levels, and 11 are continuous.

1. Select **DOE > Classical > Screening Design.**

   In the Responses panel, there is a single default response called Y. Keep this as the default response.

2. In the Factors panel, click **Categorical** and select **4 Level**.

   This adds the variable X1 with levels L1 through L4.

3. Click **Categorical** and select **6 Level**.

   This adds the variable X2, with levels L1 through L6.

4. Enter 11 next to **Add N Factors**.

5. Click **Continuous**.

   This adds 11 factors, X3 to X13, each at two levels, -1 and 1.

6. Click **Continue**.

   The Design Generation panel appears.

   There is no option to select a design from the Design List since there are no available standard designs in this situation.

   Keep the default number of runs, which is 24.

Figure 9.9 Screening Design Window for 13-Factor Design with Design Generation Panel

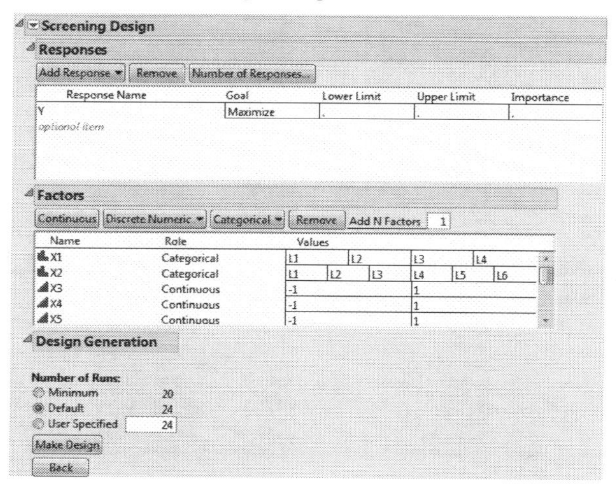

7. Click **Make Design**.

   A Design and a Design Evaluation outline appear.

8. Open the Design outline to see the randomized design.

   Note: The algorithm that generates the design uses a random starting design. To reproduce this design, save the script with the random seed by selecting **Save Script to Script Window** from the red triangle menu next to the report title.

   Next, examine the Color Map on Correlations to see that this specific design is orthogonal.

9. Open the **Design Evaluation > Color Map on Correlations** outline.

   The color map (Figure 9.10) shows red entries (using JMP default colors) on the main diagonal, indicating correlations of one. This is because each diagonal cell corresponds to the correlation of a term with itself, which is one. Off-diagonal correlations are all deep blue, indicating that correlations between distinct terms are zero. Hold your mouse pointer over any cell to see the relevant terms and their absolute correlation.

Figure 9.10 Color Map on Correlations

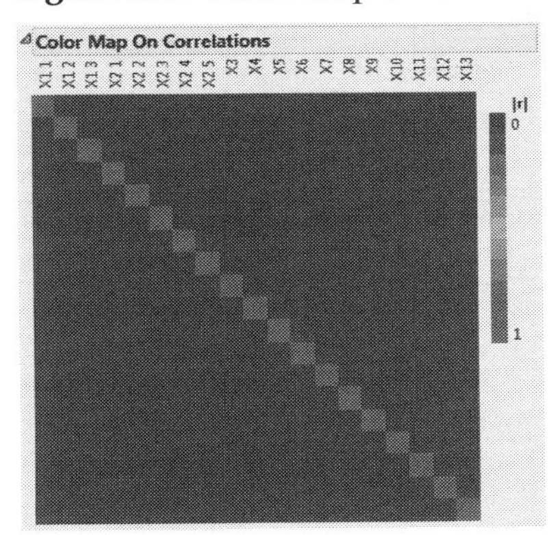

10. Click **Make Table** to construct the design table.

    The table contains the runs for your experiment in random order. Conduct the experiment in this randomized order and insert the results in column Y. Run the **Model** script in the data table to analyze your results.

Screening Design Window

The Screening Design window is updated as you work through the design steps. For more information, see "The DOE Workflow: Describe, Specify, Design" on page 54. The outlines that appear are separated by buttons that update the window. These follow the flow in the figures below.

Figure 9.11 Screening Design Flow when a Standard Design Exists

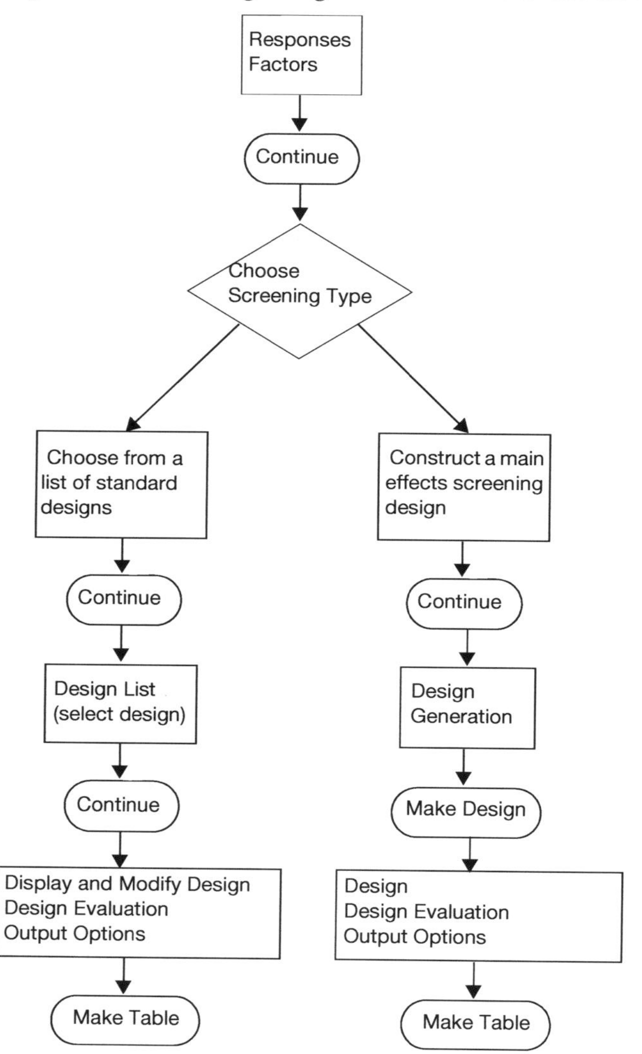

Figure 9.12 Screening Design Flow when No Standard Design Exists

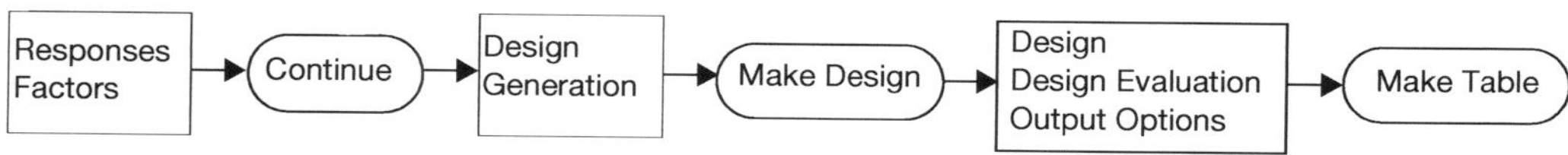

Responses

Use the Responses outline to specify one or more responses.

Tip: When you have completed the Responses outline, consider selecting **Save Responses** from the red triangle menu. This option saves the response names, goals, limits, and importance values in a data table that you can later reload in DOE platforms.

Figure 9.13 Responses Outline

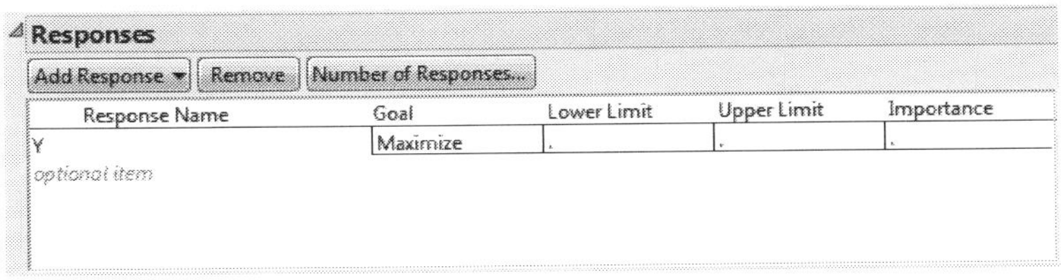

Responses

Add Response ▾ | Remove | Number of Responses...

Response Name	Goal	Lower Limit	Upper Limit	Importance
Y	Maximize	.	.	.
optional item				

Add Response Enters a single response with a goal type of Maximize, Match Target, Minimize, or None. If you select Match Target, enter limits for your target value. If you select Maximize or Minimize, entering limits is not required but can be useful if you intend to use desirability functions.

Remove Removes the selected responses.

Number of Responses Enters additional responses so that the number that you enter is the total number of responses. If you have entered a response other than the default Y, the Goal for each of the additional responses is the Goal associated with the last response entered. Otherwise, the Goal defaults to Match Target. Click the Goal type in the table to change it.

The Responses outline contains the following columns:

Response Name The name of the response. When added, a response is given a default name of Y, Y2, and so on. To change this name, double-click it and enter the desired name.

Goal, Lower Limit, Upper Limit The Goal tells JMP whether you want to maximize your response, minimize your response, match a target, or that you have no response goal. JMP assigns a Response Limits column property, based on these specifications, to each response column in the design table. It uses this information to define a desirability function for each response. The Profiler and Contour Profiler use these desirability functions to find optimal factor settings. For further details, see the Profiler chapter in the *Profilers* book and "Response Limits" on page 653 in the "Column Properties" appendix.

- A Goal of Maximize indicates that the best value is the largest possible. If there are natural lower or upper bounds, you can specify these as the Lower Limit or Upper Limit.
- A Goal of Minimize indicates that the best value is the smallest possible. If there are natural lower or upper bounds, you can specify these as the Lower Limit or Upper Limit.

- A Goal of Match Target indicates that the best value is a specific target value. The default target value is assumed to be midway between the Lower Limit and Upper Limit.
- A Goal of None indicates that there is no goal in terms of optimization. No desirability function is constructed.

Note: If your target response is not midway between the Lower Limit and the Upper Limit, you can change the target after you generate your design table. In the data table, open the Column Info window for the response column (**Cols** > **Column Info**) and enter the desired target value.

Importance When you have several responses, the Importance values that you specify are used to compute an overall desirability function. These values are treated as weights for the responses. If there is only one response, then specifying the Importance is unnecessary because it is set to 1 by default.

Editing the Responses Outline

In the Responses outline, note the following:

- Double-click a response to edit the response name.
- Click the goal to change it.
- Click on a limit or importance weight to change it.
- For multiple responses, you might want to enter values for the importance weights.

Response Limits Column Property

The Goal, Lower Limit, Upper Limit, and Importance that you specify when you enter a response are used in finding optimal factor settings. For each response, the information is saved in the generated design data table as a Response Limits column property. JMP uses this information to define the desirability function. The desirability function is used in the Prediction Profiler to find optimal factor settings. For further details about the Response Limits column property and examples of its use, see "Response Limits" on page 653 in the "Column Properties" appendix.

If you do not specify a Lower Limit and Upper Limit, JMP uses the range of the observed data for the response to define the limits for the desirability function. Specifying the Lower Limit and Upper Limit gives you control over the specification of the desirability function. For more details about the construction of the desirability function, see the Profiler chapter in the *Profilers* book.

Factors

Add factors in the Factors outline.

Tip: When you have completed the Factors outline, consider selecting **Save Factors** from the red triangle menu. This option saves the factor names, roles, changes, and values in a data table that you can later reload in DOE platforms.

Figure 9.14 Factors Outline

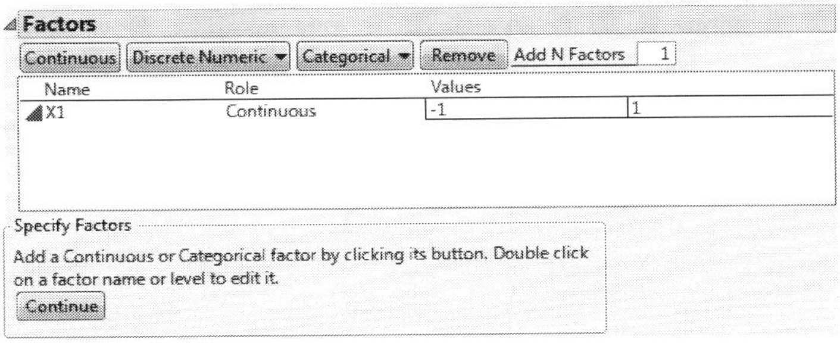

Continuous Adds a Continuous factor. The data type in the resulting data table is Numeric. A continuous factor is a factor that you can conceptually set to any value between the lower and upper limits you supply, given the limitations of your process and measurement system.

Discrete Numeric Adds a Discrete Numeric factor. A discrete numeric factor can assume only a discrete number of numeric values. These values have an implied order. The data type in the resulting data table is Numeric.

A screening design includes all levels of a discrete numeric factor and attempts to balance the levels. Fit Model treats a discrete numeric factor as a continuous predictor.

The default values for a discrete numeric factor with k levels, where $k > 2$, are the integers $1, 2, \ldots, k$. The default values for a discrete numeric factor with $k = 2$ levels are -1 and 1. Replace the default values with the settings that you plan to use in your experiment.

Categorical Adds a Categorical factor. Click to select or specify the number of levels. The data type in the resulting data table is Character. The value ordering of the levels is the order of the values, as entered from left to right. This ordering is saved in the Value Ordering column property after the design data table is created.

The default values for a categorical factor are L1, L2, ..., Lk, where k is the number of levels that you specify. Replace the default values with level names that are relevant for your experiment.

Remove Removes the selected factors.

Add N Factors Adds multiple factors. Enter the number of factors to add, click **Add Factor,** and then select the factor type. Repeat **Add N Factors** to add multiple factors of different types.

Factors Outline

The Factors outline contains the following columns:

Name The name of the factor. When added, a factor is given a default name of X1, X2, and so on. To change this name, double-click it and enter the desired name.

Role The Design Role of the factor. The Design Role column property for the factor is saved to the data table. This property ensures that the factor type is modeled appropriately. The Role of the factor determines other factor properties that are saved to the data table. For details, see "Factor Column Properties" on page 296.

Values The experimental settings for the factors.

Editing the Factors Outline

In the Factors outline, note the following:

- To edit a factor name, double-click the factor name.
- To edit a value, click the value in the Values column.

Factor Column Properties

For each factor, various column properties are saved to the design table after you create the design by selecting Make Table in the Screening Design window. These properties are also saved automatically to the data table that is created when you select the Save Factors option. You can find details about these column properties and related examples in Appendix A, "Column Properties".

Coding If the Role is Continuous or Discrete Numeric, the Coding column property for the factor is saved. This property transforms the factor values so that the low and high values correspond to –1 and +1, respectively. For details, see "Coding" on page 663 in the "Column Properties" appendix.

Value Ordering If the Role is Categorical or if a Block variable is constructed, the Value Ordering column property for the factor is saved. This property determines the order in which levels of the factor appear. For details, see "Value Ordering" on page 679 in the "Column Properties" appendix.

Design Role Each factor is given the Design Role column property. The Role that you specify in defining the factor determines the value of its Design Role column property. When you select a design with a block, that Block factor is assigned the Blocking value. The Design Role property reflects how the factor is intended to be used in modeling the experimental data. Design Role values are used in the Augment Design platform. For details, see "Design Role" on page 661 in the "Column Properties" appendix.

Factor Changes Each factor is assigned the Factor Changes column property with the value of Easy. The Factor Changes property reflects how the factor is used in modeling the experimental data. Factor Changes values are used in the Augment Design and Evaluate Design platforms. For details, see “Factor Changes” on page 676 in the “Column Properties” appendix.

RunsPerBlock For a blocking factor, indicates the maximum allowable number of runs in each block. When a Blocking factor is specified in the Factors outline, the RunsPerBlock column property is saved for that factor. For details, see “RunsPerBlock” on page 685 in the “Column Properties” appendix.

Choose Screening Type

After you enter your responses and factors and click **Continue**, one of the following results occurs:

- If a standard design can accommodate your factors and levels, two options appear in the Choose Screening Type panel. See “Choose Screening Type Options” on page 297.
- If no listed standard design exists for your factors and levels, then the Choose Screening Type panel does not appear. The Design Generation outline for constructing a main effects screening design opens. See “Design Generation” on page 306.

Choose Screening Type Options

Choose from a list of fractional factorial designs Enables you to select from a list of designs. This option is the default. For details, see “Choose from a List of Fractional Factorial Designs” on page 297.

Construct a main effects screening design Opens the Design Generation outline where you can specify the number of runs in the main effects screening design. For details about main effects screening designs, see “Main Effects Screening Designs” on page 304.

Choose from a List of Fractional Factorial Designs

The list of screening designs that you can choose from includes designs that group the experimental runs into blocks of equal sizes where the size is a power of two. Select the type of screening design that you want to use and click **Continue**.

Figure 9.15 Choosing a Type of Fractional Factorial Design

Design List

Choose a design by clicking on its row in the list.

Number Of Runs	Block Size	Design Type	Resolution - what is estimable
4		Fractional Factorial	3 - Main Effects Only
8		Full Factorial	>6 - Full Resolution
8	4	Full Factorial	5+ - All 2-factor interactions
8	2	Full Factorial	4 - Some 2-factor interactions

optional item

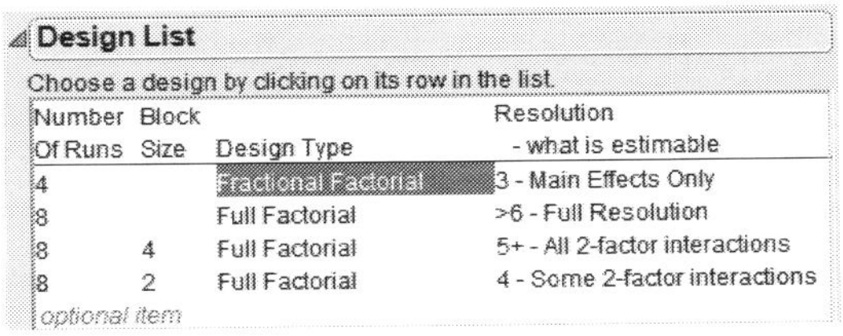

The Design List contains the following columns:

Number of Runs Total number of runs in the design.

Block Size Number of runs in a block. The number of blocks is the Number of Runs divided by Block Size.

Design Type Description of the type of design. See "Design Type" on page 298.

Resolution Gives the resolution of the design and a brief description of the type of aliasing. See"Resolution as a Measure of Confounding" on page 301.

Design Type

The Design List provides the following types of designs:

- "Two-Level Full Factorial" on page 298
- "Two-Level Regular Fractional Factorial" on page 298
- "Plackett-Burman Designs" on page 299
- "Mixed-Level Designs" on page 299
- "Cotter Designs" on page 300

Two-Level Full Factorial

A full factorial design has runs for all combinations of the levels of the factors. The sample size is the product of the levels of the factors. For two-level designs, this is 2^k where k is the number of factors.

Full factorial designs are orthogonal for all effects. It follows that estimates of the effects are uncorrelated. Also, if you remove an effect from the analysis, the values of the other estimates do not change. Their p-values change slightly, because the estimate of the error variance and the degrees of freedom are different.

Full factorial designs allow the estimation of interactions of all orders up to the number of factors. However, most empirical modeling involves only first- or second-order approximations to the true functional relationship between the factors and the responses. From this perspective, full factorial designs are an inefficient use of experimental runs.

Two-Level Regular Fractional Factorial

A regular fractional factorial design also has a sample size that is a power of two. For two-level designs, if k is the number of factors, the number of runs in a regular fractional factorial design is 2^{k-p} where $p < k$. A 2^{k-p} fractional factorial design is a 2^{-p} fraction of the k-factor full factorial design. Like full factorial designs, regular fractional factorial designs are orthogonal.

A full factorial design for k factors provides estimates of all interaction effects up to degree k. But because experimental runs are typically expensive, smaller designs are preferred. In a smaller design, some of the higher-order effects are *confounded* with other effects, meaning that the effects cannot be distinguished from each other. Although a linear combination of the confounded effects is estimable, it is not possible to attribute the variation to a specific effect or effects.

In fact, fractional factorials are designed by deciding in advance which interaction effects are confounded with other interaction effects. Experimenters are usually not concerned with interactions involving more than two factors. Three-way and higher-order interaction effects are often assumed to be negligible.

Plackett-Burman Designs

Plackett-Burman designs are an alternative to regular fractional factorials for screening. The number of runs in a Plackett-Burman design is a multiple of four rather than a power of two. There are no two-level fractional factorial designs with run sizes between 16 and 32. However, there are 20-run, 24-run, and 28-run Plackett-Burman designs.

In a Plackett-Burman design, main effects are orthogonal and two-factor interactions are only partially confounded with main effects. By contrast, in a regular Resolution 3 fractional factorial design, some two-factor interactions are indistinguishable from main effects. Plackett-Burman designs are useful when you are interested in detecting large main effects among many factors and where interactions are considered negligible.

Mixed-Level Designs

For most designs that involve categorical or discrete numeric factors at three or more levels, standard designs do not exist. In such cases, the screening platform generates *main effects screening designs*. These designs are orthogonal or near orthogonal for main effects.

For cases where standard mixed-level designs exist, the possible designs are given in the Design List. The Design List provides fractional factorial designs for pure three-level factorials with up to 13 factors. For mixed two-level and three-level designs, the Design list includes the complete factorials and the orthogonal-array designs listed in Table 9.1.

If your number of factors does not exceed the number for a design listed in the table, you can adapt that design by using an appropriate subset of its columns. Some of these designs are not balanced, even though they are all orthogonal.

Table 9.1 Table of Mixed-Level Designs

	Number of Factors	
Design	**Two–Level**	**Three–Level**
L18 John and L18 Taguchi	1	7
L18 Chakravarty	3	6
L18 Hunter	8	4
L36 Taguchi	11	12

Cotter Designs

Note: By default, Cotter designs are not included in the Design List. To include Cotter designs, deselect **Suppress Cotter Designs** in the Screening Design red triangle menu. To always show Cotter designs, select **File > Preferences > Platforms > DOE** and deselect **Suppress Cotter Designs**.

Cotter designs are useful when you must test many factors, some of which might interact, in a very small number of runs. Cotter designs rely on the principle of effect sparsity. They assume that the sum of effects shows an effect if one of the components of the sum has an active effect. The drawback is that several active effects with mixed signs might sum to near zero, thereby failing to signal an effect. Because of this false-negative risk, many statisticians discourage their use.

For k factors, a Cotter design has $2k + 2$ runs. The design structure is similar to the "vary one factor at a time" approach.

The Cotter design is constructed as follows:

- A run is defined with all factors set to their high level.
- For each of the next k runs, one factor in turn is set at its low level and the others high.
- The next run sets all factors at their low level.
- For each of the next k runs, one factor in turn is set at its high level and the others low.
- The runs are randomized.

When you construct a Cotter design, the design data table includes a set of columns to use as regressors. The column names are of the form <factor name> Odd and <factor name> Even. They are constructed by summing the odd-order and even-order interaction terms, respectively, that contain the given factor.

For example, suppose that there are three factors, A, B, and C. Table 9.2 shows how the values in the regressor columns are calculated.

Table 9.2 Cotter Design Table

Effects Summed for Odd and Even Regressor Columns	
AOdd = A + ABC	AEven = AB + AC
BOdd = B + ABC	BEven = AB + BC
COdd = C + ABC	CEven = BC + AC

The Odd and Even columns define an orthogonal transformation. For this reason, tests for the parameters of the odd and even columns are equivalent to testing the combinations on the original effects.

Resolution as a Measure of Confounding

The *resolution* of a design is a measure of the degree of confounding in the design. The trade-off in screening designs is between the number of runs and the resolution of the design.

Experiments are classified by *resolution number* into these groups:

- Resolution 3 means that some main effects are confounded with one or more two-factor interactions. In order for the main effects to be meaningful, these interactions must be assumed to be negligible.
- Resolution 4 means that main effects are not confounded with other main effects or two-factor interactions. However, some two-factor interactions are confounded with other two-factor interactions.
- Resolution 5 means that there is no confounding between main effects, between main effects and two-factor interactions, and between pairs of two-factor interactions. Some two-factor interactions are confounded with three-factor interactions.
- Resolution 5+ means that the design has resolution greater than 5 but is not a full factorial design.
- Resolution 6 means that there is no confounding between effects of any order. The design is a full factorial design.

A *minimum aberration* design is one that minimizes the number of confoundings for a given resolution. A minimum aberration design of a given resolution minimizes the number of *words* in the defining relation that are of minimum length. For a description of *words*, see “Change Generating Rules” on page 302. For a discussion of minimum aberration designs, see Fries and Hunter (1984).

Display and Modify Design

In the Design List, if you select a fractional factorial design with all continuous or two-level categorical factors, and possibly a blocking factor, the Display and Modify Design outline opens after you click Continue. Modify your design using the reports in this outline. See "Modify Generating Rules in a Fractional Factorial Design" on page 311 for an example of changing the generating rules to construct a design.

Note: The Change Generating Rules and Aliasing of Effects outlines do not appear for Plackett-Burman designs or Cotter designs, because interactions are not identically equal to main effects.

Change Generating Rules Specify the defining relation for the design. The defining relation determines which fraction of the full fractional factorial design that JMP provides. See "Change Generating Rules" on page 302.

Aliasing of Effects Shows the confounding pattern for the fractional factorial design. Click the red arrow at the bottom of the panel to see interactions to a specified order. The interactions and their aliases are presented in a data table.

Coded Design Shows the pattern of high and low values for the factors in each run.

Note: For Cotter designs, the Change Generating Rules and Aliasing of Effects outlines do not apply and are not shown.

Change Generating Rules

The generating rules define the relation used to construct a specific fractional factorial design. Your experimental situation might require that you define a fraction of the design that provides a coding or aliasing structure that is different from the standard fraction. You can do this by changing the generating rules in the Display and Modify Design outline. For details about defining relations and generating rules, see Montgomery (2009).

The defining relation for a design is determined by the *words* in the generating rules. A *word* is represented by a product of factors, but it is interpreted as the elementwise product of the entries in the design matrix for those columns. A defining relation consists of words whose product is a column of ones, called the *identity*.

Figure 9.16 shows the default-generating rules for a 2^{5-3} design (five factors and eight runs).

Figure 9.16 Generating Rules for the Standard 2^{5-3} Design

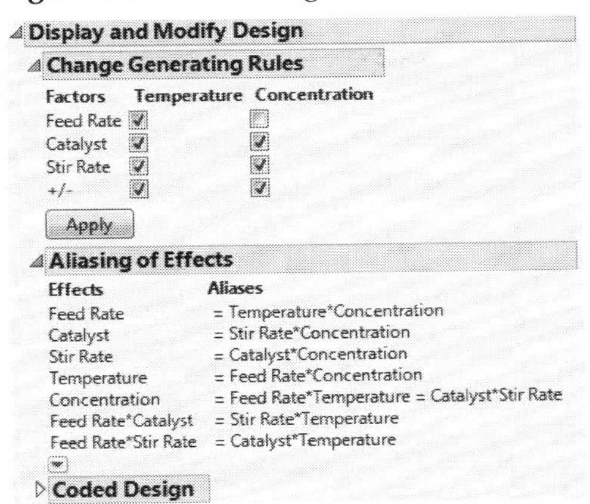

In each column of the Change Generating Rules panel, the factor listed at the top and the factors in the column whose boxes are selected form a word in the defining relation. For example, the first column indicates that Temperature = Feed Rate*Catalyst*Stir Rate is a word in the defining relation.

- If the +/- box is selected, the sign associated with the generating rule is positive and the corresponding word equals the identity.
- If the +/- box is not selected, the sign associated with the generating rule is negative and the corresponding word equals minus the identity.

See "Obtain the Defining Relations in the 2^{5-3} Design" on page 304.

The *principal fraction* of a full factorial design is the fractional factorial design obtained by setting all the defining relations equal to the identity. By default, the factorial design that JMP provides is the principal fraction. Notice that the +/- box is selected by default for all generating rules, so that each word in the defining relation equals the identity.

Generating rules determine the coding and aliasing of effects for the design. In some cases, you might want to use a fraction that results in a coding or an aliasing structure that differs from that of the standard fraction.

- To change the generating rules, select the appropriate boxes.
- To see the effect of your selections on the Aliasing of Effects results and on the Coded Design, click **Apply**.

For an example, see "Change the Generating Rules to Obtain a Different Fraction" on page 314.

Obtain the Defining Relations in the 2^{5-3} Design

Figure 9.16 shows two columns of check boxes:

- The first column represents the word Temperature = Feed Rate*Catalyst*Stir Rate.
- The second column represents the word Concentration = Catalyst*Stir Rate.

Define I to represent a column consisting of the values +1. Because all factor levels are -1 or +1, the word in the first column is equivalent to Temperature*Feed Rate*Catalyst*Stir Rate = I. The word in the second column is equivalent to Concentration*Catalyst*Stir Rate = I. Together, these give the defining relations for the 2^{5-3} design:

I = Temperature*Feed Rate*Catalyst*Stir Rate = Concentration*Catalyst*Stir Rate

Obtain the Aliasing of Effects Relations in the 2^{5-3} Design

The aliasing structure in the Aliasing of Effects outline is determined by the defining relations and the fact that factor levels are +1 and -1. Recall that the first generating rule is Temperature = Feed Rate*Catalyst*Stir Rate and the second is Concentration = Catalyst*Stir Rate.

To obtain the first relation in the Aliasing of Effects outline, notice that applying these two generating rules gives the expression:

Temperature = Feed Rate*Catalyst*Stir Rate = Feed Rate*Concentration

The second equality follows from replacing Catalyst*Stir Rate by Concentration using the second generating rule.

Now, post-multiply the first and third expressions by Concentration to obtain the following expression:

Temperature*Concentration = Feed Rate*Concentration*Concentration

Because the column for Concentration in the design matrix contains values of -1 and +1, the term Concentration*Concentration represents a column of +1 values. The expression becomes the first alias relation shown in the Aliasing of Effects outline:

Temperature*Concentration = Feed Rate*I = Feed Rate

The other alias relations can be obtained using similar calculations.

Main Effects Screening Designs

If an experiment involves categorical or discrete numeric factors, or if the number of runs is constrained, it might not be possible to construct an orthogonal design for screening main effects. However, a main effects screening design can be constructed. See Lekivetz et al. (2015).

A main effects screening design is a design with good balance properties as described by a Chi-square criterion. See "Chi-Square Efficiency" on page 305. Such designs have desirable statistical properties for main effect models.

The algorithm used to generate the design attempts to construct an orthogonal array of strength two. Strength-two orthogonal arrays permit orthogonal estimation of main effects when interactions are negligible. These arrays are ideal for screening designs. Regular fractional factorial designs of Resolution 3 and Plackett-Burman designs are examples of strength-two orthogonal arrays.

Consider all possible pairs of levels for factors in the design. The algorithm attempts to balance the number of pairs of levels as far as possible. Given that a fixed number of columns has been generated, a new balanced column is randomly constructed. A measure is defined that reflects the degree of balance achieved for pairs that involve the new column. The algorithm attempts to minimize this measure by interchanging levels within the new column.

Chi-Square Efficiency

Suppose that a design has n runs and p factors corresponding to the columns of the design matrix.

- Denote the levels of factors k and l by $a = 0, 1, \ldots, s_k - 1$ and $b = 0, 1, \ldots, s_l - 1$, respectively.
- Denote the number of times that the combination of levels (a,b) appears in columns k and l by $n_{kl}(a, b)$.

A measure of the lack of orthogonality evidenced by columns k and l is given by the following expression:

$$\chi^2_{kl} = \sum_{a=0}^{s_k-1} \sum_{b=0}^{s_l-1} \frac{[n_{kl}(a,b) - n/(s_k s_l)]^2}{n/(s_k s_l)}$$

A measure of the average non-orthogonality of the design is given by this expression:

$$\chi^2 = \sum_{1 \le k < l \le p} \chi^2_{kl} / [p(p-1)/2]$$

The maximum possible value of χ^2, denoted χ^2_{max}, is obtained. The chi-square efficiency of a design is defined as follows:

$$\text{Chi-Square Efficiency} = 100(1 - \chi^2 / \chi^2_{max})$$

Chi-square efficiency indicates how close χ^2 is to zero, relative to a design in which pairs of levels show extreme lack of balance.

Design Generation

When you construct a main effects screening design, the Design Generation outline enables you to specify the number of runs. To generate the design, click **Make Design**.

Minimum A lower bound on the number of runs necessary to avoid failures in design generation. When you select Minimum, the resulting design is saturated. There are no degrees of freedom for error.

Note: If you select the Minimum number of runs, there is no error term for testing. You cannot test parameter estimates. This choice is appropriate only when the cost of additional runs is prohibitive.

Default Suggests the number of runs. This value is based on heuristics for creating a balanced design with at least four runs more than the Minimum number of runs.

User Specified Specify the number of runs that you want. Enter that value into the Number of Runs text box. This option enables you to balance the cost of additional runs against the potential gain in information.

Design

The Design outline shows the runs for the main effects screening design. To change the run order for your design table, you can select Run Order options in the Output Options panel before generating the table.

Design Evaluation

The Design Evaluation outline provides a number of ways to evaluate the properties of the generated design. Open the Design Evaluation outline to see the following options:

Power Analysis Enables you to explore your ability to detect effects of given sizes.

Prediction Variance Profile Shows the prediction variance over the range of factor settings.

Fraction of Design Space Plot Shows how much of the model prediction variance lies below (or above) a given value.

Prediction Variance Surface Shows a surface plot of the prediction variance for any two continuous factors.

Estimation Efficiency For each parameter, gives the fractional increase in the length of a confidence interval compared to that of an ideal (orthogonal) design, which might not exist. Also gives the relative standard error of the parameters.

Alias Matrix Gives coefficients that indicate the degree by which the model parameters are biased by effects that are potentially active, but not in the model.

Color Map on Correlations Shows the absolute correlation between effects on a plot using an intensity scale.

Design Diagnostics Indicates the optimality criterion used to construct the design. Also gives efficiency measures for your design.

For more details about the Design Evaluation panel, see “Design Evaluation” on page 433 in the “Evaluate Designs” chapter.

Note: The Design Evaluation outline is not shown for Cotter designs.

Output Options

Specify details for the output data table in the Output Options panel. When you have finished, click **Make Table** to construct the data table for the design. Figure 9.17 shows the Output Options panel for a standard design selected from the Design List. For a main effects screening design, only Run Order is available.

Figure 9.17 Select the Output Options

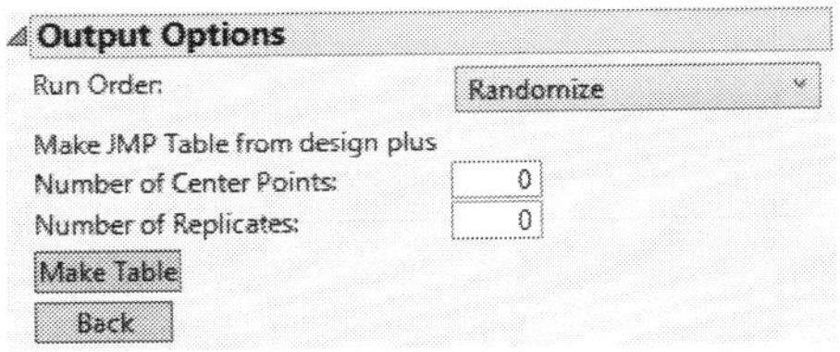

Run Order

The **Run Order** options determine the order of the runs in the design table. Choices include the following:

Keep the Same Rows in the design table are in the same order as in the Coded Design or Design outlines.

Sort Left to Right Columns in the design table are sorted from left to right.

Randomize Rows in the design table are in random order.

Sort Right to Left Columns in the design table are sorted from right to left.

Randomize within Blocks Rows in the design table are in random order within the blocks. (Not available if you select **Construct a main effects screening design.**)

Center Points and Replicates

Number of Center Points Specifies how many additional runs to add as center points to the design. A center point is a run where every continuous factor is set at the center of the factor’s range. This option is not available if you select **Construct a main effects screening design**.

Suppose that your design includes both continuous and categorical factors. If you request center points in the Output Options panel, the center points are distributed as follows:

1. The settings for the categorical factors are ordered using the value ordering specified in the Factors outline.
2. One center point is assigned to each combination of the settings of the categorical factors in order, and this is repeated, until all center points are assigned.

Number of Replicates For designs in the Design List, specify the number of times to replicate the entire design, including center points. One replicate doubles the number of runs. This option is not available if you select **Construct a main effects screening design**.

Note: If you request center points or replicates and click Make Table repeatedly, these actions are applied to the most recently constructed design table.

Make Table

Click **Make Table** to create a data table that contains the runs for your experiment. In the table, the high and low values that you specified appear for each run.

Figure 9.18 The Design Data Table

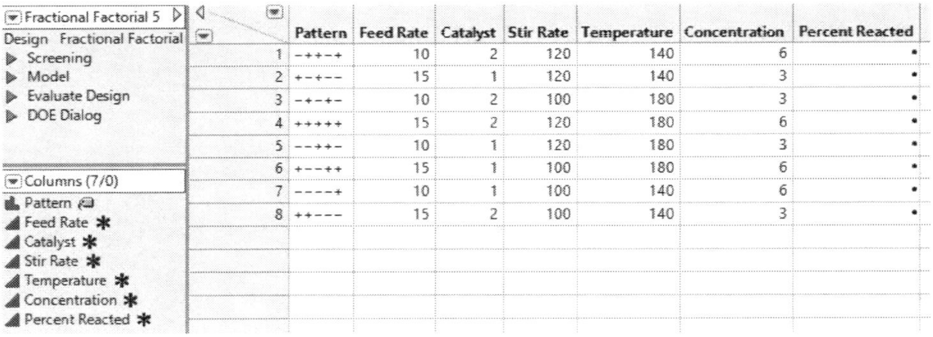

Fractional Factorial 5
Design Fractional Factorial
▶ Screening
▶ Model
▶ Evaluate Design
▶ DOE Dialog

Columns (7/0)
Pattern
Feed Rate *
Catalyst *
Stir Rate *
Temperature *
Concentration *
Percent Reacted *

	Pattern	Feed Rate	Catalyst	Stir Rate	Temperature	Concentration	Percent Reacted
1	-++-+	10	2	120	140	6	•
2	+-+--	15	1	120	140	3	•
3	-+-+-	10	2	100	180	3	•
4	+++++	15	2	120	180	6	•
5	--++-	10	1	120	180	3	•
6	+--++	15	1	100	180	6	•
7	----+	10	1	100	140	6	•
8	++---	15	2	100	140	3	•

The name of the table is the design type that generated it.

The design table includes the following scripts:

Screening Runs the Analyze > Specialized Modeling > Specialized DOE Models > Fit Two Level Screening platform. Only provided when all factors are at two levels.

Model Runs the Analyze > Fit Model platform.

Evaluate Design Runs the DOE > Design Diagnostics > Evaluate Design platform.

DOE Dialog Re-creates the Screening Design window that you used to generate the design table. The script also contains the random seed used to generate your design.

Run the **Screening** or **Model** scripts to analyze the data.

If the design was selected from the Design List, the design table contains a Pattern column. The Pattern column contains entries that summarize the run in the given row. Low settings are denoted by "–", high settings by "+", and center points by "0". Pattern can be useful as a label variable in plots.

Screening Design Options

The Screening Design red triangle menu contains the following options:

Save Responses Creates a data table containing a row for each response with a column called Response Name that identifies the responses. Four additional columns contain the Lower Limit, Upper Limit, Response Goal, and Importance. Saving responses enables you to quickly load them into a DOE window.

Load Responses Loads responses from a data table that you have saved using the Save Responses option.

Save Factors Creates a data table containing a column for each factor that contains its factor levels. A factor's column contains column properties associated with the factor. Saving factors enables you to quickly load them into a DOE window.

Note: It is possible, but not recommended, to create a factors table by entering data into an empty table, but remember to assign each column an appropriate Design Role. Do this by right-clicking on the column name in the data grid and selecting **Column Properties > Design Role**. In the Design Role area, select the appropriate role.

Load Factors Loads factors from a data table that you have saved using the Save Factors option.

Save Constraints Unavailable for this platform because constraints are not supported.

Load Constraints Unavailable for this platform because constraints are not supported.

Set Random Seed Sets the random seed that JMP uses to control certain actions that have a random component. These actions include:

– simulating responses using the Simulate Responses option
– randomizing Run Order for design construction
– selecting a starting design for designs based on random starts.

To reproduce a design or simulated responses, enter the random seed used to generate them.

Note: To reproduce a screening design, you must enter the random seed *before* you select an option under Choose Screening Type. If you select a screening type, click Continue, and *then* enter a random seed, the resulting design might not match the design you previously obtained using that random seed.

Notice that the random seed associated with a design is included in the DOE Dialog script that is saved to the design data table.

Simulate Responses Adds response values and a column containing a simulation formula to the design table. Select this option before you click Make Table.

When you click Make Table, the following occur:

- A set of simulated response values is added to each response column.
- For each response, a new a column that contains a simulation model formula is added to the design table. The formula and values are based on the model that is specified in the design window.
- A Model window appears where you can set the values of coefficients for model effects and specify one of three distributions: Normal, Binomial, or Poisson.
- A script called **DOE Simulate** is saved to the design table. This script re-opens the Model window, enabling you to re-simulate values or to make changes to the simulated response distribution.

Make selections in the Model window to control the distribution of simulated response values. When you click Apply, a formula for the simulated response values is saved in a new column called <Y> Simulated, where Y is the name of the response. Clicking Apply again updates the formula and values in <Y> Simulated.

For additional details, see "Simulate Responses" on page 106 in the "Custom Designs" chapter.

Note: JMP PRO You can use Simulate Responses to conduct simulation analyses using the JMP Pro Simulate feature. For information about Simulate and some DOE examples, see the Simulate chapter in the *Basic Analysis* book.

Suppress Cotter Designs Excludes Cotter designs from the Design List. This option is selected by default. Deselect it to show Cotter designs in the Design List.

Note: You can set a preference to always show Cotter designs. Select File > Preferences > Platforms > DOE and deselect Suppress Cotter Designs.

Number of Starts (Main Effects Screening Designs only.) Specify the maximum number of times that the algorithm regenerates entire designs from scratch, attempting to optimize the final design.

Design Search Time (Main Effects Screening Designs only.) Specify the maximum number of seconds spent searching for a design. The default search time is 15 seconds.

If the iterations of the algorithm require more than a few seconds, a Computing Design progress window appears. The progress bar displays *Chi2 Efficiency*. See "Chi-Square Efficiency" on page 305. If you click **Cancel** in the progress window, the calculation stops and gives the best design found at that point.

Note: You can set a preference for Design Search Time. Select **File > Preferences > Platforms > DOE**. Select **Design Search Time** and enter the maximum number of seconds. If an orthogonal array is found, the search terminates. In certain situations where more time is required, JMP automatically extends the search time.

Number of Column Starts (Main Effects Screening Designs only.) Specify the maximum number of times that the algorithm attempts to optimize a given column before moving on to constructing the next column. The default number of column starts is 50. For details, see "Main Effects Screening Designs" on page 304.

Advanced Options > Set Delta for Power Specify the difference in the mean response that you want to detect for model effects. See "Set Delta for Power" on page 112 in the "Custom Designs" chapter.

Save Script to Script Window Creates the script for the design that you specified in the Screening Design window and places it in an open script window.

Additional Examples of Screening Designs

This section contains the following examples:

- "Modify Generating Rules in a Fractional Factorial Design" on page 311
- "Plackett-Burman Design" on page 317

Modify Generating Rules in a Fractional Factorial Design

The following example, adapted from Meyer, *et al.* (1996), shows how to use the Screening Design platform when you have many factors. In this example, a chemical engineer investigates the effects of five factors on the percent reaction of a chemical process. The factors are:

- Feed Rate - the amount of raw material added to the reaction chamber in liters per minute
- Catalyst (as a percent)
- Stir Rate - the RPMs of a propeller in the chamber
- Temperature (in degrees Celsius)

- Concentration of reactant

Production constraints limit the size of the experiment to no more than twelve runs. You decide to consider the 8-run fractional factorial design and the 12-run Plackett-Burman design. Also, you suspect the following statements to be true:

- The Temperature*Concentration interaction is active, so you want a design that does not alias this interaction with a main effect.
- The Catalyst*Temperature* interaction is not likely to be active.
- The Stir Rate*Concentration interaction is not likely to be active.

Use this information in constructing your design.

Create the Standard Fractional Factorial Design

To create the standard fractional factorial design, do the following:

- "Specify the Response" on page 312
- "Specify the Factors" on page 312 or "Specify Factors Manually" on page 313
- "Choose a Design" on page 313

Specify the Response

1. Select **DOE > Classical > Screening Design**.
2. Double-click Y under Response Name and type Percent Reacted.

   Note that the default Goal is Maximize. The Goal is to maximize the response, but the minimum acceptable reaction percentage is 90 (Lower Limit) and the upper limit is 100 (Upper Limit).
3. Click under Lower Limit and type 90.
4. Click under Upper Limit and type 100.
5. Leave the area under Importance blank.

   Because there is only one response, that response is given Importance 1 by default.

   See Figure 9.19 for the completed Responses outline. Now, specify the factors.

Specify the Factors

You can enter the factors manually or automatically:

- To enter the factors automatically, use the Reactor Factors.jmp data table:
  1. Select **Help > Sample Data Library** and open Design Experiment/Reactor Factors.jmp.
  2. From the Screening Design red triangle menu, select **Load Factors**. Proceed to "Choose a Design" on page 313.
- To enter the factors manually, follow the steps below.

Specify Factors Manually

1. Add five continuous factors by entering 5 in the **Add N Factors** box and clicking **Continuous**.
2. Change the default factor names (X1-X5) to Feed Rate, Catalyst, Stir Rate, Temperature, and Concentration.
3. Enter the low and high values, as follows:
   – Feed Rate: 10, 15
   – Catalyst: 1, 2
   – Stir Rate: 100, 120
   – Temperature: 140, 180
   – Concentration: 3, 6

Figure 9.19 Responses and Factors Outlines

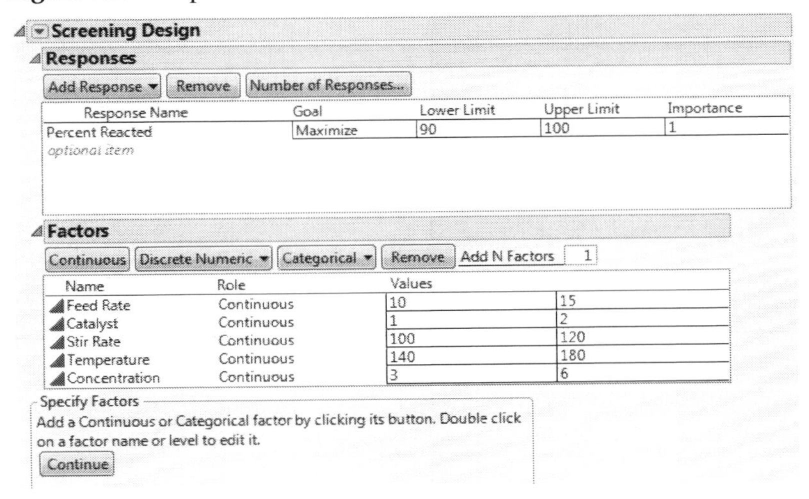

Choose a Design

1. Click **Continue**.
2. From the Choose Screening Type panel, accept the default selection to **Choose from a list of fractional factorial designs** and click **Continue**.

   Designs for the factors and levels that you specified are listed in the Design List (Figure 9.20).

Figure 9.20 Fractional Factorial Designs for Five Continuous Factors

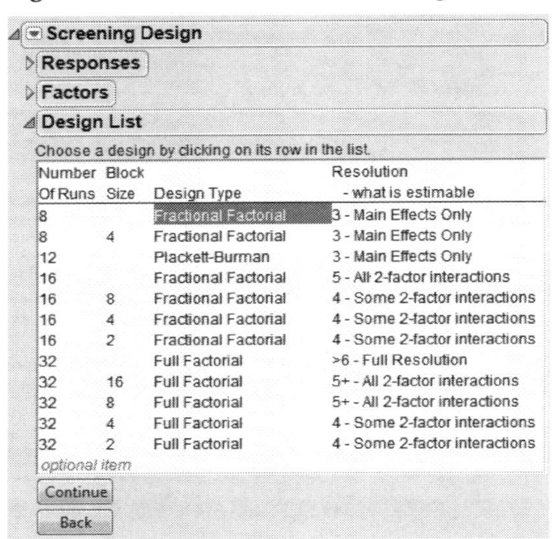

3. The design that you want is the first in the list and happens to be selected by default (Figure 9.20). Accept that selection and click **Continue**.

   Because you are limited to eight runs and have no blocking factor, your best design option is the 8-run fractional factorial design with no blocks. This design is a 2^{5-2} fractional factorial design. It is one quarter of the full factorial design for five factors.

Change the Generating Rules to Obtain a Different Fraction

In this example, you want to know whether the Temperature*Concentration interaction is confounded with a main effect. Use the Display and Modify Design outline to view the aliasing structure for the design that you selected and to change it, if appropriate.

1. Open the Aliasing of Effects outline.

Figure 9.21 Aliasing of Effects Outline

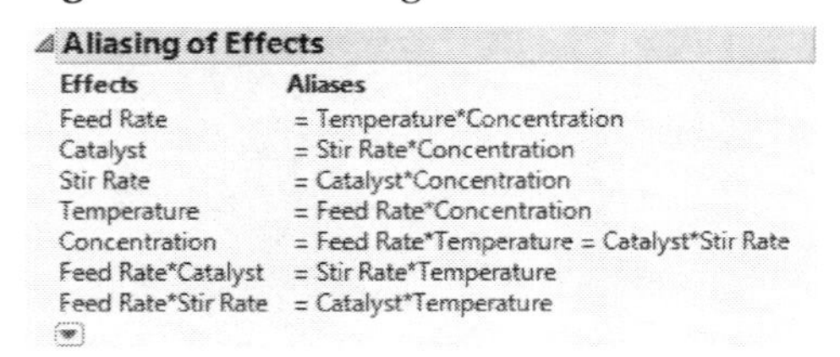

The Temperature*Concentration interaction, which you suspect is active, is confounded with Feed Rate, a main effect. You want to change the generating rules to construct a design where **Feed Rate** is aliased with effects that you suspect are inactive, and where the **Temperature*Concentration** interaction is not aliased with a main effect.

2. Open the Change Generating Rules outline.

   The default-generating rules give you the standard (or *principal*) one-quarter fraction of the full factorial design. Recall that you suspect that the **Catalyst*Temperature** and **Stir Rate*Concentration** interactions are not likely to be active. Redefine the generating rules so that these two interactions are confounded with **Feed Rate**. The redefined generating rules give you a different one-quarter fraction of the full factorial design.

3. Do the following:

   – Deselect **Stir Rate** in the **Temperature** column.

   – Deselect **Catalyst** in the **Concentration** column.

   – Select **Feed Rate** in the **Concentration** column.

Figure 9.22 New Generating Rules

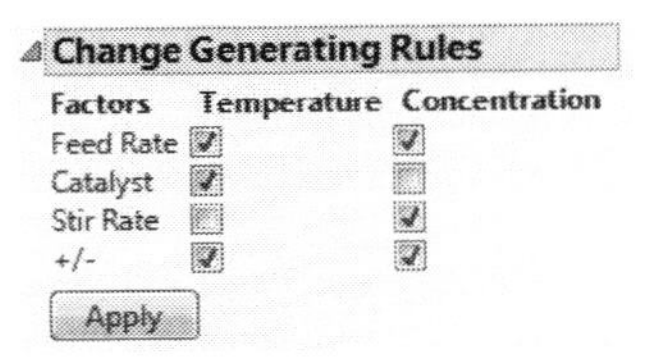

4. Click **Apply**.

Figure 9.23 Aliasing of Effects Outline for Modified Generating Rules

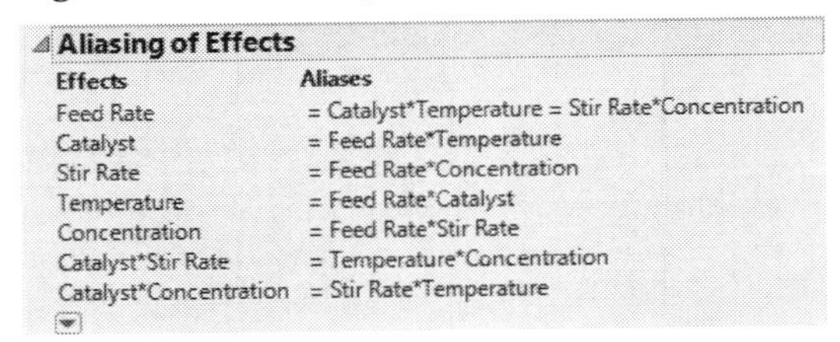

In the design that you have defined, **Feed Rate** is confounded with **Catalyst*Temperature** and **Stir Rate*Concentration**. Also, the **Temperature*Concentration** interaction is now confounded with the two-way interaction **Catalyst*Stir Rate**.

5. In the Output Options outline, accept the default Run Order setting of **Randomize** and click **Make Table**.

Figure 9.24 Eight-Run Fractional Factorial Design Table

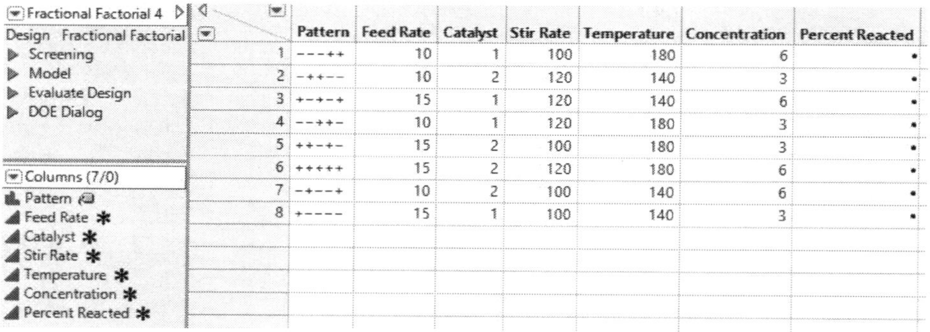

Fractional Factorial 4
Design Fractional Factorial
▶ Screening
▶ Model
▶ Evaluate Design
▶ DOE Dialog

Columns (7/0)
Pattern
Feed Rate *
Catalyst *
Stir Rate *
Temperature *
Concentration *
Percent Reacted *

	Pattern	Feed Rate	Catalyst	Stir Rate	Temperature	Concentration	Percent Reacted
1	---++	10	1	100	180	6	•
2	-++--	10	2	120	140	3	•
3	+-+-+	15	1	120	140	6	•
4	--++-	10	1	120	180	3	•
5	++-+-	15	2	100	180	3	•
6	+++++	15	2	120	180	6	•
7	-+--+	10	2	100	140	6	•
8	+----	15	1	100	140	3	•

The design table shows the design that you constructed. Notice that the table contains a column for the response that you defined in the Screening window, **Percent Reacted**, where you can record your experimental results.

The **Screening**, **Model**, and **DOE Dialog** scripts are also included. For details about these scripts, see "Make Table" on page 308.

Analyze the Results

Next you conduct the experiment, record your data, and proceed to analyze the results.

1. Select **Help > Sample Data Library** and open Design Experiment/Reactor 8 Runs.jmp.

   You can estimate seven effects with your eight runs. Of these, you expect only a few to be active. Because you want to estimate seven effects, there are no degrees of freedom for error. For these reasons, you use the Screening platform to analyze the results.

2. Run the **Screening** script in the data table.

   The **Screening** script launches the Screening platform (**Analyze > Specialized Modeling > Specialized DOE Models > Fit Two Level Screening**) for your response and factors.

   Figure 9.25 shows the report.

Figure 9.25 Report for Screening Example

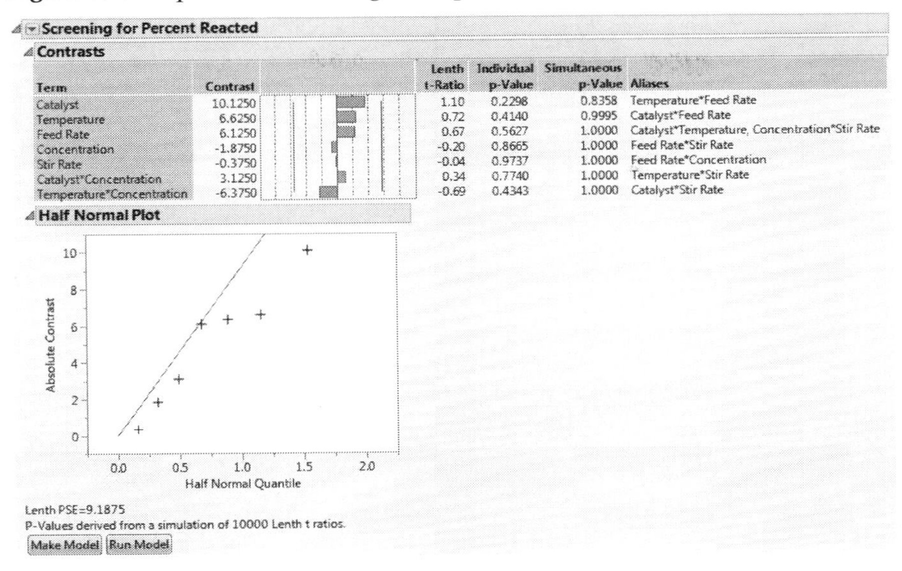
Screening for Percent Reacted

Contrasts

Term	Contrast	Lenth t-Ratio	Individual p-Value	Simultaneous p-Value	Aliases
Catalyst	10.1250	1.10	0.2298	0.8358	Temperature*Feed Rate
Temperature	6.6250	0.72	0.4140	0.9995	Catalyst*Feed Rate
Feed Rate	6.1250	0.67	0.5627	1.0000	Catalyst*Temperature, Concentration*Stir Rate
Concentration	-1.8750	-0.20	0.8665	1.0000	Feed Rate*Stir Rate
Stir Rate	-0.3750	-0.04	0.9737	1.0000	Feed Rate*Concentration
Catalyst*Concentration	3.1250	0.34	0.7740	1.0000	Temperature*Stir Rate
Temperature*Concentration	-6.3750	-0.69	0.4343	1.0000	Catalyst*Stir Rate

Half Normal Plot

Lenth PSE=9.1875
P-Values derived from a simulation of 10000 Lenth t ratios.
Make Model | Run Model

Note: Since the p-values are obtained using a simulation-based technique, your *p*-values might not precisely match those shown here.

The report shows both Individual and Simultaneous *p*-values based on Lenth t-ratios. None of the effects are significant, even with respect to the Individual *p*-values. The Half Normal Plot suggests that the effects reflect only random noise.

For more details about the Screening report, see "The Screening Report" on page 328 in the "The Fit Two Level Screening Platform" chapter.

Plackett-Burman Design

The Fractional Factorial example shows an 8-run fractional factorial design for five continuous factors. But suppose you can afford 4 additional runs. In this example, construct a 12-run Plackett-Burman design. To facilitate completing the Screening window, use the Load Responses and Load Factors commands.

Create the Plackett-Burman Design

1. Select **DOE > Classical > Screening Design**.
2. Select **Help > Sample Data Library** and open Design Experiment/Reactor Response.jmp.
3. Select **Load Responses** from the Screening Design red triangle menu.
4. Select **Help > Sample Data Library** and open Design Experiment/Reactor Factors.jmp.

5. Select **Load Factors** from the Screening Design red triangle menu.

   The **Load Responses** and **Load Factors** commands fill in the Responses and Factors outlines with the response and factor names, goal and limits for the response, and values for the factors. See Figure 9.19 for the completed Responses and Factors outlines.

6. Click **Continue.**

Note: Setting the random seed in the next step reproduces the run order shown in this example. In constructing a design on your own, this step is not necessary.

7. (Optional) From the Screening Design red triangle menu, select **Set Random Seed**, type 34567, and click **OK**.

8. From the Choose Screening Type panel, accept the default selection to **Choose from a list of fractional factorial designs** and click **Continue**.

9. Select the Plackett-Burman design, as shown in Figure 9.26.

   Plackett-Burman designs with run sizes that are not a power of two tend to have complex aliasing structures. In particular, main effects can be partially aliased with several two-way interactions. See "Evaluate the Design" on page 319. Notice that the 12-run Plackett-Burman design is designated as having Resolution 3.

Figure 9.26 Design List Showing Plackett-Burman Screening Design

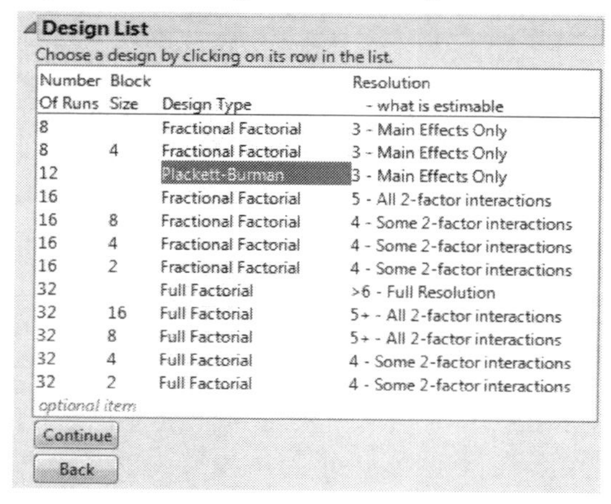

Number Of Runs	Block Size	Design Type	Resolution - what is estimable
8		Fractional Factorial	3 - Main Effects Only
8	4	Fractional Factorial	3 - Main Effects Only
12		Plackett-Burman	3 - Main Effects Only
16		Fractional Factorial	5 - All 2-factor interactions
16	8	Fractional Factorial	4 - Some 2-factor interactions
16	4	Fractional Factorial	4 - Some 2-factor interactions
16	2	Fractional Factorial	4 - Some 2-factor interactions
32		Full Factorial	>6 - Full Resolution
32	16	Full Factorial	5+ - All 2-factor interactions
32	8	Full Factorial	5+ - All 2-factor interactions
32	4	Full Factorial	4 - Some 2-factor interactions
32	2	Full Factorial	4 - Some 2-factor interactions

10. Click **Continue.**

11. Click **Make Table.**

Figure 9.27 Design Table for Placket-Burman Design

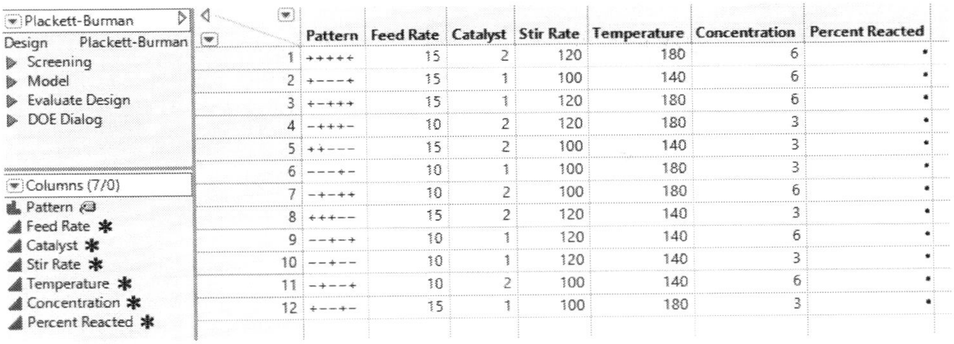

Plackett-Burman
Design Plackett-Burman
Screening
Model
Evaluate Design
DOE Dialog

Columns (7/0)
Pattern
Feed Rate
Catalyst
Stir Rate
Temperature
Concentration
Percent Reacted

	Pattern	Feed Rate	Catalyst	Stir Rate	Temperature	Concentration	Percent Reacted
1	+++++	15	2	120	180	6	•
2	+---+	15	1	100	140	6	•
3	+-+++	15	1	120	180	6	•
4	-+++-	10	2	120	180	3	•
5	++---	15	2	100	140	3	•
6	---+-	10	1	100	180	3	•
7	-+-++	10	2	100	180	6	•
8	+++--	15	2	120	140	3	•
9	--+-+	10	1	120	140	6	•
10	--+--	10	1	120	140	3	•
11	-+--+	10	2	100	140	6	•
12	+--+-	15	1	100	180	3	•

A column called Percent Reacted is included in the design table. You should conduct your experimental runs in the order shown in the table, recording your results in the Percent Reacted column.

Evaluate the Design

1. Return to your Screening Design window. If you have closed this window, run the DOE Dialog script in your design table.
2. Open the **Design Evaluation > Color Map on Correlations** outline.

Figure 9.28 Color Map for Absolute Correlations

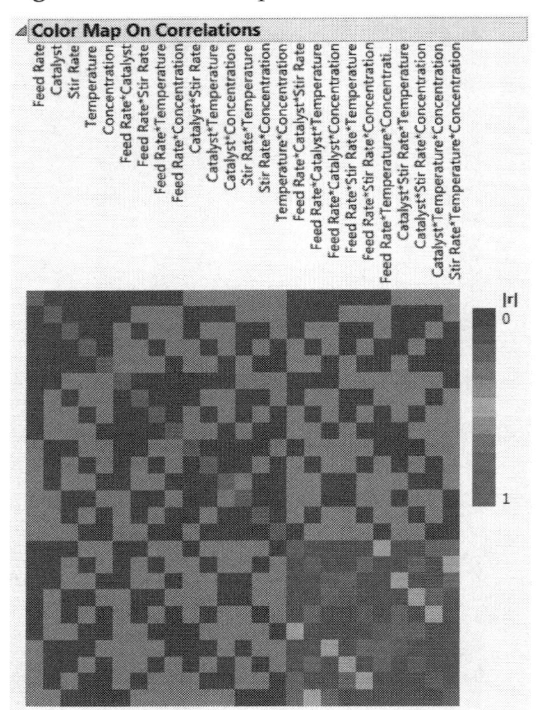

The diagonal cells have correlations of one, as expected. Cells with the deep blue color correspond to effects that have correlations equal to 0. The light blue and gray shaded cells correspond to effects that have correlations greater than zero. Place your cursor over a few of these cells with your cursor to see the effects involved and their absolute correlations. For example, notice that Feed Rate is correlated with several two-way and three-way interactions.

3. Open the Alias Matrix outline.

Figure 9.29 Alias Matrix - Partial View Showing Up to Two-Way Interactions

Alias Matrix

Effect	Feed Rate*Catalyst	Feed Rate*Stir Rate	Feed Rate*Temperature	Feed Rate*Concentration	Catalyst*Stir Rate	Catalyst*Temperature	Catalyst*Concentration	Stir Rate*Temperature	Stir Rate*Concentration	Temperature*Concentration
Intercept	0	0	0	0	0	0	0	0	0	0
Feed Rate	0	0	0	0	0.333	-0.33	-0.33	0.333	0.333	0.333
Catalyst	0	0.333	-0.33	-0.33	0	0	0	0.333	-0.33	0.333
Stir Rate	0.333	0	0.333	0.333	0	0.333	-0.33	0	0	0.333
Temperature	-0.33	0.333	0	0.333	0.333	0	0.333	0	0.333	0
Concentration	-0.33	0.333	0.333	0	-0.33	0.333	0	0.333	0	0

Because the design is orthogonal for the main effects, the Alias Matrix gives the numerical values of the correlations between effects. See "Alias Matrix" on page 446 in the "Evaluate Designs" chapter. For example, notice that Feed Rate is partially aliased with six two-way interactions and with four three-way interactions. These are the interactions corresponding to the entries of 0.333 and -0.33 in the row for Feed Rate.

Analyze the Results

The data table Plackett-Burman.jmp contains the results of the designed experiment. Recall that you suspect that the Temperature*Concentration interaction is active. You proceed under the assumption that this is the only potentially active interaction.

1. Select **Help > Sample Data Library** and open Design Experiment/Plackett-Burman.jmp.
2. Run the **Model** script by clicking the icon to its left.
3. Select Temperature in the Select Columns list and Concentration in the Construct Model Effects list.
4. Click **Cross**.
5. Click **Run**.

Figure 9.30 Parameter Estimates for Full Model

Parameter Estimates

Term	Estimate	Std Error	t Ratio	Prob>|t|
Intercept	66.166667	2.639444	25.07	<.0001*
Feed Rate(10,15)	-2.333333	2.850926	-0.82	0.4503
Catalyst(1,2)	10.5	2.850926	3.68	0.0142*
Stir Rate(100,120)	1.3333333	2.850926	0.47	0.6597
Temperature(140,180)	5	2.639444	1.89	0.1167
Concentration(3,6)	-0.5	2.639444	-0.19	0.8572
Concentration*Temperature	-6.5	3.232646	-2.01	0.1006

The Actual by Predicted Plot indicates no lack of model fit. The Parameter Estimates report shows that Catalyst is significant at the 0.05 level and that the Concentration*Temperature interaction is almost significant at the 0.10 level.

Reduce the Model

You want to identify those effects that have the most impact on the response. To see these active effects more clearly, remove insignificant effects using the Effect Summary outline.

Figure 9.31 Effect Summary Outline for Full Model

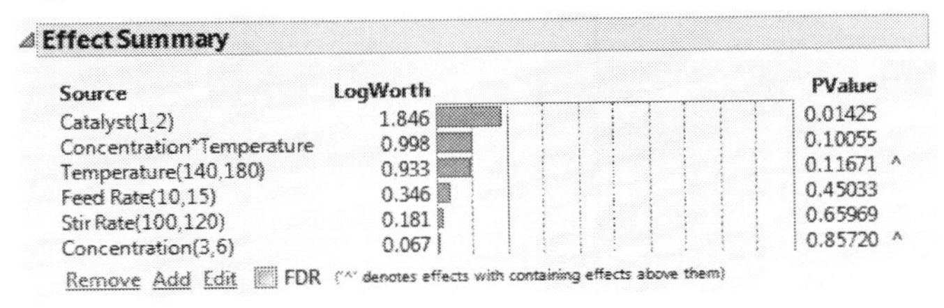

Effect Summary

Source	LogWorth	PValue
Catalyst(1,2)	1.846	0.01425
Concentration*Temperature	0.998	0.10055
Temperature(140,180)	0.933	0.11671 ^
Feed Rate(10,15)	0.346	0.45033
Stir Rate(100,120)	0.181	0.65969
Concentration(3,6)	0.067	0.85720 ^

Remove Add Edit FDR ('^' denotes effects with containing effects above them)

Although Concentration is the least significant effect, it is involved in a higher-order interaction (Concentration*Temperature), as indicated by the caret to the right of its PValue.

Based on the principle of effect heredity, Concentration should not be removed from the model while the Concentration*Temperature interaction remains in the model. See "Effect Heredity" on page 59 in the "Starting Out with DOE" chapter. The next least significant effect is Stir Rate.

1. In the Effect Summary outline, select Stir Rate and click **Remove**.

   Feed Rate is the next least significant effect that can be removed.

2. In the Effect Summary outline, select Feed Rate and click **Remove**.

Figure 9.32 Effect Summary Outline for Reduced Model

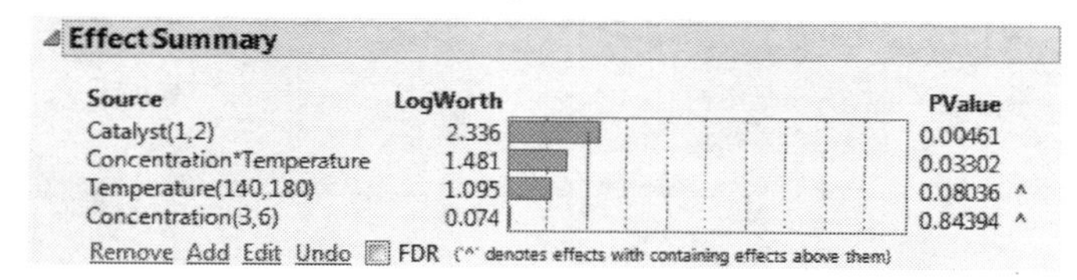

The PValue column indicates that the Catalyst main effect and the Concentration*Temperature interaction are both significant at the 0.05 level. The model should not be reduced any further. If all other interactions are inactive or negligible, then you can conclude that Catalyst and the Concentration*Temperature interaction are active effects.

Chapter 10

The Fit Two Level Screening Platform

Analyze Data from Screening Experiments

The Fit Two Level Screening platform is a modeling platform that you can use to analyze experimental data that results from a screening design. The Fit Two Level Screening platform helps you select a model by identifying effects that have a large impact on the response.

The Fit Two Level Screening platform is based on the principle of effect sparsity (Box and Meyer, 1986). This principle asserts that relatively few of the effects that you study in a screening design are active. Most are inactive, meaning that their true effects are zero and that their estimates are random error.

A screening design often provides no degrees of freedom for error. Consequently, classical tests for effects are not available. In such cases, the Fit Two Level Screening platform is particularly useful.

Figure 10.1 Half Normal Plot from Fit Two Level Screening Report

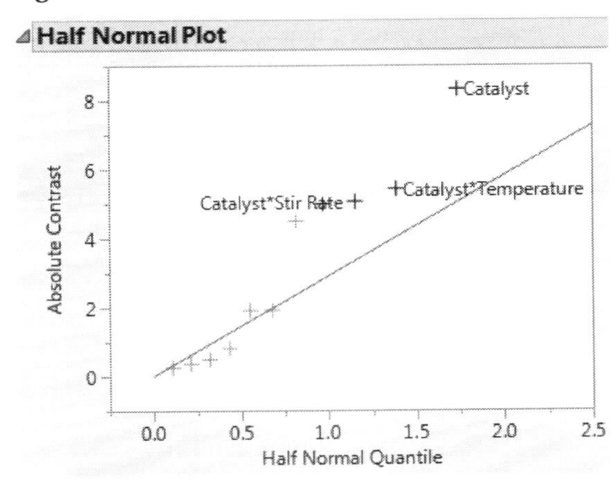

Overview of the Fit Two Level Screening Platform

The analysis of screening designs depends on *effect sparsity*, where most effects are assumed to be inactive. Using this assumption, effects with small estimates can help estimate the error in the model and determine whether the larger effects are active. Basically, if all the effects are inactive, they should vary randomly, with no effect deviating substantially from the other effects.

Data from a screening experiment can be analyzed using Fit Model (Analyze > Fit Model) or Fit Two Level Screening (Analyze > Specialized Modeling > Specialized DOE Models > Fit Two Level Screening). Use the Fit Two Level Screening platform to analyze data from screening experiments In accordance with the following guidelines:

- If your factors are all two-level and orthogonal, then all of the statistics in the Fit Two Level Screening platform are appropriate.
- For highly supersaturated main effect designs, the Fit Two Level Screening platform is effective in selecting factors, but is not as effective at estimating the error or the significance. The Monte Carlo simulation to produce *p*-values uses assumptions that are not valid for this case.
- If you have a categorical or a discrete numeric factor with more than two levels, then the Fit Two Level Screening platform is not an appropriate way to analyze the data. JMP treats the associated model terms as continuous. The variation for the factor is scattered across main and polynomial effects for that term.
- If your data are not orthogonal, then the constructed estimates are different from standard regression estimates. JMP can pick out big effects, but it does not effectively test each effect. This is because later effects are artificially orthogonalized, making earlier effects look more significant.
- The Fit Two Level Screening platform is not appropriate for mixture designs.

An Example Comparing Fit Two Level Screening and Fit Model

Consider the Reactor Half Fraction.jmp sample data table. The data are derived from a design discussed in Box, Hunter, and Hunter (1978). You are interested in a model with main effects and two-way interactions. This example uses a model with fifteen parameters for a design with sixteen runs.

For this example, select all continuous factors, except the response, Percent Reacted, as the screening effects, **X**. Select Percent Reacted as the response **Y**. The screening platform constructs interactions automatically. This is in contrast to Fit Model, where you manually specify the interactions that you want to include in your model.

Figure 10.2 shows the result of using the Fit Model platform, where a factorial to degree 2 model is specified. Since there are not enough observations to estimate an error term, it is not possible to conduct standard tests.

Figure 10.2 Traditional Saturated Half Reactor.jmp Design Output

Response Percent Reacted

Summary of Fit

RSquare	1
RSquare Adj	.
Root Mean Square Error	.
Mean of Response	65.25
Observations (or Sum Wgts)	16

Parameter Estimates

Term	Estimate	Std Error	t Ratio	Prob>|t|
Intercept	65.25	.	.	.
Feed Rate(10,15)	-1	.	.	.
Catalyst(1,2)	10.25	.	.	.
Stir Rate(100,120)	0	.	.	.
Temperature(140,180)	6.125	.	.	.
Concentration(3,6)	-3.125	.	.	.
Feed Rate*Catalyst	0.75	.	.	.
Feed Rate*Stir Rate	0.25	.	.	.
Feed Rate*Temperature	-0.375	.	.	.
Feed Rate*Concentration	0.625	.	.	.
Catalyst*Stir Rate	0.75	.	.	.
Catalyst*Temperature	5.375	.	.	.
Catalyst*Concentration	0.625	.	.	.
Stir Rate*Temperature	0.125	.	.	.
Stir Rate*Concentration	1.125	.	.	.
Temperature*Concentration	-4.75	.	.	.

JMP can calculate parameter estimates, but because there are no degrees of freedom for error, standard errors, *t*-ratios, and *p*-values are all missing. Rather than use Fit Model, you can use the Fit Two Level Screening platform, which specializes in getting the most information out of these situations, leading to a better model. The report from the Fit Two Level Screening platform for the same data is shown in Figure 10.3.

Figure 10.3 Half Reactor.jmp Fit Two Level Screening Design Report

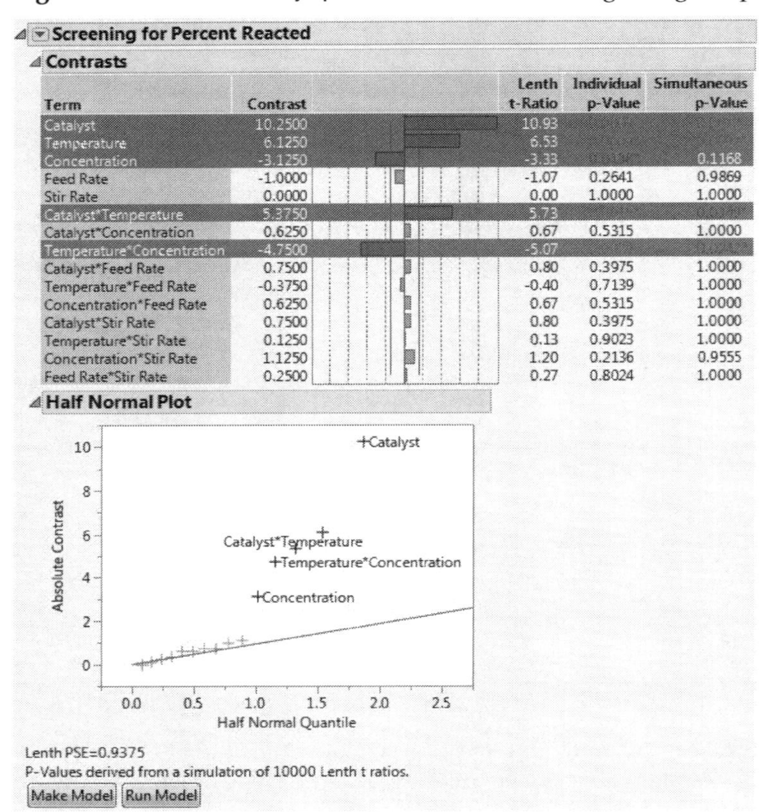

Screening for Percent Reacted

Contrasts

Term	Contrast	Lenth t-Ratio	Individual p-Value	Simultaneous p-Value
Catalyst	10.2500	10.93	[illegible]	[illegible]
Temperature	6.1250	6.53	[illegible]	[illegible]
Concentration	-3.1250	-3.33	[illegible]	0.1168
Feed Rate	-1.0000	-1.07	0.2641	0.9869
Stir Rate	0.0000	0.00	1.0000	1.0000
Catalyst*Temperature	5.3750	5.73	[illegible]	[illegible]
Catalyst*Concentration	0.6250	0.67	0.5315	1.0000
Temperature*Concentration	-4.7500	-5.07	[illegible]	[illegible]
Catalyst*Feed Rate	0.7500	0.80	0.3975	1.0000
Temperature*Feed Rate	-0.3750	-0.40	0.7139	1.0000
Concentration*Feed Rate	0.6250	0.67	0.5315	1.0000
Catalyst*Stir Rate	0.7500	0.80	0.3975	1.0000
Temperature*Stir Rate	0.1250	0.13	0.9023	1.0000
Concentration*Stir Rate	1.1250	1.20	0.2136	0.9555
Feed Rate*Stir Rate	0.2500	0.27	0.8024	1.0000

Half Normal Plot

Lenth PSE=0.9375

P-Values derived from a simulation of 10000 Lenth t ratios.

Make Model Run Model

Note the following features of the Screening report:

- Estimates labeled **Contrast**. Effects whose individual p-value is less than 0.10 are highlighted.
- A t-ratio is calculated using Lenth's PSE (pseudo-standard error). The Lenth PSE is shown below the Half Normal Plot.
- Both individual and simultaneous p-values are shown. Those that are less than 0.05 are shown with an asterisk.
- The Half Normal Plot enables you to quickly examine the effects. Effects initially highlighted in the effects list are also labeled in this plot.
- You can highlight effects by clicking on them in the Contrasts outline.
- The **Make Model** button opens the Fit Model window and populates it with the selected effects. The **Run Model** button runs the model based on the selected effects.

For this example, Catalyst, Temperature, and Concentration, along with two of their two-factor interactions, are selected.

Launch the Fit Two Level Screening Platform

Open the data table called Plackett-Burman.jmp, found in Design Experiment folder in the Sample Data installed with JMP. This table contains the design runs and the Percent Reacted experimental results for the 12-run Plackett-Burman design created in the previous section.

The data table has two scripts called **Screening** and **Model**, showing in the upper-left corner of the table, that were created by the DOE Screening designer. You can use these scripts to analyze the data, however it is simple to run the analyses yourself.

1. Select **Help > Sample Data Library** and open Design Experiment/Plackett-Burman.jmp.
2. Select **Analyze > Specialized Modeling > Specialized DOE Models > Fit Two Level Screening.**

   The populated launch window appears. When you use the Screening Design platform to create a design, a Screening script is saved to the design table. This allows JMP to populate the screening launch window.

Figure 10.4 Launch Window for the Fit Two Level Screening Platform

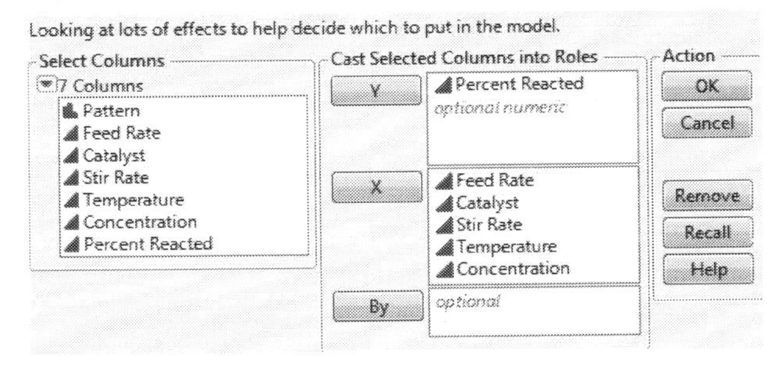

3. Click **OK.**

   The Screening report, shown in Figure 10.5, appears.

Figure 10.5 Screening Report

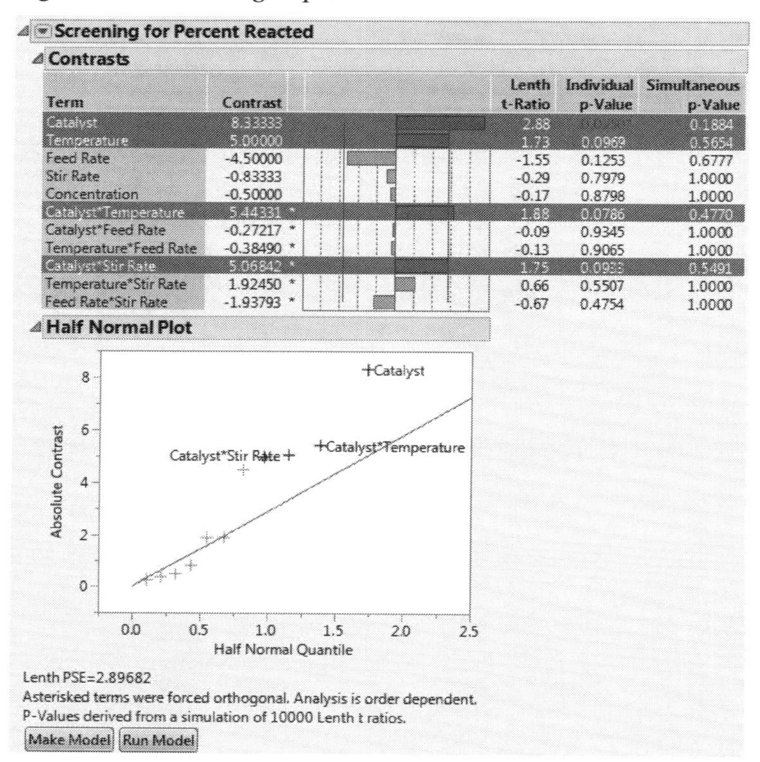

Screening for Percent Reacted

Contrasts

Term	Contrast	Lenth t-Ratio	Individual p-Value	Simultaneous p-Value
Catalyst	8.33333	2.88	[illegible]	0.1884
Temperature	5.00000	1.73	0.0969	0.5654
Feed Rate	-4.50000	-1.55	0.1253	0.6777
Stir Rate	-0.83333	-0.29	0.7979	1.0000
Concentration	-0.50000	-0.17	0.8798	1.0000
Catalyst*Temperature	5.44331 *	1.88	0.0786	0.4770
Catalyst*Feed Rate	-0.27217 *	-0.09	0.9345	1.0000
Temperature*Feed Rate	-0.38490 *	-0.13	0.9065	1.0000
Catalyst*Stir Rate	5.06842 *	1.75	0.0933	0.5491
Temperature*Stir Rate	1.92450 *	0.66	0.5507	1.0000
Feed Rate*Stir Rate	-1.93793 *	-0.67	0.4754	1.0000

Half Normal Plot

Lenth PSE=2.89682
Asterisked terms were forced orthogonal. Analysis is order dependent.
P-Values derived from a simulation of 10000 Lenth t ratios.
Make Model Run Model

If all effects are inactive, their estimates are random normal noise. Their estimates (contrasts) should fall close to the line shown in the Half Normal plot in Figure 10.5. Effects that fall far from the line are likely not noise, and so may represent active effects. Note that effects with Individual p-Values that fall below 0.10 are highlighted in the Contrasts outline. These effects are labeled in the Half Normal Plot and then tend to fall far from the line.

The Screening Report

The Screening report has two parts: The Contrasts outline and the Half Normal Plot outline.

Contrasts

The Contrasts outline lists model effects, a contrast value for each effect, Lenth *t*-ratios (calculated as the contrast value divided by the Lenth PSE (pseudo-standard error), individual and simultaneous *p*-values, and aliases if there are any. Effects are entered into the analysis following a hierarchical ordering. See "Order of Effect Entry" on page 334 for details. Effects with Individual p-Value less than 0.10 are highlighted.

Term Name of the factor.

Contrast Estimate for the factor. For orthogonal designs, this number is the same as the regression parameter estimate. This is not the case for non-orthogonal designs. An asterisk might appear next to the contrast, indicating a lack of orthogonality.

Bar Chart Shows the Lenth *t*-ratios with blue vertical lines indicating a value that is significant at the 0.10 level.

Lenth t-Ratio Lenth's *t*-ratio, calculated as $\text{Contrast}/PSE$, where PSE is Lenth's Pseudo-Standard Error. See "Lenth's Pseudo-Standard Error" on page 335 for details.

Individual p-Value Analogous to the standard *p*-values for a linear model. Small values of this value indicate a significant effect. Refer to "Lenth's Pseudo-Standard Error" on page 335 for details.

Do not expect the *p*-values to be exactly the same if the analysis is re-run. The Monte Carlo method should give similar, but not identical, values if the same analysis is repeated.

Simultaneous p-Value Similar to the individual *p*-value, but multiple-comparison adjusted.

Aliases Appears only if there are exact aliases of later effects to earlier effects.

Half Normal Plot

The Half Normal Plot shows the absolute value of the contrasts plotted against the absolute value of quantiles for the half-normal distribution. Significant effects appear separated from the line towards the upper right of the graph.

The Half Normal Plot is interactive. Select different model effects by dragging a rectangle around the effects of interest, or hold down CTRL and click on an effect name in the report.

Using the Fit Model Platform

The **Make Model** button beneath the Half Normal Plot creates a Fit Model dialog that includes all the highlighted effects.

1. Open the Plackett-Burman.jmp sample data table, found in Design Experiment folder.
2. Select **Analyze > Specialized Modeling > Specialized DOE Models > Fit Two Level Screening**.
3. Click **OK**.
4. Click the **Make Model** Button beneath the Half Normal Plot.

Note that the Catalyst*Stir Rate interaction is highlighted, but the Stir Rate main effect is not. In accordance with the principle of Effect Heredity, add the Stir Rate main effect to the model. See "Effect Heredity" on page 59 in the "Starting Out with DOE" chapter.

5. Select Stir Rate and click **Add** in the Fit Model window.
6. Click **Run**.

The Actual-by-Predicted Plot

The Whole Model actual-by-predicted plot, shown in Figure 10.6, appears at the top of the Fit Model report. You see at a glance that this model fits well. The blue line falls outside the bounds of the 95% confidence curves (red-dotted lines), which tells you the model is significant. The model *p*-value (p = 0.0208), R^2, and RMSE appear below the plot. The RMSE is an estimate of the standard deviation, assuming that the unestimated effects are negligible.

Figure 10.6 An Actual-by-Predicted Plot

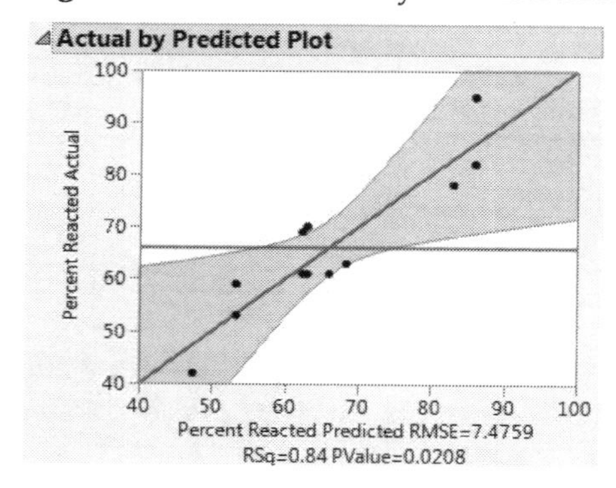

The Scaled Estimates Report

To see a scaled estimates report, use **Effect Screening > Scaled Estimates** found in the red triangle menu on the **Response Percent Reacted** title bar. When there are quadratic or polynomial effects, the coefficients and the tests for them are more meaningful if effects are scaled and coded. The Scaled Estimates report includes a bar chart of the individual effects embedded in a table of parameter estimates. The last column of the table has the *p*-values for each effect.

Figure 10.7 Example of a Scaled Estimates Report

Scaled Estimates

Term	Scaled Estimate	Std Error	t Ratio	Prob>|t|
Intercept	66.166667	2.158103	30.66	<.0001*
Catalyst(1,2)	8.3333333	2.158103	3.86	0.0083*
Temperature(140,180)	3.5	2.289014	1.53	0.1771
Catalyst*Temperature	6.5	2.289014	2.84	0.0296*
Catalyst*Stir Rate	4.5	2.289014	1.97	0.0969
Stir Rate(100,120)	-3	2.289014	-1.31	0.2379

A Power Analysis

Open the Effect Details disclosure icon to see outline nodes for the Catalyst and Temperature effects. To run a power analysis for an effect, click the red triangle icon on its title bar and select **Power Analysis**.

This example shows a power analysis for the Catalyst variable, using default value for α (0.05), the root mean square error and parameter estimate for Catalyst, for a sample size of 12. The resulting power is 0.8926, which means that in similar experiments, you can expect an 89% chance of detecting a significant effect for Catalyst.

Figure 10.8 Example of a Power Analysis

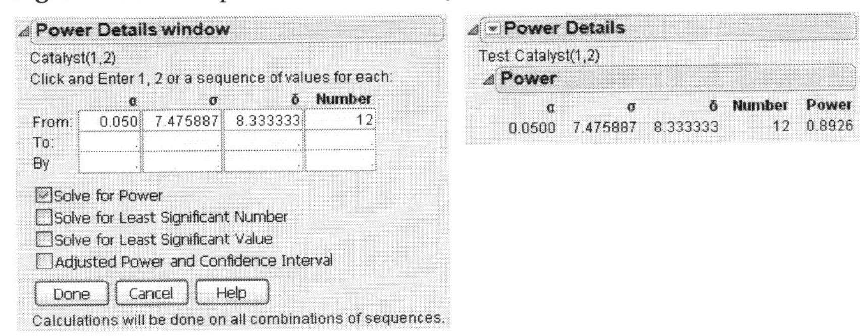

Refer to the Standard Least Squares chapter in the *Fitting Linear Models* book for details.

Additional Fit Two Level Screening Analysis Examples

This section provides examples of using the Fit Two Level Screening platform.

Analyzing a Plackett-Burman Design

Plackett-Burman designs are an alternative to fractional-factorial screening designs. Two-level fractional factorial designs must, by their nature, have a number of runs that are a power of two. Plackett-Burman designs exist for 12-, 24-, and 28-run designs.

The Weld-Repaired Castings.jmp sample data table uses a Plackett-Burman design, and is found in textbooks such as Giesbrecht and Gumpertz (2004) and Box, Hunter, and Hunter (1978). Seven factors are thought to be influential on weld quality. The seven factors include Initial Structure, Bead Size, Pressure Treatment, Heat Treatment, Cooling Rate, Polish, and Final Treatment. A Plackett-Burman design with 12 runs is used to investigate the importance of the seven factors. The response is $100 \times \log(\text{lifetime})$. (There are also four terms that were used to model error that are not used in this analysis.)

Using the Fit Two Level Screening platform, select the seven effects as *X* and Log Life as *Y*. (If terms are automatically populated in the Fit Two Level Screening platform launch window, remove the four error terms listed as effects.) Click **OK**. Figure 10.9 appears, showing only a single significant effect.

Figure 10.9 Screening Report for Weld-Repaired Castings.jmp

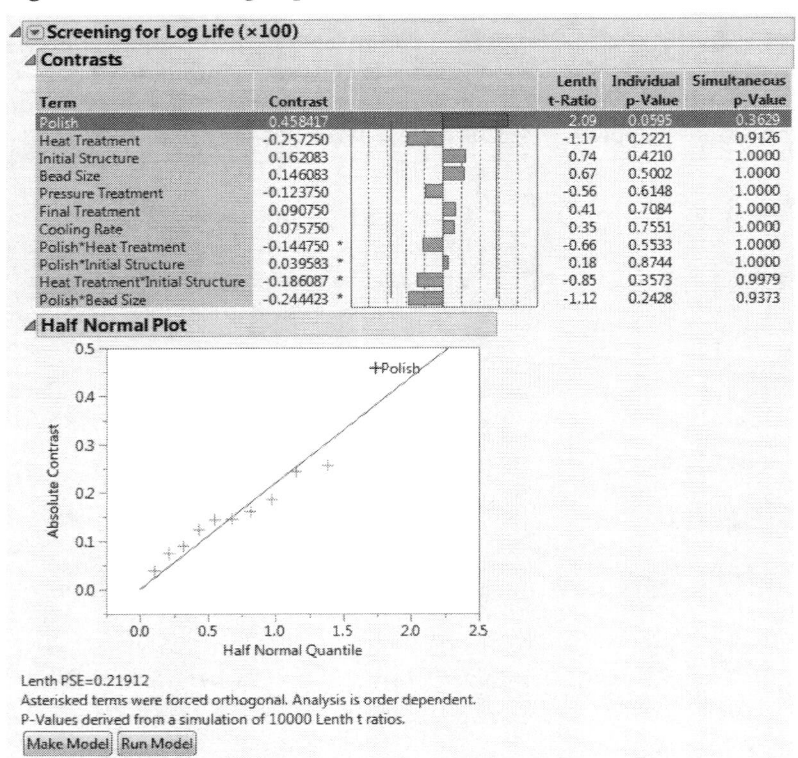

Screening for Log Life (×100)

Contrasts

Term	Contrast	Lenth t-Ratio	Individual p-Value	Simultaneous p-Value
Polish	0.458417	2.09	0.0595	0.3629
Heat Treatment	-0.257250	-1.17	0.2221	0.9126
Initial Structure	0.162083	0.74	0.4210	1.0000
Bead Size	0.146083	0.67	0.5002	1.0000
Pressure Treatment	-0.123750	-0.56	0.6148	1.0000
Final Treatment	0.090750	0.41	0.7084	1.0000
Cooling Rate	0.075750	0.35	0.7551	1.0000
Polish*Heat Treatment	-0.144750 *	-0.66	0.5533	1.0000
Polish*Initial Structure	0.039583 *	0.18	0.8744	1.0000
Heat Treatment*Initial Structure	-0.186087 *	-0.85	0.3573	0.9979
Polish*Bead Size	-0.244423 *	-1.12	0.2428	0.9373

Note asterisks mark four terms, indicating that they are not orthogonal to effects preceding them, and the obtained contrast value was after orthogonalization. So, they would not match corresponding regression estimates.

Analyzing a Supersaturated Design

Supersaturated designs have more factors than runs. The objective is to determine which effects are active. They rely heavily on effect sparsity for their analysis, so the Fit Two Level Screening platform is ideal for their analysis.

As an example, look at Supersaturated.jmp, from the sample data folder, a simulated data set with 18 factors but only 12 runs. Y is generated by

$$Y = 2(X7) + 5(X10) - 3(X15) + \varepsilon$$

where $\varepsilon \sim N(0,1)$. So, Y has been constructed with three active factors.

To detect the active factors, run the Fit Two Level Screening platform with X1–X18 as *X* and Y as *Y*. The report shown in Figure 10.10 appears.

Figure 10.10 Screening Report for Supersaturated.jmp

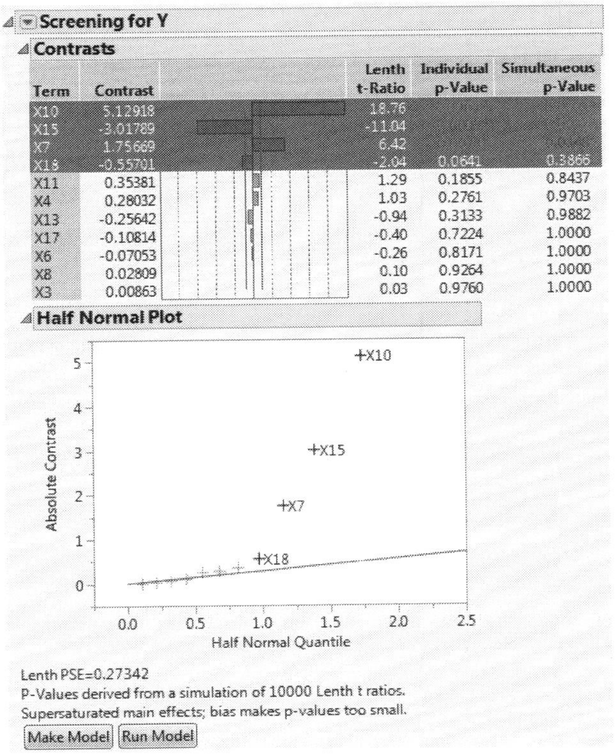

Screening for Y

Contrasts

Term	Contrast	Lenth t-Ratio	Individual p-Value	Simultaneous p-Value
X10	5.12918	18.76	[illegible]	[illegible]
X15	-3.01789	-11.04	[illegible]	[illegible]
X7	1.75669	6.42	[illegible]	[illegible]
X18	-0.55701	-2.04	0.0641	0.3866
X11	0.35381	1.29	0.1855	0.8437
X4	0.28032	1.03	0.2761	0.9703
X13	-0.25642	-0.94	0.3133	0.9882
X17	-0.10814	-0.40	0.7224	1.0000
X6	-0.07053	-0.26	0.8171	1.0000
X8	0.02809	0.10	0.9264	1.0000
X3	0.00863	0.03	0.9760	1.0000

Half Normal Plot

Absolute Contrast
Half Normal Quantile
+X10
+X15
+X7
+X18

Lenth PSE=0.27342
P-Values derived from a simulation of 10000 Lenth t ratios.
Supersaturated main effects; bias makes p-values too small.
Make Model Run Model

Note that the three active factors have been highlighted. One other factor, X18, has also been highlighted. It shows in the Half Normal plot close to the blue line, indicating that it is close to the 0.1 cutoff significance value. The 0.1 critical value is generous in its selection of factors so you don't miss those that are possibly active.

The contrasts of 5.1, –3, and 1.8 are close to their simulated values (5, –3, 2). However, the similarity of these values can be increased by using a regression model, without the effect of orthogonalization.

The *p*-values, while useful, are not entirely valid statistically, since they are based on a simulation that assumes orthogonal designs, which is not the case for supersaturated designs.

Technical Details

Order of Effect Entry

The Fit Two Level Screening platform has a carefully defined order of operations.

- First, the main effect terms enter according to the absolute size of their contrast. All effects are orthogonalized to the effects preceding them in the model. The method assures that their order is the same as it would be in a forward stepwise regression. Ordering by main effects also helps in selecting preferred aliased terms later in the process.
- After main effects, all second-order interactions are brought in, followed by third-order interactions, and so on. The second-order interactions cross with all earlier terms before bringing in a new term. For example, with size-ordered main effects A, B, C, and D, B*C enters before A*D. If a factor has more than two levels, square and higher-order polynomial terms are also considered.
- An effect that is an exact alias for an effect already in the model shows in the alias column. Effects that are a linear combination of several previous effects are not displayed. If there is partial aliasing (a lack of orthogonality) the effects involved are marked with an asterisk.
- The process continues until *n* effects are obtained, where *n* is the number of rows in the data table, thus fully saturating the model. If complete saturation is not possible with the factors, JMP generates random orthogonalized effects to absorb the rest of the variation. They are labeled Null n where *n* is a number. For example, this situation occurs if there are exact replicate rows in the design.

Fit Two Level Screening as an Orthogonal Rotation

Mathematically, the Fit Two Level Screening platform takes the *n* values in the response vector and rotates them into *n* new values. The rotated values are then mapped by the space of the factors and their interactions.

$$\text{Contrasts} = \mathbf{T'} \times \text{Responses}$$

where **T** is an orthonormalized set of values starting with the intercept, main effects of factors, two-way interactions, three-way interactions, and so on, until n values have been obtained. Since the first column of **T** is an intercept, and all the other columns are orthogonal to it, these other columns are all contrasts, that is, they sum to zero. Since **T** is orthogonal, it can serve as **X** in a linear model. It does not need inversion, since **T'** is also $\mathbf{T}^{-1}$ and **(T'T)T'**. The contrasts are the parameters estimated in a linear model.

- If no effect in the model is active other than the intercept, the contrasts are just an orthogonal rotation of random independent variates into different random independent variates. These newly orthogonally rotated variates have the same variance as the original random independent variates. To the extent that some effects are active, the inactive effects still represent the same variation as the error in the model. The hope is that the effects and the design are strong enough to separate the active effects from the random error effects.

Lenth's Pseudo-Standard Error

Lenth's method (Lenth, 1989), known as the *Lenth Pseudo Standard Error* (*PSE*), constructs an estimate of the residual standard error using effects that appear to be inactive. Lenth's PSE can be used to estimate the standard error for experiments where contrasts are independent and have a common variance.

If there are n rows, the platform constructs $n - 1$ contrasts. Denote these contrasts by $\hat{C}_i$, where $i = 1, \ldots, n - 1$.

To obtain Lenth's PSE, first calculate the following:

$$v = 1.5[median_{i = 1, \ldots, n-1}|\hat{C}_i|]$$

Lenth's PSE is based on the effects that are likely to be inactive and is given by:

$$PSE = 1.5\left[median_{|\hat{C}_i| < 2.5v}|\hat{C}_i|\right]$$

The value for Lenth's PSE is shown at the bottom of the Screening report.

Lenth t-Ratios

For each contrast, a t-ratio is computed by dividing the contrast by the PSE. The reference distribution of these t-ratios under the null hypothesis is not computationally tractable, and so it is obtained by simulation. The method, described below, is based on a discussion in Ye and Hamada (2000).

Denote the t-ratio for the ith contrast by t_i:

$$t_i = \hat{C}_i/(PSE)$$

Of primary importance in screening experiments is the *individual error rate,* namely the probability of declaring a given effect as active when it is not. For the ith effect, this occurs when $|t_i|$ is large, falling in the upper tail of it's reference distribution.

Because the platform constructs a relatively large number of effects, the *experimentwise error rate* is also of importance. The experimentwise error rate is the probability of declaring *any* effect as active when no effects are active. An experimentwise error occurs when no effects are active and the maximum of the absolute *t*-ratios, $\max|t_i|$, is large and falls in the upper tail of its reference distribution.

The Fit Two Level Screening platform obtains reference distributions for both types of error rates using Monte Carlo simulation. Consider a set of $n - 1$ values that is simulated from a normal distribution with mean 0 and standard deviation equal to PSE. This set of values represents potential contrast values for the experiment under the null hypothesis of no active effects. In all, 10,000 sets of contrast values are generated.

Individual p-Values

Because the contrast distributions are identical, all of the 10,000*($n - 1$) values obtained in the simulation are generated from the distribution of values for any specific contrast under the null hypothesis that the contrast is not active.

Consider the ith contrast. Lenth *t*-ratios are constructed using each simulated value. The reference distribution for the individual error rate is approximated by the absolute values of these *t*-ratios. The *p*-value given in the Individual *p*-Value column of the report is the interpolated fractional position of the observed absolute Lenth t-Ratio among the 10,000*($n - 1$) simulated absolute *t*-ratios arranged in descending order. This approximates the area to the right of the absolute value of the observed absolute Lenth t-Ratio with respect to the reference distribution.

Simultaneous p-Values

An experimentwise error occurs if any t-ratio leads to rejecting the null hypothesis when all effects are inactive. Equivalently, an experimentwise error occurs if the maximum of the absolute *t*-ratios, $\max|t_i|$, leads to rejecting the null hypothesis.

To obtain a reference distribution in this case, in each of the 10,000 simulations, the maximum of the absolute *t*-ratios is computed. These 10,000 maximum values form the reference distribution. The *p*-value given in the Simultaneous *p*-Value column of the report is the interpolated fractional position of the observed absolute Lenth t-Ratio among the 10,000 simulated maximum absolute *t*-ratios arranged in descending order. This approximates the area to the right of the absolute value of the absolute Lenth t-Ratio with respect to the reference distribution based on the simulated maximum absolute *t*-ratios.

Controlling the Monte Carlo Simulation

To change the number of default sets of simulations from 10,000, you must assign a value to a global JSL variable named `LenthSimN`. As an example, do the following:

1. Select **Help > Sample Data Library** and open Reactor Half Fraction.jmp.
2. Select **Analyze > Specialized Modeling > Specialized DOE Models > Fit Two Level Screening.**
3. Select Percent Reacted as the response variable, *Y*.
4. Select all the other continuous variables as effects, *X*.
5. Click **OK**.
6. Click the Screening for Percent Reacted red triangle and select **Save Script > To Script Window**.
7. At the top of the Script Window (above the code), type: **LenthSimN=50000;**
8. Highlight `LenthSimN=50000;` and the remaining code.
9. Right-click in the script window and select **Run Script**.

Note: If `LenthSimN=0`, the standard *t*-distribution is used and simultaneous *p*-values are not provided (not recommended).

Chapter 11

Response Surface Designs

Response surface designs are useful for modeling quadratic surfaces. A response surface model can identify a point where a minimum or maximum value of the response occurs, if one exists inside the design region. Three distinct values for each factor are necessary to fit a quadratic function, so standard two-level designs are not appropriate for fitting curved surfaces.

Response surface designs are capable of fitting a second-order prediction equation for the response. The quadratic terms in these equations model the curvature in the true response function. If a maximum or minimum exists inside the design region, the point where that value occurs can be estimated.

Figure 11.1 Model for Response Surface Design Results

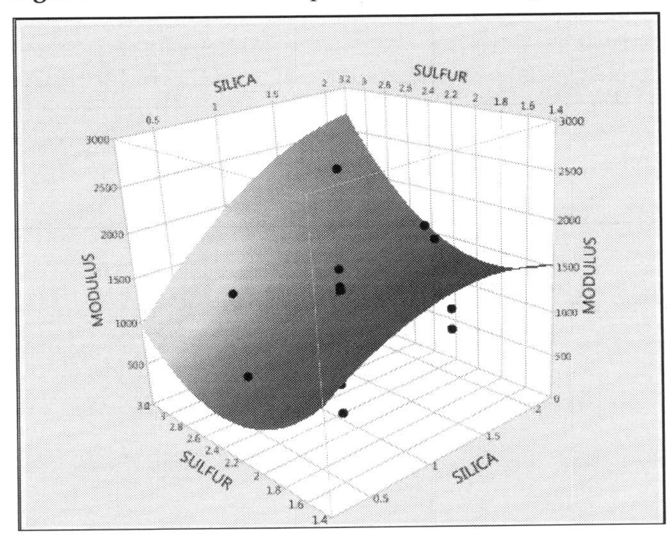

Overview of Response Surface Designs

The Response Surface Design platform provides the classical central composite and Box-Behnken designs, including blocked versions of these designs. For central composite designs, you can control the placement of axial points and other aspects of the design. Response surface designs are available for continuous factors only and are provided for up to eight factors.

Tip: You can use DOE > Custom Design to construct optimal response surface designs that accommodate your specific experimental situation. Custom Design constructs response surface designs that are much more flexible than classical response surface designs. In particular, you can use the Custom Design platform to create response surface designs that involve categorical factors or more than eight continuous factors. You can also specify the number of runs and restrictions on the design space. For examples, see "Response Surface Experiments" on page 150 in the "Examples of Custom Designs" chapter.

A *central composite design* (Figure 11.2) combines a two-level fractional factorial design and two other types of points:

- *Center points*, where all the factor values are set to the midrange value.
- *Axial points*, where one factor is set to a high or low value (an *axial* value) and all other factors are set to the midrange value.

Depending on your selections relative to axial points, a central composite design can have as many as five distinct settings for each factor and the axial points can extend beyond the specified range of the factors.

Figure 11.2 Central Composite Design for Three Factors

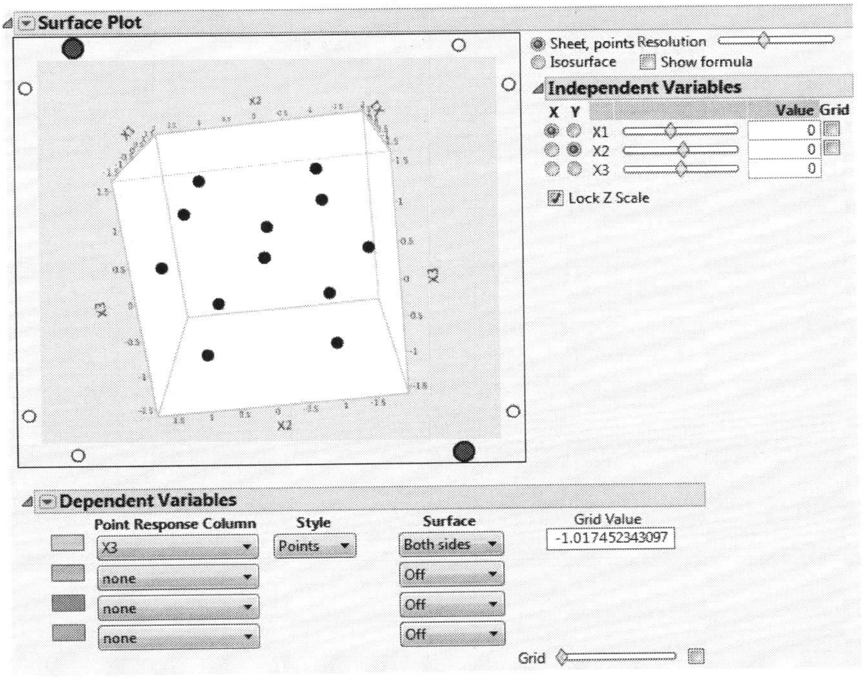

A *Box-Behnken design* (Figure 11.3) has only three levels per factor and has no design points at the vertices of the cube defined by the ranges of the factors. This type of design can be useful when you must avoid these points due to engineering considerations. But, the lack of design points at the vertices of the cube means that a Box-Behnken design has higher prediction variance, and so less precision, near the vertices compared to a central composite design.

Figure 11.3 Box-Behnken Design for Three Factors

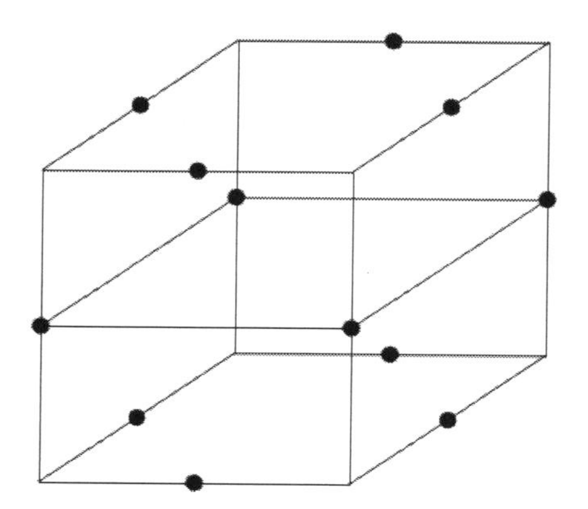

In JMP, you can construct a response surface design in two ways:

- Using the Response Surface Design platform (for up to eight continuous factors)
- Using the Custom Design platform (and clicking the RSM button in the Model outline)

In both cases, the design table contains a Model script that you can run to fit a model. The Model script applies the Response Surface Effect attribute to each main effect, so that the main effects appear with a **&RS** suffix in the Fit Model window. This attribute ensures that the Fit Least Squares report contains a Response Surface report. For details about this report, see the Standard Least Squares Report and Options chapter in the *Fitting Linear Models* book.

Note: The Response Surface outline in the Standard Least Squares report is not shown for response surface designs that contain more than 20 continuous factors.

Example of a Response Surface Design

In this example, you construct a Box-Behnken design for a tire tread experiment. Your objective is to match a target value of 450 for a measure of elongation (Stretch). The stretch varies as a function of the amounts of Silica, Silane, and Sulfur used to manufacture the tire tread compound. You want to experiment over a wide range of factor settings to find the settings that achieve the target.

Construct a Box-Behnken Design

In this example, for convenience, you load the responses and factors from existing tables. When designing a new experiment on your own, enter the responses and factors manually. See "Responses" on page 350 and "Factors" on page 352.

1. Select **DOE > Classical > Response Surface Design.**
2. Select **Help > Sample Data Library** and open Design Experiment/Bounce Response.jmp.
3. Click the Response Surface Design red triangle menu and select **Load Responses.**
4. Select **Help > Sample Data Library** and open Design Experiment/Bounce Factors.jmp.
5. Click the Response Surface Design red triangle menu and select **Load Factors.**

Figure 11.4 Responses and Factors Outlines for Tire Tread Design

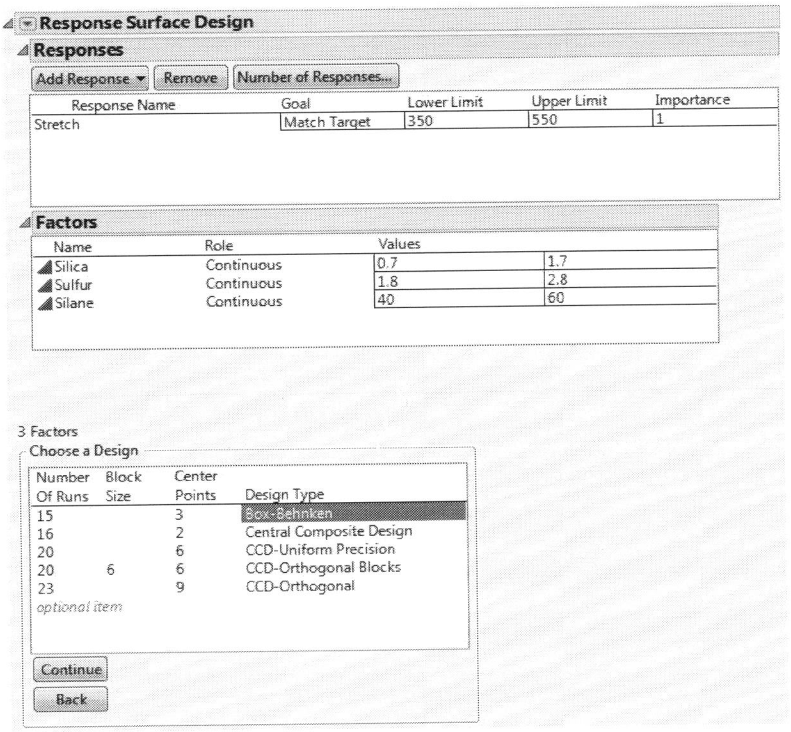

Response Surface Design

Responses

Add Response Remove Number of Responses...

Response Name	Goal	Lower Limit	Upper Limit	Importance
Stretch	Match Target	350	550	1

Factors

Name	Role	Values	
Silica	Continuous	0.7	1.7
Sulfur	Continuous	1.8	2.8
Silane	Continuous	40	60

3 Factors

Choose a Design

Number Of Runs	Block Size	Center Points	Design Type
15		3	Box-Behnken
16		2	Central Composite Design
20		6	CCD-Uniform Precision
20	6	6	CCD-Orthogonal Blocks
23		9	CCD-Orthogonal
optional item			

Continue

Back

In the Responses outline, notice that the Goal for Stretch is set to Match Target.

In the Choose a Design panel, possible designs appear.

Note: Setting the Random Seed in step 6 reproduces the exact results shown in this example. In constructing a design on your own, this step is not necessary.

6. (Optional) Click the Response Surface Design red triangle menu and select **Set Random Seed**. Type 12345 and click **OK**.
7. Click **Continue** to retain the Box-Behnken design selection.
8. Click **Make Table**.

Figure 11.5 Box-Behnken Design Table

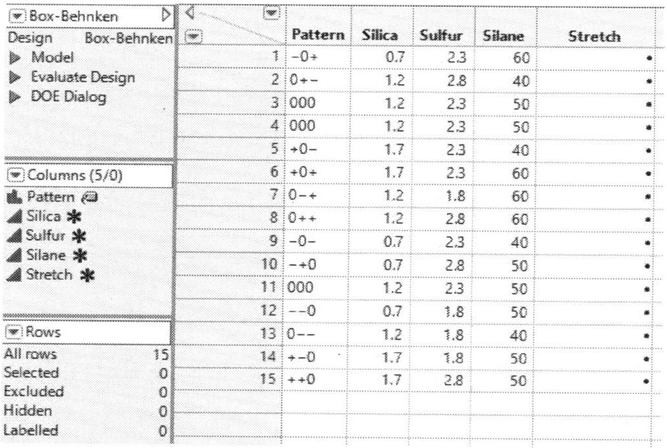

Box-Behnken
Design Box-Behnken
Model
Evaluate Design
DOE Dialog

Columns (5/0)
Pattern
Silica
Sulfur
Silane
Stretch

Rows	
All rows	15
Selected	0
Excluded	0
Hidden	0
Labelled	0

	Pattern	Silica	Sulfur	Silane	Stretch
1	-0+	0.7	2.3	60	•
2	0+-	1.2	2.8	40	•
3	000	1.2	2.3	50	•
4	000	1.2	2.3	50	•
5	+0-	1.7	2.3	40	•
6	+0+	1.7	2.3	60	•
7	0-+	1.2	1.8	60	•
8	0++	1.2	2.8	60	•
9	-0-	0.7	2.3	40	•
10	-+0	0.7	2.8	50	•
11	000	1.2	2.3	50	•
12	--0	0.7	1.8	50	•
13	0--	1.2	1.8	40	•
14	+-0	1.7	1.8	50	•
15	++0	1.7	2.8	50	•

At this point, conduct the experiment and enter the responses into the data table.

Analyze the Experimental Data

1. Select **Help > Sample Data Library** and open Design Experiment/Bounce Data.jmp.

   The file Bounce Data.jmp contains your experiment results.

2. Run the **Model** script.

   Notice that the main effects in the Construct Model Effects list are followed by the **& RS** suffix. This suffix indicates that these are response surface effects, which produce a Response Surface report in the Standard Least Squares report.

3. Click **Run**.

Figure 11.6 Lack of Fit and Effect Tests Reports

Lack Of Fit

Source	DF	Sum of Squares	Mean Square	F Ratio
Lack Of Fit	3	11.750000	3.91667	0.9792
Pure Error	2	8.000000	4.00000	Prob > F
Total Error	5	19.750000		0.5411
				Max RSq
				0.9999

Box-Cox Transformations

Parameter Estimates

Effect Tests

Source	Nparm	DF	Sum of Squares	F Ratio	Prob > F
Silica(0.7,1.7)	1	1	12880.125	3260.791	<.0001*
Sulfur(1.8,2.8)	1	1	5778.125	1462.816	<.0001*
Silane(40,60)	1	1	968.000	245.0633	<.0001*
Silica*Sulfur	1	1	52441.000	13276.20	<.0001*
Silica*Silane	1	1	0.250	0.0633	0.8114
Sulfur*Silane	1	1	8556.250	2166.139	<.0001*
Silica*Silica	1	1	2592.923	656.4362	<.0001*
Sulfur*Sulfur	1	1	4653.231	1178.033	<.0001*
Silane*Silane	1	1	0.231	0.0584	0.8186

There is no indication of lack of fit and the Effect Tests report indicates that all but two higher-order terms (**Silica*Silane** and **Silane*Silane**) have *p*-values below 0.0001. See the Standard Least Squares chapter in the *Fitting Linear Models* book for more information about interpretation of the tables in Figure 11.6.

4. Click the disclosure icon next to Response Surface to open the report.
5. Click the disclosure icon next to Canonical Curvature.

Figure 11.7 Response Surface Report

Response Surface

Coef

	Silica(0.7,1.7)	Sulfur(1.8,2.8)	Silane(40,60)	Stretch
Silica(0.7,1.7)	26.5	114.5	-0.25	-40.125
Sulfur(1.8,2.8)	.	-35.5	46.25	-26.875
Silane(40,60)	.	.	0.25	11

Solution

Variable	Critical Value
Silica(0.7,1.7)	1.7912411
Sulfur(1.8,2.8)	2.1986422
Silane(40,60)	23.424426

Solution is a SaddlePoint
Critical values outside data range
Predicted Value at Solution 360.38388

Canonical Curvature

Eigenvalues and Eigenvectors

Eigenvalue	62.9095	3.2989	-74.9584
Silica(0.7,1.7)	0.82779	-0.29879	-0.47486
Sulfur(1.8,2.8)	0.52687	0.12315	0.84097
Silane(40,60)	0.19280	0.94634	-0.25937

The Coef table shown as the first part of the report gives a concise summary of the estimated model parameters. The first three columns give the coefficients of the second-order terms. The last column gives the coefficients of the linear terms. To see the prediction expression in its entirely, select **Estimates > Show Prediction Expression** from the Response Stretch red triangle menu.

The Solution report gives the coordinates of the point where the single critical value occurs. In this instance, that point is a saddle point (a point where neither a maximum nor a minimum occurs) and falls outside the range of the design space.

The Canonical Curvature report shows eigenvalues and eigenvectors of the effects. These give information about the nature and direction of the surface's curvature. The large positive eigenvalue of 62.9095 indicates positive curvature and the eigenvector values indicate that the curvature is primarily in the Silica direction. The large negative eigenvalue of -74.9584 indicates negative curvature and the eigenvector values indicate that the curvature is primarily in the Sulfur direction.

See the Standard Least Squares chapter in the *Fitting Linear Models* book for details about the response surface analysis tables in Figure 11.7.

Next, use the prediction profiler and the contour profiler to find optimal settings.

Explore Optimal Settings

1. Click the Prediction Profiler red triangle menu and select **Optimization and Desirability > Maximize Desirability**.

Figure 11.8 Prediction Profiler for Bounce Data.jmp with Desirability Maximized

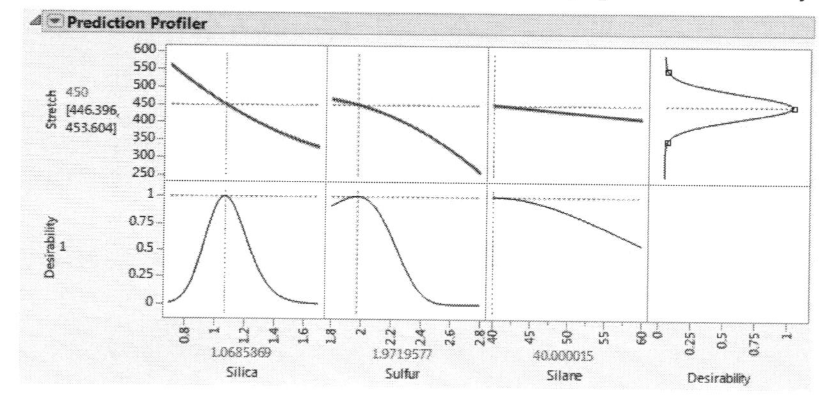

Note: Your optimal settings might differ. This is because there are many points for which the predicted Stretch is 450.

When you specified the response, the goal was set to match a target of 450, with lower and upper limits of 350 and 550. This goal was carried over to the design table and these limits were put in a Response Limits column property for Stretch. A desirability function is constructed from these response limits (top right cell in Figure 11.8). For details, see "Response Limits" on page 653 in the "Column Properties" appendix.

When you maximize the desirability function, JMP identifies one combination of factor level settings (usually out of many possible combinations) that results in a predicted Stretch of 450. Figure 11.8 shows these settings as Silica = 1.069, Sulfur = 1.972, and Silane = 40.000. Next, you use the Contour Profiler to identify other points that maximize the desirability function.

For more information about the Prediction Profiler, see the Profiler chapter in the *Profilers* book.

2. Click the Response Stretch red triangle menu and select **Factor Profiling > Contour Profiler.** Suppose that you want to achieve your target while setting Sulfur to the value 2.0. Also, you want to make sure that the settings that you choose for Silane and Silica maintain predicted Stretch within 5 units of 450.
3. In the plot controls area above the plot, click the radio button under **Vert** to the left of Silane.

Figure 11.9 Contour Profiler for Bounce Data.jmp

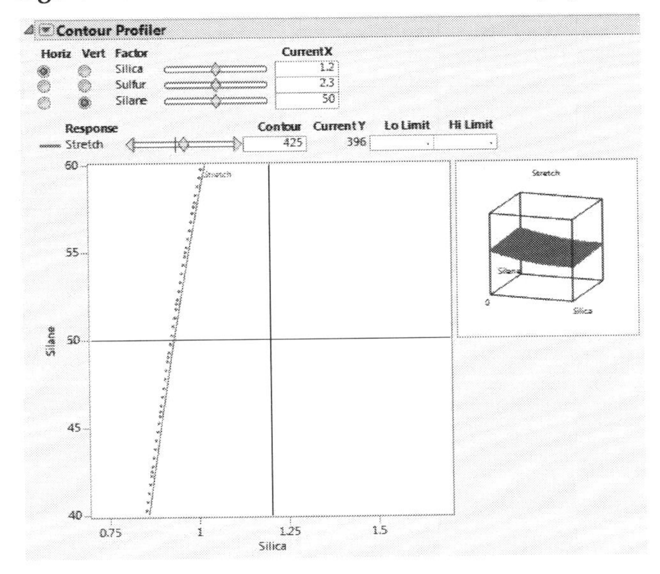

The plot shows the contour of values of Silane and Silica for Stretch at 425 and Sulfur at 2.3.

4. Set the **Current X** for Sulfur to 2.
5. Set the **Contour** for Stretch to 450.
6. Set the **Lo Limit** and **Hi Limit** for Stretch to 445 and 455, respectively. Press Enter.

Figure 11.10 Contour Profiler Showing Optimal Settings for Silica and Silane

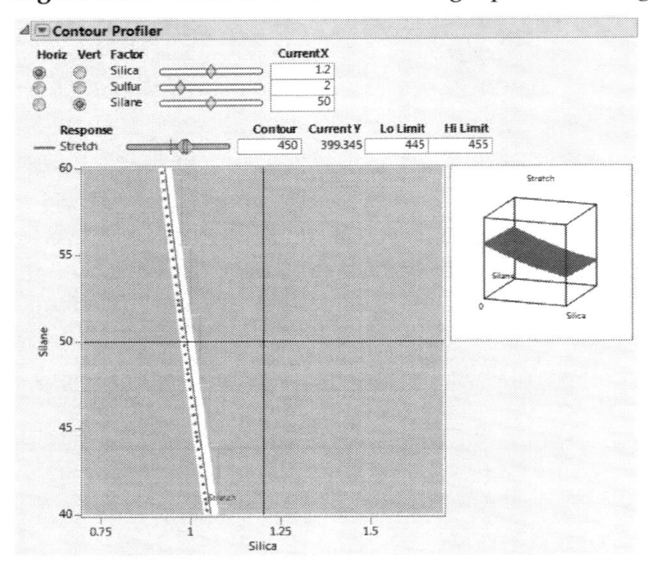

The unshaded band of Silica and Silane values gives predicted Stretch between 445 and 455 when Sulfur is set at 2.0. The values on the solid red curve give predicted Stretch of 450.

7. Drag the crosshairs that appear in the plot to the unshaded band to find settings for Silica and Silane that are best for your process from a practical perspective.

   Suppose that your process is known to be more robust at low levels of Silane than at high levels. Then you might consider the settings in Figure 11.11.

Figure 11.11 Contour Profiler Showing Specific Settings for Silica and Silane

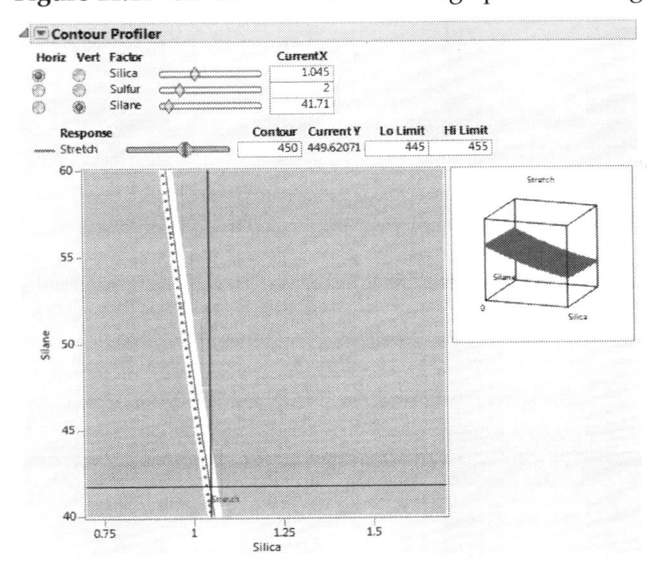

For Sulfur = 2.0, the factor settings identified by the crosshairs are Silica = 1.045 and Silane = 41.71. These settings are shown under Current X. At these settings, the predicted Stretch is 449.62071, shown next to Current Y.

For further information about the Contour Profiler, see the Contour Profiler chapter in the *Profilers* book.

Response Surface Design Window

The Response Surface Design window walks you through the steps to construct a design for modeling a quadratic surface. You can select a central composite design, a Box-Behnken design, or a blocked version of one of these design types. If you select a central composite design, you can adjust the axial points.

The Response Surface Design window is updated as you work through the design steps. The outlines, separated by buttons that update the outlines, follow the flow in Figure 11.12.

Figure 11.12 Response Surface Design Flow

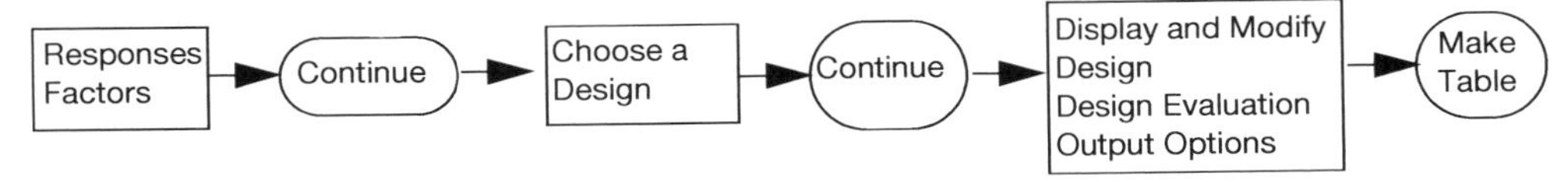

The following sections describe the steps in creating a response surface design:

- "Responses" on page 350
- "Factors" on page 352
- "Choose a Design" on page 353
- "Specify Output Options" on page 355
- "Make Table" on page 356

Responses

Use the Responses outline to specify one or more responses.

Tip: When you have completed the Responses outline, consider selecting **Save Responses** from the red triangle menu. This option saves the response names, goals, limits, and importance values in a data table that you can later reload in DOE platforms.

Figure 11.13 Responses Outline

Add Response Enters a single response with a goal type of Maximize, Match Target, Minimize, or None. If you select Match Target, enter limits for your target value. If you select Maximize or Minimize, entering limits is not required but can be useful if you intend to use desirability functions.

Remove Removes the selected responses.

Number of Responses Enters additional responses so that the number that you enter is the total number of responses. If you have entered a response other than the default Y, the Goal for each of the additional responses is the Goal associated with the last response entered. Otherwise, the Goal defaults to Match Target. Click the Goal type in the table to change it.

The Responses outline contains the following columns:

Response Name The name of the response. When added, a response is given a default name of Y, Y2, and so on. To change this name, double-click it and enter the desired name.

Goal, Lower Limit, Upper Limit The Goal tells JMP whether you want to maximize your response, minimize your response, match a target, or that you have no response goal. JMP assigns a Response Limits column property, based on these specifications, to each response column in the design table. It uses this information to define a desirability function for each response. The Profiler and Contour Profiler use these desirability functions to find optimal factor settings. For further details, see the Profiler chapter in the *Profilers* book and "Response Limits" on page 653 in the "Column Properties" appendix.

– A Goal of Maximize indicates that the best value is the largest possible. If there are natural lower or upper bounds, you can specify these as the Lower Limit or Upper Limit.

– A Goal of Minimize indicates that the best value is the smallest possible. If there are natural lower or upper bounds, you can specify these as the Lower Limit or Upper Limit.

– A Goal of Match Target indicates that the best value is a specific target value. The default target value is assumed to be midway between the Lower Limit and Upper Limit.

– A Goal of None indicates that there is no goal in terms of optimization. No desirability function is constructed.

Note: If your target response is not midway between the Lower Limit and the Upper Limit, you can change the target after you generate your design table. In the data table, open the Column Info window for the response column (**Cols** > **Column Info**) and enter the desired target value.

Importance When you have several responses, the Importance values that you specify are used to compute an overall desirability function. These values are treated as weights for the responses. If there is only one response, then specifying the Importance is unnecessary because it is set to 1 by default.

Editing the Responses Outline

In the Responses outline, note the following:

- Double-click a response to edit the response name.
- Click the goal to change it.
- Click on a limit or importance weight to change it.
- For multiple responses, you might want to enter values for the importance weights.

Response Limits Column Property

The Goal, Lower Limit, Upper Limit, and Importance that you specify when you enter a response are used in finding optimal factor settings. For each response, the information is saved in the generated design data table as a Response Limits column property. JMP uses this information to define the desirability function. The desirability function is used in the

Prediction Profiler to find optimal factor settings. For further details about the Response Limits column property and examples of its use, see "Response Limits" on page 653 in the "Column Properties" appendix.

If you do not specify a Lower Limit and Upper Limit, JMP uses the range of the observed data for the response to define the limits for the desirability function. Specifying the Lower Limit and Upper Limit gives you control over the specification of the desirability function. For more details about the construction of the desirability function, see the Profiler chapter in the *Profilers* book.

Factors

Factors in a response surface design can only be continuous.

Tip: Use **DOE > Custom Design** to create response surface designs that involve categorical factors.

The initial Factors panel for a response surface design appears with two continuous factors.

Figure 11.14 Factors Outline

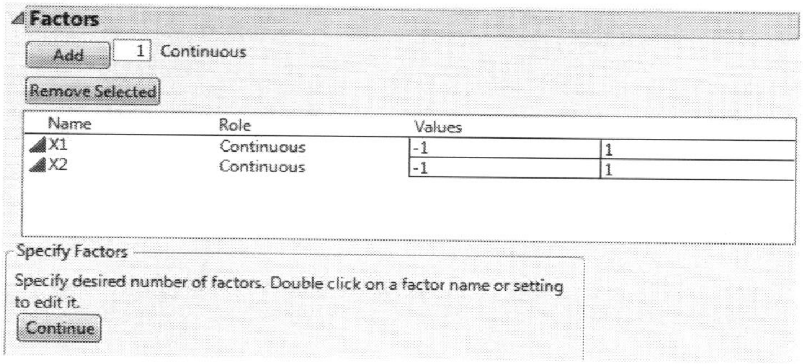

The factors outline contains the following buttons.

Add Enters the number of continuous factors specified.

Remove Selected Removes the selected factors.

Tip: When you have completed your Factors panel, select Save Factors from the red triangle menu. This saves the factor names and values in a data table that you can later reload. See "Response Surface Design Options" on page 357.

The Factors outline contains the following columns:

Name The name of the factor. When added, a factor is given a default name of X1, X2, and so on. To change this name, double-click it and enter the desired name.

Role Specifies the Design Role of the factor as Continuous. The Design Role column property for the factor is saved to the data table. This property ensures that the factor type is modeled appropriately.

Values The experimental settings for the factors. To insert Values, click on the default values and enter the desired values.

Factor Column Properties

For each factor, various column properties are saved to the data table for the completed design.

Design Role Each factor is assigned the Design Role column property. The Role that you specify in defining the factor determines the value of its Design Role column property. When you select a design with a block, that factor is assigned the Blocking value. The Design Role property reflects how the factor is intended to be used in modeling the experimental data. Design Role values are used in the Augment Design platform.

Factor Changes Each factor is assigned the Factor Changes column property with a setting of Easy. In the Response Surface Design platform, it is assumed that factor levels can be changed for each experimental run. Factor Changes values are used in the Evaluate Design and Augment Design platforms.

Coding If the Design Role is Continuous, the Coding column property for the factor is saved. This property transforms the factor values so that the low and high values correspond to –1 and +1, respectively. The estimates and tests in the Fit Least Squares report are based on the transformed values.

RunsPerBlock Indicates the number of runs in each block. When you select a design with a block and then click Make Table, a factor with the default name Block is added to the Factors list. The RunsPerBlock column property is saved for that factor.

Choose a Design

After you enter your responses and factors and click **Continue**, you select from a list of designs. The designs include two types:

- "Box-Behnken Designs" on page 354
- "Central Composite Designs" on page 354

Select the design that you want to use and click **Continue**.

Figure 11.15 Choose a Design Panel for Four Factors

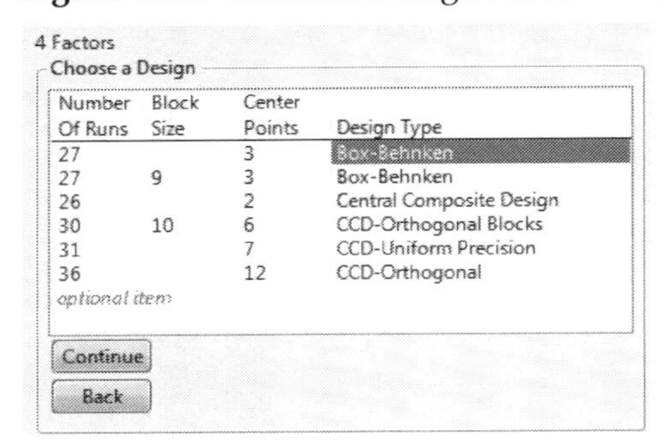

Box-Behnken Designs

Box-Behnken designs have only three levels for each factor and have no design points at the vertices of the cube defined by the ranges of the factors. These designs can be useful when it is desirable to avoid extreme settings for engineering considerations. However, these designs result in higher prediction variance near the vertices than do central composite designs.

Central Composite Designs

Central composite designs have center points and axial points. An *axial point* is a point where one factor is set to a high or low value (an *axial* value) and all other factors are set to the midrange, or center, value.

A central composite design can have axial points that fall beyond the faces of the hypercube defined by the specified factor ranges. This means that each factor might require five distinct settings, including two that fall beyond the range of values specified in the Factors outline. However, JMP enables you to place design points on the face.

The following types of central composite designs are available:

Central Composite Design The usual central composite design for the specified number of factors.

CCD-Uniform Precision The number of center points is chosen so that the prediction variance near the center of the design space is very flat.

CCD-Orthogonal The number of center points and the axial values are chosen so that the second-order parameter estimates are minimally correlated with the other parameter estimates.

CCD-Orthogonal Blocks The second-order parameter estimates and block effects are minimally correlated with the other parameter estimates.

Axial Value

When you select a central composite design and then click **Continue**, you have the option to provide axial scaling information. In placing axial values, the values shown are used to multiply half of the specified range of a factor. If you specify a value of 1.0 next to Axial Value, then axial points in the resulting design are placed on the faces of the cube defined by the factors. You can set the axial value according to the following options:

Figure 11.16 Axial Value Panel

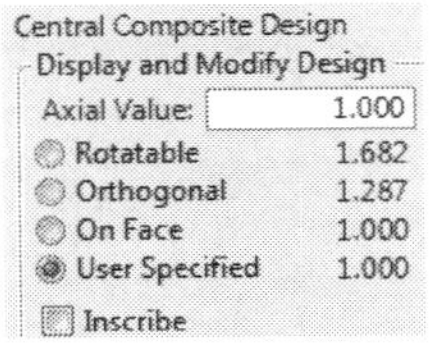

Central Composite Design
Display and Modify Design
Axial Value: 1.000
Rotatable 1.682
Orthogonal 1.287
On Face 1.000
User Specified 1.000
Inscribe

Rotatable The prediction variance depends only on the scaled distance from the center of the design. The axial points are more extreme than the factor ranges. If this factor range cannot be practically achieved, select **On Face** or specify your own value.

Orthogonal The effects are orthogonal. The axial points are more extreme than the factor ranges. If this factor range cannot be practically achieved, select **On Face** or specify your own value.

On Face Places the axial points at the extremes of the specified factor ranges.

User Specified Places the axial points at a distance specified by the value that you enter in the Axial Value text box.

Inscribe Rescales the design so that the axial points are at the low and high ends of the factor range. The factorial design points are shrunken based on that scaling.

Specify Output Options

You can specify details for the output data table in the Output Options panel. When you finish, click **Make Table** to construct the data table for the design.

Run Order

The **Run Order** options determine the order of the runs in the design table. Choices include the following:

Keep the Same Rows in the design table are in the same order as in the Design and Anticipated Coefficients outline.

Sort Left to Right Columns in the design table are sorted from left to right.

Randomize Rows in the design table are in random order.

Sort Right to Left Columns in the design table are sorted from right to left.

Randomize within Blocks Rows in the design table are in random order within the blocks.

Center Points and Replicates

Number of Center Points The number of center points that appear in the design. A center point is a run where every continuous factor is set at the center of the factor's range. The initial value shown is the number of center points in the design that you selected.

Number of Replicates The number of times to replicate the entire design, including center points. One replicate doubles the number of runs.

Make Table

Click **Make Table** to create a design table that contains the runs for your experiment.

Figure 11.17 Orthogonal Central Composite Design for Bounce Factors and Response

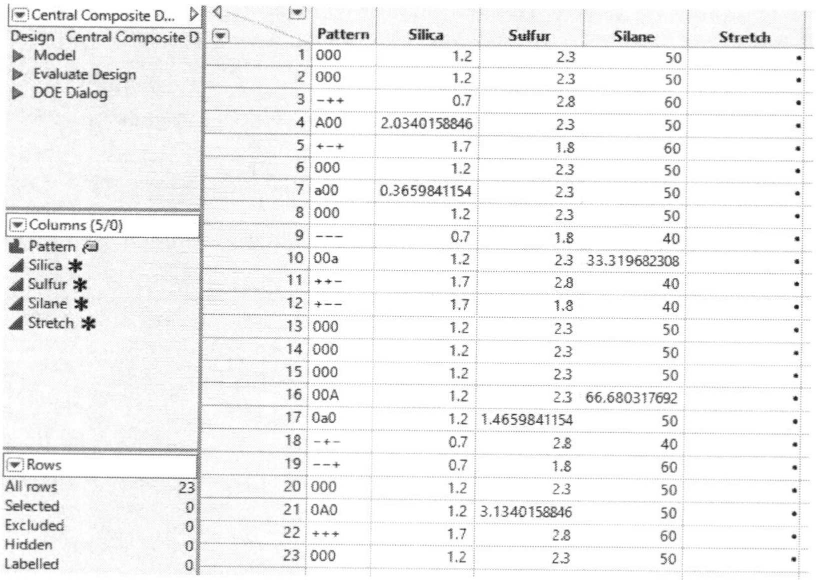

Central Composite D...
Design Central Composite D
▶ Model
▶ Evaluate Design
▶ DOE Dialog

Columns (5/0)
Pattern
Silica
Sulfur
Silane
Stretch

Rows
All rows 23
Selected 0
Excluded 0
Hidden 0
Labelled 0

	Pattern	Silica	Sulfur	Silane	Stretch
1	000	1.2	2.3	50	•
2	000	1.2	2.3	50	•
3	−++	0.7	2.8	60	•
4	A00	2.0340158846	2.3	50	•
5	+−+	1.7	1.8	60	•
6	000	1.2	2.3	50	•
7	a00	0.3659841154	2.3	50	•
8	000	1.2	2.3	50	•
9	−−−	0.7	1.8	40	•
10	00a	1.2	2.3	33.319682308	•
11	++−	1.7	2.8	40	•
12	+−−	1.7	1.8	40	•
13	000	1.2	2.3	50	•
14	000	1.2	2.3	50	•
15	000	1.2	2.3	50	•
16	00A	1.2	2.3	66.680317692	•
17	0a0	1.2	1.4659841154	50	•
18	−+−	0.7	2.8	40	•
19	−−+	0.7	1.8	60	•
20	000	1.2	2.3	50	•
21	0A0	1.2	3.1340158846	50	•
22	+++	1.7	2.8	60	•
23	000	1.2	2.3	50	•

The **Design** note in the Table panel at the upper left gives the design type that generated the table (Central Composite Design). This information can be helpful if you are comparing multiple designs.

Pattern Column

A Pattern column gives a symbolic description of the run in each row in terms of the factor values.

Tip: Pattern can be a useful label variable in plots.

Table 11.1 Pattern Column Description

-	Low value
+	High value
0	Midrange (center) value
a	Low axial value
A	High axial value

Design Table Scripts

The design table includes the following scripts:

Model Runs the **Analyze** > **Fit Model** platform.

Evaluate Design Runs the DOE > Design Diagnostics > Evaluate Design platform.

DOE Dialog Re-creates the Response Surface Design window that you used to generate the design table. The script also contains the random seed used to generate your design.

Response Surface Design Options

The red triangle menu in the Response Surface Design platform contains these options:

Save Responses Saves the information in the Responses panel to a new data table. You can then quickly load the responses and their associated information into most DOE windows. This option is helpful if you anticipate re-using the responses.

Load Responses Loads responses that you saved using the Save Responses option.

Save Factors Saves the information in the Factors panel to a new data table. Each factor's column contains its levels. Other information is stored as column properties. You can then quickly load the factors and their associated information into most DOE windows.

Note: It is possible to create a factors table by entering data into an empty table, but remember to assign each column an appropriate Design Role. Do this by right-clicking on the column name in the data grid and selecting **Column Properties > Design Role**. In the Design Role area, select the appropriate role.

Load Factors Loads factors that you saved using the Save Factors option.

Save Constraints (Unavailable for some platforms) Saves factor constraints that you defined in the Define Factor Constraints or Linear Constraints outline into a data table, with a column for each constraint. You can then quickly load the constraints into most DOE windows.

In the constraint table, the first rows contain the coefficients for each factor. The last row contains the inequality bound. Each constraint's column contains a column property called ConstraintState that identifies the constraint as a "less than" or a "greater than" constraint. See "ConstraintState" on page 686 in the "Column Properties" appendix.

Load Constraints (Unavailable for some platforms) Loads factor constraints that you saved using the Save Constraints option.

Set Random Seed Sets the random seed that JMP uses to control certain actions that have a random component. These actions include the following:

- simulating responses using the Simulate Responses option
- randomizing Run Order for design construction
- selecting a starting design for designs based on random starts

To reproduce a design or simulated responses, enter the random seed that generated them. For designs using random starts, set the seed before clicking Make Design. To control simulated responses or run order, set the seed before clicking Make Table.

Note: The random seed associated with a design is included in the DOE Dialog script that is saved to the design data table.

Simulate Responses Adds response values and a column containing a simulation formula to the design table. Select this option before you click Make Table.

When you click Make Table, the following occur:

- A set of simulated response values is added to each response column.
- For each response, a new a column that contains a simulation model formula is added to the design table. The formula and values are based on the model that is specified in the design window.
- A Model window appears where you can set the values of coefficients for model effects and specify one of three distributions: Normal, Binomial, or Poisson.
- A script called **DOE Simulate** is saved to the design table. This script re-opens the Model window, enabling you to re-simulate values or to make changes to the simulated response distribution.

Make selections in the Model window to control the distribution of simulated response values. When you click Apply, a formula for the simulated response values is saved in a new column called <Y> Simulated, where Y is the name of the response. Clicking Apply again updates the formula and values in <Y> Simulated.

For additional details, see "Simulate Responses" on page 106 in the "Custom Designs" chapter.

Note: JMP PRO You can use Simulate Responses to conduct simulation analyses using the JMP Pro Simulate feature. For information about Simulate and some DOE examples, see the Simulate chapter in the *Basic Analysis* book.

Advanced Options > Set Delta for Power Specifies the difference in the mean response that you want to detect for model effects. See "Evaluate Design Options" on page 451 in the "Evaluate Designs" chapter.

Save Script to Script Window Creates the script for the design that you specified in the Response Surface Design window and saves it in an open script window.

Chapter 12

Full Factorial Designs

A full factorial design defines an experiment where trials are run at all possible combinations of factor settings. A full factorial design allows the estimation of all possible interactions. Full factorial designs are large compared to screening designs, and since high-level interactions are often not active, they can be inefficient. They are typically used when you have a small number of factors and levels and want information about all possible interactions. For example, full factorial designs often form the basis for a measurement systems analysis (MSA).

Figure 12.1 Full Factorial Design for Three Two-Level Factors

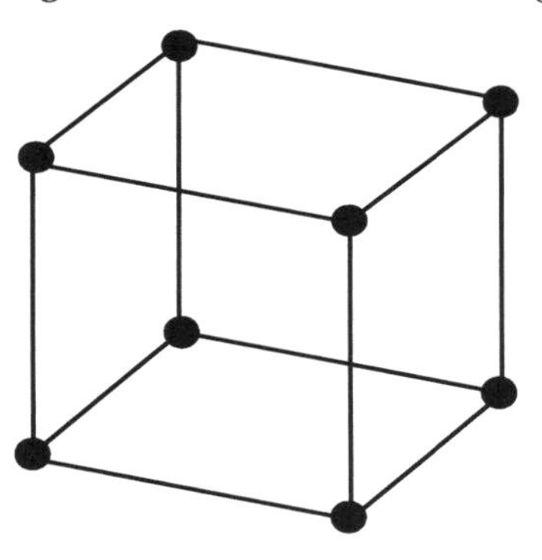

Overview of Full Factorial Design

In a full factorial design, you perform an experimental run at every combination of the factor levels. The sample size is the product of the numbers of levels of the factors. For example, a factorial experiment with a two-level factor, a three-level factor, and a four-level factor has $2 \times 3 \times 4 = 24$ runs.

The Full Factorial Design platform supports both continuous factors and categorical factors with arbitrary numbers of levels. It is assumed that you can run the trials in a completely random fashion.

Full factorial designs are the most conservative of all design types. Unfortunately, because the sample size grows exponentially with the number of factors, full factorial designs are often too expensive to run. Custom designs, definitive screening designs, and screening designs are less conservative but more efficient and cost-effective.

Example of a Full Factorial Design

In this example, you construct a full factorial design to study the effects of five two-level factors (**Feed Rate**, **Catalyst**, **Stir Rate**, **Temperature**, and **Concentration**) on the yield of a reactor. Because there are five factors, each at two levels, the full factorial design includes at least $2^5 = 32$ runs. For smaller screening designs involving this experimental situation, see "Additional Examples of Screening Designs" on page 311 in the "Screening Designs" chapter.

In this example you load the responses and factors from existing tables. When designing a new experiment on your own, enter the responses and factors manually. See "Responses" on page 369 and "Factors" on page 371.

Construct the Design

1. Select **DOE > Classical > Full Factorial Design**.
2. Select **Help > Sample Data Library** and open Design Experiment/Reactor Response.jmp.
3. Click the Full Factorial Design red triangle and select **Load Responses**.
4. Select **Help > Sample Data Library** and open Design Experiment/Reactor Factors.jmp.
5. Click the Full Factorial Design red triangle and select **Load Factors**.
6. Click **Continue**.

Figure 12.2 Full Factorial Example Response and Factors Panels

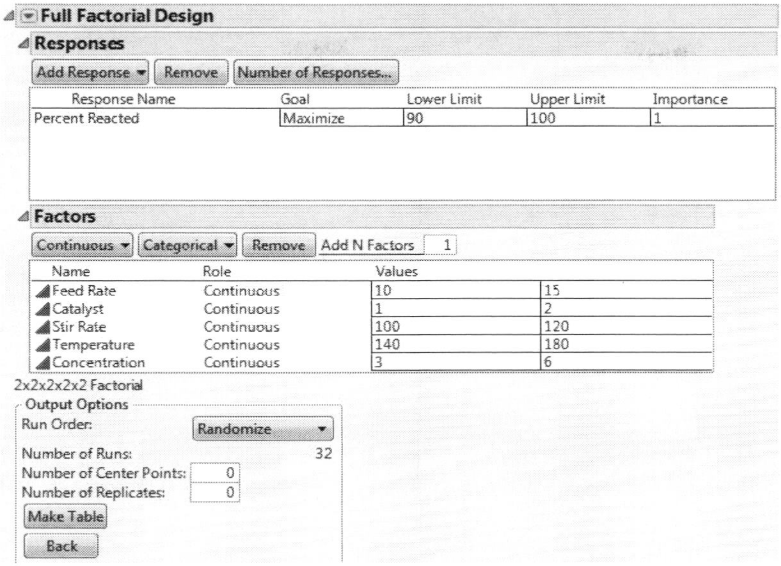

Full Factorial Design

Responses

Add Response | Remove | Number of Responses...

Response Name	Goal	Lower Limit	Upper Limit	Importance
Percent Reacted	Maximize	90	100	1

Factors

Continuous | Categorical | Remove | Add N Factors 1

Name	Role	Values	
Feed Rate	Continuous	10	15
Catalyst	Continuous	1	2
Stir Rate	Continuous	100	120
Temperature	Continuous	140	180
Concentration	Continuous	3	6

2x2x2x2x2 Factorial

Output Options

Run Order: Randomize

Number of Runs: 32

Number of Center Points: 0

Number of Replicates: 0

Make Table

Back

Note: Setting the Random Seed in step 7 ensures that the runs in your design table appear in the same order as in this example. In constructing a design on your own, this step is not necessary.

7. (Optional) Click the Full Factorial Design red triangle menu and select **Set Random Seed**. Type 12345 and click **OK**.

   The Run Order in the Output Options panel is set to Randomize. The order of runs in the design table will be random, as determined by the random seed.

   The Number of Runs is set to 32. This is the number of all possible factor level combinations.

8. Click **Make Table**.

   The first column in the design data table shows the factor level combination for each run in terms of + and - signs, indicating high and low factor settings. The table also has an empty Y column named Percent Reacted for entering response values as you conduct the experiment.

Figure 12.3 Full Factorial Design for Reactor Experiment

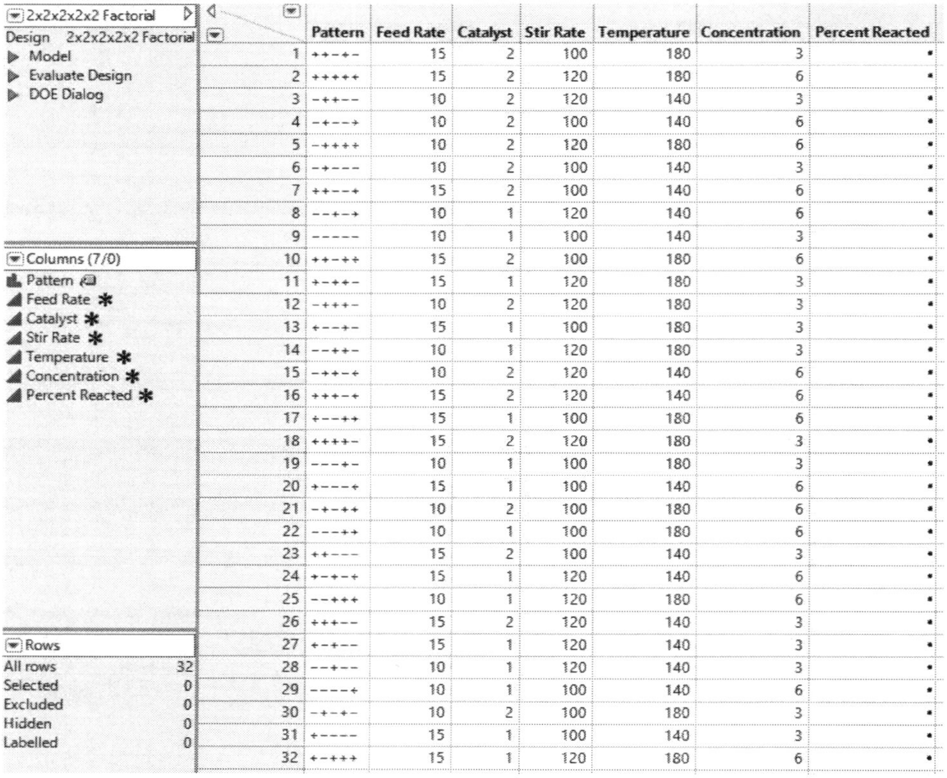

2x2x2x2x2 Factorial
Design 2x2x2x2x2 Factorial
Model
Evaluate Design
DOE Dialog

Columns (7/0)
Pattern
Feed Rate *
Catalyst *
Stir Rate *
Temperature *
Concentration *
Percent Reacted *

Rows
All rows 32
Selected 0
Excluded 0
Hidden 0
Labelled 0

	Pattern	Feed Rate	Catalyst	Stir Rate	Temperature	Concentration	Percent Reacted
1	++-+-	15	2	100	180	3	•
2	+++++	15	2	120	180	6	•
3	-++--	10	2	120	140	3	•
4	-+--+	10	2	100	140	6	•
5	-++++	10	2	120	180	6	•
6	-+---	10	2	100	140	3	•
7	++--+	15	2	100	140	6	•
8	--+-+	10	1	120	140	6	•
9	-----	10	1	100	140	3	•
10	++-++	15	2	100	180	6	•
11	+-++-	15	1	120	180	3	•
12	-+++-	10	2	120	180	3	•
13	+--+-	15	1	100	180	3	•
14	--++-	10	1	120	180	3	•
15	-++-+	10	2	120	140	6	•
16	+++-+	15	2	120	140	6	•
17	+--++	15	1	100	180	6	•
18	++++-	15	2	120	180	3	•
19	---+-	10	1	100	180	3	•
20	+---+	15	1	100	140	6	•
21	-+-++	10	2	100	180	6	•
22	---++	10	1	100	180	6	•
23	++---	15	2	100	140	3	•
24	+-+-+	15	1	120	140	6	•
25	--+++	10	1	120	180	6	•
26	+++--	15	2	120	140	3	•
27	+-+--	15	1	120	140	3	•
28	--+--	10	1	120	140	3	•
29	----+	10	1	100	140	6	•
30	-+-+-	10	2	100	180	3	•
31	+----	15	1	100	140	3	•
32	+-+++	15	1	120	180	6	•

Analyze the Experimental Data

Next, proceed to analyze the data from the completed experiment. You will use two methods to analyze the results: Screening and Stepwise Regression. Then you will find optimal settings using the Prediction Profiler.

Analysis Using Screening Platform

1. Select **Help > Sample Data Library** and open Design Experiment/Reactor 32 Runs.jmp.
2. Run the **Screening** script.

   The Screening report shows a Contrasts report and a Half Normal Plot. The Contrasts report shows estimates for all 31 potential effects, up to the five-way interaction.

Figure 12.4 Contrasts Report for Reactor 32 Runs.jmp

Contrasts

Term	Contrast	Lenth t-Ratio	Individual p-Value	Simultaneous p-Value
Catalyst	9.75000	14.86	[illegible]	[illegible]
Temperature	5.37500	8.19	[illegible]	[illegible]
Concentration	-3.12500	-4.76	[illegible]	[illegible]
Feed Rate	-0.68750	-1.05	0.2883	0.9998
Stir Rate	-0.31250	-0.48	0.6494	1.0000
Catalyst*Temperature	6.62500	10.10	[illegible]	[illegible]
Catalyst*Concentration	1.00000	1.52	0.1335	0.9553
Temperature*Concentration	-5.50000	-8.38	[illegible]	[illegible]
Catalyst*Feed Rate	0.68750	1.05	0.2883	0.9998
Temperature*Feed Rate	-0.43750	-0.67	0.4941	1.0000
Concentration*Feed Rate	0.06250	0.10	0.9266	1.0000
Catalyst*Stir Rate	0.43750	0.67	0.5212	1.0000
Temperature*Stir Rate	1.06250	1.62	0.1139	0.9205
Concentration*Stir Rate	0.43750	0.67	0.4941	1.0000
Feed Rate*Stir Rate	0.37500	0.57	0.5859	1.0000
Catalyst*Temperature*Concentration	-0.12500	-0.19	0.8557	1.0000
Catalyst*Temperature*Feed Rate	0.68750	1.05	0.2883	0.9998
Catalyst*Concentration*Feed Rate	-0.93750	-1.43	0.1587	0.9784
Temperature*Concentration*Feed Rate	0.31250	0.48	0.6494	1.0000
Catalyst*Temperature*Stir Rate	0.56250	0.86	0.3772	1.0000
Catalyst*Concentration*Stir Rate	0.06250	0.10	0.9266	1.0000
Temperature*Concentration*Stir Rate	0.06250	0.10	0.9266	1.0000
Catalyst*Feed Rate*Stir Rate	0.75000	1.14	0.2478	0.9989
Temperature*Feed Rate*Stir Rate	-0.37500	-0.57	0.5859	1.0000
Concentration*Feed Rate*Stir Rate	-1.25000	-1.90	0.0705	0.7592
Catalyst*Temperature*Concentration*Feed Rate	0.31250	0.48	0.6494	1.0000
Catalyst*Temperature*Concentration*Stir Rate	-0.31250	-0.48	0.6494	1.0000
Catalyst*Temperature*Feed Rate*Stir Rate	0.00000	0.00	1.0000	1.0000
Catalyst*Concentration*Feed Rate*Stir Rate	0.75000	1.14	0.2478	0.9989
Temperature*Concentration*Feed Rate*Stir Rate	0.50000	0.76	0.4328	1.0000
Catalyst*Temperature*Concentration*Feed Rate*Stir Rate	-0.25000	-0.38	0.7150	1.0000

Note: Because the p-values in the Contrasts report are obtained using a Monte Carlo simulation, you will not obtain the same values as shown in Figure 12.4. For more information, see "Lenth's Pseudo-Standard Error" on page 335 in the "The Fit Two Level Screening Platform" chapter.

The six highlighted effects in the Contrasts outline are labeled in the Half Normal Plot.

Figure 12.5 Half Normal Plot for Reactor 32 Runs.jmp

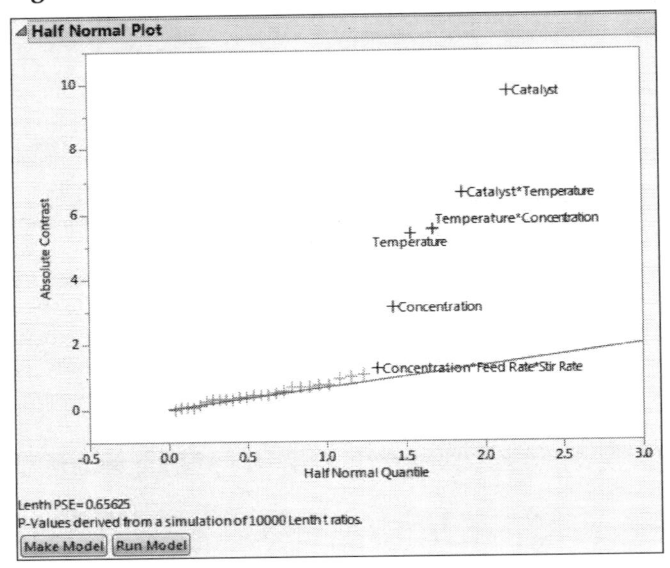

The Half Normal Plot provides strong evidence that at least five of the labeled effects are larger than would be expected if they were the result of random variation. This suggests that these effects are active. The plot does *not* provide a clear indication that the three-way Concentration*Feed Rate*Stir Rate interaction is active.

In the Contrasts outline in Figure 12.4, the Individual p-Value for the three-way Concentration*Feed Rate*Stir Rate interaction is 0.0705 and its Simultaneous p-Value is 0.7592. Because the effect does not stand out on the Half Normal Plot and because its *p*-values are large, you decide not to include this effect in your model.

3. In the Half Normal Plot, drag a rectangle to select all labeled effects except for Concentration*Feed Rate*Stir Rate.
4. Click **Make Model** to open a Fit Model window containing the five effects.
5. Click **Run.**

   The Actual by Predicted plot shows no evidence of lack of fit and the Effect Summary outline shows that all five effects are significant.

Analysis Using Stepwise Regression

1. Return to the Reactor 32 Runs.jmp data table, or reopen it by selecting **Help > Sample Data Library** and opening Design Experiment/Reactor 32 Runs.jmp.
2. Run the **Model** script.

The Construct Model Effects list contains only up to two-way interactions. You want to consider all interactions.

3. From the Select Columns list, select Feed Rate through Concentration.
4. Click **Macros** > **Full Factorial**.

   All possible effects are added in the Construct Model Effects list.
5. Change the **Personality** to **Stepwise**.
6. Click **Run**.
7. Change the **Stopping Rule** to **Minimum AICc.**

   For designed experiments, AICc is preferred to BIC. This is because BIC is typically a more lenient stopping rule than AICc as it tends to allow inactive effects into the model.
8. Click **Go**.

   The Stepwise procedure selects seven effects as potentially active.
9. Click **Run Model.**

   This fits a model using the seven effects. The Effect Summary outline indicates that Catalyst*Temperature*Concentration is not significant, because its *p*-value is 0.83042.
10. Select Catalyst*Temperature*Concentration in the Effect Summary outline and click **Remove**.

    The effect Catalyst*Concentration has a *p*-value of 0.08960. You decide to remove it too.
11. Select Catalyst*Concentration in the Effect Summary outline and click **Remove**.

    The five remaining effects are all highly significant. These are the same five effects that you identified using the Screening platform ("Analysis Using Screening Platform" on page 364).

Optimal Settings Using the Prediction Profiler

Now, find optimal settings for the three active factors involved in the five significant effects that you retained in your model.

1. In the Reactor 32 Runs.jmp data table, run the **Reduced Model** script.

   The Reduced Model script opens a Fit Model window for the five-effect model that you identified in "Analysis Using Screening Platform" on page 364 and "Analysis Using Stepwise Regression" on page 366.
2. Click **Run**.

   The Prediction Profiler report displays Desirability because in the Full Factorial window, you specified a Goal of Maximize when you defined your response. The desirability function shown in the rightmost cell in the top row of the profiler shows that a value of 100 is most desirable and a value of 90 or below is least desirable.

3. Click the Prediction Profiler red triangle and select **Optimization and Desirability** > **Maximize Desirability**.

Figure 12.6 Prediction Profiler Showing Settings That Optimize Desirability

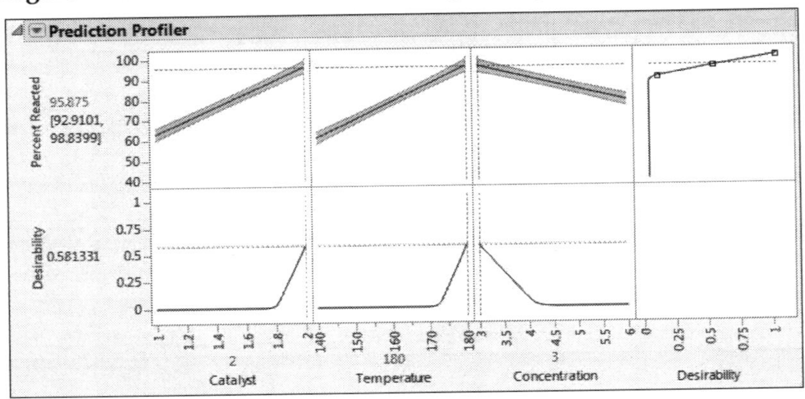

The predicted mean **Percent Reacted** at the settings that are shown is 95.875, with a confidence interval of 92.91 to 98.84. Note that, for all three factors, the settings that are identified are at the extremes of the ranges used in the experiment. In a future experiment, you should explore the process behavior beyond these settings.

Full Factorial Design Window

The Full Factorial design window is updated as you work through the design steps. For more information, see "The DOE Workflow: Describe, Specify, Design" on page 54. The outlines, separated by buttons that update the outlines, follow the flow in Figure 12.7.

Figure 12.7 Full Factorial Design Flow

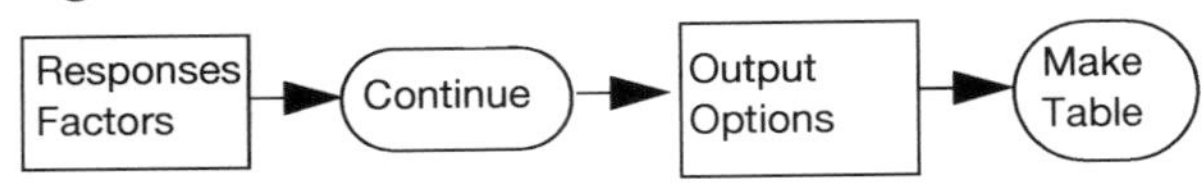

The following sections describe the steps in creating a full factorial design:

- "Responses" on page 369
- "Factors" on page 371
- "Select Output Options" on page 373
- "Make Table" on page 374

Responses

Use the Responses outline to specify one or more responses.

Tip: When you have completed the Responses outline, consider selecting **Save Responses** from the red triangle menu. This option saves the response names, goals, limits, and importance values in a data table that you can later reload in DOE platforms.

Figure 12.8 Responses Outline

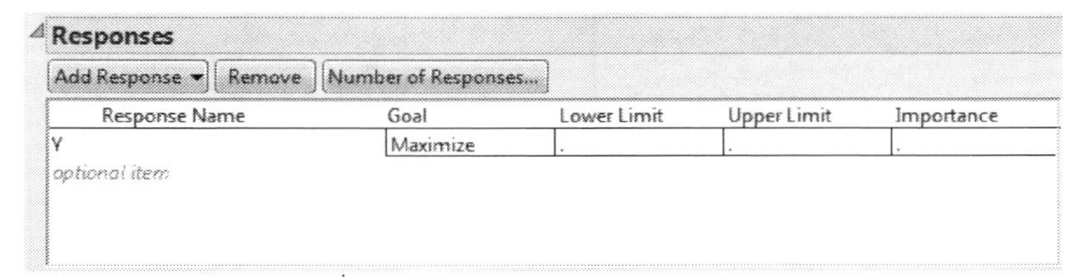

Add Response Enters a single response with a goal type of Maximize, Match Target, Minimize, or None. If you select Match Target, enter limits for your target value. If you select Maximize or Minimize, entering limits is not required but can be useful if you intend to use desirability functions.

Remove Removes the selected responses.

Number of Responses Enters additional responses so that the number that you enter is the total number of responses. If you have entered a response other than the default Y, the Goal for each of the additional responses is the Goal associated with the last response entered. Otherwise, the Goal defaults to Match Target. Click the Goal type in the table to change it.

The Responses outline contains the following columns:

Response Name The name of the response. When added, a response is given a default name of Y, Y2, and so on. To change this name, double-click it and enter the desired name.

Goal, Lower Limit, Upper Limit The Goal tells JMP whether you want to maximize your response, minimize your response, match a target, or that you have no response goal. JMP assigns a Response Limits column property, based on these specifications, to each response column in the design table. It uses this information to define a desirability function for each response. The Profiler and Contour Profiler use these desirability functions to find optimal factor settings. For further details, see the Profiler chapter in the *Profilers* book and "Response Limits" on page 653 in the "Column Properties" appendix.

- A Goal of Maximize indicates that the best value is the largest possible. If there are natural lower or upper bounds, you can specify these as the Lower Limit or Upper Limit.

- A Goal of Minimize indicates that the best value is the smallest possible. If there are natural lower or upper bounds, you can specify these as the Lower Limit or Upper Limit.
- A Goal of Match Target indicates that the best value is a specific target value. The default target value is assumed to be midway between the Lower Limit and Upper Limit.
- A Goal of None indicates that there is no goal in terms of optimization. No desirability function is constructed.

Note: If your target response is not midway between the Lower Limit and the Upper Limit, you can change the target after you generate your design table. In the data table, open the Column Info window for the response column (**Cols** > **Column Info**) and enter the desired target value.

Importance When you have several responses, the Importance values that you specify are used to compute an overall desirability function. These values are treated as weights for the responses. If there is only one response, then specifying the Importance is unnecessary because it is set to 1 by default.

Editing the Responses Outline

In the Responses outline, note the following:

- Double-click a response to edit the response name.
- Click the goal to change it.
- Click on a limit or importance weight to change it.
- For multiple responses, you might want to enter values for the importance weights.

Response Limits Column Property

The Goal, Lower Limit, Upper Limit, and Importance that you specify when you enter a response are used in finding optimal factor settings. For each response, the information is saved in the generated design data table as a Response Limits column property. JMP uses this information to define the desirability function. The desirability function is used in the Prediction Profiler to find optimal factor settings. For further details about the Response Limits column property and examples of its use, see "Response Limits" on page 653 in the "Column Properties" appendix.

If you do not specify a Lower Limit and Upper Limit, JMP uses the range of the observed data for the response to define the limits for the desirability function. Specifying the Lower Limit and Upper Limit gives you control over the specification of the desirability function. For more details about the construction of the desirability function, see the Profiler chapter in the *Profilers* book.

Factors

Factors in a full factorial design can be continuous or categorical.

Tip: When you have completed the Factors outline, consider selecting **Save Factors** from the red triangle menu. This option saves the factor names, roles, changes, and values in a data table that you can later reload in DOE platforms.

Figure 12.9 Factors Outline

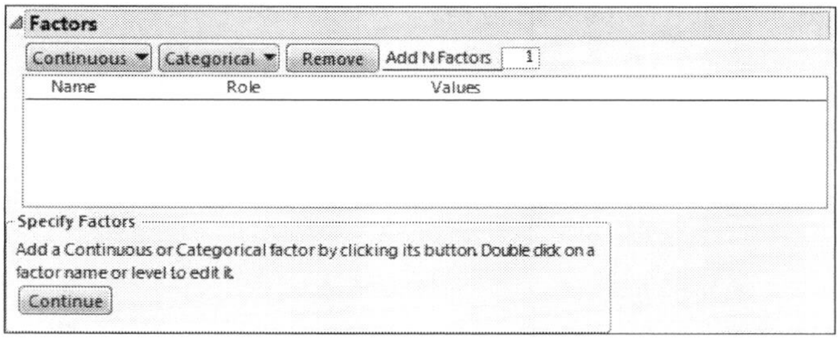

Factors
Continuous Categorical Remove Add N Factors 1
Name Role Values
Specify Factors
Add a Continuous or Categorical factor by clicking its button. Double click on a factor name or level to edit it.
Continue

Continuous Adds a Continuous factor. The data type in the resulting data table is Numeric. A continuous factor is a factor that you can conceptually set to any value between the lower and upper limits you supply, given the limitations of your process and measurement system.

Categorical Adds a Categorical factor. Click to select or specify the number of levels. The data type in the resulting data table is Character. The value ordering of the levels is the order of the values, as entered from left to right. This ordering is saved in the Value Ordering column property after the design data table is created.

The default values for a categorical factor are L1, L2, ..., Lk, where *k* is the number of levels that you specify. Replace the default values with level names that are relevant for your experiment.

Remove Removes the selected factors.

Add N Factors Adds multiple factors. Enter the number of factors to add, click **Add Factor,** and then select the factor type. Repeat **Add N Factors** to add multiple factors of different types.

Factors Outline

The Factors outline contains the following columns:

Name The name of the factor. When added, a factor is given a default name of X1, X2, and so on. To change this name, double-click it and enter the desired name.

Role The Design Role of the factor. The Design Role column property for the factor is saved to the data table. This property ensures that the factor type is modeled appropriately. The Role of the factor determines other factor properties that are saved to the data table. For details, see "Factor Column Properties" on page 372.

Values The experimental settings for the factors.

Editing the Factors Outline

In the Factors outline, note the following:

- To edit a factor name, double-click the factor name.
- To edit a value, click the value in the Values column.

Factor Column Properties

For each factor, various column properties are saved to the design table after you create the design by selecting Make Table in the Screening Design window. These properties are also saved automatically to the data table that is created when you select the Save Factors option. You can find details about these column properties and related examples in Appendix A, "Column Properties".

Coding If the Role is Continuous, the Coding column property for the factor is saved. This property transforms the factor values so that the low and high values correspond to –1 and +1, respectively. For details, see "Coding" on page 663 in the "Column Properties" appendix.

Value Ordering If the Role is Categorical or if a Block variable is constructed, the Value Ordering column property for the factor is saved. This property determines the order in which levels of the factor appear. For details, see "Value Ordering" on page 679 in the "Column Properties" appendix.

Design Role Each factor is given the Design Role column property. The Role that you specify in defining the factor determines the value of its Design Role column property. The Design Role property reflects how the factor is intended to be used in modeling the experimental data. Design Role values are used in the Augment Design platform. For details, see "Design Role" on page 661 in the "Column Properties" appendix.

Factor Changes Each factor is assigned the Factor Changes column property with the value of Easy. The Factor Changes property reflects how the factor is used in modeling the experimental data. Factor Changes values are used in the Augment Design and Evaluate Design platforms. For details, see "Factor Changes" on page 676 in the "Column Properties" appendix.

Select Output Options

After you enter your responses and factors and click **Continue**, you can make selections for your design table in the Output Options outline. The structure of the full factorial design appears at the top of the outline.

Figure 12.10 Output Options Panel

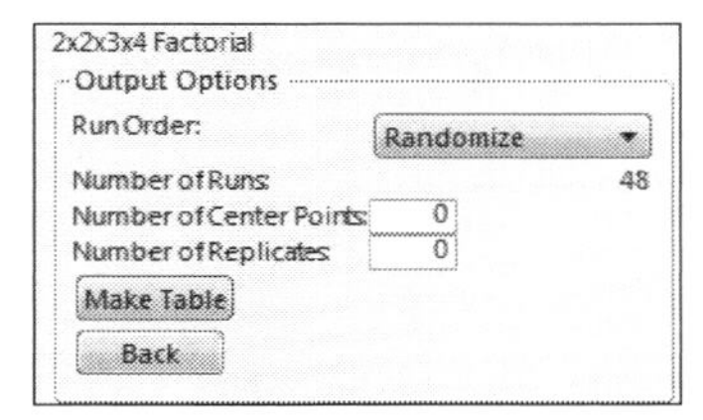

Run Order

The **Run Order** options determine the order of the runs in the design table. Choices include the following:

Keep the Same Rows in the design table are sorted from left to right.

Sort Left to Right Columns in the design table are sorted from left to right.

Randomize Rows in the design table are in random order.

Sort Right to Left Columns in the design table are sorted from right to left.

Center Points and Replicates

Number of Runs Shows the number of runs in the design before you add center points or replicates.

Number of Center Points Specifies how many additional runs to add as center points to the design. A center point is a run where every continuous factor is set at the center of the factor's range.

Suppose that your design includes both continuous and categorical factors. If you request center points in the Output Options panel, the center points are distributed as follows:

1. The settings for the categorical factors are ordered using the value ordering specified in the Factors outline.
2. One center point is assigned to each combination of the settings of the categorical factors in order, and this is repeated until all center points are assigned.

Number of Replicates The number of times to replicate the entire design, including center points. One replicate doubles the number of runs.

Make Table

Clicking Make Table creates a data table that contains the runs for your experiment. The example in Figure 12.11 shows a full factorial design with five center points for three factors: X1 (a two-level continuous factor), X2 (a three-level continuous factor), and X3 (a two-level categorical factor). The design uses the default values for the factor levels. The center points are in rows 1, 5, 8, 10, and 15. See "Pattern Column" on page 374.

Figure 12.11 Design Data Table

2x3x2 Factorial
Design 2x3x2 Factorial
Model
Evaluate Design
DOE Dialog

Columns (5/0)
Pattern
X1 *
X2 *
X3 *
Y *

Rows
All rows 17
Selected 0
Excluded 0
Hidden 0
Labelled 0

	Pattern	X1	X2	X3	Y
1	-22	-1	2	L2	•
2	-11	-1	1	L1	•
3	-32	-1	3	L2	•
4	+11	1	1	L1	•
5	+32	1	3	L2	•
6	+22	1	2	L2	•
7	001	0	2	L1	•
8	002	0	2	L2	•
9	-31	-1	3	L1	•
10	+31	1	3	L1	•
11	-21	-1	2	L1	•
12	001	0	2	L1	•
13	+12	1	1	L2	•
14	002	0	2	L2	•
15	-12	-1	1	L2	•
16	+21	1	2	L1	•
17	001	0	2	L1	•

The name of the table, shown in the upper left corner, is the design type that generated it.

Design Table Scripts

The design table includes the following scripts:

Model Runs the Analyze > Fit Model platform.

Evaluate Design Runs the DOE > Design Diagnostics > Evaluate Design platform.

DOE Dialog Re-creates the Full Factorial Design window that you used to generate the design table. The script also contains the random seed used to generate your design.

Run the **Screening** or **Model** scripts to analyze the data.

Pattern Column

The Pattern column contains entries that summarize the run in the given row. You can use Pattern as a label variable in plots.

- For a two-level continuous factor, the low setting is denoted by "–", the high setting by "+", and a center point by "0".

- For a continuous factor with more than two levels:
  - For a non-center point, the factor setting is denoted by an integer that corresponds to the value level for the run.
  - For a center point, the factor setting is denoted by a "0".
- For a categorical factor, the factor setting is denoted by an integer that corresponds to the value level for the run.

Full Factorial Design Options

The red triangle menu in the Full Factorial Design platform contains these options:

Save Responses Saves the information in the Responses panel to a new data table. You can then quickly load the responses and their associated information into most DOE windows. This option is helpful if you anticipate re-using the responses.

Load Responses Loads responses that you saved using the Save Responses option.

Save Factors Saves the information in the Factors panel to a new data table. Each factor's column contains its levels. Other information is stored as column properties. You can then quickly load the factors and their associated information into most DOE windows.

Note: It is possible to create a factors table by entering data into an empty table, but remember to assign each column an appropriate Design Role. Do this by right-clicking on the column name in the data grid and selecting **Column Properties > Design Role**. In the Design Role area, select the appropriate role.

Load Factors Loads factors that you saved using the Save Factors option.

Save Constraints (Unavailable for some platforms) Saves factor constraints that you defined in the Define Factor Constraints or Linear Constraints outline into a data table, with a column for each constraint. You can then quickly load the constraints into most DOE windows.

In the constraint table, the first rows contain the coefficients for each factor. The last row contains the inequality bound. Each constraint's column contains a column property called ConstraintState that identifies the constraint as a "less than" or a "greater than" constraint. See "ConstraintState" on page 686 in the "Column Properties" appendix.

Load Constraints (Unavailable for some platforms) Loads factor constraints that you saved using the Save Constraints option.

Set Random Seed Sets the random seed that JMP uses to control certain actions that have a random component. These actions include the following:

- simulating responses using the Simulate Responses option
- randomizing Run Order for design construction

– selecting a starting design for designs based on random starts

To reproduce a design or simulated responses, enter the random seed that generated them. For designs using random starts, set the seed before clicking Make Design. To control simulated responses or run order, set the seed before clicking Make Table.

Note: The random seed associated with a design is included in the DOE Dialog script that is saved to the design data table.

Simulate Responses Adds response values and a column containing a simulation formula to the design table. Select this option before you click Make Table.

When you click Make Table, the following occur:

– A set of simulated response values is added to each response column.

– For each response, a new a column that contains a simulation model formula is added to the design table. The formula and values are based on the model that is specified in the design window.

– A Model window appears where you can set the values of coefficients for model effects and specify one of three distributions: Normal, Binomial, or Poisson.

– A script called **DOE Simulate** is saved to the design table. This script re-opens the Model window, enabling you to re-simulate values or to make changes to the simulated response distribution.

Make selections in the Model window to control the distribution of simulated response values. When you click Apply, a formula for the simulated response values is saved in a new column called <Y> Simulated, where Y is the name of the response. Clicking Apply again updates the formula and values in <Y> Simulated.

For additional details, see “Simulate Responses” on page 106 in the “Custom Designs” chapter.

Note: JMP PRO You can use Simulate Responses to conduct simulation analyses using the JMP Pro Simulate feature. For information about Simulate and some DOE examples, see the Simulate chapter in the *Basic Analysis* book.

Advanced Options None available.

Save Script to Script Window Creates the script for the design that you specified in the Full Factorial Design window and saves it in an open script window.

Chapter 13

Mixture Designs

The Mixture Design platform supports experiments with factors that are ingredients in a mixture. Choose from several classical mixture design approaches, such as simplex, extreme vertices, and lattice. For the extreme vertices approach, you can supply a set of linear inequality constraints limiting the geometry of the mixture factor space.

The properties of a mixture are almost always a function of the relative proportions of the ingredients rather than their absolute amounts. In experiments with mixtures, a factor's value is its proportion in the mixture, which falls between zero and one. The sum of the proportions in any mixture recipe is one (100%).

Designs for mixture experiments are fundamentally different from those for screening experiments. Screening experiments are orthogonal. That is, over the course of an experiment, the setting of one factor varies independently of any other factor. Thus, the interpretation of screening experiments is relatively simple, because the effects of the factors on the response are separable.

With mixtures, it is impossible to vary one factor independently of all the others. When you change the proportion of one ingredient, the proportion of one or more other ingredients must also change to compensate. This simple fact has a profound effect on every aspect of experimentation with mixtures: the factor space, the design properties, and the interpretation of the results.

Because the proportions sum to one, mixture designs have an interesting geometry. The feasible region for the response in a mixture design takes the form of a simplex. For example, consider three factors in a 3-D graph. The plane where the sum of the three factors sum to one is a triangle-shaped slice. You can rotate the plane to see the triangle face-on and see the points in the form of a ternary plot.

Figure 13.1 Mixture Design

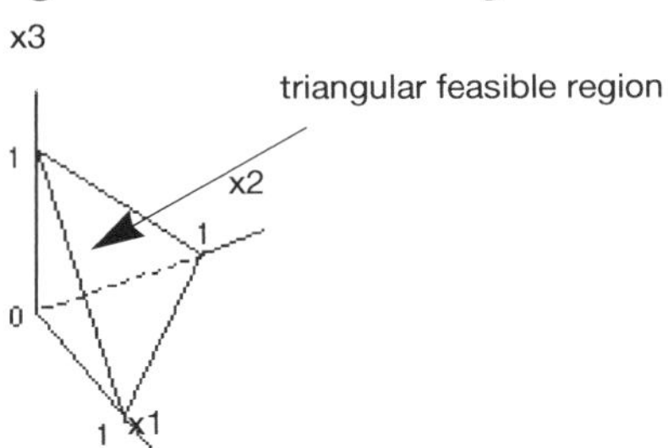

Overview of Mixture Designs

You can choose from the following types of mixture designs:

Optimal invokes the custom designer with all the mixture variables already defined.

Simplex Centroid Specify the degree to which the factor combinations are made.

Simplex Lattice Specify how many levels you want on each edge of the grid.

Extreme Vertices Specify linear constraints or restrict the upper and lower bounds to be within the 0 to 1 range.

ABCD Design Creates a screening design for mixtures devised by Snee (1975).

Space Filling Constructs a design that accommodates linear constraints. Design points are spread throughout the design space.

Linear Constraints

You can specify linear constraints in the following design types:

- Optimal
- Extreme Vertices
- Space Filling
- Design Workflow

Mixture Design Window

The mixture design window updates as you work through the design steps. For more information, see "The DOE Workflow: Describe, Specify, Design" on page 54. The workflow and outlines vary depending on design type.

Figure 13.2 Workflow for Optimal Mixture Designs

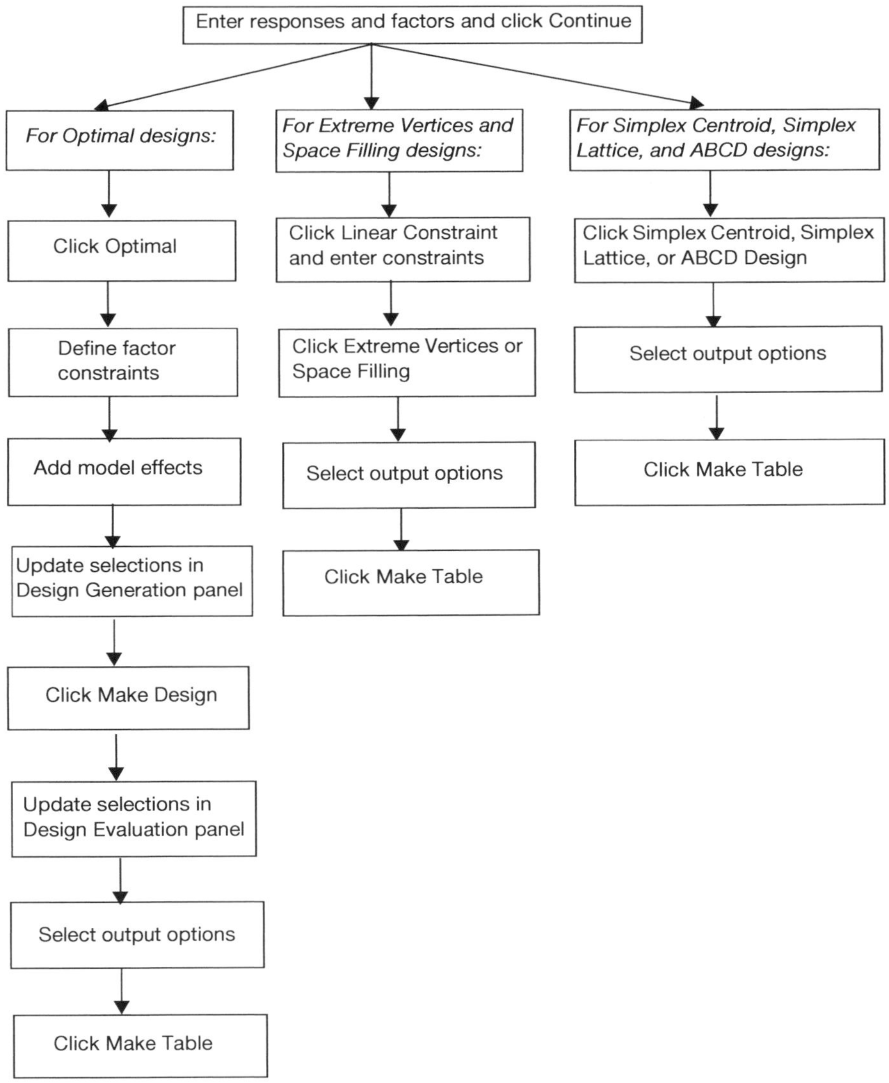

Enter responses and factors and click Continue

For Optimal designs:
- Click Optimal
- Define factor constraints
- Add model effects
- Update selections in Design Generation panel
- Click Make Design
- Update selections in Design Evaluation panel
- Select output options
- Click Make Table

For Extreme Vertices and Space Filling designs:
- Click Linear Constraint and enter constraints
- Click Extreme Vertices or Space Filling
- Select output options
- Click Make Table

For Simplex Centroid, Simplex Lattice, and ABCD designs:
- Click Simplex Centroid, Simplex Lattice, or ABCD Design
- Select output options
- Click Make Table

Responses

Use the Responses outline to specify one or more responses.

Tip: When you have completed the Responses outline, consider selecting **Save Responses** from the red triangle menu. This option saves the response names, goals, limits, and importance values in a data table that you can later reload in DOE platforms.

Figure 13.3 Responses Outline

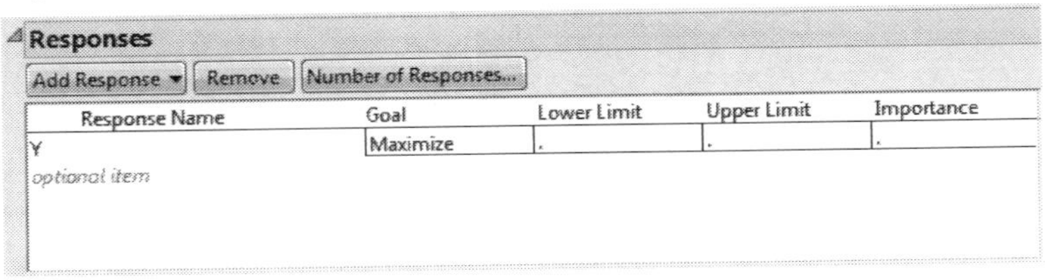

Responses

Add Response | Remove | Number of Responses...

Response Name	Goal	Lower Limit	Upper Limit	Importance
Y	Maximize	.	.	.
optional item				

Add Response Enters a single response with a goal type of Maximize, Match Target, Minimize, or None. If you select Match Target, enter limits for your target value. If you select Maximize or Minimize, entering limits is not required but can be useful if you intend to use desirability functions.

Remove Removes the selected responses.

Number of Responses Enters additional responses so that the number that you enter is the total number of responses. If you have entered a response other than the default Y, the Goal for each of the additional responses is the Goal associated with the last response entered. Otherwise, the Goal defaults to Match Target. Click the Goal type in the table to change it.

The Responses outline contains the following columns:

Response Name The name of the response. When added, a response is given a default name of Y, Y2, and so on. To change this name, double-click it and enter the desired name.

Goal, Lower Limit, Upper Limit The Goal tells JMP whether you want to maximize your response, minimize your response, match a target, or that you have no response goal. JMP assigns a Response Limits column property, based on these specifications, to each response column in the design table. It uses this information to define a desirability function for each response. The Profiler and Contour Profiler use these desirability functions to find optimal factor settings. For further details, see the Profiler chapter in the *Profilers* book and "Response Limits" on page 653 in the "Column Properties" appendix.

– A Goal of Maximize indicates that the best value is the largest possible. If there are natural lower or upper bounds, you can specify these as the Lower Limit or Upper Limit.

– A Goal of Minimize indicates that the best value is the smallest possible. If there are natural lower or upper bounds, you can specify these as the Lower Limit or Upper Limit.

– A Goal of Match Target indicates that the best value is a specific target value. The default target value is assumed to be midway between the Lower Limit and Upper Limit.

– A Goal of None indicates that there is no goal in terms of optimization. No desirability function is constructed.

Note: If your target response is not midway between the Lower Limit and the Upper Limit, you can change the target after you generate your design table. In the data table, open the Column Info window for the response column (**Cols > Column Info**) and enter the desired target value.

Importance When you have several responses, the Importance values that you specify are used to compute an overall desirability function. These values are treated as weights for the responses. If there is only one response, then specifying the Importance is unnecessary because it is set to 1 by default.

Editing the Responses Outline

In the Responses outline, note the following:

- Double-click a response to edit the response name.
- Click the goal to change it.
- Click on a limit or importance weight to change it.
- For multiple responses, you might want to enter values for the importance weights.

Response Limits Column Property

The Goal, Lower Limit, Upper Limit, and Importance that you specify when you enter a response are used in finding optimal factor settings. For each response, the information is saved in the generated design data table as a Response Limits column property. JMP uses this information to define the desirability function. The desirability function is used in the Prediction Profiler to find optimal factor settings. For further details about the Response Limits column property and examples of its use, see "Response Limits" on page 653 in the "Column Properties" appendix.

If you do not specify a Lower Limit and Upper Limit, JMP uses the range of the observed data for the response to define the limits for the desirability function. Specifying the Lower Limit and Upper Limit gives you control over the specification of the desirability function. For more details about the construction of the desirability function, see the Profiler chapter in the *Profilers* book.

Factors

Add factors in the Factors outline.

Tip: When you have completed the Factors outline, consider selecting **Save Factors** from the red triangle menu. This saves the factor names, roles, changes, and values in a data table that you can later reload.

Figure 13.4 Factors Outline

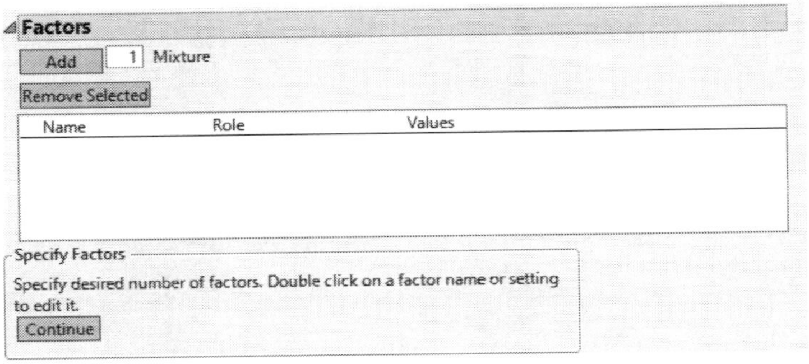

Add Enter the number of factors to add and click Add.

Remove Selected Removes the selected factors.

Factors List

The Factors list contains the following columns:

Name The name of the factor. When added, a factor is given a default name of X1, X2, and so on. To change this name, double-click it and enter the desired name.

Role Specifies the Design Role of the factor. The Design Role column property for the factor is saved to the data table. This property ensures that the factor type is modeled appropriately.

Values The experimental settings for the factors. To insert Values, click on the default values and enter the desired values.

Editing the Factors List

In the Factors list, do the following:

- To edit a factor name, double-click the factor name.
- To edit a value, click the value in the **Values** column.

Linear Constraints

Click the **Linear Constraint** button to enter one or more linear inequality constraints. A template appears for a linear expression involving all the continuous factors in your design. Enter coefficient values for the factors and select the direction of the inequality to reflect your linear constraint. Specify the constraining value in the box to the right of the inequality. To add more constraints, click **Linear Constraint** again.

Note: When you save a script for a design that involves a linear constraint, the script expresses the linear constraint as a *less than or equal to* inequality ($\leq$).

Examples of Mixture Design Types

This section presents examples of each of the different mixture design types:

- "Optimal Mixture Design" on page 383
- "Simplex Centroid Design" on page 384
- "Simplex Lattice Design" on page 387
- "Extreme Vertices Design" on page 389
- "ABCD Design" on page 395
- "Space Filling Design" on page 395

Optimal Mixture Design

Optimal mixture designs use the Custom Design platform with the mixture variables entered into the response and factors panels.

To create an optimal mixture design:

1. Select **DOE > Classical > Mixture Design**.
2. Enter factors and responses. The steps for entering responses are outlined in "Responses" on page 83 in the "Custom Designs" chapter.
3. After you enter responses and factors, click **Continue**.
4. Click **Optimal** on the Choose Mixture Design Type panel.
5. Open the **Define Factor Constraints** node and click **Add Constraint** to add linear constraints, if you have any.
6. Add effects to the model using the instructions below.
7. In the **Design Generation** panel, make selections relative to blocks, center points, replicates, and the number of runs.
8. Click **Make Design** to generate the Mixture Design report, which displays the design and Design Evaluation report.
9. Click **Make Table** in the Output Options panel of the Mixture Design report to generate the data table.

Adding Effects to the Model

Initially, the Model panel lists only the main effects corresponding to the factors that you entered, as shown in Figure 13.5.

Figure 13.5 The Model Panel

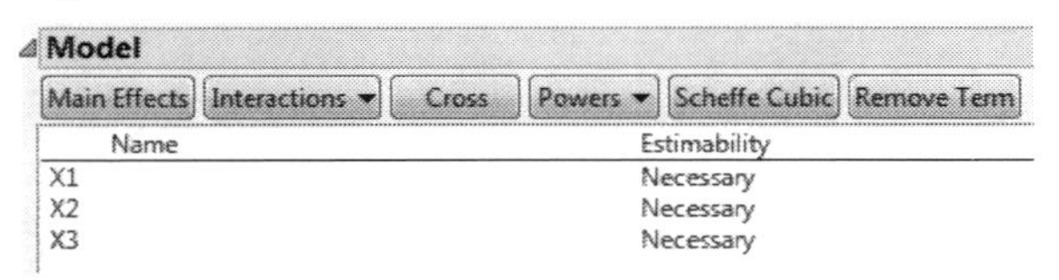

However, you can add factor interactions, specific crossed factor terms, powers, or Scheffé Cubic terms to the model.

- To add interaction terms to a model, click the **Interactions** button and select **2nd**, **3rd**, **4th**, or **5th**. For example, if you have factors X1 and X2, click **Interactions > 2nd** and X1*X2 is added to the list of model effects.
- To add crossed effects to a model, highlight the factors and effects that you want to cross and click the **Cross** button.
- To add powers of continuous factors to the model, click the **Powers** button and select **2nd**, **3rd**, **4th**, or **5th**.
- When you want a mixture model with third-degree polynomial terms, the **Scheffé Cubic** button provides a polynomial specification of the surface by adding terms of the form X1*X2*(X1-X2).

Simplex Centroid Design

A simplex centroid design of degree k with n factors consists of mixture runs with the following characteristics:

- all one factor
- all combinations of two factors at equal levels
- all combinations of three factors at equal levels
- and so on, up to k factors at a time combined at k equal levels.

A center point run with equal amounts of all the ingredients is always included.

Creating the Design

To create a simplex centroid design:

1. Select **DOE > Classical > Mixture Design**.
2. Enter factors and responses. The steps for entering responses are outlined in "Responses" on page 83 in the "Custom Designs" chapter.
3. After you enter responses and factors, click **Continue**.
4. Enter the number of ingredients in the box under **K**. JMP creates runs for each ingredient without mixing. JMP also creates runs that mix equal proportions of K ingredients at a time to the specified limit.
5. Click the **Simplex Centroid** button.
6. View Design and Output Options, as illustrated in Figure 13.6.

Figure 13.6 Example of Factor Settings and Output Options

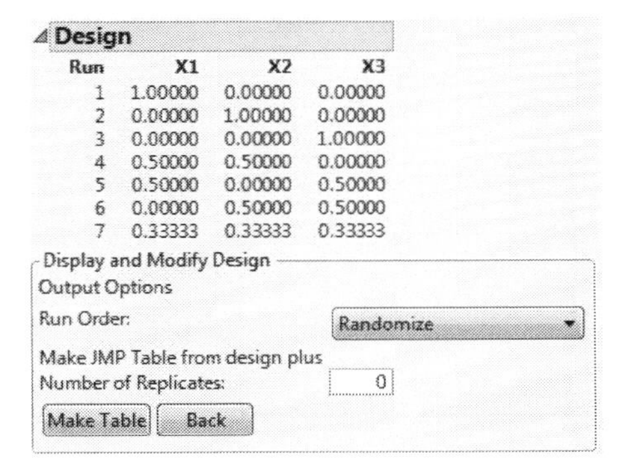

7. Specify Run Order, which is the order in which the runs that appear in the data table that you will create. Run order choices are:

   Keep the Same: The rows (runs) in the output table appear as they do in the Factor Settings panel.

   Sort Left to Right: The rows (runs) in the output table appear sorted from left to right.

   Randomize: The rows (runs) in the output table appear in a random order.

   Sort Right to Left: The rows (runs) in the output table appear sorted from right to left.

   Randomize within Blocks: The rows (runs) in the output table appear in random order within the blocks that you set up.

8. Specify **Number of Replicates.** The number of replicates is the number of times to replicate the entire design, including center points. Enter the number of times you want to replicate the design in the associated text box. One replicate doubles the number of runs.
9. Click **Make Table**.

Simplex Centroid Design Examples

The table of runs for a design of degree 1 with three factors (left in Figure 13.7) shows runs for each single ingredient followed by the center point. The table of runs to the right is for three factors of degree 2. The first three runs are for each single ingredient, the second set shows each combination of two ingredients in equal parts, and the last run is the center point.

Figure 13.7 Three-Factor Simplex Centroid Designs of Degrees 1 and 2

Run	X1	X2	X3
1	1	0	0
2	0	1	0
3	0	0	1
4	0.333	0.333	0.333

Run	X1	X2	X3
1	1	0	0
2	0	1	0
3	0	0	1
4	0.5	0.5	0
5	0.5	0	0.5
6	0	0.5	0.5
7	0.333	0.333	0.333

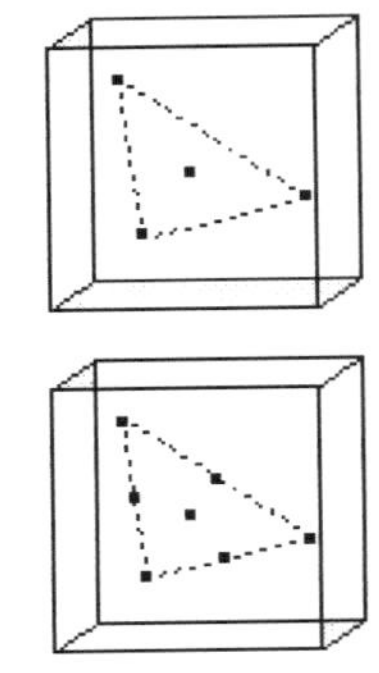

To generate the two sets of runs in Figure 13.7:

1. Choose **DOE > Classical > Mixture Design.**
2. Enter three mixture factors.
3. Click **Continue.**
4. Enter 1 in the **K** box, and click **Simplex Centroid** to see the design on the left in Figure 13.8.
5. Click the **Back** button, click **Continue**, and then enter 2 in the **K** box. Then click **Simplex Centroid** to see the design on the right in Figure 13.8.

Figure 13.8 Create Simplex Centroid Designs of Degrees 1 and 2

Design

Run	X1	X2	X3
1	1.00000	0.00000	0.00000
2	0.00000	1.00000	0.00000
3	0.00000	0.00000	1.00000
4	0.33333	0.33333	0.33333

Design

Run	X1	X2	X3
1	1.00000	0.00000	0.00000
2	0.00000	1.00000	0.00000
3	0.00000	0.00000	1.00000
4	0.50000	0.50000	0.00000
5	0.50000	0.00000	0.50000
6	0.00000	0.50000	0.50000
7	0.33333	0.33333	0.33333

As another example:

1. Choose **DOE > Classical > Mixture Design.**
2. Enter five factors and click **Continue.**
3. Use the default value, 4, in the **K** box.

4. Click **Simplex Centroid**.
5. Click **Make Table**.

   Figure 13.9 shows part of the 31-run design. Note that your table might look different because the design was created with Run Order set to Randomize.

Figure 13.9 Partial Listing of Factor Settings for Five-Factor Simplex Centroid Design

Simplex Centroid
Design Simplex Centroid
Screening
Model
DOE Dialog

Columns (6/0)
X1 *
X2 *
X3 *
X4 *
X5 *
Y *

Rows
All rows 31
Selected 0
Excluded 0
Hidden 0
Labelled 0

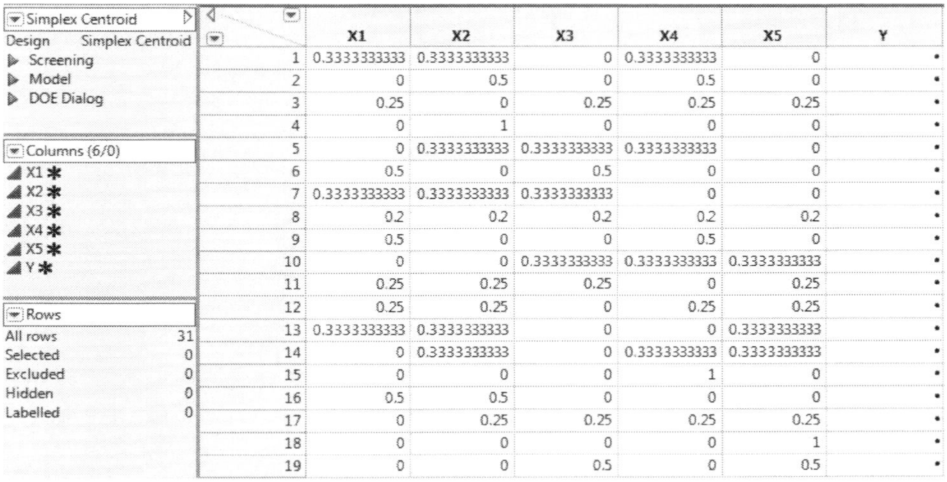

	X1	X2	X3	X4	X5	Y
1	0.3333333333	0.3333333333	0	0.3333333333	0	•
2	0	0.5	0	0.5	0	•
3	0.25	0	0.25	0.25	0.25	•
4	0	1	0	0	0	•
5	0	0.3333333333	0.3333333333	0.3333333333	0	•
6	0.5	0	0.5	0	0	•
7	0.3333333333	0.3333333333	0.3333333333	0	0	•
8	0.2	0.2	0.2	0.2	0.2	•
9	0.5	0	0	0.5	0	•
10	0	0	0.3333333333	0.3333333333	0.3333333333	•
11	0.25	0.25	0.25	0	0.25	•
12	0.25	0.25	0	0.25	0.25	•
13	0.3333333333	0.3333333333	0	0	0.3333333333	•
14	0	0.3333333333	0	0.3333333333	0.3333333333	•
15	0	0	0	1	0	•
16	0.5	0.5	0	0	0	•
17	0	0.25	0.25	0.25	0.25	•
18	0	0	0	0	1	•
19	0	0	0.5	0	0.5	•

Simplex Lattice Design

The simplex lattice design is a space filling design that creates a triangular grid of runs. The design is the set of all combinations where the factors' values are i / m, where i is an integer that varies from 0 to m such that the sum of the factor values is 1.

To create simplex lattice designs:

1. Select **DOE > Classical > Mixture Design**.
2. Enter factors and responses. The steps for entering responses are outlined in "Responses" on page 83 in the "Custom Designs" chapter.
3. Click **Continue**.
4. Specify the number of levels that you want in the Mixture Design Type panel and click **Simplex Lattice**.

Figure 13.10 shows the runs for three-factor simplex lattice designs of degrees 3, 4, and 5, with their corresponding geometric representations. In contrast to the simplex centroid design, the simplex lattice design does not necessarily include the centroid.

Figure 13.10 Three-Factor Simplex Lattice Designs for Factor Levels 3, 4, and 5

Design

Run	X1	X2	X3
1	0.00000	0.00000	1.00000
2	0.00000	0.33333	0.66667
3	0.00000	0.66667	0.33333
4	0.00000	1.00000	0.00000
5	0.33333	0.00000	0.66667
6	0.33333	0.33333	0.33333
7	0.33333	0.66667	0.00000
8	0.66667	0.00000	0.33333
9	0.66667	0.33333	0.00000
10	1.00000	0.00000	0.00000

Design

Run	X1	X2	X3
1	0.00000	0.00000	1.00000
2	0.00000	0.25000	0.75000
3	0.00000	0.50000	0.50000
4	0.00000	0.75000	0.25000
5	0.00000	1.00000	0.00000
6	0.25000	0.00000	0.75000
7	0.25000	0.25000	0.50000
8	0.25000	0.50000	0.25000
9	0.25000	0.75000	0.00000
10	0.50000	0.00000	0.50000
11	0.50000	0.25000	0.25000
12	0.50000	0.50000	0.00000
13	0.75000	0.00000	0.25000
14	0.75000	0.25000	0.00000
15	1.00000	0.00000	0.00000

Design

Run	X1	X2	X3
1	0.00000	0.00000	1.00000
2	0.00000	0.20000	0.80000
3	0.00000	0.40000	0.60000
4	0.00000	0.60000	0.40000
5	0.00000	0.80000	0.20000
6	0.00000	1.00000	0.00000
7	0.20000	0.00000	0.80000
8	0.20000	0.20000	0.60000
9	0.20000	0.40000	0.40000
10	0.20000	0.60000	0.20000
11	0.20000	0.80000	0.00000
12	0.40000	0.00000	0.60000
13	0.40000	0.20000	0.40000
14	0.40000	0.40000	0.20000
15	0.40000	0.60000	0.00000
16	0.60000	0.00000	0.40000
17	0.60000	0.20000	0.20000
18	0.60000	0.40000	0.00000
19	0.80000	0.00000	0.20000
20	0.80000	0.20000	0.00000
21	1.00000	0.00000	0.00000

Figure 13.11 lists the runs for a simplex lattice of degree 3 for five effects. In the five-level example, the runs creep across the hyper-triangular region and fill the space in a grid-like manner.

Figure 13.11 JMP Design Table for Simplex Lattice with Five Variables, Order (Degree) 3

Design

Run	X1	X2	X3	X4	X5
1	0.00000	0.00000	0.00000	0.00000	1.00000
2	0.00000	0.00000	0.00000	0.33333	0.66667
3	0.00000	0.00000	0.00000	0.66667	0.33333
4	0.00000	0.00000	0.00000	1.00000	0.00000
5	0.00000	0.00000	0.33333	0.00000	0.66667
6	0.00000	0.00000	0.33333	0.33333	0.33333
7	0.00000	0.00000	0.33333	0.66667	0.00000
8	0.00000	0.00000	0.66667	0.00000	0.33333
9	0.00000	0.00000	0.66667	0.33333	0.00000
10	0.00000	0.00000	1.00000	0.00000	0.00000
11	0.00000	0.33333	0.00000	0.00000	0.66667
12	0.00000	0.33333	0.00000	0.33333	0.33333
13	0.00000	0.33333	0.00000	0.66667	0.00000
14	0.00000	0.33333	0.33333	0.00000	0.33333
15	0.00000	0.33333	0.33333	0.33333	0.00000
16	0.00000	0.33333	0.66667	0.00000	0.00000
17	0.00000	0.66667	0.00000	0.00000	0.33333
18	0.00000	0.66667	0.00000	0.33333	0.00000
19	0.00000	0.66667	0.33333	0.00000	0.00000
20	0.00000	1.00000	0.00000	0.00000	0.00000
21	0.33333	0.00000	0.00000	0.00000	0.66667
22	0.33333	0.00000	0.00000	0.33333	0.33333
23	0.33333	0.00000	0.00000	0.66667	0.00000
24	0.33333	0.00000	0.33333	0.00000	0.33333
25	0.33333	0.00000	0.33333	0.33333	0.00000
26	0.33333	0.00000	0.66667	0.00000	0.00000
27	0.33333	0.33333	0.00000	0.00000	0.33333
28	0.33333	0.33333	0.00000	0.33333	0.00000
29	0.33333	0.33333	0.33333	0.00000	0.00000
30	0.33333	0.66667	0.00000	0.00000	0.00000
31	0.66667	0.00000	0.00000	0.00000	0.33333
32	0.66667	0.00000	0.00000	0.33333	0.00000
33	0.66667	0.00000	0.33333	0.00000	0.00000
34	0.66667	0.33333	0.00000	0.00000	0.00000
35	1.00000	0.00000	0.00000	0.00000	0.00000

Extreme Vertices Design

The extreme vertices design can be selected only if you have modified the ranges on the factors in the Factors panel or if you have specified a linear constraint. The extreme vertices design accounts for factor limits and selects vertices and their averages (formed by factor limits) as design points. Additional limits are usually in the form of range constraints, upper bounds, and lower bounds on the factor values.

The extreme vertices design finds the corners (vertices) of a factor space constrained by limits specified for one or more of the factors. The property that the factors must be nonnegative and must add up to one is the basic mixture constraint that makes a triangular-shaped region.

Sometimes other ingredients need range constraints that confine their values to be greater than a lower bound or less than an upper bound. Range constraints chop off parts of the triangular-shaped (simplex) region to make additional vertices. It is also possible to have a linear constraint, which defines a linear combination of factors to be greater or smaller than some constant.

The geometric shape of a region bound by linear constraints is called a simplex. Because the vertices represent extreme conditions of the operating environment, they are often the best places to use as design points in an experiment.

You usually want to add points between the vertices. The average of points that share a constraint boundary is called a *centroid* point, and centroid points of various degrees can be added. The centroid point for two neighboring vertices joined by a line is a second degree centroid because a line is two dimensional. The centroid point for vertices sharing a plane is a third degree centroid because a plane is three dimensional, and so on.

Creating the Design

Follow these steps to create an extreme vertices design. The next sections show examples with specific constraints.

1. Select **DOE > Classical > Mixture Design.**
2. Enter factors and responses. These steps are outlined in "Custom Design Window" on page 82 in the "Custom Designs" chapter. If your factor ranges are constrained, enter the limits as upper and lower limits in the Factors panel (see Figure 13.12).
3. Click **Continue.**
4. If you have linear constraints, click **Linear Constraints** and enter them.

   Note: The extreme vertices design can be selected only if you have modified the ranges on the factors in the Factors panel or if you have specified a linear constraint.

5. In the **Degree** text box, enter the degree of the centroid point that you want to add. The centroid point is the average of points that share a constraint boundary.
6. If you have linear constraints, click the **Linear Constraints** button for each constraint that you want to add. Use the text boxes that appear to define a linear combination of factors to be greater or smaller than some constant.
7. Click **Extreme Vertices** to see the factor settings.
8. Specify the Run Order. This determines the order of the runs in the data table when it is created. Run order choices are:

   Keep the Same: The rows (runs) in the output table appear as they do in the Design panel.

   Sort Left to Right: The rows (runs) in the output table appear sorted from left to right.

   Randomize: The rows (runs) in the output table appear in a random order.

   Sort Right to Left: The rows (runs) in the output table appear sorted from right to left.

   Randomize within Blocks: The rows (runs) in the output table appear in random order within the blocks that you set up.

9. Enter the sample size that you want in the Choose desired sample size text box.
10. (Optional) Click **Find Subset** to generate the optimal subset having the number of runs specified in sample size box described in Step 8. The **Find Subset** option uses the row exchange method (not coordinate exchange) to find the optimal subset of rows.

11. Click **Make Table**.

An Extreme Vertices Example with Range Constraints

The following example design table is for five factors with the range constraints shown in Figure 13.12, where the ranges are smaller than the default 0 to 1 range.

1. Select **DOE > Classical > Mixture Design**.
2. Add two additional factors (for a total of 5 factors) and give them the values shown in Figure 13.12.
3. Click **Continue**.
4. Enter 4 in the **Degree** text box (Figure 13.12).

Figure 13.12 Example of Five-factor Extreme Vertices

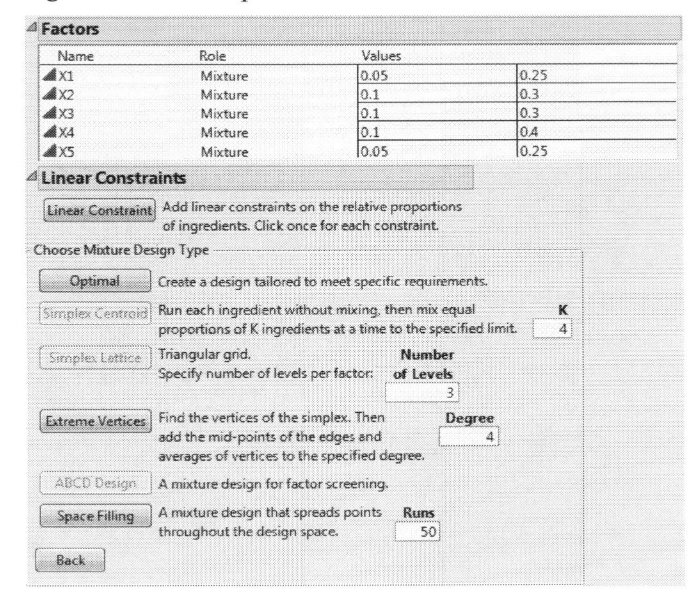

5. Click **Extreme Vertices**.
6. Select **Sort Left to Right** from the Run Order menu.
7. Click **Make Table**.

   Figure 13.13 shows a partial listing of a resulting design. Note that the Rows panel in the data table shows that the table has the default 116 runs.

Figure 13.13 JMP Design Table for Extreme Vertices with Range Constraints

Extreme Vertices
Design Extreme Vertices
▶ Screening
▶ Model
▶ DOE Dialog

Columns (6/0)
X1 *
X2 *
X3 *
X4 *
X5 *
Y *

Rows
All rows 116
Selected 0
Excluded 0
Hidden 0
Labelled 0

	X1	X2	X3	X4	X5	Y
1	0.25	0.3	0.2	0.1	0.15	•
2	0.25	0.2	0.2	0.3	0.05	•
3	0.05	0.22	0.22	0.34	0.17	•
4	0.1833333333	0.2333333333	0.2333333333	0.1	0.25	•
5	0.1166666667	0.3	0.3	0.1666666667	0.1166666667	•
6	0.1833333333	0.3	0.2333333333	0.1	0.1833333333	•
7	0.1	0.1	0.3	0.4	0.1	•
8	0.15	0.2	0.1	0.3	0.25	•
9	0.1166666667	0.3	0.1666666667	0.1666666667	0.25	•
10	0.25	0.1	0.3	0.2	0.15	•
11	0.25	0.1	0.2	0.2	0.25	•
12	0.25	0.3	0.2	0.2	0.05	•
13	0.25	0.25	0.1	0.35	0.05	•
14	0.25	0.1	0.1	0.4	0.15	•
15	0.05	0.3	0.2	0.4	0.05	•
16	0.05	0.3	0.2	0.3	0.15	•
17	0.05	0.3	0.1	0.3	0.25	•
18	0.17	0.1	0.22	0.34	0.17	•
19	0.15	0.1	0.2	0.3	0.25	•
20	0.25	0.3	0.3	0.1	0.05	•

Suppose you want fewer runs. You can go back and enter a different sample size (number of runs).

8. Click **Back** and then click **Continue**.
9. Enter 4 in the **Degree** text box and click **Extreme Vertices**.
10. Choose desired sample size text box, enter 10.
11. Click **Find Subset** to generate an optimal subset having the number of runs specified.

   The resulting design (Figure 13.14) is an optimal 10-run subset of the 116 current runs. This is useful when the extreme vertices design generates a large number of vertices. Your design might look different, because there are different subsets that achieve the same D-efficiency.

Figure 13.14 JMP Design Table for 10-Run Subset of the 116 Current Runs

Extreme Vertices 2
Design Extreme Vertices
▶ Screening
▶ Model
▶ DOE Dialog

Columns (6/0)
X1 *
X2 *
X3 *
X4 *
X5 *
Y *

	X1	X2	X3	X4	X5	Y
1	0.05	0.3	0.3	0.1	0.25	•
2	0.15	0.1	0.3	0.4	0.05	•
3	0.25	0.3	0.1	0.3	0.05	•
4	0.25	0.3	0.1	0.1	0.25	•
5	0.05	0.3	0.3	0.3	0.05	•
6	0.25	0.1	0.1	0.4	0.15	•
7	0.25	0.3	0.3	0.1	0.05	•
8	0.25	0.1	0.3	0.1	0.25	•
9	0.05	0.1	0.2	0.4	0.25	•
10	0.05	0.3	0.1	0.3	0.25	•

Note: The **Find Subset** option uses the row exchange method (not coordinate exchange) to find the optimal subset of rows.

An Extreme Vertices Example with Linear Constraints

Consider the classic example presented by Snee (1979) and Piepel (1988). This example has three factors, X1, X2, and X3, with five individual factor bound constraints and three additional linear constraints:

Table 13.1 Linear Constraints for the Snee and Piepel Example

```
X1 ≥ 0.1        90 ≤ 85*X1 + 90*X2 + 100*X3
X1 ≤ 0.5           85*X1 + 90*X2 + 100*X3 ≤ 95
X2 ≥ 0.1        .4 ≤ 0.7*X1 + X3
X2 ≤ 0.7
X3 ≤ 0.7
```

To enter these constraints:

1. Enter the upper and lower limits in the factors panel.
2. Click **Continue**.
3. Click the **Linear Constraint** button three times. Enter the constraints as shown in Figure 13.15.
4. Click the **Extreme Vertices** button.
5. Change the run order to **Sort Right to Left,** and keep the sample size at 13. See Figure 13.15 for the default Design and completed Output Options.
6. Click **Make Table**.

Figure 13.15 Constraints

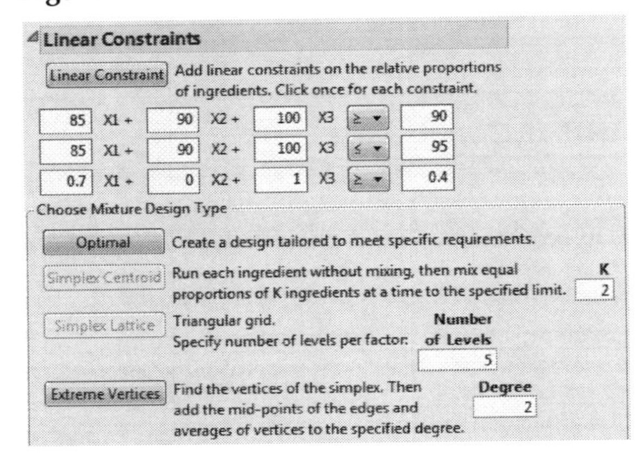

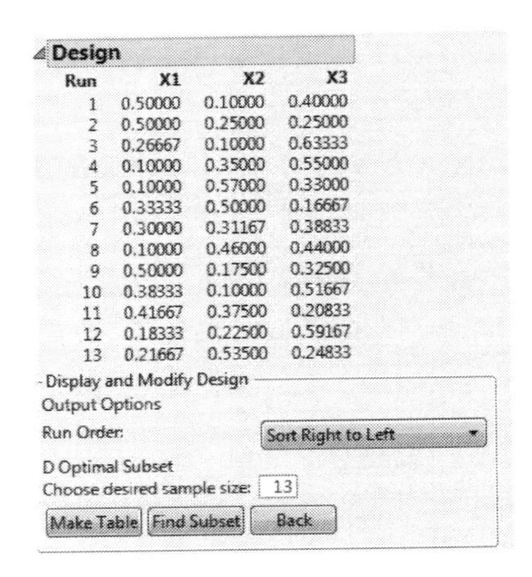

Run	X1	X2	X3
1	0.50000	0.10000	0.40000
2	0.50000	0.25000	0.25000
3	0.26667	0.10000	0.63333
4	0.10000	0.35000	0.55000
5	0.10000	0.57000	0.33000
6	0.33333	0.50000	0.16667
7	0.30000	0.31167	0.38833
8	0.10000	0.46000	0.44000
9	0.50000	0.17500	0.32500
10	0.38333	0.10000	0.51667
11	0.41667	0.37500	0.20833
12	0.18333	0.22500	0.59167
13	0.21667	0.53500	0.24833

This example is best understood by viewing the design as a ternary plot, as shown at the end of this chapter, in Figure 13.18. The ternary plot shows how close to one a given component is by how close it is to the vertex of that variable in the triangle. See "Creating Ternary Plots" on page 398 for details.

Extreme Vertices Method: How It Works

If there are linear constraints, JMP uses the CONSIM algorithm developed by R.E. Wheeler, described in Snee (1979) and presented by Piepel (1988) as CONVRT. The method is also described in Cornell (1990, Appendix 10a). The method combines constraints and checks to see whether vertices violate them. If so, it drops the vertices and calculates new ones. The method named CONAEV for doing centroid points is by Piepel (1988).

If there are no linear constraints (only range constraints), the extreme vertices design is constructed using the XVERT method developed by Snee and Marquardt (1974) and Snee (1975). After the vertices are found, a simplex centroid method generates combinations of vertices up to a specified order.

The XVERT method first creates a full 2^{nf-1} design using the given low and high values of the $nf-1$ factors with smallest range. Then, it computes the value of the one factor left out based on the restriction that the factors' values must sum to one. It keeps points that are not in factor's range. If not, it increments or decrements the value to bring it within range, and decrements or increments each of the other factors in turn by the same amount. This method keeps the points that still satisfy the initial restrictions.

The above algorithm creates the vertices of the feasible region in the simplex defined by the factor constraints. However, Snee (1975) has shown that it can also be useful to have the centroids of the edges and faces of the feasible region. A generalized n-dimensional face of the feasible region is defined by $nf-n$ of the boundaries and the centroid of a face defined to be the average of the vertices lying on it. The algorithm generates all possible combinations of the boundary conditions and then averages over the vertices generated on the first step.

ABCD Design

This approach by Snee (1975) generates a screening design for mixtures. To create an ABCD design:

1. Select **DOE > Classical > Mixture Design**.
2. Enter factors and responses. The steps for entering responses are outlined in "Responses" on page 83 in the "Custom Designs" chapter.
3. After you enter responses and factors, click **Continue**.
4. Click the **ABCD Design** button.
5. View factor settings and Output Options.
6. Specify Run Order, which is the order you want the runs to appear in the data table when it is created. Run order choices are:

   Keep the Same: The rows (runs) in the output table appear as they do in the Factor Settings panel.

   Sort Left to Right: The rows (runs) in the output table appear sorted from left to right.

   Randomize: The rows (runs) in the output table appear in a random order.

   Sort Right to Left: The rows (runs) in the output table appear sorted from right to left.

   Randomize within Blocks: The rows (runs) in the output table appear in random order within the blocks that you set up.
7. Specify Number of Replicates. The number of replicates is the number of times to replicate the entire design, including center points. Enter the number of times you want to replicate the design in the associated text box. One replicate doubles the number of runs.
8. Click **Make Table**.

Space Filling Design

The Space Filling mixture design type spreads design points fairly uniformly throughout the design region. It accommodates linear constraints. The design is generated in a fashion similar to the Fast Flexible Filling design method found under DOE > Special Purpose > Space Filling Design ("Fast Flexible Filling Designs" on page 594).

Two red triangle options relate specifically to Space Filling Designs:

FFF Optimality Criterion For the Fast Flexible Filling mixture design type, enables you to select between the MaxPro criterion (the default) and the Centroid criterion. See "FFF Optimality Criterion" on page 396.

Advanced Options > Set Average Cluster Size For the Fast Flexible Filling mixture design type, enables you to specify the average number of randomly generated points used to define each cluster or, equivalently, each design point. See "Set Average Cluster Size" on page 397.

FFF Optimality Criterion

The algorithms for Fast Flexible Filling designs begin by generating a large number of random points within the specified design region. These points are then clustered using a Fast Ward algorithm into a number of clusters that equals the Number of Runs that you specified.

The final design points can be obtained by using the default MaxPro (*maximum projection*) optimality criterion or by selecting the Centroid criterion. You can find these options under FFF Optimality Criterion in the report's red triangle menu. MaxPro

For p factors and n equal to the specified Number of Runs, the MaxPro criterion strives to find points in the clusters that minimize the following criterion:

$$C_{MaxPro} = \sum_{i}^{n-1} \sum_{j=i+1}^{n} \left[1 / \prod_{k=1}^{p} (x_{ik} - x_{jk})^2 \right]$$

The MaxPro criterion maximizes the product of the distances between potential design points in a way that involves all factors. This supports the goal of providing good space-filling properties on projections of factors. See Joseph et al. (2015). The Max Pro option is the default.

Centroid This method places a design point at the centroid of each cluster. It has the property that the average distance from an arbitrary point in the design space to its closest neighboring design point is smaller than for other designs.

Note: You can set a preference to always use a given optimality criterion. Select **File > Preferences > Platforms > DOE**. Select FFF Optimality Criterion and select your preferred criterion.

Set Average Cluster Size

The Set Average Cluster Size option is found under Advanced Options in the Mixture Design red triangle menu. This option enables you to specify the average number of uniformly generated points used to define each cluster or, equivalently, each design point.

By default, if the number of Runs for the Space Filling design type is 200 or smaller, a total of 10,000 random uniformly generated points are used as the basis for the clustering algorithm. When the number of Runs exceeds 200, the default value is 50. Increasing this value can be particularly useful in designs with a large number of factors.

Note: Depending on the number of factors and the specified value for Runs, you might want to increase the average number of initial points per design point by selecting **Advanced Options > Set Average Cluster Size.**

Linear Constraints

The design region can be restricted by selecting the **Linear Constraint** option in the Linear Constraints outline.

When you specify linear constraints, the random points that form the basis for the clustering algorithm are randomly distributed within the constrained design region. The clustering algorithm uses these points.

Space Filling Example

To create a space filling design:

1. Select **DOE > Classical > Mixture Design**.
2. Enter factors and responses.
3. Click **Continue**.
4. Add **Linear Constraints**, if you have any.
5. Specify the number of runs you want in the **Runs** box to the right of the Space Filling button in the Mixture Design Type panel.
6. Click **Space Filling**.

A Space Filling Example with a Linear Constraint

Consider a three-factor mixture design with the single linear constraint: $0.7*X1 + X3 \geq 0.4$. Figure 13.16 shows a ternary plot for a 30-run Space Filling design that satisfies this constraint. (For a discussion of ternary plots, see “Creating Ternary Plots” on page 398.) This design is constructed using the Centroid FFF Optimality Criterion. Note that the points fall in the constrained design region and are fairly well spread throughout this region.

Figure 13.16 Space Filling Design with One Linear Constraint

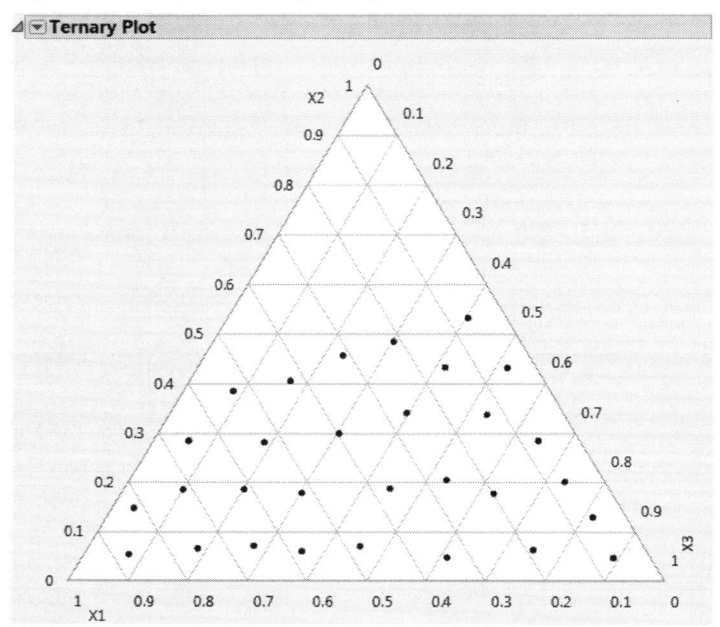

Creating Ternary Plots

A mixture problem in three components can be represented in two dimensions because the third component is a linear function of the others. The *ternary plot* in Figure 13.18 shows how close to one (1) a given component is by how close it is to the vertex of that variable in the triangle. The plot in Figure 13.17 illustrates a ternary plot.

Figure 13.17 Ternary Plot for Mixture Design

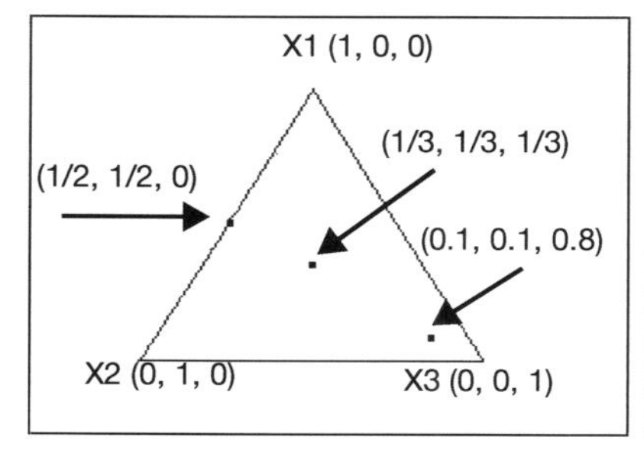

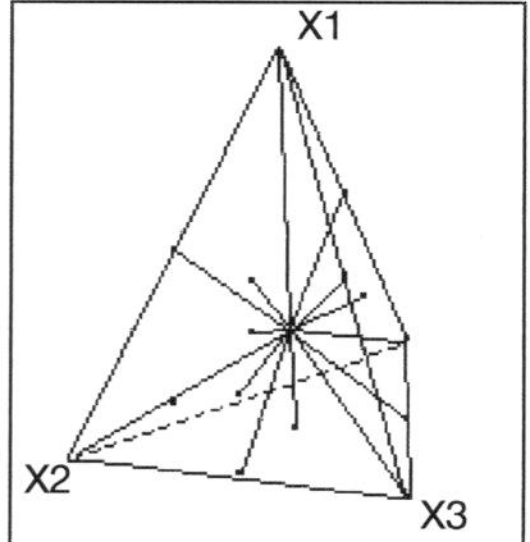

The Piepel (1979) example referenced in “An Extreme Vertices Example with Linear Constraints” on page 393 is best understood by the ternary plot shown in Figure 13.18.

To view a mixture design as a ternary plot:

1. Create the Piepel mixture data as shown previously, or open the table called Piepel.jmp, found in the Design Experiments folder of the Sample Data Library.
2. Choose **Graph > Ternary Plot**.
3. In the Ternary Plot launch window, specify the three mixture components and click **OK**.

The ternary plot platform recognizes the three factors as mixture factors, and also considers the upper and lower constraints entered into the Factors panel when the design was created. The ternary plot uses shading to exclude the unfeasible areas excluded by those constraints.

The Piepel data had additional constraints, entered as linear constraints for the extreme vertices design. There are six active constraints, six vertices, and six centroid points shown on the plot, as well as three inactive (redundant) constraints. The feasible area is the inner white polygon delimited by the design points and constraint lines.

Figure 13.18 Diagram of Ternary Plot Showing Piepel Example Constraints

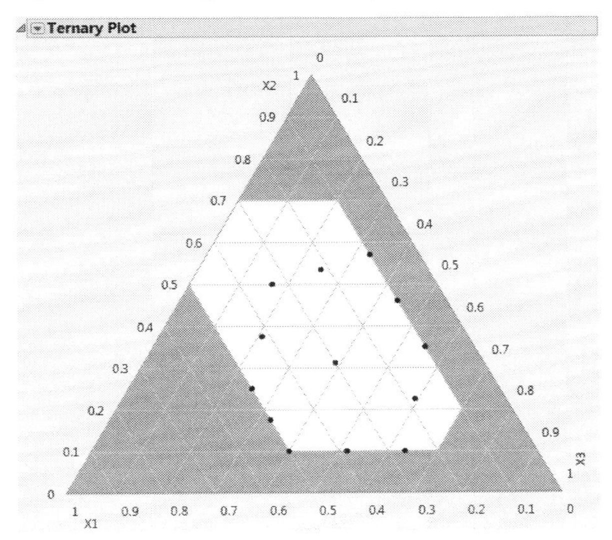

Fitting Mixture Designs

When fitting a model for mixture designs, take into account that the factors sum to a constant; a traditional full linear model are not fully estimable.

The recommended response surface model is called the Scheffé polynomial (Scheffé 1958). See the discussion of Cox Mixtures and the Scheffé Cubic macro in the Standard Least Squares chapter in the *Fitting Linear Models* book. The Scheffé polynomial model does the following:

- suppresses the intercept
- includes all the linear main-effect terms
- excludes all the square terms (such as X1*X1)
- includes all the cross terms (such as X1*X2)

To fit a model:

1. Choose **DOE > Classical > Mixture Design** and make the design data table. To fit a model, the Y column in the data table must contain values. Either assign responses or select **Simulate Responses** from the red triangle menu before you click **Make Table**.
2. The design data table stores the model in the data table as a table property. This table property is a JSL script called **Model**, located in the left panel of the table.
3. Click the green triangle next to the **Model** script to open the Fit Model launch window, which is automatically filled with the saved model.
4. Click **Run** on the Fit Model window.

In this model, the parameters are easy to interpret (Cornell 1990). The coefficients on the linear terms are the fitted response at the extreme points where the mixture consists of a single factor. The coefficients on the cross terms indicate the curvature across each edge of the factor space.

The model report usually has several sections of interest, including the whole model tests, Analysis of Variance reports, and response surface reports, which are described below.

Whole Model Tests and Analysis of Variance Reports

In a whole-model Analysis of Variance table, JMP traditionally tests that all the parameters are zero except for the intercept. In a mixture model without an intercept, JMP looks for a hidden intercept, in the sense that a linear combination of effects is a constant. If it finds a hidden intercept, it does the whole model test with respect to the intercept model rather than a zero-intercept model. This test is equivalent to testing that all the parameters are zero except the linear parameters, and testing that they are equal.

The hidden-intercept property also causes the R^2 to be reported with respect to the intercept model rather than reported as missing.

Understanding Response Surface Reports

When there are effects marked as response surface effects “&RS,” JMP creates additional reports that analyze the fitted response surface. These reports were originally designed for full response surfaces, not mixture models. However, JMP might encounter a no-intercept model and find a hidden intercept with linear response surface terms, but no square terms. Then it *folds* its calculations, collapsing on the last response surface term to calculate critical values for the optimum. This can be done for any combination that yields a constant and involves the last response surface term.

A Chemical Mixture Example

Three plasticizers (p1, p2, and p3) comprise 79.5% of the vinyl used for automobile seat covers (Cornell, 1990). Within this 79.5%, the individual plasticizers are restricted by the following constraints: $0.409 \leq x1 \leq 0.849$, $0 \leq x2 \leq 0.252$, and $0.151 \leq x3 \leq 0.274$.

Create the Design

To create Cornell's mixture design in JMP:

1. Select **DOE > Classical > Mixture Design**.
2. In the Factors panel, use the three default factors but name them **p1**, **p2**, and **p3**, and enter the high and low constraints as shown in Figure 13.19. Or, load the factors with the **Load Factors** command in the red triangle on the Mixture Design title bar. To import the factors, open Plastifactors.jmp, found in the Design Experiment sample data folder that was installed with JMP.

Figure 13.19 Factors and Factor Constraints for the Plasticizer Experiment

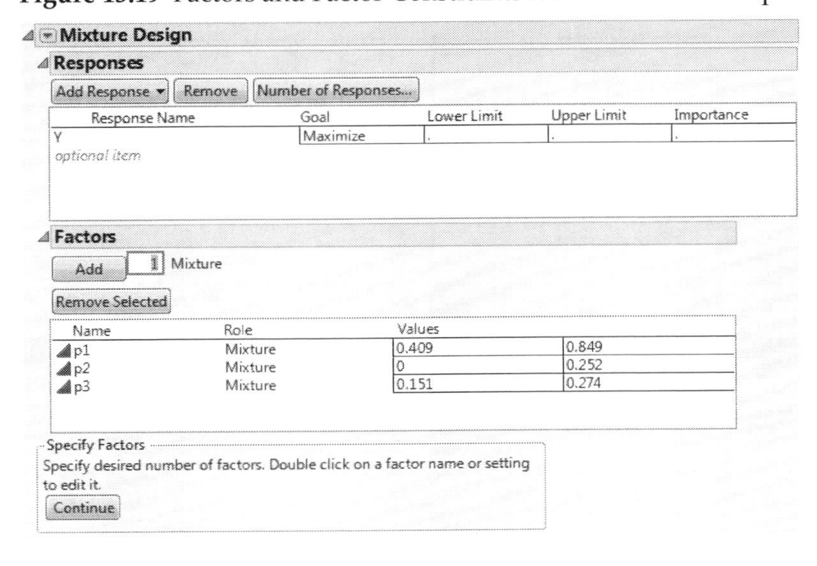

3. Click **Continue**.

4. Enter 3 in the **Degree** text box.
5. Click **Extreme Vertices.**
6. Click **Make Table.** JMP uses the 9 factor settings to generate a JMP table (Figure 13.20).

Figure 13.20 Extreme Vertices Mixture Design

Extreme Vertices 2
Design Extreme Vertices
▶ Screening
▶ Model
▶ DOE Dialog

Columns (4/0)
p1 *
p2 *
p3 *
Y *

Rows
All rows 9
Selected 0
Excluded 0
Hidden 0
Labelled 0

	p1	p2	p3	Y
1	0.597	0.252	0.151	•
2	0.726	0	0.274	•
3	0.6615	0.126	0.2125	•
4	0.6	0.126	0.274	•
5	0.7875	0	0.2125	•
6	0.849	0	0.151	•
7	0.5355	0.252	0.2125	•
8	0.474	0.252	0.274	•
9	0.723	0.126	0.151	•

Next, you add an extra five design runs by duplicating the vertex points and interior point, to give a total of 14 rows in the table.

Note: To identify the vertex points and the center (or interior) point, use the sample data script called LabelMixturePoints.jsl in the Sample Scripts folder installed with JMP.

7. Run the LabelMixturePoints.jsl to see the results in Figure 13.21, and highlight the vertex points and the interior point as shown.

Figure 13.21 Identify Vertices and Center Point with Sample Script

Extreme Vertices
Design Extreme Vertices
▶ Model
▶ Evaluate Design
▶ DOE Dialog
▶ Ternary Plot

Columns (5/0)
p1 *
p2 *
p3 *
Y *
Point Type

	p1	p2	p3	Y	Point Type
1	0.597	0.252	0.151	•	Vertex
2	0.726	0	0.274	•	Vertex
3	0.6615	0.126	0.2125	•	Interior
4	0.6	0.126	0.274	•	Edge
5	0.7875	0	0.2125	•	Edge
6	0.849	0	0.151	•	Vertex
7	0.5355	0.252	0.2125	•	Edge
8	0.474	0.252	0.274	•	Vertex
9	0.723	0.126	0.151	•	Edge

8. Select **Edit > Copy**, to copy the selected rows to the clipboard.
9. Select **Rows > Add Rows** and type **5** as the number of rows to add.
10. Click the At End radio button on the dialog and then click **OK**.

11. Highlight the new rows and select **Edit > Paste** to add the duplicate rows to the table.

The Plasticizer design with the results (*Y* values) that Cornell obtained are available in the sample data. Open Plasticizer.jmp in the sample data folder installed with JMP to see this table (Figure 13.22). You can check that, up to run order, the design that you created is identical to Cornell's.

Figure 13.22 Plasticizer.jmp Data Table from the Sample Data Library

Plasticizer
Design Extreme Vertices
Model
Ternary Plot

Columns (6/0)
p1
p2
p3
Y
Point Type
Pred Formula Y

Rows
All rows 14
Selected 0
Excluded 0
Hidden 0
Labelled 0

	p1	p2	p3	Y	Point Type	Pred Formula Y
1	0.474	0.252	0.274	12	Vertex	10.8923077
2	0.726	0	0.274	4	Vertex	5.09230769
3	0.849	0	0.151	8	Vertex	7.49230769
4	0.597	0.252	0.151	13	Vertex	11.2923077
5	0.6	0.126	0.274	13	Edge	13.0307692
6	0.6615	0.126	0.2125	18	Interior	19.2692308
7	0.7875	0	0.2125	12	Edge	11.8307692
8	0.723	0.126	0.151	14	Edge	14.4307692
9	0.5355	0.252	0.2125	16	Edge	16.6307692
10	0.474	0.252	0.274	10	Vertex	10.8923077
11	0.726	0	0.274	6	Vertex	5.09230769
12	0.849	0	0.151	7	Vertex	7.49230769
13	0.597	0.252	0.151	10	Vertex	11.2923077
14	0.6615	0.126	0.2125	21	Interior	19.2692308

Analyze the Mixture Model

Use the Plasticizer.jmp data from the sample data library (Figure 13.22) to run the mixture model:

1. Click the green triangle next to the **Model** script in the upper left corner of the data table.

   A completed Fit Model launch window appears.

2. Click **Run** to see the response surface analysis.
3. Plasticizer.jmp contains a column called Pred Formula Y. This column was added after the analysis by selecting **Save Columns > Prediction Formula** from the Response Y red triangle menu.
4. To see the prediction formula, right-click (press Ctrl and click on Macintosh) the column name and select **Formula**:

   0-50.1465*$p1$ - 282.198*$p2$ - 911.648*$p3$ + $p2$*$p1$*317.363 + $p2$*$p1$*1464.330 + $p3$*$p2$*1846.218

Note: These results correct the coefficients reported in Cornell (1990).

The Response Surface Solution report (Figure 13.23) shows that a maximum predicted value of 19.570299 occurs at point (0.63505, 0.15568, 0.20927).

Figure 13.23 Mixture Response Surface Analysis

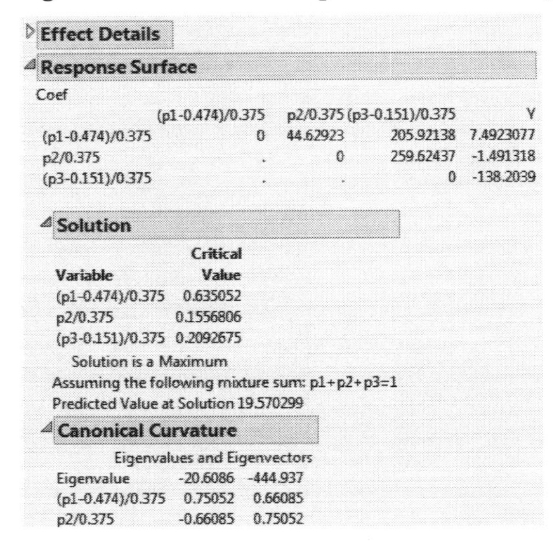

Effect Details

Response Surface

Coef

	(p1-0.474)/0.375	p2/0.375	(p3-0.151)/0.375	Y
(p1-0.474)/0.375	0	44.62923	205.92138	7.4923077
p2/0.375	.	0	259.62437	-1.491318
(p3-0.151)/0.375	.	.	0	-138.2039

Solution

Variable	Critical Value
(p1-0.474)/0.375	0.635052
p2/0.375	0.1556806
(p3-0.151)/0.375	0.2092675

Solution is a Maximum
Assuming the following mixture sum: p1+p2+p3=1
Predicted Value at Solution 19.570299

Canonical Curvature

Eigenvalues and Eigenvectors

Eigenvalue	-20.6086	-444.937
(p1-0.474)/0.375	0.75052	0.66085
p2/0.375	-0.66085	0.75052

The Prediction Profiler

The report contains a prediction profiler.

1. If the profiler is not visible, click the red triangle in the Response Y title bar and select **Factor Profiling > Profiler**. You should see the initial profiler shown in Figure 13.24.

   The crossed effects show as curvature in the prediction traces. Drag one of the vertical reference lines, and the other two lines move in the opposite direction maintaining their ratio.

   Note: The axes of prediction profiler traces range from the upper and lower bounds of the factors, p1, p2, and p3, entered to create the design and the design table. When you experiment moving a variable trace, you see the other traces move such that their ratio is preserved. As a result, when the limit of a variable is reached, it cannot move further and only the third variable changes.

2. To limit the visible profile curves to bounds that use all three variables, select **Profile at Boundary > Stop at Boundaries** from Prediction Profiler red triangle menu.
3. If needed, select the **Optimization and Desirability > Desirability Functions** command to display the desirability function showing to the right of the prediction profile plots in Figure 13.25.

4. Select **Optimization and Desirability > Maximize Desirability** from the Prediction Profiler menu to see the best factor settings.

The profiler in Figure 13.25, displays optimal settings (rounded) of 0.6350 for **p1**, 0.1557 for **p2**, and 0.2093 for **p3**, which give an estimated response of 19.5703.

Figure 13.24 Initial Prediction Profiler

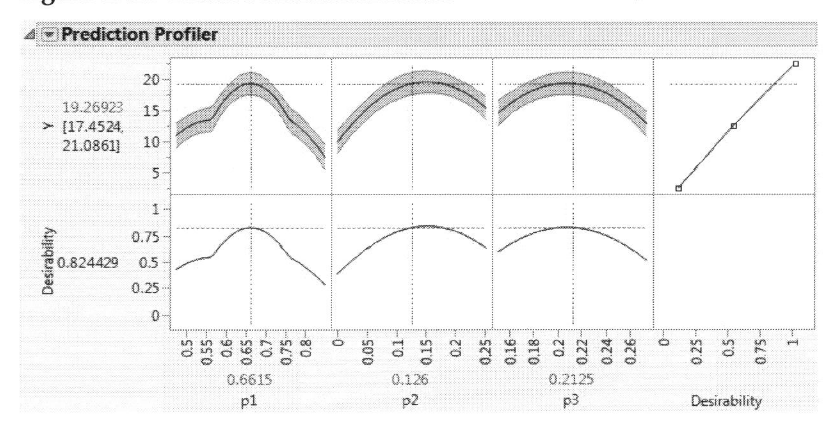

Figure 13.25 Maximum Desirability in Profiler for Mixture Analysis Example

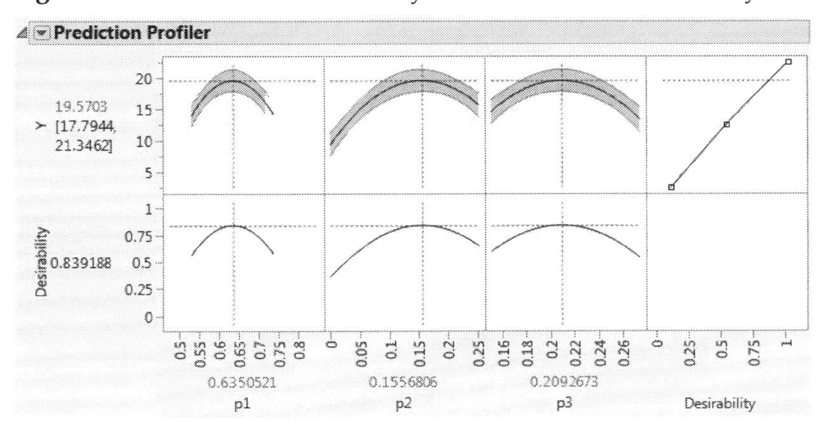

The Mixture Profiler

The Fit Model report also has a **Mixture Profiler** that is useful for visualizing and optimizing response surfaces from mixture experiments.

Many of the features are the same as those of the Contour Profiler. However, some are unique to the Mixture Profiler:

- A ternary plot is used instead of a Cartesian plot, which enables you to view three mixture factors at a time.
- If you have more than three factors, radio buttons let you choose which factors to plot.
- If the factors have constraints, you can enter their low and high limits in the Lo Limit and Hi Limit columns. This setting shades non-feasible regions in the profiler.

Select **Factor Profiling** > **Mixture Profiler** from Response Y red triangle menu to see the mixture profiler for the plasticizer data, shown in Figure 13.26.

Figure 13.26 Mixture Profiler for Plasticizer Example

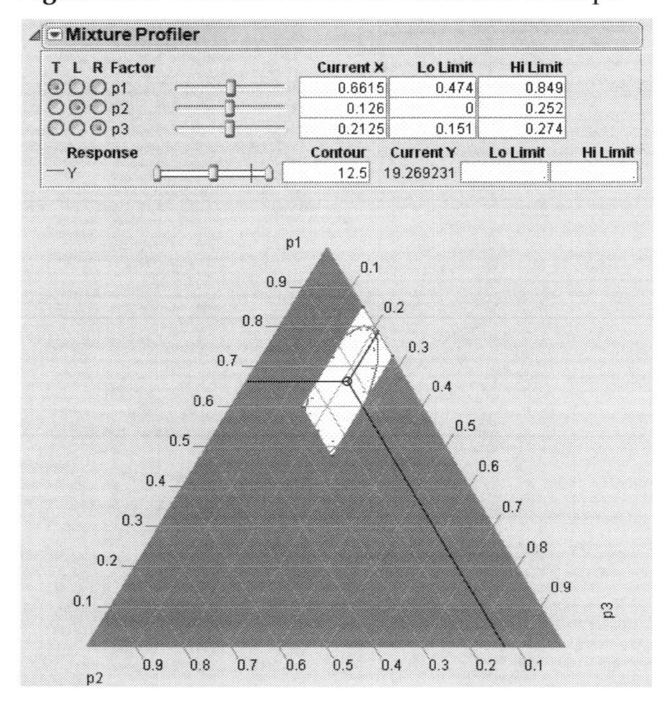

A Ternary Plot of the Mixture Response Surface

You can also plot the response surface of the plasticizer data as a ternary plot using the Ternary Plot platform and contour the plot with information from an additional variable:

1. Choose **Graph** > **Ternary Plot**.
2. Specify plot variables (p1, p2, and p3) and click **X, Plotting**, as shown in Figure 13.27.

3. To identify the contour variable (the prediction equation), select Pred Formula Y and click the **Contour Formula** button. The contour variable must have a prediction formula to form the contour lines, as shown by the ternary plots at the bottom in Figure 13.28. If there is no prediction formula, the ternary plot only shows points.

Figure 13.27 Ternary Plot Launch Window

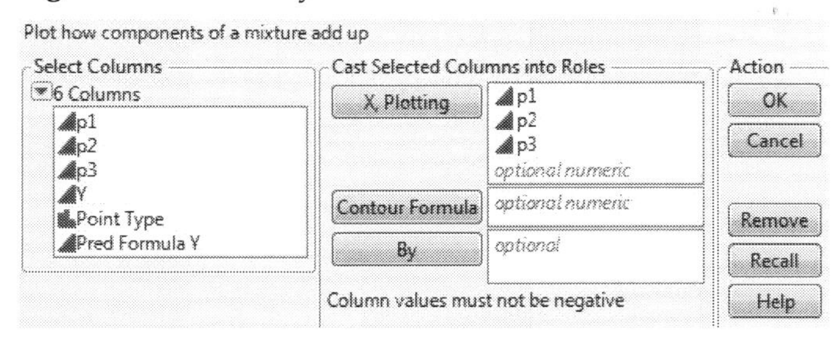

4. Click **OK** and view the results, as shown in Figure 13.28.

   By default, the ternary plot shows contour lines only. Add a fill by selecting **Contour Fill** from the Ternary Plot red triangle menu and then selecting **Fill Above** or **Fill Below**.

Figure 13.28 Ternary Plot of a Mixture Response Surface

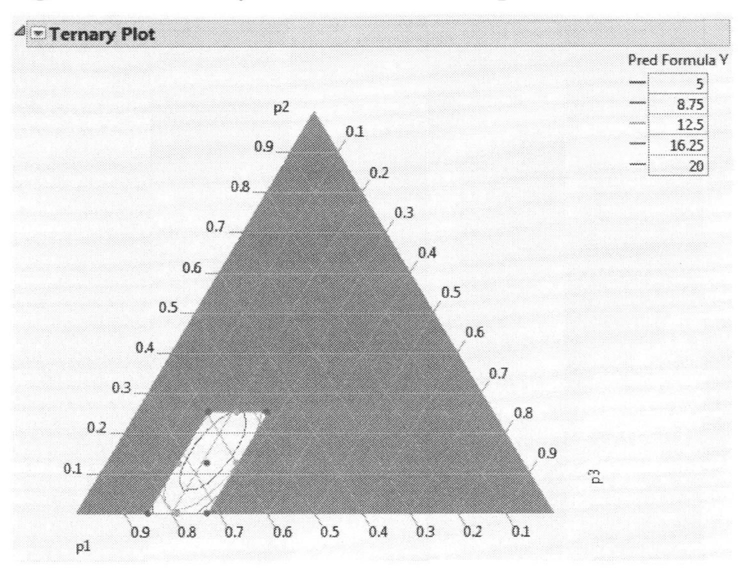

Chapter 14

Taguchi Designs

The goal of the Taguchi method is to find control factor settings that generate acceptable responses despite natural environmental and process variability. In each experiment, Taguchi's design approach employs two designs called the *inner* and *outer* array. The Taguchi experiment is the cross product of these two arrays.

- The *control* factors, used to tweak the process, form the inner array.
- The *noise* factors, associated with process or environmental variability, form the outer array.

Taguchi's *signal-to-noise ratios* are functions of the observed responses over an outer array. The Taguchi platform supports all these features of the Taguchi method. You choose from inner and outer array designs, which use the traditional Taguchi orthogonal arrays, such as L4, L8, and L16.

Dividing system variables according to their signal and noise factors is a key ingredient in robust engineering. Signal factors are system control inputs. Noise factors are variables that are typically difficult or expensive to control.

The inner array is a design in the signal factors and the outer array is a design in the noise factors. A signal-to-noise ratio is a statistic calculated over an entire outer array. Its formula depends on whether the experimental goal is to maximize, minimize or match a target value of the quality characteristic of interest.

A Taguchi experiment repeats the outer array design for each run of the inner array. The response variable in the data analysis is not the raw response or quality characteristic; it is the signal-to-noise ratio.

The Taguchi platform in JMP supports signal and noise factors, inner and outer arrays, and signal-to-noise ratios as Taguchi specifies.

Overview of Taguchi Designs

The Taguchi method defines two types of factors: control factors and noise factors:

- An *inner* design constructed over the control factors finds optimum settings.
- An *outer* design over the noise factors looks at how the response behaves for a wide range of noise conditions.

The experiment is performed on all combinations of the inner and outer design runs. A performance statistic is calculated across the outer runs for each inner run. This becomes the response for a fit across the inner design runs. Table 14.1 lists the recommended performance statistics.

Table 14.1 Recommended Performance Statistics

Goal	S/N Ratio Formula
nominal is best	$\frac{S}{N} = 10\log\left(\frac{\bar{Y}^2}{s^2}\right)$
larger-is-better (maximize)	$\frac{S}{N} = -10\log\left(\frac{1}{n}\sum_i \frac{1}{Y_i^2}\right)$
smaller-is-better (minimize)	$\frac{S}{N} = -10\log\left(\frac{1}{n}\sum_i Y_i^2\right)$

Example of a Taguchi Design

The following example is an experiment described by Byrne and Taguchi (1986). The objective of the experiment is to find settings of predetermined control factors that simultaneously maximize the adhesiveness (pull-off force) and minimize the assembly costs of nylon tubing.

Here are the signal and noise factors in the Byrne Taguchi Data for this example:

Interfer Tubing and connector interference. Control factor with 3 levels.

Wall Wall thickness of the connector. Control factor with 3 levels.

Depth Insertion depth of the tubing into the connector. Control factor with 3 levels.

Adhesive Percent of adhesive. Control factor with 3 levels.

Time Conditioning time. Noise factor with 2 levels.

Temperature Temperature. Noise factor with 2 levels.

Humidity Relative humidity. Noise factor with 2 levels.

Create the Design

1. Select **DOE > Classical > Taguchi Arrays**.
2. Select **Help > Sample Data Library** and open Design Experiment/Byrne Taguchi Factors.jmp.
3. In the Taguchi Design window, click the Taguchi Design red triangle and select **Load Factors**.

   The factors panel shows the four three-level control (signal) factors and three noise factors.

Figure 14.1 Response, and Signal and Noise Factors for the Byrne-Taguchi Example

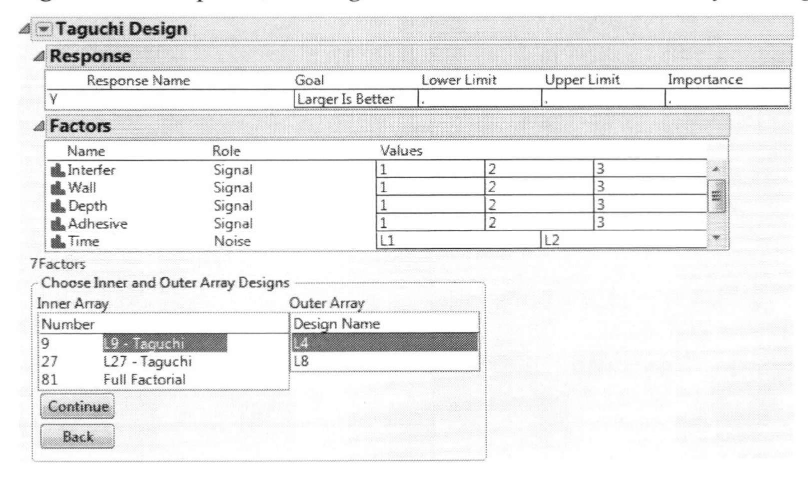

4. Ensure that **L9-Taguchi** is selected to give the L9 orthogonal array for the inner design.
5. Click **L8** to give an eight-run outer array design.
6. Click **Continue**.
7. Click **Make Table** to create the design table shown in Figure 14.2.

   The outer design has three two-level factors. A full factorial in eight runs is generated. However, it is only used as a guide to identify a new set of eight columns in the final JMP data table—one for each combination of levels in the outer design.

Figure 14.2 Taguchi Design Before Data Entry

L9
Design L9
Model

Columns (15/0)
Interfer *
Wall *
Depth *
Adhesive *
Pattern

--+

	Interfer	Wall	Depth	Adhesive	Pattern	---	--+	-+-	-++	+--	+-+	++-	+++	Mean	SN Ratio
1	1	1	1	1	----	•	•	•	•	•	•	•	•	•	•
2	1	2	2	2	-000	•	•	•	•	•	•	•	•	•	•
3	1	3	3	3	-+++	•	•	•	•	•	•	•	•	•	•
4	2	1	2	3	0-0+	•	•	•	•	•	•	•	•	•	•
5	2	2	3	1	00+-	•	•	•	•	•	•	•	•	•	•
6	2	3	1	2	0+-0	•	•	•	•	•	•	•	•	•	•
7	3	1	3	2	+-+0	•	•	•	•	•	•	•	•	•	•
8	3	2	1	3	+0-+	•	•	•	•	•	•	•	•	•	•
9	3	3	2	1	++0-	•	•	•	•	•	•	•	•	•	•

Now, suppose the pull-off adhesive force measures are collected and entered into the columns containing missing data, as shown in Figure 14.3. The missing data column names are appended with the letter Y before the levels (+ or –) of the noise factors for that run. For example, **Y---** is the column of measurements taken with the three noise factors set at their low levels.

8. Select **Help > Sample Data Library** and openDesign Experiment/Byrne Taguchi Data.jmp. Figure 14.3 shows the completed design.

Figure 14.3 Complete Taguchi Design Table (Byrne Taguchi Data.jmp)

	Interfer	Wall	Depth	Adhesive	Pattern	Y---	Y--+	Y-+-	Y-++	Y+--	Y+-+	Y++-	Y+++	Mean Y	SN Ratio Y
1	1	1	1	1	----	15.6	9.5	16.9	19.9	19.6	19.6	20	19.1	17.525	24.02534
2	1	2	2	2	-000	15	16.2	19.4	19.6	19.7	19.8	24.2	21.9	19.475	25.52164
3	1	3	3	3	-+++	16.3	16.7	19.1	15.6	22.6	18.2	23.3	20.4	19.025	25.33476
4	2	1	2	3	0-0+	18.3	17.4	18.9	18.6	21	18.9	23.2	24.7	20.125	25.90425
5	2	2	3	1	00+-	19.7	18.6	19.4	25.1	25.6	21.4	27.5	25.3	22.825	26.90753
6	2	3	1	2	0+-0	16.2	16.3	20	19.8	14.7	19.6	22.5	24.7	19.225	25.32574
7	3	1	3	2	+-+0	16.4	19.1	18.4	23.6	16.8	18.6	24.3	21.6	19.85	25.71081
8	3	2	1	3	+0-+	14.2	15.6	15.1	16.8	17.8	19.6	23.2	24.4	18.3375	24.83231
9	3	3	2	1	++0-	16.1	19.9	19.3	17.3	23.1	22.7	22.6	28.6	21.2	26.15198

The column named SN Ratio Y is the performance statistic computed with the formula shown below. In this case, it is the "larger-the-better" (LTB) formula, which is –10 times the common logarithm of the average squared reciprocal:

$$-10Log10\left[Mean\left[\frac{1}{(\mathrm{Y}---)^2}, \frac{1}{(\mathrm{Y}--+)^2}, \frac{1}{(\mathrm{Y}-+-)^2}, \frac{1}{(\mathrm{Y}-++)^2}, \frac{1}{(\mathrm{Y}+--)^2},\right.\right.$$
$$\left.\left.\frac{1}{(\mathrm{Y}+-+)^2}, \frac{1}{(\mathrm{Y}++-)^2}, \frac{1}{(\mathrm{Y}+++)^2}\right]\right]$$

This expression is large when all of the individual y values are large.

Analyze the Data

The data are now ready to analyze. The goal of the analysis is to find factor settings that maximize both the mean and the signal-to-noise ratio.

1. In the Byrne Taguchi Data.jmp data table, click the green arrow to run the Model script.

Figure 14.4 Fit Model Window for Taguchi Data

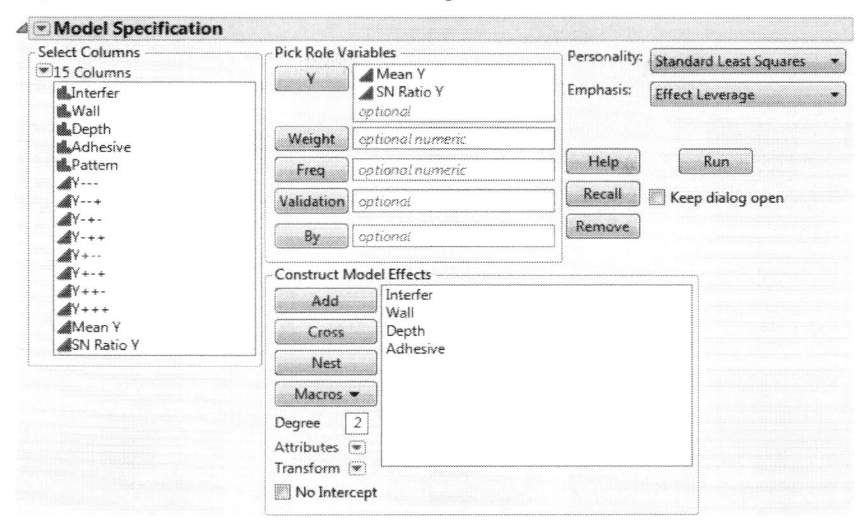

The Fit Model window automatically has the appropriate effects. It includes the main effects of the four signal factors. The two responses are the mean (**Mean Y**) and signal-to-noise ratio (**SN Ratio Y**) over the outer array.

2. Click **Run**.

   The prediction profiler is a quick way to find settings that give the highest signal-to-noise ratio for this experiment.

3. From the Response Mean Y red triangle menu, select **Factor Profiling > Profiler**.

Figure 14.5 The Prediction Profiler

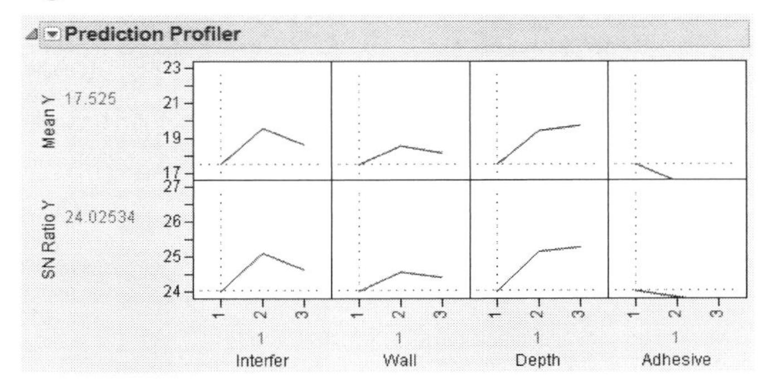

The profile traces (Figure 14.5) indicate that different settings of the first three factors would increase **SN Ratio Y**. Find the optimal settings.

4. From the Prediction Profiler red triangle menu, select **Optimization and Desirability > Desirability Functions.**

   This adds the row of traces and a column of function settings to the profiler, as shown in Figure 14.6. The default desirability functions are set to larger-is-better, which is what you want in this experiment. See the Profiler chapter in the *Profilers* book for more details about the prediction profiler.

5. From the Prediction Profiler red triangle menu, select **Optimization and Desirability > Maximize Desirability.**

   This automatically sets the prediction traces that give the best results according to the desirability functions.

Figure 14.6 Best Factor Settings for Byrne Taguchi Data

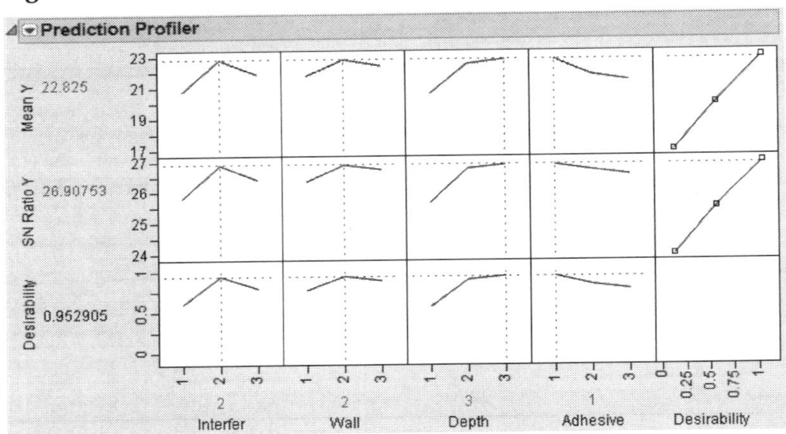

In this example, the settings for **Interfer** and **Wall** changed from 1 to 2. (See Figure 14.5 and Figure 14.6). The **Depth** setting changed from 1 to 3. The settings for **Adhesive** did not change. These new settings increased the signal-to-noise ratio from 24.0253 to 26.9075.

Taguchi Design Window

The Taguchi design window updates as you work through the design steps. For more information, see "The DOE Workflow: Describe, Specify, Design" on page 54. The outlines, separated by buttons that update the outlines, follow the flow in Figure 14.7.

Figure 14.7 Taguchi Design Flow

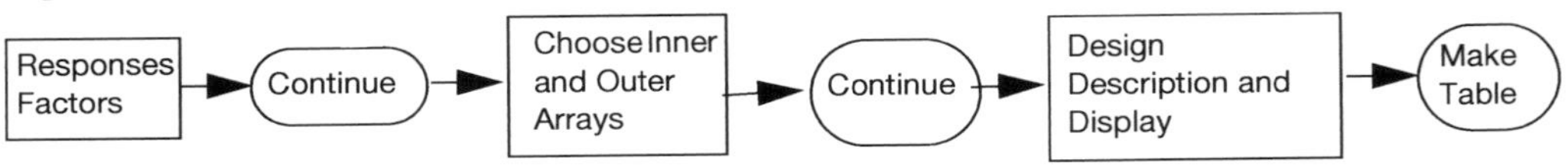

Responses

Use the Responses outline to specify a response.

Figure 14.8 Responses Outline

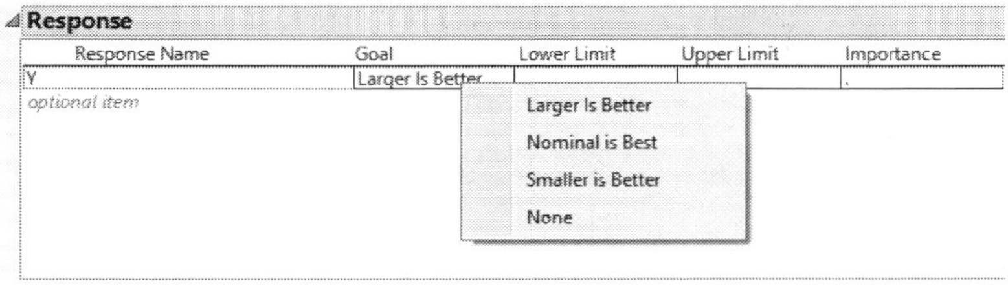

The Responses outline contains the following columns:

Response Name The name of the response. The default response name is Y. To change this name, double-click it and enter the desired name.

Goal, Lower Limit, Upper Limit Select one of the following Goals: Larger Is Better, Nominal is Best, Smaller is Better, or None. When you create your design, JMP saves a formula for the SN Ratio to the data table that reflects your selected goal.

Importance Because there is only one response, specifying the Importance is unnecessary because it is set to 1 by default.

Editing the Responses Outline

In the Responses outline, note the following:

- Double-click a response to edit the response name.
- Click the goal to change it.

Factors

Add factors in the Factors outline.

Signal Specify one or more 2- or 3-level signal factors. Signal factors are system control inputs. These are factors that you can control in production.

Noise Specify one or more noise factors. Noise factors are variables that are difficult or expensive to control in production. However, you must be able to control noise factors during the experiment.

Remove Removes the selected factors.

The steps for specifying factors are given in Figure 14.9.

1. Click to add a signal, then select a signal type: **2 Level**, or **3 Level**.

Or click to add a noise.

2. Double-click to edit the factor name.
3. To change the value of a signal or noise, click and then type the new value.

Figure 14.9 Entering Factors

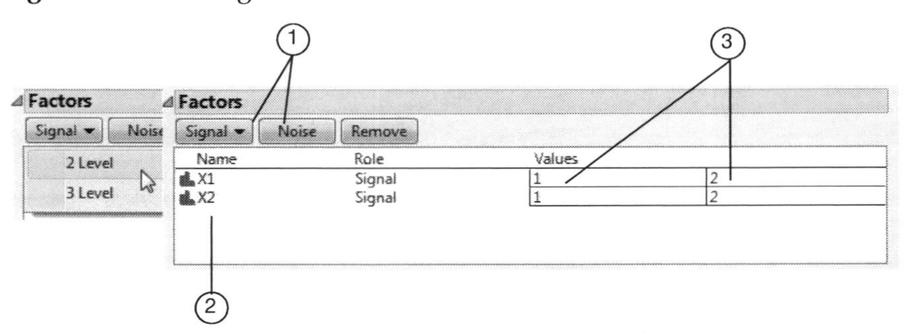

When you finish adding factors, click **Continue**.

Choose Inner and Outer Array Designs

Your choice for inner and outer arrays depends on the number and type of factors you have. Figure 14.10 shows the available inner array choices when you have eight signal factors. If you also have noise factors, choices include designs for the outer array. To follow along, enter eight two-level **Signal** factors and click **Continue**. Then highlight the design you want and again click **Continue**. This example uses the L12 design.

Figure 14.10 Selecting a Design for Eight Signal Factors

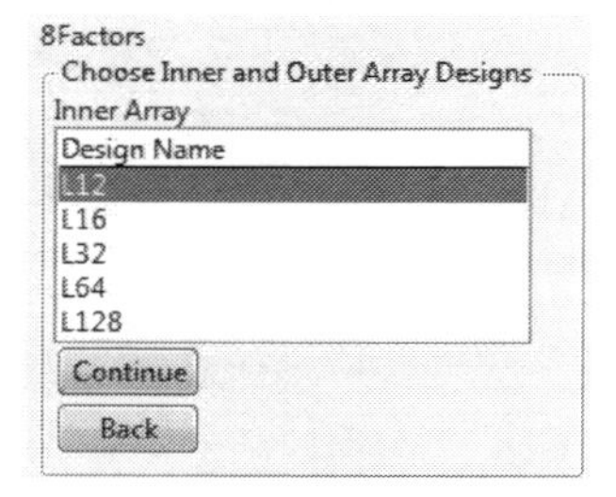

If you did not specify a noise factor, after you click **Continue**, a dialog appears that asks you to specify how many times you want to perform each inner array run. Specify two (2) for this example.

Display Coded Design

After you select a design type, the Coded Design (Figure 14.11) is shown below the Factors panel.

Figure 14.11 Coding for Eight Factor L12 Design

The Coded Design shows the pattern of high and low values for the factors in each run.

Make the Design Table

When you click **Make Table**, a table similar to that shown in Figure 14.12 appears. In the data table, each row represents a run. In the values for the Pattern variable, plus signs designate high levels and minus signs represent low levels.

Figure 14.12 Taguchi Design Table for Eight Factor L12 Design

Taguchi Array L12
Design Taguchi Array L12
Model
Columns (13/0)
X1, X2, X3, X4, X5, X6, X7, X8, Pattern, run 1
Rows

	X1	X2	X3	X4	X5	X6	X7	X8	Pattern	run 1
1	1	1	1	1	1	1	1	1	--------	•
2	1	1	1	1	1	2	2	2	-----+++	•
3	1	1	2	2	2	1	1	1	--+++---	•
4	1	2	1	2	2	1	2	2	-+-++-++	•
5	1	2	2	1	2	2	1	2	-++-++-+	•
6	1	2	2	2	1	2	2	1	-+++-++-	•
7	2	1	2	2	1	1	2	2	+-++--++	•
8	2	1	2	1	2	2	2	1	+-+-+++-	•
9	2	1	1	2	2	2	1	2	+--+++-+	•
10	2	2	2	1	1	1	1	2	+++----+	•
11	2	2	1	2	1	2	1	1	++-+-+--	•
12	2	2	1	1	2	1	2	1	++--+-+-	•

Chapter 15

Evaluate Designs

Explore Properties of Your Design

Using the Evaluate Design platform, you can:

- See the strengths and limitations of your existing experimental design.
- Determine your design's ability to detect effects associated with meaningful changes in the response.
- Address prediction variance and the precision of your estimates.
- Gain insight on aliasing.
- Obtain efficiency measures.

The Evaluate Design platform generates the results that appear in the Design Evaluation outlines provided by several DOE platforms. Diagnostics provided by both Evaluate Design and the Design Evaluation outlines include the following:

- power analysis
- a prediction variance profiler and surface plot
- a fraction of design space plot, showing how much of the design space has prediction variance above a given value
- estimation efficiency measures for parameter estimates
- the alias matrix, showing the bias structure for model effects
- a color map showing absolute correlations among effects
- design efficiency values

Figure 15.1 Comparison of Two Fraction of Design Space Plots

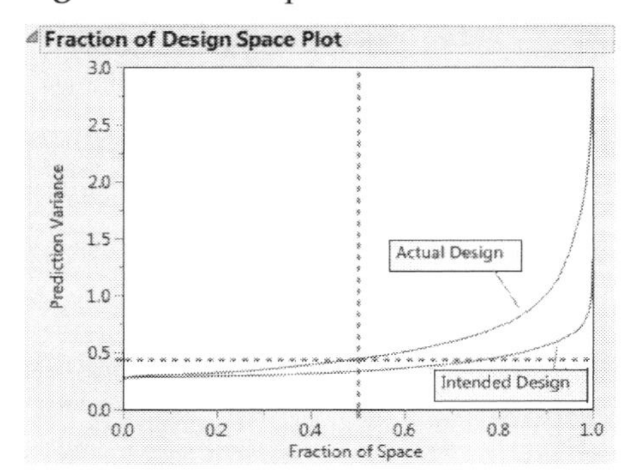

Overview of Evaluate Design

The Evaluate Design platform provides powerful tools that enable you to assess the properties of your design, whether it is created by JMP or another tool. You can use the platform before conducting an experiment to choose from several designs. You can also assess the impact of incorrect settings or lost runs in a design that you have conducted. You can modify the terms in your assumed model to see the impact of estimating a modified model. You can also modify the terms that appear in the Alias Terms outline to see the impact on the Alias Matrix.

You start by entering information about the design in the launch window. Then you can modify the assumed model and specify which effects not included in the model are of potential interest. Based on your specifications, the Design Evaluation platform then provides a number of ways to evaluate the properties of the generated design:

Power Analysis Enables you to explore your ability to detect effects of given sizes.

Prediction Variance Profile Shows the prediction variance over the range of factor settings.

Fraction of Design Space Plot Shows how much of the model prediction variance lies below or above a given value.

Prediction Variance Surface Shows a surface plot of the prediction variance for any two continuous factors.

Estimation Efficiency For each parameter, gives the fractional increase in the length of a confidence interval compared to that of an ideal (orthogonal) design, which might not exist. Also gives the relative standard error of the parameters.

Alias Matrix Gives coefficients that indicate the degree to which the model parameters are biased by effects that are potentially active, but not in the model.

Color Map on Correlations Shows the absolute correlations between effects on a plot using an intensity scale.

Design Diagnostics Gives efficiency measures for your design.

Note: In several DOE platforms, when you construct a design, a Design Evaluation outline appears. This outline shows results provided by the Evaluate Design platform. The platforms that provide a Design Evaluation outline are: Custom Design, Definitive Screening Design, Screening Design, Response Surface Design, and Mixture Design with Optimal design type.

Example of Evaluate Design

This section presents two examples. The first example assesses the impact of lost runs. The second example evaluates the power to detect a quadratic effect in a specified model.

Assessing the Impact of Lost Runs

An experiment was conducted to explore the effect of three factors (Silica, Sulfur, and Silane) on tennis ball bounciness (Stretch). The goal of the experiment is to develop a predictive model for Stretch. A 15-run Box-Behnken design was selected using the Response Surface Design platform. After the experiment, the researcher learned that the two runs where Silica = 0.7 and Silane = 50 were not processed correctly. These runs could not be included in the analysis of the data.

Use the Evaluate Design platform to assess the impact of not including those two runs. Obtain diagnostics for the intended 15-run design and compare these to the actual 13-run design that is missing the two runs.

Construct the Intended and Actual Designs

Intended Design

1. Select **Help > Sample Data Library** and open Design Experiment/Bounce Data.jmp.
2. Select **DOE > Design Diagnostics > Evaluate Design**.
3. Select Silica, Sulfur, and Silane and click **X, Factor**.

   You can add Stretch as **Y, Response** if you wish. But specifying the response has no effect on the properties of the design.
4. Click **OK**.

Leave your Evaluate Design window for the intended design open.

Tip: Place the Evaluate Design window for the intended design in the left area of your screen. After the next steps, you will place the corresponding window for the actual design to its right.

Actual Design with Missing Runs

In this section, you will exclude the two runs where Silica = 0.7 and Silane = 50. These are rows 3 and 7 in the data table.

1. In Bounce Data.jmp, select rows 3 and 7, right click in the highlighted area, and select **Hide and Exclude**.
2. Select **DOE > Design Diagnostics > Evaluate Design**.
3. Click **Recall**.
4. Click **OK**.
5. Click **OK** in the JMP Alert that appears.

Leave your Evaluate Design window for the actual design open.

Tip: Place the Evaluate Design window for the actual design to the *right* of the Evaluate Design window for the intended design to facilitate comparing the two designs.

Comparison of Intended and Actual Designs

You can now compare the two designs using these methods:

- "Power Analysis" on page 422
- "Prediction Variance Profile" on page 423
- "Fraction of Design Space Plot" on page 425
- "Estimation Efficiency" on page 426
- "Color Map on Correlations" on page 427
- "Design Diagnostics" on page 427

Power Analysis

In each window, do the following:

1. Open the Power Analysis outline.

   The outline shows default values of 1 for all Anticipated Coefficients. These values correspond to detecting a change in the anticipated response of 2 units across the levels of main effect terms, assuming that the interaction and quadratic terms are not active.

   The power calculations assume an error term (Anticipated RMSE) of 1. From previous studies, you believe that the RMSE is approximately 2.

2. Type 2 next to **Anticipated RMSE**.

   When you click outside the text box, the power values are updated.

   You are interested in detecting differences in the anticipated response that are on the order of 6 units across the levels of main effects, assuming that interaction and quadratic terms are not active. To set these uniformly, use a red triangle option.

3. From the red triangle menu next to Evaluate Design, select **Advanced Options > Set Delta for Power**.
4. Type 6 as your value for delta.
5. Click **OK**.

Figure 15.2 shows both outlines, with the Design and Anticipated Responses outline closed.

Figure 15.2 Power Analysis Outlines, Intended Design (Left) and Actual Design (Right)

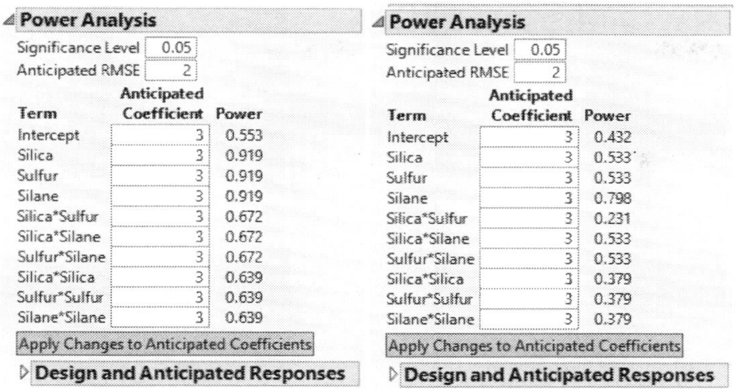

Power Analysis (Intended Design)

Significance Level 0.05
Anticipated RMSE 2

Term	Anticipated Coefficient	Power
Intercept	3	0.553
Silica	3	0.919
Sulfur	3	0.919
Silane	3	0.919
Silica*Sulfur	3	0.672
Silica*Silane	3	0.672
Sulfur*Silane	3	0.672
Silica*Silica	3	0.639
Sulfur*Sulfur	3	0.639
Silane*Silane	3	0.639

Apply Changes to Anticipated Coefficients

Design and Anticipated Responses

Power Analysis (Actual Design)

Significance Level 0.05
Anticipated RMSE 2

Term	Anticipated Coefficient	Power
Intercept	3	0.432
Silica	3	0.533
Sulfur	3	0.533
Silane	3	0.798
Silica*Sulfur	3	0.231
Silica*Silane	3	0.533
Sulfur*Silane	3	0.533
Silica*Silica	3	0.379
Sulfur*Sulfur	3	0.379
Silane*Silane	3	0.379

Apply Changes to Anticipated Coefficients

Design and Anticipated Responses

The power values for the actual design are uniformly smaller than for the intended design. For **Silica** and **Sulfur**, the power of the tests in the intended design is almost twice the power in the actual design. For the **Silica*Sulfur** interaction, the power of the test in the actual design is 0.231, compared to 0.672 in the intended design. The actual design results in substantial loss of power in comparison with the intended design.

Prediction Variance Profile

1. In each window, open the Prediction Variance Profile outline.
2. In the window for the actual design, place your cursor on the scale for the vertical axis. When your cursor becomes a hand, right click. **Select Edit > Copy Axis Settings**.

   This action creates a script containing the axis settings. Next, apply these axis settings to the Prediction Variance Profile plot for the intended design.
3. In the Evaluate Design window for the intended design, locate the Prediction Variance Profile outline. When your cursor becomes a hand, right click. **Select Edit > Paste Axis Settings**.

   The plots are shown in Figure 15.5, with the plot for the intended design at the top and for the actual design at the bottom.

Figure 15.3 Prediction Variance Profile, Intended Design (Top) and Actual Design (Bottom)

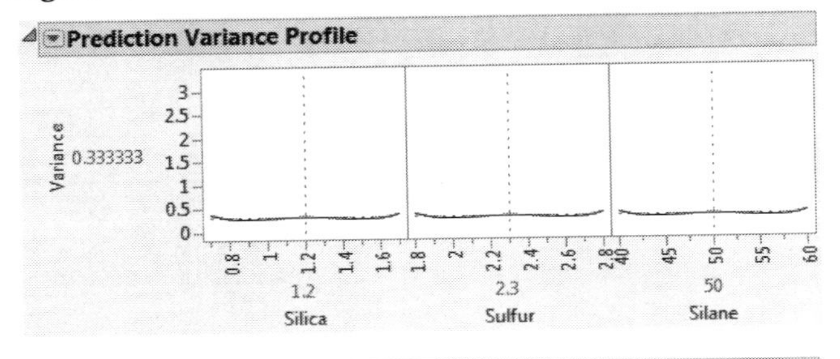

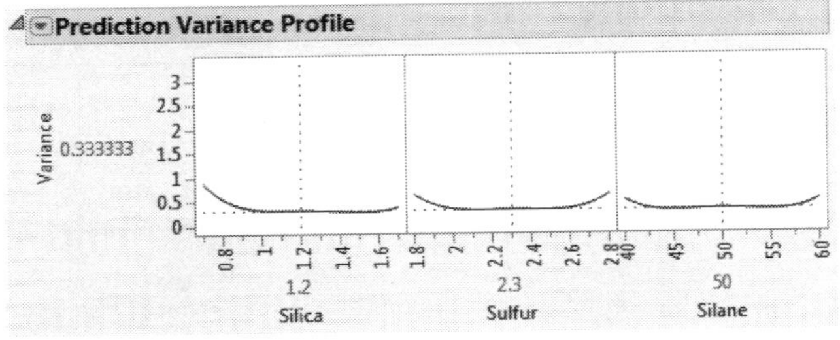

The Prediction Variance Profile plots are profiler views of the relative prediction variance. You can explore the relative prediction variance in various regions of design space.

Both plots show the same relative prediction variance in the center of the design space. However, the variance for points near the edges of the design space appears greater than for the same points in the intended design. Explore this phenomenon by moving all three vertical lines to points near the edges of the factor settings.

4. In both windows, select **Optimization and Desirability > Maximize Desirability** from the Prediction Variance Profile red triangle menu.

   Figure 15.4 shows the maximum relative prediction variance for the intended and actual designs.

Figure 15.4 Prediction Variance Profile Maximized, Intended Design (Top) and Actual Design (Bottom)

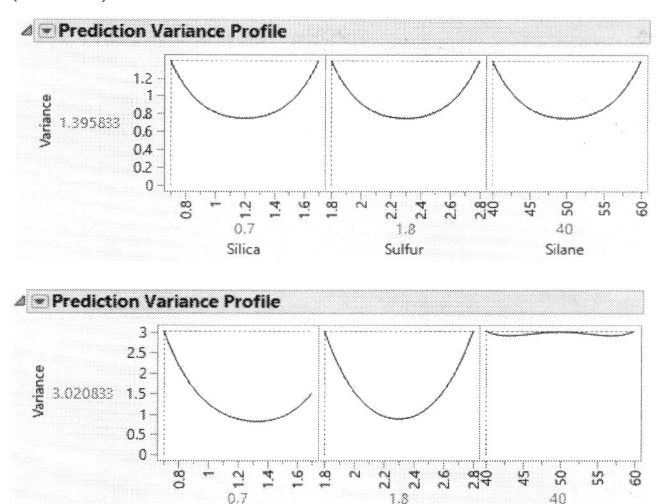

For both designs, the profilers identify the same point as one of the design points where the maximum prediction variance occurs: Silica=0.7, Sulfur=1.8, and Silane=40. The maximum prediction variance is 1.396 for the intended design, and 3.021 for the actual design. Note that there are other points where the prediction variance is maximized. The larger maximum prediction variance for the actual design means that predictions in parts of the design space are less accurate than they would have been had the intended design been conducted.

Fraction of Design Space Plot

1. In each window, open the Fraction of Design Space Plot outline.
2. In the window for the intended design, right-click in the plot and select **Edit > Copy Frame Contents**.
3. In the window for the actual design, locate the Fraction of Design Space Plot outline.
4. Right-click in the plot and select **Edit > Paste Frame Contents**

   Figure 15.5 shows the plot with annotations. Each Fraction of Design Space Plot shows the proportion of the design space for which the relative prediction variance falls below a specific value.

Figure 15.5 Fraction of Design Space Plots

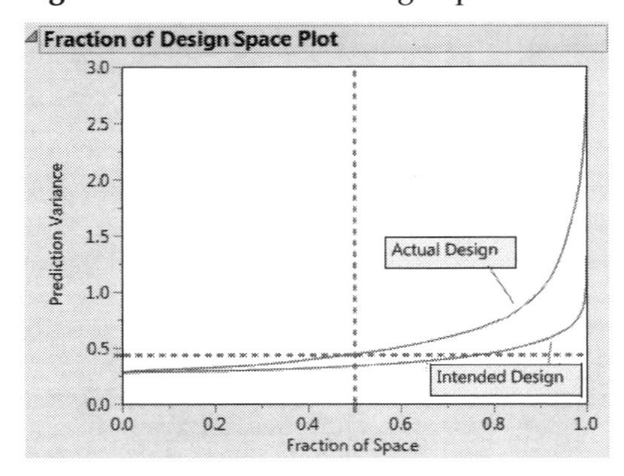

The relative prediction variance for the actual design is greater than that of the intended design over the entire design space. The discrepancy increases with larger design space coverage.

Estimation Efficiency

In each window, open the Estimation Efficiency outline.

Figure 15.6 Estimation Efficiency Outlines, Intended Design (Left) and Actual Design (Right)

Estimation Efficiency (Intended Design)

Term	Fractional Increase in CI Length	Relative Std Error of Estimate
Intercept	1.236	0.577
Silica	0.369	0.354
Sulfur	0.369	0.354
Silane	0.369	0.354
Silica*Sulfur	0.936	0.5
Silica*Silane	0.936	0.5
Sulfur*Silane	0.936	0.5
Silica*Silica	1.016	0.52
Sulfur*Sulfur	1.016	0.52
Silane*Silane	1.016	0.52

Estimation Efficiency (Actual Design)

Term	Fractional Increase in CI Length	Relative Std Error of Estimate
Intercept	1.082	0.577
Silica	0.803	0.5
Sulfur	0.803	0.5
Silane	0.275	0.354
Silica*Sulfur	2.122	0.866
Silica*Silane	0.803	0.5
Sulfur*Silane	0.803	0.5
Silica*Silica	1.268	0.629
Sulfur*Sulfur	1.268	0.629
Silane*Silane	1.268	0.629

In the actual design (right), the relative standard errors for all parameters either exceed or equal the standard errors for the intended design (left). For all except three of the non-intercept parameters, the relative standard errors in the actual design exceed those in the intended design.

The Fractional Increase in CI Length compares the length of a parameter's confidence interval as given by the current design to the length of such an interval given by an ideal design of the same run size. The length of the confidence interval, and consequently the Fractional Increase in CI Length, is affected by the number of runs. See "Fractional Increase in CI Length" on page 444. Despite the reduction in run size, for the actual design, the terms Silane, Silica*Silane, and Sulfur*Silane have a smaller increase than for the intended design. This is because the two runs that were removed to define the actual design had Silane set to its center point. By removing these runs, the widths of the confidence intervals for these parameters more closely resemble those of an ideal orthogonal design, which has no center points.

Color Map on Correlations

In each report, do the following:

1. Open the **Color Map On Correlations** outline.

   The two color maps show the effects in the Model outline. Each plot shows the absolute correlations between effects colored using the JMP default blue to red intensity scale. Ideally, you would like zero or very small correlations between effects.

Figure 15.7 Color Map on Correlations, Intended Design (Left) and Actual Design (Right)

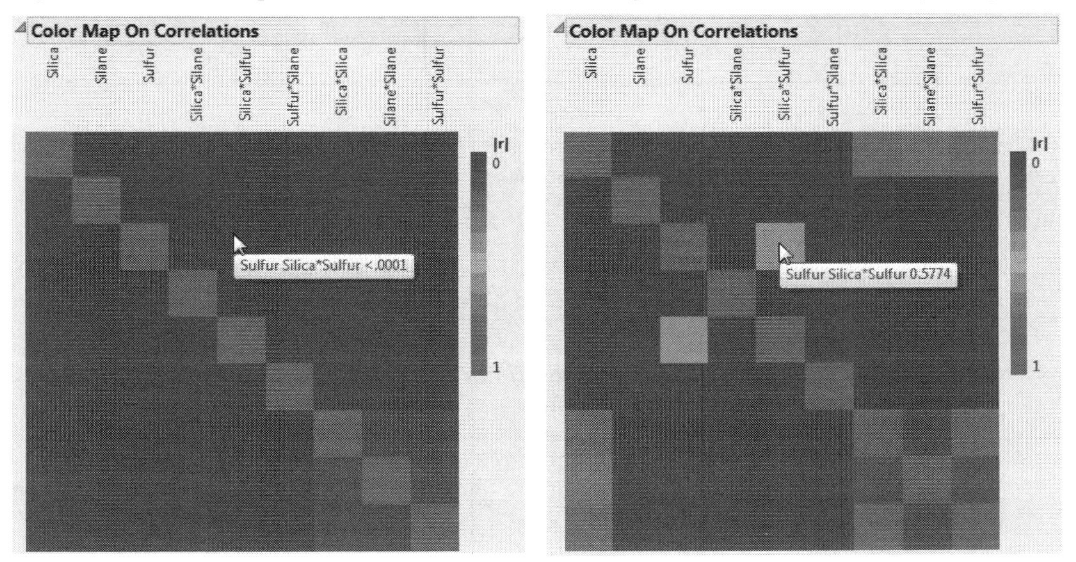

The absolute values of the correlations range from 0 (blue) to 1 (red). Hover over a cell to see the value of the absolute correlation. The color map for the actual design shows more absolute correlations that are large than does the color map for the intended design. For example, the correlation between Sulfur and Silica*Sulfur is < .0001 for the intended design, and 0.5774 for the actual design.

Design Diagnostics

In each report, open the Design Diagnostics outline.

Figure 15.8 Design Diagnostics, Intended Design (Left) and Actual Design (Right)

Design Diagnostics	
D Efficiency	36.6429
G Efficiency	69.10947
A Efficiency	29.3578
Average Variance of Prediction	0.384722
Design Creation Time (seconds)	0

Design Diagnostics	
D Efficiency	36.6429
G Efficiency	69.10947
A Efficiency	29.3578
Average Variance of Prediction	0.384722
Design Creation Time (seconds)	0

The intended design (left) has higher efficiency values and a lower average prediction variance than the actual design (right). The results of the Design Evaluation analysis indicate that the two lost runs have had a negative impact on the design.

Note that both the number of runs and the model matrix factor into the calculation of efficiency measures. In particular, the *D*-, *G*-, and *A*- efficiencies are calculated relative to the ideal design for the run size of the given design. It is not necessarily true that larger designs are more efficient than smaller designs. However, for a given number of factors, larger designs tend to have smaller Average Variance of Prediction values than do smaller designs. For details on how efficiency measures are defined, see "Design Diagnostics" on page 449.

Evaluating Power Relative to a Specified Model

For this example, you have constructed a definitive screening design to determine which of six factors have an effect on the yield of an extraction process. The data are given in the Extraction Data.jmp sample data table, located in the Design Experiment folder. Because the design is a *definitive* screening design, each factor has three levels. See the "Definitive Screening Designs" chapter on page 231.

You are interested in the power of tests to detect a strong quadratic effect. You consider a strong effect to be one whose magnitude is at least three times as large as the error variation.

Although the experiment studies six factors, effect sparsity suggests that only a small subset of factors is active. Consequently, you feel comfortable investigating power in a model based on a smaller number of factors. Also, past studies on a related process provide strong evidence to suggest that three of the factors, Propanol, Butanol, and pH, have negligible main effects, do not interact with other factors, and do not have quadratic effects. This leads you to believe that the likely model contains main, interaction, and quadratic effects only for Methanol, Ethanol, and Time. You decide to investigate power in the context of a three-factor response surface model.

Use the Evaluate Design platform to determine the power of your design to detect strong quadratic effects for Methanol, Ethanol, or Time.

1. Select **Help > Sample Data Library** and open Design Experiment/Extraction Data.jmp.
2. Select **DOE > Design Diagnostics > Evaluate Design**.
3. Select Methanol, Ethanol, and Time and click **X, Factor**.

You can add Yield as **Y, Response** if you wish. But specifying the response has no effect on the properties of the design.

4. Click **OK**.
5. In the Model outline, click **RSM**.

   This adds the interaction and quadratic terms for the three factors.
6. Open the Power Analysis outline.

   Note that the Anticipated RMSE is set to 1 by default. Although you have an estimate of the RMSE from past studies, you need not enter it. This is because the magnitude of the effect of interest is three times the error variation.
7. Under **Anticipated Coefficient**, type 3 next to Methanol*Methanol, Ethanol*Ethanol, and Time*Time.
8. Click **Apply Changes to Anticipated Coefficients**.

Figure 15.9 Power Analysis Outline after Applying Changes to Coefficients

Power Analysis

Significance Level 0.05
Anticipated RMSE 1

Term	Anticipated Coefficient	Power
Intercept	1	0.111
Methanol	1	0.573
Ethanol	1	0.573
Time	1	0.573
Methanol*Methanol	3	0.737
Methanol*Ethanol	-1	0.331
Ethanol*Ethanol	3	0.737
Methanol*Time	-1	0.331
Ethanol*Time	1	0.331
Time*Time	3	0.737

Apply Changes to Anticipated Coefficients

The power of detecting a quadratic effect whose magnitude is three times the error variation is 0.737. This assumes a final model that is a response surface in three factors. It also assumes a 0.05 significance level for the test.

Evaluate Design Launch Window

To launch the Evaluate Design platform, open the data table of interest and select **DOE > Design Diagnostics > Evaluate Design**. The example in Figure 15.10 uses the Bounce Data.jmp sample data table, located in the Design Experiment folder.

Figure 15.10 Evaluate Design Launch Window

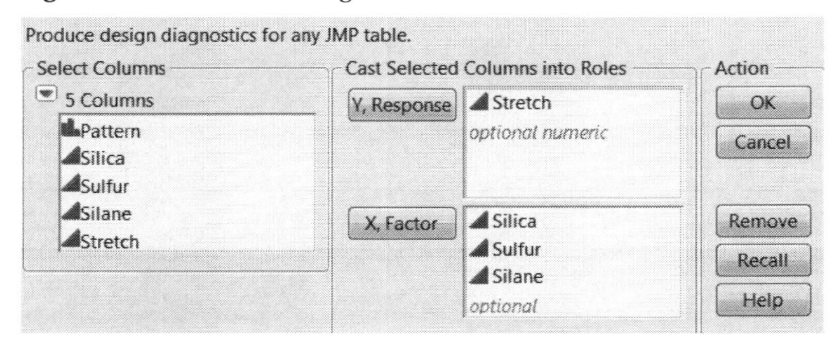

The launch window contains the following buttons:

Y, Response Enter the response column or columns. Entering a response is optional. Response values are not used in evaluating the design. Responses must be numeric.

X, Factor Enter the factor columns. Factors can be of any Data Type or Modeling Type.

Evaluate Design Window

The Evaluate Design window consists of two parts. See Figure 15.11, where all outline nodes are closed.

- The Factors, Model, Alias Terms, and Design outlines define the model and design.
- The Design Evaluation outline provides results that describe the properties of your design.

Figure 15.11 Evaluate Design Window Showing All Possible Outlines

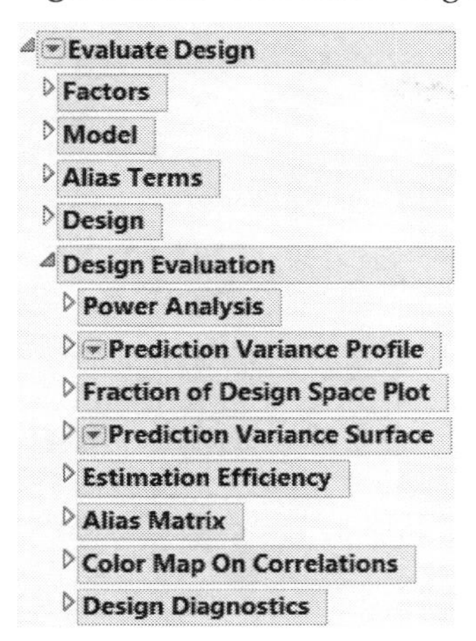

The Factors, Model, Alias Terms, and Design outlines contain information that you enter about the factors, assumed model, potentially aliased effects of interest, and the actual design. JMP populates these outlines using your selections in the launch window and the design table. However, you can modify the effects in the Model and Alias Terms outlines. These outlines are described in the following sections:

- "Factors" on page 432
- "Model" on page 432
- "Alias Terms" on page 433
- "Design" on page 433

Once you have made your specifications, the Design Evaluation outlines are updated. You can open these outlines to see reports or control windows that provide information about your design. These outlines are described in the following sections:

- "Power Analysis" on page 434
- "Prediction Variance Profile" on page 440
- "Fraction of Design Space Plot" on page 442
- "Prediction Variance Surface" on page 442
- "Estimation Efficiency" on page 444
- "Alias Matrix" on page 446
- "Color Map on Correlations" on page 448
- "Design Diagnostics" on page 449

Factors

The factors outline lists the factors entered in the launch window. You can select factors to construct effects in the Model outline.

Model

If the data table contains a script called Model or Fit Model, the Model outline contains the effects specified in that script. Otherwise, the Model outline contains only main effects.

Figure 15.12 shows the Model outline for the Bounce Data.jmp data table, found in the Design Experiment folder. The Model script in the data table contains response surface effects for the three factors Silica, Silane, and Sulfur. Consequently, the Model outline contains the main effects, two-way interactions, and quadratic effects for these three factors.

Figure 15.12 Model Outline for Bounce Data.jmp

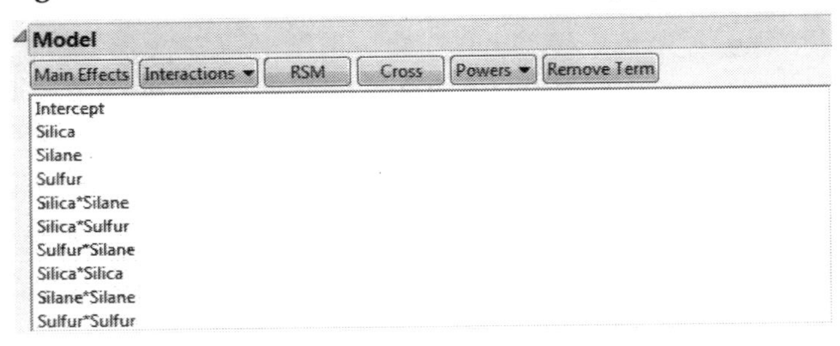

You can add effects to the Model outline using the following buttons:

Main Effects Adds main effects for all factors in the model.

Interactions Adds interaction effects. If no factors are selected in the Factors outline, select 2nd, 3rd, 4th, or 5th to add all appropriate interactions up to that order. Add interactions up to a given order for specific factors by selecting the factor names in the Factors outline, selecting Interactions, and then specifying the appropriate order. Interactions between non-mixture and mixture factors, and interactions with blocking and constant factors, are not added.

RSM Adds interaction and quadratic terms up to the second order (response surface model terms) for continuous factors. Categorical factors are not included in RSM terms. Main effects for non-mixture factors that interact with all the mixture factors are removed.

Cross Adds specific interaction terms. Select factor names in the Factors outline and effect names in the Model outline. Click Cross to add the crossed terms to the Model outline.

Powers Adds polynomial terms. If no factor names are selected in the Factors outline, adds polynomial terms for all continuous factors. If factor names are selected in the Factors outline, adds polynomial terms for only those factors. Select 2nd, 3rd, 4th, or 5th to add polynomial terms of that order.

Scheffé Cubic Adds Scheffé cubic terms for all mixture factors. These terms are used to specify a mixture model with third-degree polynomial terms.

Remove Term Removes selected effects.

Alias Terms

It is possible that effects *not* included in your assumed model are active. In the Alias Terms outline, list potentially active effects that are not in your assumed model but might bias the estimates of model terms. The Alias Matrix entries represent the degree of bias imparted to model parameters by the effects that you specified in the Alias Terms outline. For details, see "The Alias Matrix" on page 690 in the "Technical Details" appendix.

By default, the Alias Terms outline includes all two-way interaction effects that are not in your Model outline (with the exception of terms involving blocking factors). Add or remove terms using the buttons. For a description of how to use these buttons to add effects to the Alias Terms table, see "Model" on page 432.

In the Evaluate Design platform, the Alias Matrix outline is immediately updated to reflect changes to Alias Matrix effects. In the Custom Design platform, you must click Make Design after modifying the effects in the Alias Terms outline. Within other DOE platforms that construct designs, there is no Alias Terms outline. However, the Alias Matrix outline, containing appropriate effects, appears under Design Evaluation after you construct the design.

Design

The Design outline shows the design runs for the factors that you have specified in the launch window. You can easily view the design as you explore its properties in the Design Evaluation outline.

Design Evaluation

Design Evaluation within the Evaluate Design platform is based on your design and the specifications that you make in the Model and Alias Terms outlines. Several DOE Design platforms provide a Design Evaluation outline: Custom, Definitive Screening, Screening, Response Surface, and Mixture with Optimal design type. Design Evaluation within these platforms is based on the design that you construct.

The Design Evaluation outline contains eight headings:

- "Power Analysis" on page 434

- "Prediction Variance Profile" on page 440
- "Fraction of Design Space Plot" on page 442
- "Prediction Variance Surface" on page 442
- "Estimation Efficiency" on page 444
- "Alias Matrix" on page 446
- "Color Map on Correlations" on page 448
- "Design Diagnostics" on page 449

Power Analysis

The Power Analysis outline calculates the power of tests for the parameters in your model. Power is the probability of detecting an active effect of a given size. The Power Analysis outline helps you evaluate the ability of your design to detect effects of practical importance. Power depends on the number of runs, the significance level, and the estimated error variation. In particular, you can determine if additional runs are necessary.

This section covers the following topics:

- "Power Analysis Overview" on page 434
- "Power Analysis Details" on page 435
- "Tests for Individual Parameters" on page 436
- "Tests for Categorical Effects with More Than Two Levels" on page 437
- "Design and Anticipated Responses Outline" on page 437
- "Power Analysis for Coffee Experiment" on page 438

Power Analysis Overview

Power is calculated for the effects listed in the Model outline. These include continuous, discrete numeric, categorical, blocking, covariate, mixture, and covariate factors. The tests are for individual model parameters and for whole effects. For details on how power is calculated, see "Power Calculations" on page 691 in the "Technical Details" appendix.

Power is the probability of rejecting the null hypothesis of no effect at specified values of the model parameters. In practice, your interest is not in the values of the model parameters, but in detecting differences in the mean response of practical importance. In the Power Analysis outline, you can compute Anticipated Responses for specified values of the Anticipated Coefficients. This helps you to determine the coefficient values associated with the differences you want to detect in the mean response.

Figure 15.13 shows the Power Analysis outline for the design in the Coffee Data.jmp sample data table, found in the Design Experiment folder. The model specified in the Model script is a main effects only model.

Figure 15.13 Power Analysis for Coffee Data.jmp

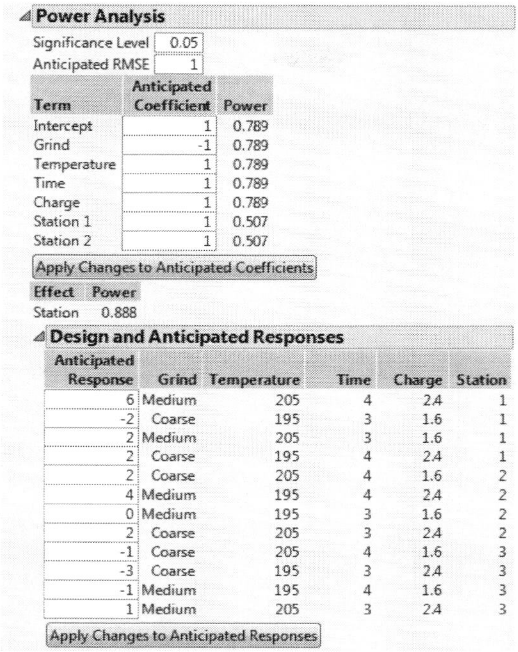

Power Analysis

Significance Level 0.05
Anticipated RMSE 1

Term	Anticipated Coefficient	Power
Intercept	1	0.789
Grind	-1	0.789
Temperature	1	0.789
Time	1	0.789
Charge	1	0.789
Station 1	1	0.507
Station 2	1	0.507

Apply Changes to Anticipated Coefficients

Effect	Power
Station	0.888

Design and Anticipated Responses

Anticipated Response	Grind	Temperature	Time	Charge	Station
6	Medium	205	4	2.4	1
-2	Coarse	195	3	1.6	1
2	Medium	205	3	1.6	1
2	Coarse	195	4	2.4	1
2	Coarse	205	4	1.6	2
4	Medium	195	4	2.4	2
0	Medium	195	3	1.6	2
2	Coarse	205	3	2.4	2
-1	Coarse	205	4	1.6	3
-3	Coarse	195	3	2.4	3
-1	Medium	195	4	1.6	3
1	Medium	205	3	2.4	3

Apply Changes to Anticipated Responses

In the Power Analysis outline, you can:

- Specify coefficient values that reflect differences that you want to detect. You enter these as Anticipated Coefficients in the top part of the outline.
- Specify anticipated response values and apply these to determine the corresponding Anticipated Coefficients. You specify Anticipated Responses in the Design and Anticipated Responses panel.

Power Analysis Details

Specify values for the Significance Level and Anticipated RMSE. These are used to calculate the power of the tests for the model parameters.

Significance Level The probability of rejecting the hypothesis of no effect, if it is true. The power calculations update immediately when you enter a value.

Anticipated RMSE An estimate of the square root of the error variation. The power calculations update immediately when you enter a value.

The top portion of the Power Analysis report opens with default values for the Anticipated Coefficients. See Figure 15.13. The default values are based on Delta. For details, see "Advanced Options > Set Delta for Power" on page 451.

Note: If the design is supersaturated, meaning that the number of parameters to be estimated exceeds the number of runs, the anticipated coefficients are set to 0.

Figure 15.14 shows the top portion of the Power Analysis report where values have been specified for the Anticipated Coefficients. These values reflect the differences you want to detect.

Figure 15.14 Possible Specification of Anticipated Coefficients for Coffee Data.jmp

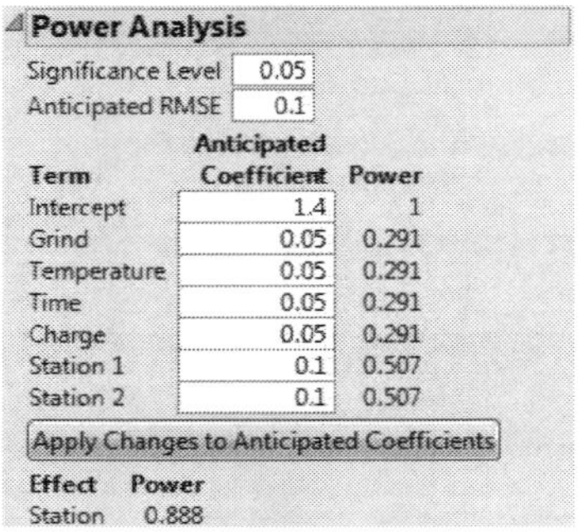

Power Analysis

Significance Level 0.05
Anticipated RMSE 0.1

Term	Anticipated Coefficient	Power
Intercept	1.4	1
Grind	0.05	0.291
Temperature	0.05	0.291
Time	0.05	0.291
Charge	0.05	0.291
Station 1	0.1	0.507
Station 2	0.1	0.507

Apply Changes to Anticipated Coefficients

Effect	Power
Station	0.888

Tests for Individual Parameters

The Term column contains a list of model terms. For each term, the Anticipated Coefficient column contains a value for that term. The value in the Power column is the power of a test that the coefficient for the term is 0 if the true value of the coefficient is given by the Anticipated Coefficient.

Term The model term associated with the coefficient being tested.

Note: The order in which model terms appear in the Power Analysis report may not be identical to their order in the Parameter Estimates report obtained using Standard Least Squares. This difference can only occur when the model contains an interaction with more than one degree of freedom.

Anticipated Coefficient A value for the coefficient associated with the model term. This value is used in the calculations for Power. These values are also used to calculate the Anticipated Response column in the Design and Anticipated Responses outline. When you set a new value in the Anticipated Coefficient column, click **Apply Changes to Anticipated Coefficients** to update the Power and Anticipated Response columns.

Note: The anticipated coefficients have default values of 1 for continuous effects. They have alternating values of 1 and –1 for categorical effects. You can specify a value for Delta be selecting **Advanced Options > Set Delta for Power** from the red triangle menu. If you change the value of Delta, the values of the anticipated coefficients are updated so that their absolute values are one-half of Delta. For details, see "Advanced Options > Set Delta for Power" on page 451.

Power Probability of rejecting the null hypothesis of no effect when the true coefficient value is given by the specified Anticipated Coefficient. For a coefficient associated with a numeric factor, the change in the mean response (based on the model) is twice the coefficient value. For a coefficient associated with a categorical factor, the change in the mean response (based on the model) across the levels of the factor equals twice the absolute value of the anticipated coefficient.

Calculations use the specified Significance Level and Anticipated RMSE. For details of the power calculation, see "Power for a Single Parameter" on page 691 in the "Technical Details" appendix.

Apply Changes to Anticipated Coefficients When you set a new value in the Anticipated Coefficient column, click **Apply Changes to Anticipated Coefficients** to update the Power and Anticipated Response columns.

Tests for Categorical Effects with More Than Two Levels

If your model contains a categorical effect with more than two levels, then the following columns appear below the Apply Changes to Anticipated Coefficients button:

Effect The categorical effect.

Power The power calculation for a test of no effect. The null hypothesis for the test is that all model parameters corresponding to the effect are zero. The difference to be detected is defined by the values in the Anticipated Coefficient column that correspond to the model terms for the effect. The power calculation reflects the differences in response means determined by the anticipated coefficients.

Calculations use the specified Significance Level and Anticipated RMSE. For details of the power calculation, see "Power for a Categorical Effect" on page 692 in the "Technical Details" appendix.

Design and Anticipated Responses Outline

The Design and Anticipated Responses outline shows the design preceded by an Anticipated Response column. Each entry in the first column is the Anticipated Response corresponding to the design settings. The Anticipated Response is calculated using the Anticipated Coefficients.

Figure 15.15 shows the Design and Anticipated Responses outline corresponding to the specification of Anticipated Coefficients given in Figure 15.14.

Figure 15.15 Anticipated Responses for Coffee Data.jmp

Design and Anticipated Responses

Anticipated Response	Grind	Temperature	Time	Charge	Station
1.6	Medium	205	4	2.4	1
1.4	Coarse	195	3	1.6	1
1.4	Medium	205	3	1.6	1
1.6	Coarse	195	4	2.4	1
1.6	Coarse	205	4	1.6	2
1.5	Medium	195	4	2.4	2
1.3	Medium	195	3	1.6	2
1.6	Coarse	205	3	2.4	2
1.3	Coarse	205	4	1.6	3
1.2	Coarse	195	3	2.4	3
1.1	Medium	195	4	1.6	3
1.2	Medium	205	3	2.4	3

Apply Changes to Anticipated Responses

In the Anticipated Response column, you can specify a value for each setting of the factors. These values reflect the differences you want to detect.

Click **Apply Changes to Anticipate Responses** to update both the Anticipated Coefficient and Power columns.

Anticipated Response The response value obtained using the Anticipated Coefficient values as coefficients in the model. When the outline first appears, the calculation of Anticipated Response values is based on the default values in the Anticipated Coefficient column. When you set new values in the Anticipated Response column, click **Apply Changes to Anticipated Responses** to update the Anticipated Coefficient and Power columns.

Design The columns to the right of the Anticipated Response column show the factor settings for all runs in your design.

Apply Changes to Anticipated Responses When you set new values in the Anticipated Response column, click **Apply Changes to Anticipated Responses** to update the Anticipated Coefficient and Power columns.

Power Analysis for Coffee Experiment

Consider the design in the Coffee Data.jmp data table. Suppose that you are interested in the power of your design to detect effects of various magnitudes on Strength. Recall that Grind is a two-level categorical factor, Temperature, Time, and Charge are continuous factors, and Station is a three-level categorical (blocking) factor.

In this example, ignore the role of Station as a blocking factor. You are interested in the effect of Station on Strength. Since Station is a three-level categorical factor, it is represented by two terms in the Parameters list: Station 1 and Station 2.

Specifically, you are interested the probability of detecting the following changes in the mean Strength:

- A change of 0.10 units as you vary Grind from Coarse to Medium.
- A change of 0.10 units or more as you vary Temperature, Time, and Charge from their low to high levels.
- An increase due to each of Stations 1 and 2 of 0.10 units beyond the overall anticipated mean. This corresponds to a decrease due to Station 3 of 0.20 units from the overall anticipated mean.

You set 0.05 as your Significance Level. Your estimate of the standard deviation of Strength for fixed design settings is 0.1 and you enter this as the Anticipated RMSE.

Figure 15.16 shows the Power Analysis node with these values entered. Specifically, you specify the Significance Level, Anticipated RMSE, and the value of each Anticipated Coefficient.

When you click Apply Changes to Anticipated Coefficients, the Anticipated Response values are updated to reflect the model you have specified.

Figure 15.16 Power Analysis Outline with User Specifications in Anticipated Coefficients Panel

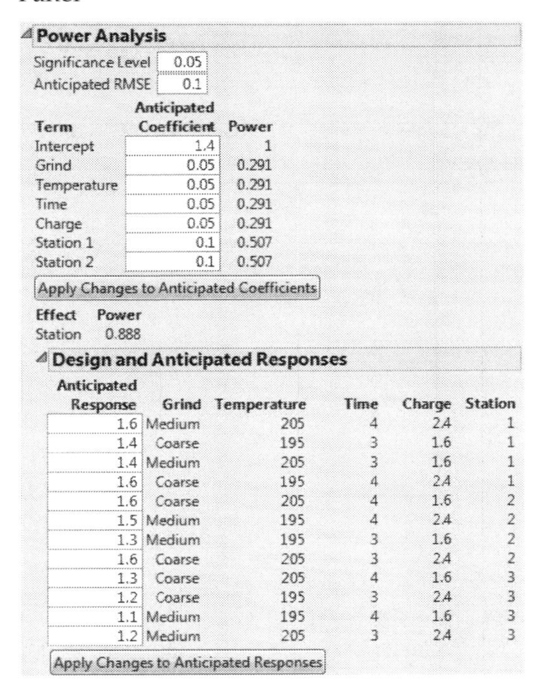

Recall that **Temperature** is a continuous factor with coded levels of -1 and 1. Consider the test whose null hypothesis is that **Temperature** has no effect on **Strength**. Figure 15.16 shows that the power of this test to detect a difference of 0.10 (=2*0.05) units across the levels of **Temperature** is only 0.291.

Now consider the test for the whole **Station** effect, where **Station** is a three-level categorical factor. Consider the test whose null hypothesis is that **Station** has no effect on **Strength**. This is the usual *F* test for a categorical factor provided in the Effect Tests report when you run **Analyze > Fit Model**. (See the Standard Least Squares chapter in the *Fitting Linear Models* book.)

The Power of this test is shown directly beneath the Apply Changes to Anticipated Coefficients button. The entries under Anticipated Coefficients for the model terms Station 1 and Station 2 are both 0.10. These settings imply that the effect of both stations is to increase **Strength** by 0.10 units above the overall anticipated mean. For these settings of the Station 1 and Station 2 coefficients, the effect of Station 3 on **Strength** is to decrease it by 0.20 units from the overall anticipated mean. Figure 15.16 shows that the power of the test to detect a difference of at least this magnitude is 0.888.

Prediction Variance Profile

The Prediction Variance Profile outline shows a profiler of the relative variance of prediction. Select the **Optimization and Desirability > Maximize Desirability** option from the red triangle next to Prediction Variance Profile to find the maximum value of the relative prediction variance over the design space. For details, see "Maximize Desirability" on page 441.

The Prediction Variance Profile plots the relative variance of prediction as a function of each factor at fixed values of the other factors. Figure 15.17 shows the Prediction Variance Profile for the **Bounce Data.jmp** data table, located in the Design Experiment folder.

Figure 15.17 Prediction Variance Profiler

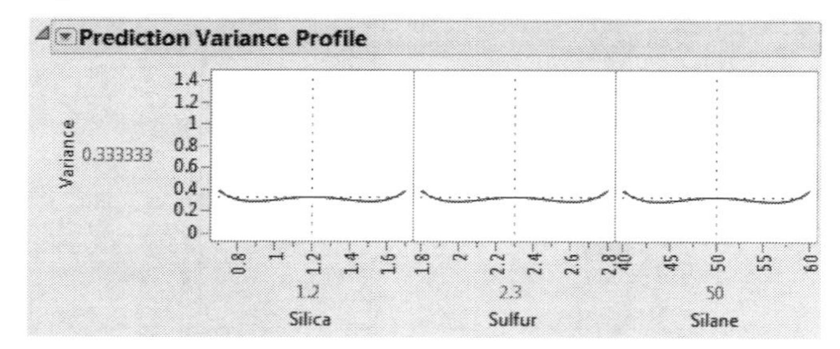

Relative Prediction Variance

For given settings of the factors, the prediction variance is the product of the error variance and a quantity that depends on the design and the factor settings. Before you run your experiment, the error variance is unknown, so the prediction variance is also unknown.

However, the ratio of the prediction variance to the error variance is not a function of the error variance. This ratio, called the *relative prediction variance*, depends only on the design and the factor settings. Consequently, the relative variance of prediction can be calculated before acquiring the data. For details, see "Relative Prediction Variance" on page 694 in the "Technical Details" appendix.

After you run your experiment and fit a least squares model, you can estimate the error variance using the mean squared error (MSE) of the model fit. You can estimate the actual variance of prediction at any setting by multiplying the relative variance of prediction at that setting.

It is ideal for the prediction variance to be small throughout the design space. Generally, the error variance drops as the sample size increases. In comparing designs, you may want to place the prediction variance profilers for two designs side-by-side. A design with lower prediction variance on average is preferred.

Maximize Desirability

You can also evaluate a design or compare designs in terms of the maximum relative prediction variance. Select the **Optimization and Desirability > Maximize Desirability** option from the red triangle next to Prediction Variance Profile. JMP uses a desirability function that maximizes the relative prediction variance. The value of the Variance displayed in the Prediction variance Profile is the worst (least desirable from a design point of view) value of the relative prediction variance.

Figure 15.18 shows the Prediction Variance Profile after Maximize Desirability was selected. The plot is for the Bounce Data.jmp sample data table, located in the Design Experiment folder. The largest value of the relative prediction variance is 1.395833. The plot also shows values of the factors that give this worst-case relative variance. However, keep in mind that many settings can lead to this same relative variance. See "Prediction Variance Surface" on page 442.

Figure 15.18 Prediction Variance Profile Showing Maximum Variance

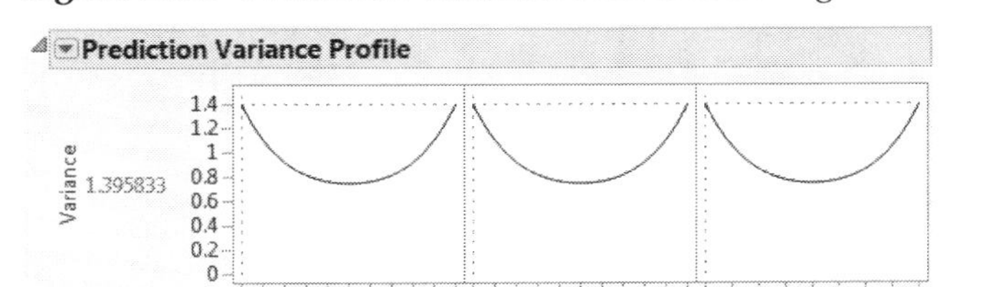

Fraction of Design Space Plot

The Fraction of Design Space Plot shows the proportion of the design space over which the relative prediction variance lies below a given value. Figure 15.19 shows the Fraction of Design Space plot for the for the Bounce Data.jmp sample data table, located in the Design Experiment folder.

Figure 15.19 Fraction of Design Space Plot

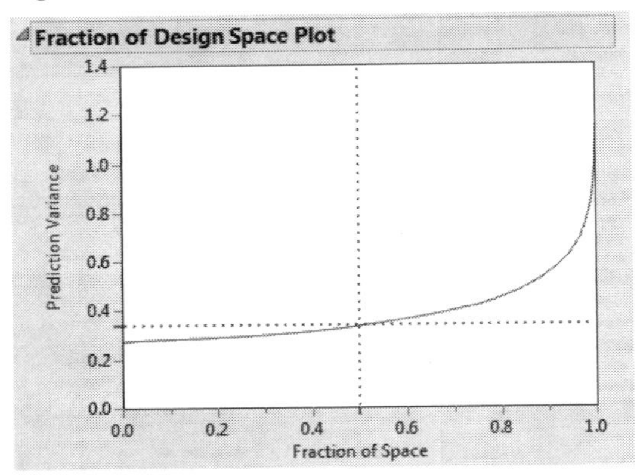

The *X* axis in the plot represents the proportion of the design space, ranging from 0 to 100%. The *Y* axis represents relative prediction variance values. For a point (x, y) that falls on the blue curve, the value x is the proportion of design space with variance less than or equal to y. Red dotted crosshairs mark the value that bounds the relative prediction variance for 50% of design space.

Figure 15.19 shows that the minimum relative prediction variance is slightly less than 0.3, while the maximum is below 1.4. (The actual maximum is 1.395833, as shown in Figure 15.18.) The red dotted crosshairs indicate that the relative prediction variance is less than 0.34 over about 50% of the design space. You can use the crosshairs tool to find the maximum relative prediction variance that corresponds to any Fraction of Space value. Use the crosshairs tool in Figure 15.19 to see that 90% of the prediction variance values are below approximately 0.55.

Note: Monte Carlo sampling of the design space is used in constructing the Fraction of Design Space Plot. Therefore, plots for the same design may vary slightly.

Prediction Variance Surface

The Prediction Variance Surface report plots the relative prediction variance surface as a function of any two design factors. Figure 15.20 shows the Prediction Variance Surface outline for the for the Bounce Data.jmp sample data table, located in the Design Experiment folder.

Show or hide the controls by selecting **Control Panel** on the red triangle menu. See "Control Panel" on page 443.

Figure 15.20 Prediction Variance Surface

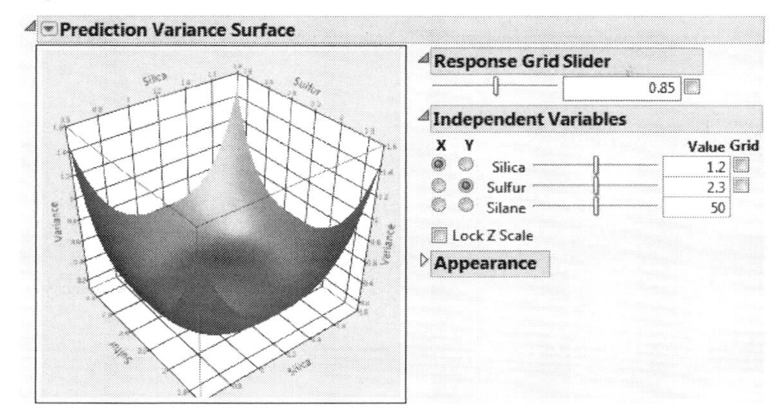

When there are two or more factors, the Prediction Variance Surface outline shows a plot of the relative prediction variance for any two variables. The Prediction Variance Surface outline plots the relative prediction variance formula. Drag on the plot to rotate and change the perspective.

Control Panel

The Control Panel consists of the following:

Response Grid Slider The **Grid** check box superimposes a grid that shows constant values of Variance. The value of the Variance is shown in the text box. The slider enables you to adjust the placement of the grid. Alternatively, you can enter a Variance value in the text box. Click outside the box to update the plot.

Independent Variables This panel enables you to select which two factors are used as axes for the plot and to specify the settings for factors not used as axes. Select a factor for each of the X and Y axes by clicking in the appropriate column. Use the sliders and text boxes to specify values for each factor *not* selected for an axis. The plot shows the three-dimensional slice of the surface at the specified values of the factors that are not used as axes in the plot. Move the sliders to see different slices.

Each grid check box activates a grid for the corresponding factor. Use the sliders to adjust the placement of each grid.

Lock Z Scale locks the *z*-axis to its current values. This is useful when moving the sliders that are not on an axis.

Appearance The **Resolution** slider affects how many points are evaluated for a formula. Too coarse a resolution means that a function with a sharp change might not be represented very well. But setting the resolution high can make evaluating and displaying the surface slower.

The **Orthographic projection** check box shows a projection of the plot in two dimensions.

The **Contour** menu controls the placement of contour curves. A contour curve is a set of points whose Response values are constant. You can select to turn the contours Off (the default) or place them contours Below, Above, or On Surface.

Estimation Efficiency

This report gives the Fractional Increase in CI (Confidence Interval) Length and Relative Std (Standard) Error of Estimate for each parameter estimate in the model. Figure 15.21 shows the Estimation Efficiency outline for the Bounce Data.jmp sample data table, located in the Design Experiment folder.

Figure 15.21 Estimation Efficiency Outline

Estimation Efficiency

Term	Fractional Increase in CI Length	Relative Std Error of Estimate
Intercept	1.236	0.577
Silica	0.369	0.354
Silane	0.369	0.354
Sulfur	0.369	0.354
Silica*Silane	0.936	0.5
Silica*Sulfur	0.936	0.5
Sulfur*Silane	0.936	0.5
Silica*Silica	1.016	0.52
Silane*Silane	1.016	0.52
Sulfur*Sulfur	1.016	0.52

Fractional Increase in CI Length

The Fractional Increase in CI Length compares the length of a parameter's confidence interval as given by the current design to the length of such an interval given an ideal design:

- The length of the ideal confidence interval for the parameter is subtracted from the length of its actual confidence interval.
- This difference is then divided by the length of the ideal confidence interval.

For an orthogonal D-optimal design, the fractional increase is zero. In selecting a design, you would like the fractional increase in confidence interval length to be as small as possible.

The Ideal Design

The covariance matrix for the ordinary least squares estimator is $\sigma^2(\mathbf{X'X})^{-1}$. The diagonal elements of $(\mathbf{X'X})^{-1}$ are the *relative variances* (the variances divided by σ^2) of the parameter estimates. For two-level designs and using the effects coding convention (see "Coding" on page 663 in the "Column Properties" appendix), the minimum value of the relative variance for any parameter estimate is $1/n$, where n is the number of runs. This occurs when all the effects for the design are orthogonal and the design is D-optimal.

Let $\hat{\beta}$ denote the vector of parameter estimates. The *ideal design*, which may not exist, is a design whose covariance matrix is given as follows:

$$\mathrm{Var}(\hat{\beta}) = (\sigma^2/n)\mathbf{I}_n$$

where $\mathbf{I}_n$ is the n by n identity matrix and σ is the standard deviation of the response.

If an orthogonal D-optimal design exists, it is the ideal design. However, the definition above extends the idea of an ideal design to situations where a design that is both orthogonal and D-optimal does not exist.

The definition is also appropriate for designs with multi-level categorical factors. The orthogonal coding used for categorical factors allows such designs to have the ideal covariance matrix. For a Custom Design, you can view the coding matrix by selecting Save X Matrix from the options in the Custom Design window, making the design table, and looking at the script Model Matrix that is saved to the design table.

Fractional Increase in Length of Confidence Interval

Note that, in the ideal design, the standard error for the parameter estimates would be given as follows:

$$SE_{Ideal}(\hat{\beta}) = (\sigma/\sqrt{n})\mathbf{I}_n$$

The length of a confidence interval is determined by the standard error. The Fractional Increase in Confidence Interval Length is the difference between the standard error of the given design and the standard error of the ideal design, divided by the standard error of the ideal design.

Specifically, for the ith parameter estimate, the Fractional Increase in Confidence Interval Length is defined as follows:

$$FI = \frac{\sigma\sqrt{(\mathbf{X'X}_{ii})^{-1}} - (\sigma/(\sqrt{n}))}{(\sigma/\sqrt{n})} = \sqrt{n(\mathbf{X'X}_{ii})^{-1}} - 1$$

where

σ^2 is the unknown response variance,

$\mathbf{X}$ is the model matrix for the given design, defined in "The Alias Matrix" on page 690 in the "Technical Details" appendix,

$(\mathbf{X'X})^{-1}_{ii}$ is the ith diagonal entry of $(\mathbf{X'X})^{-1}$, and

n is the number of runs.

Relative Std Error of Estimate

The Relative Std Error of Estimate gives the ratio of the standard deviation of a parameter's estimate to the error standard deviation. These values indicate how large the standard errors of the model's parameter estimates are, relative to the error standard deviation. For the ith parameter estimate, the Relative Std Error of Estimate is defined as follows:

$$SE = \sqrt{(\mathbf{X'X})^{-1}_{ii}}$$

where

$\mathbf{X}$ is the model matrix defined in "The Alias Matrix" on page 690 in the "Technical Details" appendix, and

$(\mathbf{X'X})^{-1}_{ii}$ is the ith diagonal entry of $(\mathbf{X'X})^{-1}$.

Alias Matrix

The Alias Matrix addresses the issue of how terms that are not included in the model affect the estimation of the model terms, if they are indeed active. In the Alias Terms outline, you list potentially active effects that are not in your assumed model but that might bias the estimates of model terms. The Alias Matrix entries represent the degree of bias imparted to model parameters by the Alias Terms effects. See "Alias Terms" on page 433.

The rows of the Alias Matrix are the terms corresponding to the model effects listed in the Model outline. The columns are terms corresponding to effects listed in the Alias Terms outline. The entry in a given row and column indicates the degree to which the alias term affects the parameter estimate corresponding to the model term.

In evaluating your design, you ideally want one of two situations to occur relative to any entry in the Alias Matrix. Either the entry is small or, if it is not small, the effect of the alias term is small so that the bias will be small. If you suspect that the alias term may have a substantial effect, then that term should be included in the model or you should consider an alias optimal design.

For details on how the Alias Matrix is computed, see "The Alias Matrix" on page 690 in the "Technical Details" appendix. See also Lekivetz, R. (2014).

Note the following:

- If the design is orthogonal for the assumed model, then the correlations in the Alias Matrix correspond to the absolute correlations in the Color Map on Correlations.
- Depending on the complexity of the design, it is possible to have alias matrix entries greater than 1 or less than -1.

Alias Matrix Examples

Consider the Coffee Data.jmp sample data table, located in the Design Experiment folder. The design assumes a main effects model. You can see this by running the Model script in the data table. Consequently, in the Evaluate Design window's Model outline, only the Intercept and five main effects appear. The Alias Terms outline contains the two-way interactions. The Alias Matrix is shown in Figure 15.22.

Figure 15.22 Alias Matrix for Coffee Data.jmp

Alias Matrix

Effect	Grind*Temperature	Grind*Time	Grind*Charge	Temperature*Time	Temperature*Charge	Time*Charge
Intercept	0	0	0	0	0	0
Grind	0	0	0	0.333	-0.33	-0.33
Temperature	0	0.333	-0.33	0	0	-0.33
Time	0.333	0	-0.33	0	-0.33	0
Charge	-0.33	-0.33	0	-0.33	0	0
Station 1	-0.41	0	0	0	0	0.816
Station 2	0.707	0	0	0	0	0

The Alias Matrix shows the Model terms in the first column defining the rows. The two-way interactions in the Alias Terms are listed across the top, defining the columns. Consider the model effect Temperature for example. If the Grind*Time interaction is the only active two-way interaction, the estimate for the coefficient of Temperature is biased by 0.333 times the true value of the Grind*Time effect. If other interactions are active, then the value in the Alias Matrix indicates the additional amount of bias incurred by the Temperature coefficient estimate.

Consider the Bounce Data.jmp sample data table, located in the Design Experiment folder. The Model script contains all two-way interactions. Consequently, the Evaluate Design window shows all main effects and two-way interactions in the Model outline. The three two-way interactions are automatically added to the list of Alias Terms. Therefore, the Alias Matrix shows a column for each of these three interactions (Figure 15.23). Notice that the only non-zero entries in the alias matrix correspond to the bias impact of the two-way interactions on themselves. These entries are 1s, which is expected because the two-way interactions are already in the model.

Figure 15.23 Alias Matrix for Bounce Data.jmp

Alias Matrix

Effect	Silica*Sulfur	Silica*Silane	Sulfur*Silane
Intercept	0	0	0
Silica	0	0	0
Silane	0	0	0
Sulfur	0	0	0
Silica*Silane	0	1	0
Silica*Sulfur	1	0	0
Sulfur*Silane	0	0	1
Silica*Silica	0	0	0
Silane*Silane	0	0	0
Sulfur*Sulfur	0	0	0

Color Map on Correlations

The Color Map on Correlations shows the absolute value of the correlation between any two effects that appear in either the Model or the Alias Terms outline. The cells of the color map are identified above the map. There is a cell for each effect in the Model outline and a cell for each effect in the Alias Terms outline.

By default, the absolute magnitudes of the correlations are represented by a blue to gray to red intensity color theme. In general terms, the color map for a good design shows a lot of blue off the diagonal, indicating orthogonality or small correlations between distinct terms. Large absolute correlations among effects inflate the standard errors of estimates.

To see the absolute value of the correlation between two effects, hover your cursor over the corresponding cell. To change the color theme for the entire plot, right click in the plot and select Color Theme.

Color Map Example

Figure 15.24 shows the Color Map on Correlations for the Bounce Data.jmp sample data table, found in the Design Experiment folder. The deep red coloring indicates absolute correlations of one. Note that there are red cells on the diagonal, showing correlations of model terms with themselves.

All other cells are either deep blue or light blue. The light blue squares correspond to correlations between quadratic terms. To see this, hover your cursor over each of the light blue squares. The absolute correlations of quadratic terms with each other are small, 0.0714.

From the perspective of correlation, this is a good design. When effects are highly correlated, it is more difficult to determine which is responsible for an effect on the response.

Figure 15.24 Color Map on Correlations

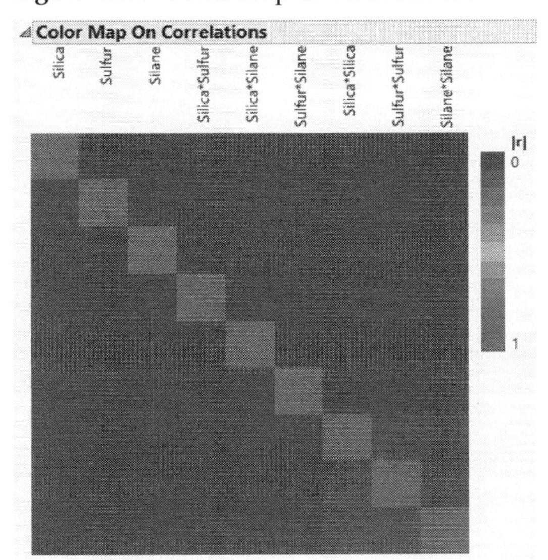

Design Diagnostics

The Design Diagnostics outline shows *D*-, *G*-, and *A*-efficiencies and the average variance of prediction. These diagnostics are not shown for designs that include factors with Changes set to Hard or Very Hard or effects with Estimability designated as If Possible.

When Evaluate Design is accessed from a DOE platform other than Evaluate Design, the Design Creation Time gives the amount of time required to create the design. When Design Diagnostics is accessed from the Evaluate Design platform, Design Creation Time gives the amount of time required for the Evaluate Design platform to calculate results.

Figure 15.25 shows the Design Diagnostics outline for the Bounce Data.jmp sample data table, found in the Design Experiment folder.

Figure 15.25 Design Diagnostics Outline

Design Diagnostics

D Efficiency	36.6429
G Efficiency	69.10947
A Efficiency	29.3578
Average Variance of Prediction	0.384722
Design Creation Time (seconds)	0

Caution: The efficiency measures should not be interpreted on their own. But they can be used to compare designs. Given two designs, the one with the higher efficiency measure is better. While the maximum efficiency is 100 for any criterion, an efficiency of 100% is impossible for many design problems.

Notation

The descriptions of the efficiency measures given below use the following notation:

- $\mathbf{X}$ is the model matrix
- n is the number of runs in the design
- p is the number of terms, including the intercept, in the model
- $Var(\hat{y}|\underset{\sim}{\mathbf{x}})$ is the relative prediction variance at the point $\underset{\sim}{\mathbf{x}}$. See "Relative Prediction Variance" on page 694 in the "Technical Details" appendix.
- $Var(\hat{y}|\underset{\sim}{\mathbf{x}})_{max}$ is the maximum relative prediction variance over the design region

D Efficiency

The efficiency of the design to that of an ideal orthogonal design in terms of the D-optimality criterion. A design is D-optimal if it minimizes the volume of the joint confidence region for the vector of regression coefficients:

$$D\text{-efficiency} = 100\left(\frac{1}{n}|\mathbf{X}'\mathbf{X}|^{1/p}\right)$$

G Efficiency

The efficiency of the design to that of an ideal orthogonal design in terms of the G-optimality criterion. A design is G-optimal if it minimizes the maximum prediction variance over the design region:

$$G\text{-efficiency} = 100p/(nVar(\hat{y}|\underset{\sim}{\mathbf{x}})_{max})$$

Letting D denote the design region,

$$Var(\hat{y}|\underset{\sim}{\mathbf{x}})_{max} = \underset{\underset{\sim}{\mathbf{x}} \text{ in D}}{maximum}[\underset{\sim}{\mathbf{x}}'(\mathbf{X}'\mathbf{X})^{-1}\underset{\sim}{\mathbf{x}}]$$

Note: G-Efficiency is calculated using Monte Carlo sampling of the design space. Therefore, calculations for the same design may vary slightly.

A Efficiency

The efficiency of the design to that of an ideal orthogonal design in terms of the A-optimality criterion. A design is A-optimal if it minimizes the sum of the variances of the regression coefficients:

$$A\text{-efficiency} = 100p/(n\mathrm{Trace}(\mathbf{X}'\mathbf{X})^{-1})$$

Average Variance of Prediction

At a point $\underset{\sim}{\mathbf{x}}$ in the design space, the relative prediction variance is defined as:

$$Var(\hat{y}|\underset{\sim}{\mathbf{x}}) = \underset{\sim}{\mathbf{x}}'(\mathbf{X}'\mathbf{X})^{-1}\underset{\sim}{\mathbf{x}}$$

Note that this is the prediction variance divided by the error variance. For details of the calculation, see Section 4.3.5 in Goos and Jones, 2011.

Design Creation Time

Design Creation Time gives the amount of time required for the Evaluate Design platform to calculate results.

Evaluate Design Options

The Evaluate Design red triangle menu contains the following options:

Advanced Options > Split Plot Variance Ratio Specify the ratio of the variance of the random whole plot and the subplot variance (if present) to the error variance. Before setting this value, you must define a hard-to-change factor for your split-plot design, or hard and very-hard-to-change factors for your split-split-plot design. Then you can enter one or two positive numbers for the variance ratios, depending on whether you have specified a split-plot or a split-split-plot design.

Advanced Options > Set Delta for Power Specify a value for the difference you want to detect that is applied to Anticipated Coefficients in the Power Analysis report. The Anticipated Coefficient values are set to Delta/2 for continuous effects. For categorical effects, they are alternating values of Delta/2 and –Delta/2. For additional details on power analysis, see "Power Analysis" on page 434.

By default, Delta is set to two. Consequently, the Anticipated Coefficient default values are 1 for continuous effects and alternating values of 1 and –1 for categorical effects. The default values that are entered as Anticipated Coefficients when Delta is 2 ensure these properties:

- The power calculation for a numeric effect assumes a change of Delta in the response mean due to linear main effects as the factor changes from the lowest setting to the highest setting in the design region.
- The power calculation for the parameter associated with a two-level categorical factor assumes a change of Delta in the response mean across the levels of the factor.
- The power calculation for a categorical effect with more than two levels is based on the multiple degree of freedom F-test for the null hypothesis that all levels have the same response mean. Power is calculated at the values of the response means that are determined by the Anticipated Coefficients. Various configurations of the Anticipated Coefficients can define a difference in levels as large as Delta. However, the power values for such configurations will differ based on the Anticipated Coefficients for the other levels.

Save Script to Script Window Creates a script that reproduces the Evaluate Design window and places it in an open script window.

Chapter 16

Compare Designs

Compare and Evaluate Designs Simultaneously

The Compare Designs platform compares two or three designs simultaneously to explore and evaluate their performance. Diagnostics show how the designs perform relative to each other and how they perform in an absolute sense. To compare designs relative to your specific needs, you can change the terms in the assumed model and in the alias terms list.

Figure 16.1 Comparing Three Designs with Different Run Sizes

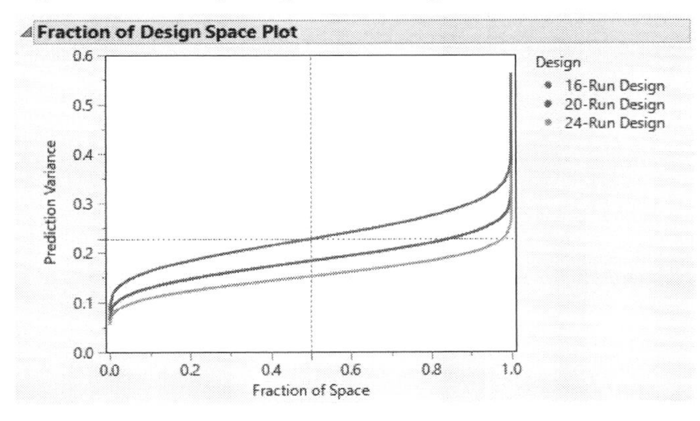

Overview of Comparing Designs

The Compare Designs platform, which is an extension of the Evaluate Design platform, enables you to easily compare two or three designs. To compare the performance of one or two designs relative to another, you select a reference design that is treated as the base design. You can specify effects in the Model outline, and effects of interest in the Alias Terms outline.

The Design Evaluation report shows diagnostic results and plots covering these areas:

- Power analysis
- Prediction variance
- Fraction of design space
- Relative estimation efficiency
- Alias matrix diagnostics
- Correlations among effects (including confounding)
- Relative efficiency measures for the overall designs

Examples of Comparing Designs

This section contains three examples:

- "Designs of Same Run Size" on page 454
- "Designs of Different Run Sizes" on page 459
- "Split Plot Designs with Different Numbers of Whole Plots" on page 463

Designs of Same Run Size

In this example, you compare two designs for six factors each with 13 runs. One is a 12-run Plackett-Burman (PB) design augmented with a single center point. The other is a Definitive Screening Design (DSD).

Comparison in Terms of Main Effects Only

First, compare the two designs assuming that the model to be estimated contains only the main effects.

1. Select **Help > Sample Data**, click **Open the Sample Scripts Directory**, and select Compare Same Run Size.jsl.
2. Right-click in the script window and select **Run Script**.

Two 13-run design tables are constructed: Definitive Screening Design and Plackett-Burman. You want to compare these two designs. Because the Plackett-Burman table is active, it is the reference design to which you compare the DSD.

3. In the Plackett-Burman data table, select **DOE > Design Diagnostics > Compare Designs**.
4. Select Definitive Screening Design from the **Compare 'Plackett-Burman' with** list.
5. Select X1 through X6 in the Plackett-Burman panel and in the Definitive Screening Panel.
6. Open the Match Columns outline and click **Match**.

Figure 16.2 Launch Window with Matched Columns

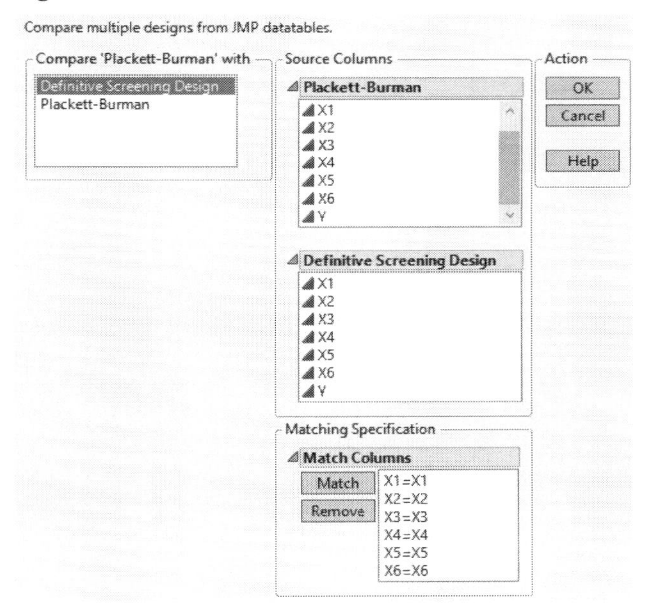

This defines the correspondence between the factors in your two designs.

7. Click **OK**.

The reference design is the Plackett-Burman design. In the Design Evaluation outline, comparison metrics compare the PB to the DSD. The designs are compared relative to power, prediction variance, estimation efficiency, aliasing, and design efficiency measures.

Figure 16.3 Power Analysis for PB and DSD Comparison

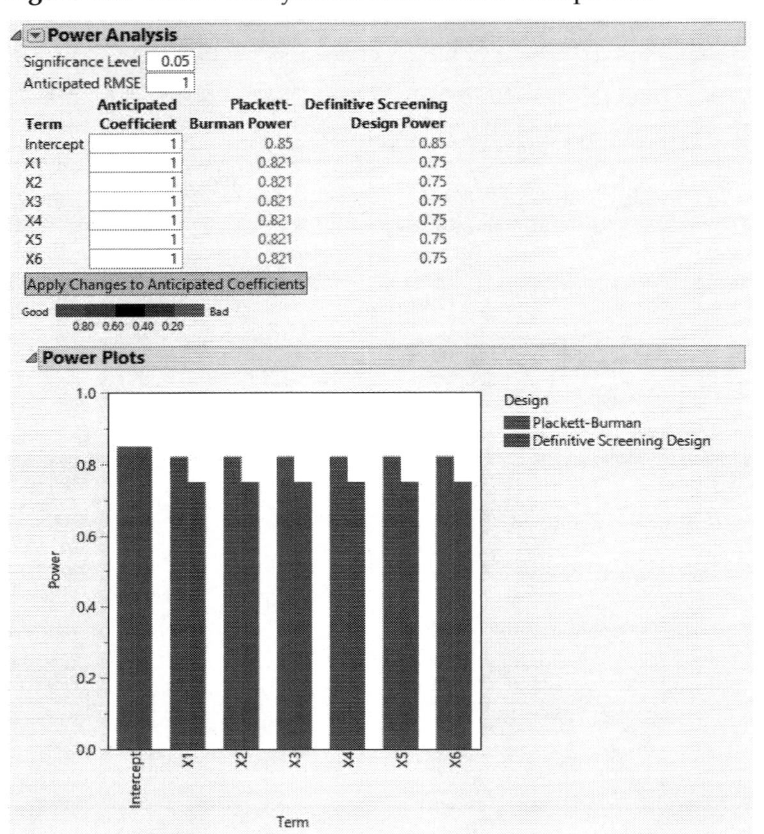

In terms of power, prediction variance, and estimation efficiency, the PB design outperforms the DSD. Figure 16.3 shows the Power Analysis report with the default settings for the significance level, Anticipated RMSE, and coefficients. For tests for the main effects, the PB design has higher power than does the DSD.

Figure 16.4 Fraction of Design Space Plot for PB and DSD Comparison

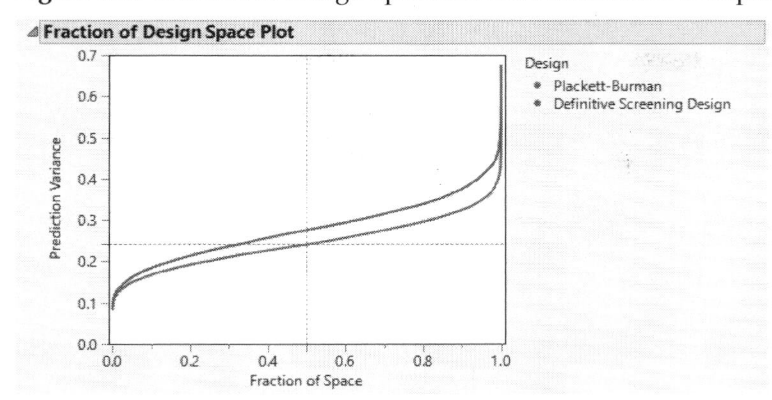

The Fraction of Design Space plot indicates that the PB design has smaller prediction variance than the DSD over the entire design space.

You conclude that, if you suspect that only main effects are active, the PB design is preferable.

Comparison in Terms of Two-Way Interactions

Now suppose you suspect that some two-way interactions might be active. The analysis below shows that if those two-way interactions are actually active, then the PB design might be less desirable than the DSD.

1. In the Absolute Correlations report, open the Color Map on Correlations report and the color map reports under it.

Figure 16.5 Color Maps for PB and DSD Comparison

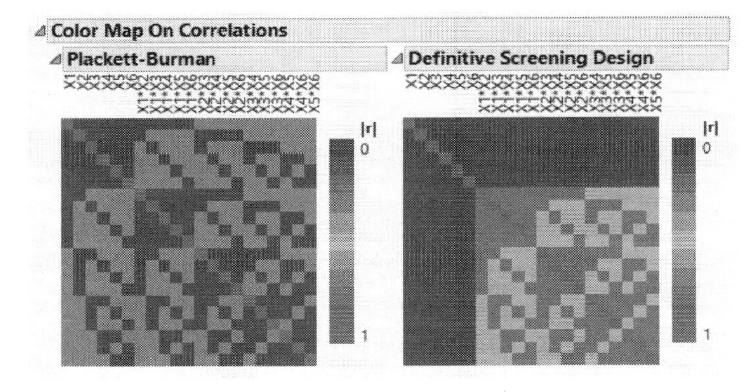

The Color Map on Correlations plots in Figure 16.5 show that the PB design aliases main effects with two-way interactions. In contrast, the DSD does not alias main effects with two-way interactions.

To gain more insight on how the designs compare if some two-way interactions are active, add two-way interactions in the Model outline.

2. In the Factors outline, select X1 through X3.
3. In the Model outline, select **Interactions > 2nd**.

Figure 16.6 Power Analysis for PB and DSD Comparison with Interactions

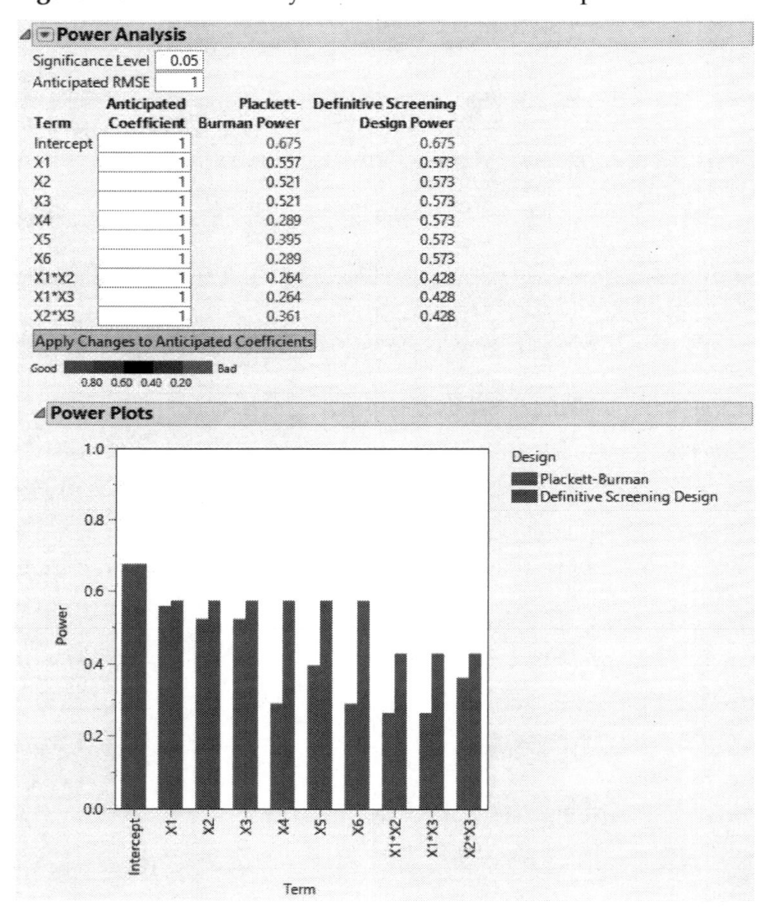

The Term list shows the three two-way interactions. If these two-way interactions are active, then the DSD has better performance in terms of power across all effects than the PB.

Figure 16.7 Prediction Variance for PB and DSD Comparison with Interactions

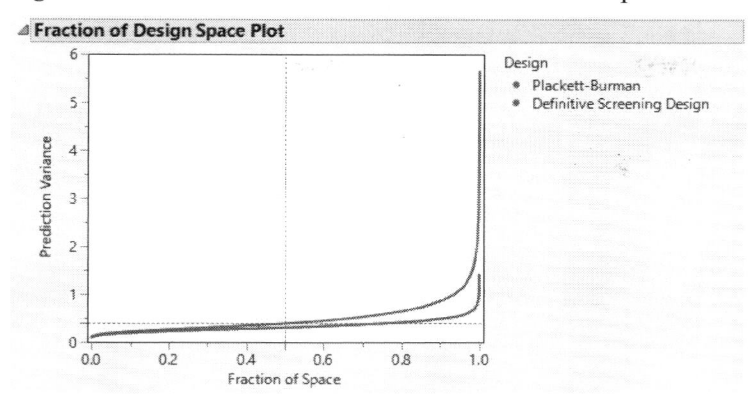

The DSD also outperforms the PB design in terms of prediction variance with the three interactions in the model. You can explore the other reports to see that the DSD is preferred when there are potentially active interactions.

Designs of Different Run Sizes

In this example, compare three designs with run sizes 16, 20, and 24. The designs are constructed for main effect models. Use the Compare Designs platform to determine whether the potential benefits of using a larger run size are worth the additional cost in resources.

1. Select **Help > Sample Data**, click **Open the Sample Scripts Directory**, and select Compare Three Run Sizes.jsl.
2. Right-click in the script window and select **Run Script**.

   Three design tables are constructed using Custom Design, with only main effects as entries in the Model outline:

   – 16-Run Design
   – 20-Run Design
   – 24-Run Design

   You want to compare these three designs. Notice that the 16-Run Design table is active.
3. In the 16-Run Design table, select **DOE > Design Diagnostics > Compare Designs**.
4. From the **Compare '16-Run Design' with** list, select 20-Run Design and 24-Run Design.

   Panels for each of these designs are added to the launch window. JMP automatically matches the columns in the order in which they appear in the three design tables.
5. Click **OK**.

Figure 16.8 Power Analysis Comparison

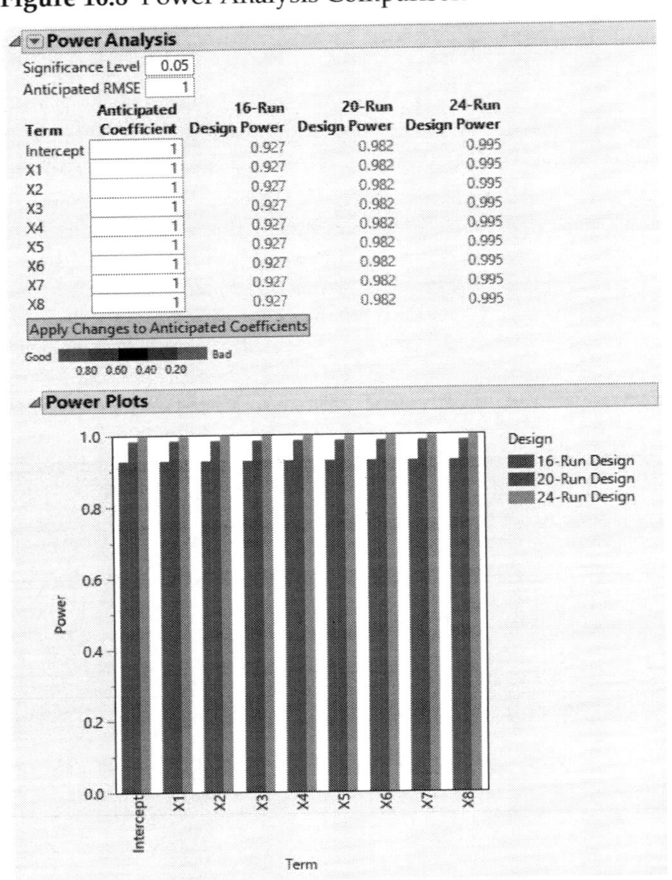
Power Analysis

Significance Level 0.05
Anticipated RMSE 1

Term	Anticipated Coefficient	16-Run Design Power	20-Run Design Power	24-Run Design Power
Intercept	1	0.927	0.982	0.995
X1	1	0.927	0.982	0.995
X2	1	0.927	0.982	0.995
X3	1	0.927	0.982	0.995
X4	1	0.927	0.982	0.995
X5	1	0.927	0.982	0.995
X6	1	0.927	0.982	0.995
X7	1	0.927	0.982	0.995
X8	1	0.927	0.982	0.995

Apply Changes to Anticipated Coefficients

All three designs have high power for detecting main effects if the coefficients are on the order of the Anticipated RMSE.

Figure 16.9 Fraction of Design Space Comparison

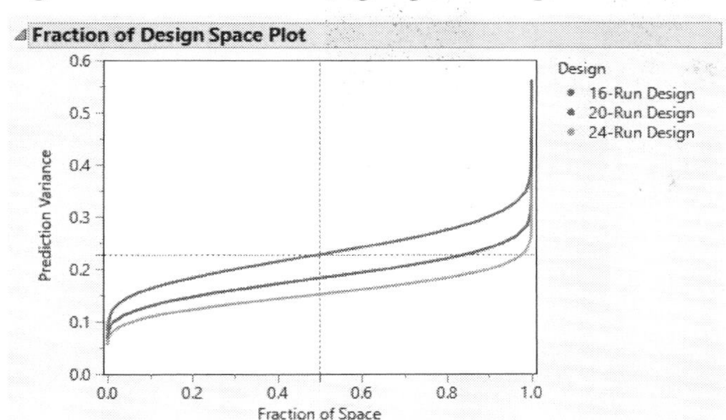

As expected, the 24-run design is superior to the other two designs in terms of prediction variance over the entire design space. The 20-run design is superior to the 16-run design.

6. In the Absolute Correlations report, open the Color Map on Correlations report and the three color map reports under it.

Figure 16.10 Color Map on Correlations Comparison

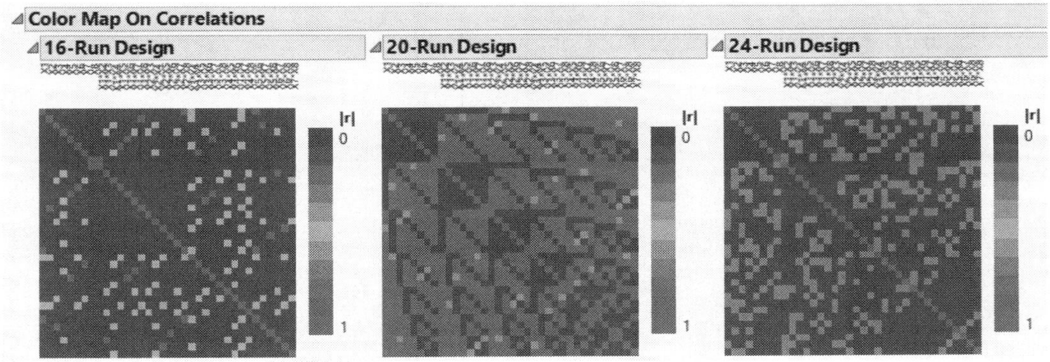

Color Map On Correlations

16-Run Design | 20-Run Design | 24-Run Design

For the 16-run design, the Color Map on Correlations indicates that there is confounding of some main effects with some two-factor interactions, and confounding of two-factor interactions.

For the 20-run design, the Color Map on Correlations indicates that there are some large correlations between some main effects and some two-factor interactions, and between some two-factor interactions.

The 24-run design shows only moderate correlations between main effects and two-factor interactions, and between two-factor interactions.

Figure 16.11 Absolute Correlations Comparison

Absolute Correlations

Model x Model	Average Correlation	Pairwise Confoundings	Pairwise Terms
16-Run Design	0	0	28
20-Run Design	0	0	28
24-Run Design	0	0	28
Model x Alias	**Average Correlation**	**Pairwise Confoundings**	**Pairwise Terms**
16-Run Design	0.08	9	224
20-Run Design	0.166	0	224
24-Run Design	0.103	0	224
Alias x Alias	**Average Correlation**	**Pairwise Confoundings**	**Pairwise Terms**
16-Run Design	0.063	6	378
20-Run Design	0.124	0	378
24-Run Design	0.066	0	378

Good 0.10 0.30 0.50 0.70 Bad

The Absolute Correlations table summarizes the information shown in the Color Maps on Correlations. Recall that the model for all three designs consists of only main effects and the Alias Matrix contains two-factor interactions.

For the 16-run design, the Model x Alias portion of the table indicates that there are nine confoundings of main effects with two-factor interactions. The Alias x Alias portion indicates that six two-factor interactions are confounded.

Figure 16.12 Design Diagnostics Comparison

Design Diagnostics

	Efficiency of 16-Run Design Relative to 20-Run Design	Efficiency of 16-Run Design Relative to 24-Run Design
D-efficiency	0.800	0.667
G-efficiency	0.894	0.816
A-efficiency	0.800	0.667
I-efficiency	0.800	0.667
Additional Run Size	-4	-8

Good 1.50 1.25 0.80 0.67 Bad

The Design Diagnostics report compares the efficiency of the 16-run design to both the 20-run and 24-run designs in terms of several efficiency measures. Relative efficiency values that exceed 1 indicate that the reference design is preferable for the given measure. Values less than 1 indicate that the design being compared to the reference design is preferable. The 16-run design has lower efficiency than the other two designs across all metrics, indicating that the larger designs are preferable.

7. In the Factors outline, select X1 through X3.
8. In the Model outline, select **Interactions > 2nd**.

   An Inestimable Terms window appears, telling you that the 16-run design cannot fit one of the effects that you just added to the model (X1*X2).
9. Click **OK**.

The other two effects, X1*X3 and X2*X3, are added to the Compare Design report. You can examine the report to compare the designs if the two interactions are active.

Split Plot Designs with Different Numbers of Whole Plots

In this example, compare two split-plot designs with different numbers of whole plots. The designs are for three factors:

- A continuous hard-to-change factor
- A continuous easy-to-change factor
- A three-level categorical easy-to-change factor

The designs include all two-factor interactions in the assumed model. You can afford 20 runs and want to compare using 4 or 8 whole plots.

Launch Compare Designs

1. Select **Help > Sample Data**, click **Open the Sample Scripts Directory**, and select Compare Split Plots.jsl.
2. Right-click in the script window and select **Run Script**.

   Two design tables are constructed using Custom Design:

   – 4 Whole Plots

   – 8 Whole Plots

   You want to compare these two designs. Notice that the 4 Whole Plots table is active.
3. In the 4 Whole Plots table, select **DOE > Design Diagnostics > Compare Designs**.
4. From the **Compare '4 Whole Plots' with** list, select 8 Whole Plots.

   A panel for this design is added to the launch window. JMP automatically matches the columns in the order in which they appear in the two design tables.

Figure 16.13 Completed Launch Window

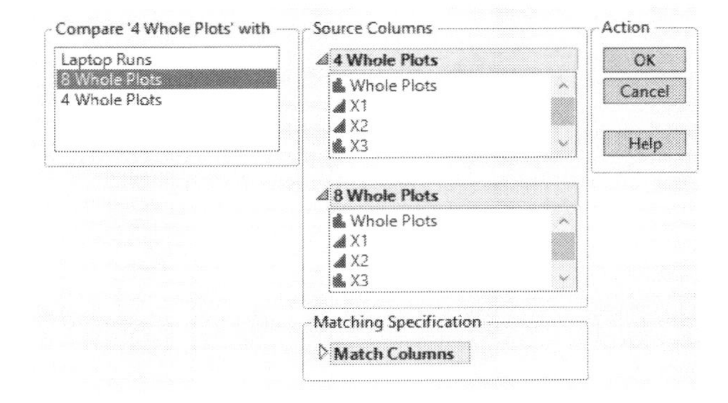

5. Click **OK**.
6. Open the **Matching Specification** outline under Reference Design: 20 run '4 Whole Plots'.

Figure 16.14 Matching Specification for Split-Plot Designs

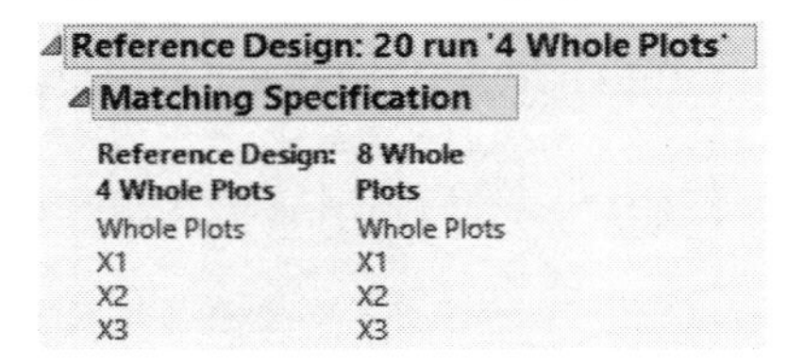

Notice that the Whole Plots column is entered as part of the design. This is necessary because Compare Designs needs to know the whole plot structure.

Examine the Report

The Design Evaluation report provides various diagnostics that compare the two designs.

Figure 16.15 Power Analysis for Two Split-Plot Designs

Power Analysis

Significance Level 0.05
Anticipated RMSE 1

Term	Anticipated Coefficient	4 Whole Plots Power	8 Whole Plots Power
Intercept	1	0.190	0.497
X1	1	0.190	0.497
X2	1	0.973	0.973
X3 1	1	0.746	0.767
X3 2	-1	0.823	0.708
X1*X2	1	0.963	0.968
X1*X3 1	1	0.746	0.767
X1*X3 2	-1	0.823	0.708
X2*X3 1	1	0.734	0.591
X2*X3 2	-1	0.833	0.526

Apply Changes to Anticipated Coefficients

Good 0.80 0.60 0.40 0.20 Bad

Effect	4 Whole Plots Power	8 Whole Plots Power
X3	0.792	0.714
X1*X3	0.792	0.714
X2*X3	0.797	0.523

Power Plot

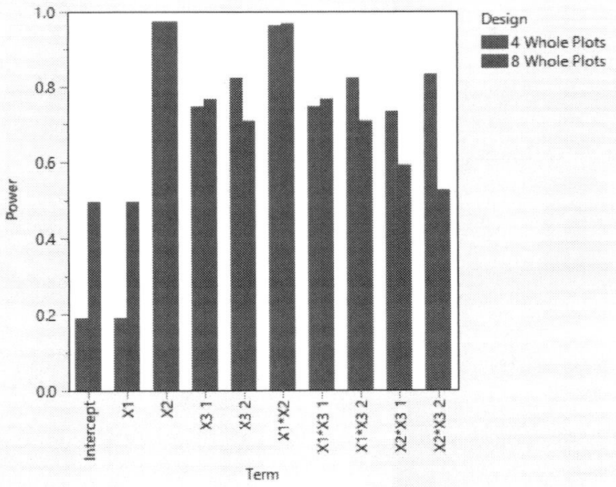

The Power Analysis report shows that the power for the whole-plot factor, X1, is much smaller for the four whole-plot design (0.19) than for the eight whole-plot design (0.497). However, the four whole-plot design has higher power to detect split-plot effects, especially the interaction of the two split-plot factors, X2*X3 (0.797 compared to 0.523). Notice that the power for the combined effect X2*X3 is given under the color bar and legend.

Figure 16.16 Relative Estimation Efficiency Comparing Split-Plot Designs

Relative Estimation Efficiency

Term	Efficiency of 4 Whole Plots Relative to 8 Whole Plots
Intercept	0.778
X1	0.778
X2	1.004
X3 1	0.924
X3 2	1.154
X1*X2	0.982
X1*X3 1	0.924
X1*X3 2	1.154
X2*X3 1	1.107
X2*X3 2	1.449

Good 1.50 1.25 0.80 0.67 Bad

Relative Std Error of Estimates

Term	4 Whole Plots	8 Whole Plots
Intercept	0.553	0.43
X1	0.553	0.43
X2	0.23	0.231
X3 1	0.25	0.231
X3 2	0.22	0.254
X1*X2	0.24	0.235
X1*X3 1	0.25	0.231
X1*X3 2	0.22	0.254
X2*X3 1	0.256	0.283
X2*X3 2	0.218	0.315

The Relative Estimation Efficiency report shows the relative estimation efficiency for X1 to be 0.778. This indicates that the standard error for X1 is notably larger for the four whole-plot design than for the eight whole-plot design.

Open the Relative Std Error of Estimates report. You can see that the relative standard error for X1 in the four whole-plot design is 0.553, compared to the eight whole-plot error of 0.43.

In the Relative Estimation Efficiency report, the relative estimation efficiency for X2*X3 2 is 1.449, indicating that the standard error for the parameter associated with X2*X3 2 is notably larger for the eight whole-plot design than for the four whole-plot design.

The Power Analysis and the Relative Estimation Efficiency reports indicate that the choice of designs revolves around the importance of detecting the whole plot effect X1. The eight whole-plots design gives you a better chance of detecting a whole plot effect. The four whole-plots design is somewhat better for detecting split-plot effects involving the categorical variable.

Compare Designs Launch Window

Launch the Compare Designs platform by selecting **DOE > Design Diagnostics > Compare Designs.** All open data tables appear in the list at the left. The active data table and its columns appear in a Source Columns panel. The design in the initial Source Columns panel is the *reference design*, namely, the design to which other designs are compared. When you add designs to compare to the reference design, their columns appear in panels under the reference design panel.

Figure 16.17 shows the launch window for the three designs in "Designs of Different Run Sizes" on page 459.

Figure 16.17 Compare Designs Launch Window

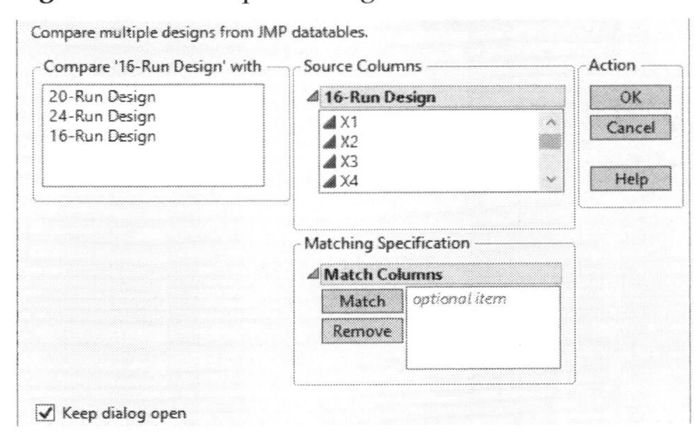

Design Table Selection

Select one or two design tables from the list on the left.

- To compare two designs to the reference design, you must select their design tables simultaneously from the list on the left.
- To replace a design (or designs) in the Source Columns list, select the desired table (or tables) from the list at the left. The design table (or tables) under the reference design table are replaced.

Note: The reference design table can be compared to itself, which can be useful when exploring the assignment of design columns to factors.

Match Columns

Specify which columns in each of the design tables correspond to each other in the Match Columns panel. To match columns, select the columns to match in each of the design table Source Columns lists, and then click Match.

Figure 16.18 Selection of Columns for Matching

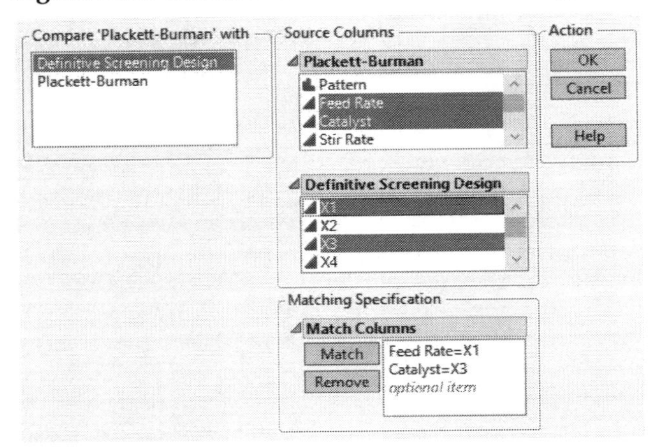

- To match single columns in each list, select the single column in each list, and then click Match.
- To match several columns that appear in the correct matching order in each list, select them in each list. Click the Match button. They are matched in their list order. See Figure 16.18. In this example, Feed Rate is matched with X1, and Catalyst is matched with X3.
- If the lists contain the same numbers of columns and your desired match order is their order of appearance in the lists, you do not have to click Match. When you click OK to run the launch window, JMP matches the columns automatically in their order of appearance. You can review the matching in the report's Matching Specification outline.

Compare Designs Window: Specify Model and Alias Terms

The Compare Designs window consists of two sets of outlines:

- Specify which effects are in the model and which effects are potentially active using the Factors, Model, and Alias Terms outlines.
- Compare the designs using the diagnostics in the Design Evaluation outlines. Changes that you make in the Model and Alias Terms outlines are updated in the Design Evaluation report.

The Compare Designs report uses the column names from the reference design.

This section describes the Reference Design, Factors, Model, and Alias Terms outlines. See "Compare Designs Window: Design Evaluation" on page 469 for a description of the Design Evaluation outlines.

Reference Design

The name of the window for the reference design appears in the outline title. The Matching Specification outline lists the specifications that you entered in the launch window.

Factors

Use the Factors outline to add effects to the Model and Alias Terms lists.

The Factors outline lists the factors, using the column names from the reference design, and coded values. Because they are not factors, whole plot and subplot columns do not appear in the Factors outline. However, they are required for the analysis.

Model

Add or remove effects to compare your designs for the effects that you believe should be in the model. The Model outline initially lists effects that are in the Model script of the reference design table and that are estimated by all designs being compared. If there is no Model script in the reference design table, the Model outline shows only the main effects that can be estimated by all designs being compared. For details about how to add and remove effects, see "Model" on page 432 in the "Evaluate Designs" chapter.

Note: If any of the designs are supersaturated, meaning that the number of parameters to be estimated exceeds the number of runs, the Model outline lists only a set of effects that can be estimated.

Alias Terms

Add or remove effects to compare your designs for effects that might be active. The Alias Terms outline initially contains all two-factor interactions that are not in the Model outline. The effects in this outline impact the calculations in the Alias Matrix Summary and Absolute Correlations outline. See "Alias Matrix Summary" on page 478 and "Absolute Correlations" on page 480.

For details about how to add and remove effects, see "Alias Terms" on page 433 in the "Evaluate Designs" chapter.

Compare Designs Window: Design Evaluation

The Design Evaluation report consists of these outlines:

- "Power Analysis" on page 470
- "Prediction Variance Profile" on page 474

- "Fraction of Design Space Plot" on page 476
- "Relative Estimation Efficiency" on page 477
- "Alias Matrix Summary" on page 478
- "Absolute Correlations" on page 480
- "Design Diagnostics" on page 483

Color Dashboard

Several of the Design Evaluation outlines show values colored according to a color bar. The colors are applied to diagnostic measures and they help you see which values (and designs) reflect good or bad behavior. You can edit the legend values to apply colors that reflect your definitions of good and bad behavior.

Figure 16.19 Color Dashboard

You can modify the color bar by selecting these two options in the red triangle menu for the outline or by right-clicking the color bar:

Show Legend Values Shows or hides the values that appear under the color bar.

Edit Legend Values Specify the values that define the colors.

Power Analysis

Power is the probability of detecting an active effect of a given size. The Power Analysis report helps you evaluate and compare the ability of your designs to detect effects of practical importance. For each of your designs, the Power Analysis report calculates the power of tests for the effects in the Model outline.

The Power Analysis report gives the power of tests for individual model parameters and for whole effects. It also provides a Power Plot and a Power versus Sample Size plot.

Power depends on the number of runs, the significance level, and the estimated error variation. For details about how power is calculated, see "Power Calculations" on page 691 in the "Technical Details" appendix.

Figure 16.20 Power Analysis Outline for Three Designs

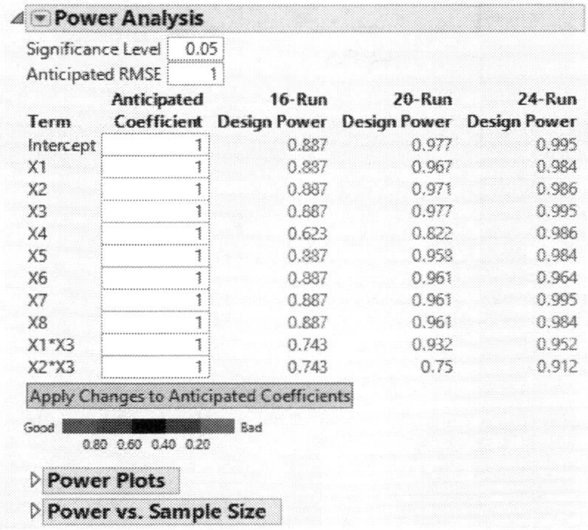

Power Analysis

Significance Level 0.05
Anticipated RMSE 1

Term	Anticipated Coefficient	16-Run Design Power	20-Run Design Power	24-Run Design Power
Intercept	1	0.887	0.977	0.995
X1	1	0.887	0.967	0.984
X2	1	0.887	0.971	0.986
X3	1	0.887	0.977	0.995
X4	1	0.623	0.822	0.986
X5	1	0.887	0.958	0.984
X6	1	0.887	0.961	0.964
X7	1	0.887	0.961	0.995
X8	1	0.887	0.961	0.984
X1*X3	1	0.743	0.932	0.952
X2*X3	1	0.743	0.75	0.912

Apply Changes to Anticipated Coefficients

Good 0.80 0.60 0.40 0.20 Bad

Power Plots

Power vs. Sample Size

Figure 16.20 shows the Power Analysis outline for the three designs constructed in "Designs of Different Run Sizes" on page 459. Two two-way interactions have been added to the Model outline.

Power Analysis Report

When you specify values for the Significance Level and Anticipated RMSE, they are used to calculate the power of the tests for the model parameters. Enter coefficient values that reflect differences that you want to detect as Anticipated Coefficients. To update the results for all designs, click Apply Changes to Anticipated Coefficients.

Significance Level The probability of rejecting the hypothesis of no effect, if it is true. The power calculations update immediately when you enter a value.

Anticipated RMSE An estimate of the square root of the error variation. The power calculations update immediately when you enter a value.

The power values are colored according to a color gradient that appears under the Apply Changes to Anticipated Coefficients button. You can control the color legend using the options in the Power Analysis red triangle menu. See "Color Dashboard" on page 470.

For details about the Power Plots, see "Power Plot" on page 473.

Note: If the design is supersaturated, meaning that the number of parameters to be estimated exceeds the number of runs, the Power Analysis outline lists only a set of effects that can be estimated.

Tests for Individual Parameters

The Term column contains a list of model terms. For each term, the Anticipated Coefficient column contains a value for that term. The Power value is the power of a test that the coefficient for the term is zero if the true value of the coefficient is given by the Anticipated Coefficient, given the design, and the terms in the Model outline.

Term The model term associated with the coefficient being tested.

Anticipated Coefficient A value for the coefficient associated with the model term. This value is used in the calculations for Power. When you set a new value in the Anticipated Coefficient column, click **Apply Changes to Anticipated Coefficients** to update the Power calculations.

Note: The anticipated coefficients have default values of 1 for continuous effects. They have alternating values of 1 and –1 for categorical effects.

Power The probability of rejecting the null hypothesis of no effect when the true coefficient value is given by the specified Anticipated Coefficient.

- For a coefficient associated with a numeric factor, the change in the mean response (based on the model) is twice the coefficient value.
- For a coefficient associated with a categorical factor, the change in the mean response (based on the model) across the levels of the factor equals twice the absolute value of the anticipated coefficient.

Calculations use the specified Significance Level and Anticipated RMSE. For details about the power calculation, see "Power for a Single Parameter" on page 691 in the "Technical Details" appendix.

Apply Changes to Anticipated Coefficients When you set a new value in the Anticipated Coefficient column, click **Apply Changes to Anticipated Coefficients** to update the Power values.

Tests for Categorical Effects with More Than Two Levels

If your model contains a categorical effect with more than two levels, then the following columns appear below the Apply Changes to Anticipated Coefficients button:

Effect The categorical effect.

Power The power calculation for a test of no effect. The null hypothesis for the test is that all model parameters corresponding to the effect are zero. The difference to be detected is defined by the values in the Anticipated Coefficient column that correspond to the model terms for the effect. The power calculation reflects the differences in response means determined by the anticipated coefficients.

Calculations use the specified Significance Level and Anticipated RMSE. For details about the power calculation, see "Power for a Categorical Effect" on page 692 in the "Technical Details" appendix.

Power Plot

The Power Plot shows the power values from the Power Analysis in graphical form. The plot shows the power for each effect and for each design in a side-by-side bar chart.

Figure 16.21 Power Plot for Three Designs

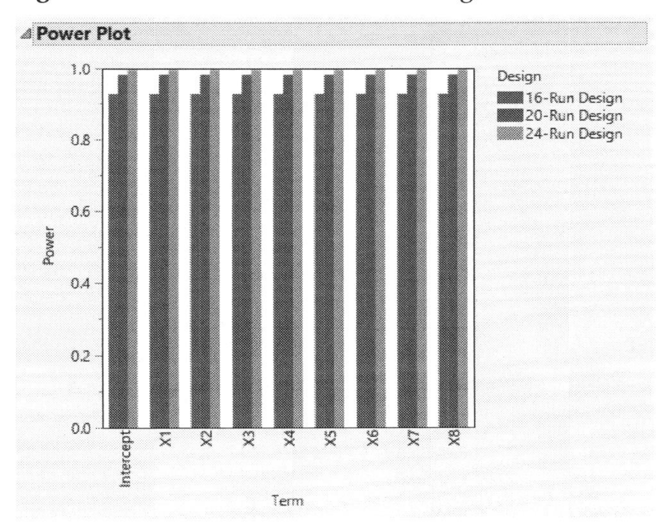

The Power Plot in Figure 16.21 is for the three designs constructed in "Designs of Different Run Sizes" on page 459. Two two-way interactions have been added to the Model outline.

Power versus Sample Size

The Power versus Sample Size profiler appears only when the designs that you are comparing differ in run size. The profiler enables you to see how sample size affects power for each effect in the model. It conveys the same information as is in the Power Plots graph, but in a different format. The power values at integer sample sizes are connected with line segments.

Figure 16.22 Power versus Sample Size Profiler for Three Designs

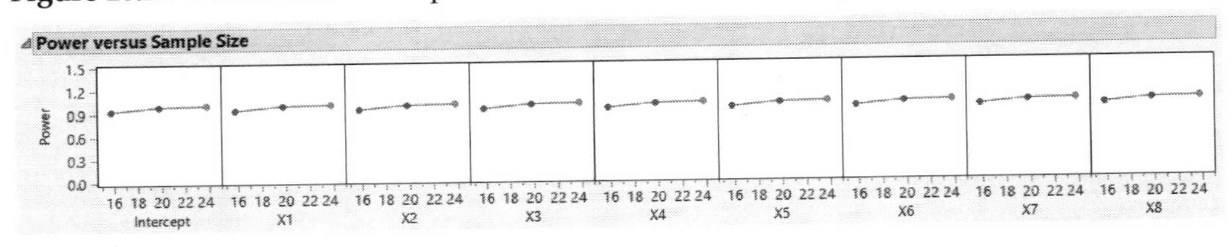

The Power versus Sample Size profiler in Figure 16.22 is for the three designs constructed in "Designs of Different Run Sizes" on page 459. Two two-way interactions have been added to the Model outline. Notice that the power for X4 increases more dramatically with sample size than does the power for other factors.

Prediction Variance Profile

The Prediction Variance Profile outline shows profilers of the relative variance of prediction for each design being compared. Each plot shows the relative variance of prediction as a function of each factor at fixed values of the other factors.

To find the maximum value of the relative prediction variance over the design space for all designs, select the **Optimization and Desirability > Maximize Desirability** option from the red triangle next to Prediction Variance Profile. For more details, see "Maximize Desirability" on page 475.

Figure 16.23 Prediction Variance Profile for Three Designs

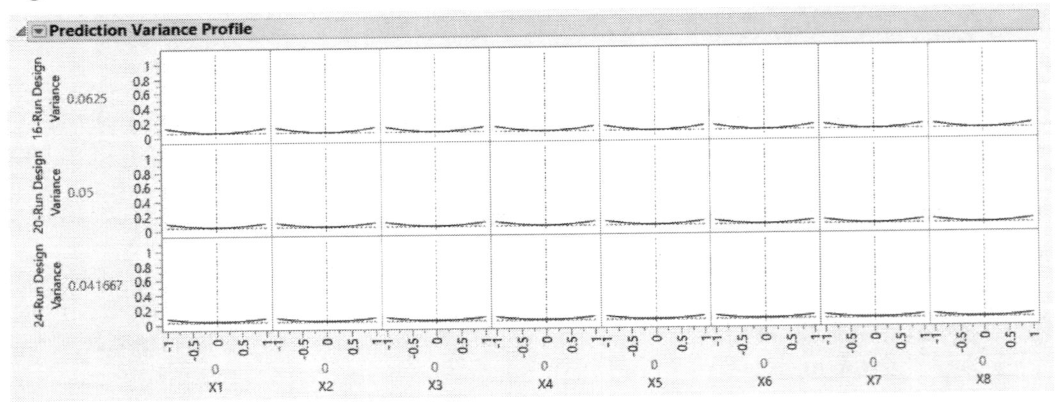

The Prediction Variance Profile plot in Figure 16.23 is for the three designs constructed in "Designs of Different Run Sizes" on page 459. Two two-way interactions, X1*X3 and X2*X3, have been added to the Model outline. The initial value for each continuous factor in the plot is the midpoint of its design settings. The Variance values to the left indicate that, as the number of runs increases, the variance decreases at the center point.

Relative Prediction Variance

For given settings of the factors, the prediction variance is the product of the error variance and a quantity that depends on the design and the factor settings. Before you run your experiment, the error variance is unknown, so the prediction variance is also unknown. However, the ratio of the prediction variance to the error variance is not a function of the error variance. This ratio, called the *relative prediction variance*, depends only on the design and the factor settings. Consequently, the relative variance of prediction can be calculated before acquiring the data. For details, see "Relative Prediction Variance" on page 694 in the "Technical Details" appendix.

After you run your experiment and fit a least squares model, you can estimate the error variance using the mean squared error (MSE) of the model fit. You can estimate the actual variance of prediction at any setting by multiplying the relative variance of prediction at that setting.

Ideally, the prediction variance is small throughout the design space. Generally, the error variance drops as the sample size increases. In comparing designs, a design with lower prediction variance on average is preferable.

Maximize Desirability

You can also evaluate a design or compare designs in terms of the maximum relative prediction variance. Select the **Optimization and Desirability > Maximize Desirability** option from the red triangle next to Prediction Variance Profile. JMP uses a desirability function that maximizes the relative prediction variance. The value of the Variance in the Prediction Variance Profile is the worst (least desirable from a design point of view) value of the relative prediction variance.

Figure 16.24 Prediction Variance Profile Showing Maximum Variance for Three Designs

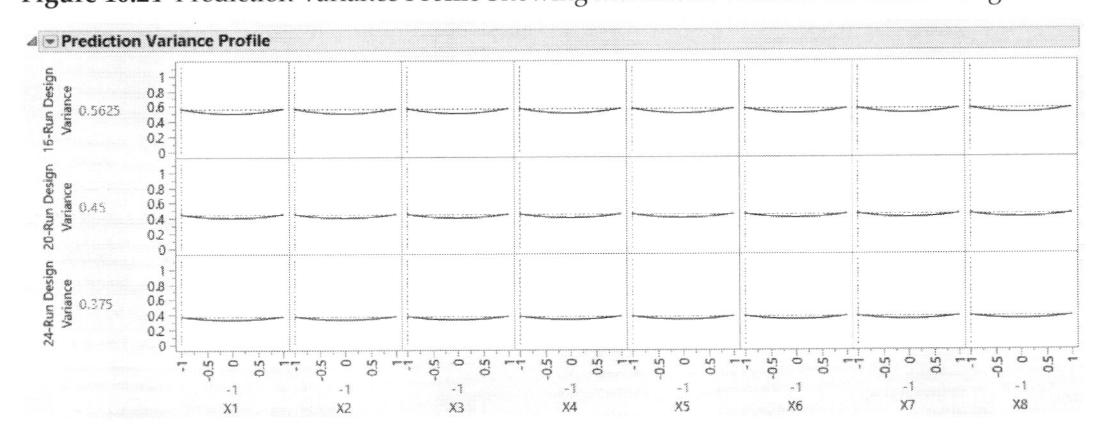

Figure 16.24 shows the Prediction Variance Profile after Maximize Desirability was selected for the three designs constructed in "Designs of Different Run Sizes" on page 459. As expected, the maximum relative prediction variance decreases as the run size increases. The plot also shows values of the factors that give this worst-case relative variance. However, keep in mind that many settings can lead to this same maximum relative variance.

Fraction of Design Space Plot

The Fraction of Design Space Plot shows the proportion of the design space over which the relative prediction variance lies below a given value.

Figure 16.25 Fraction of Design Space Plot for Three Designs

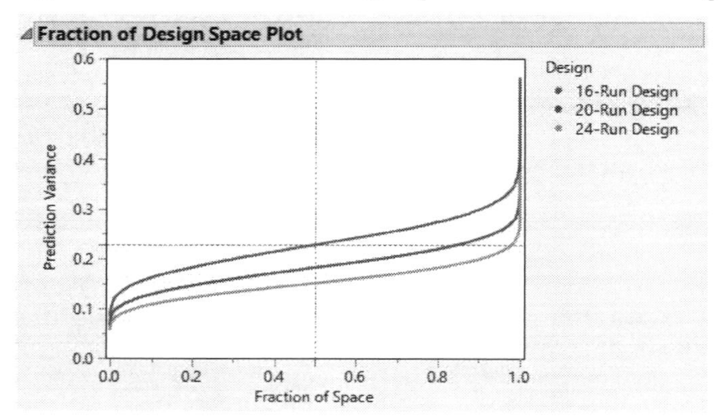

Figure 16.25 shows the Fraction of Design Space plot for the three designs constructed in "Designs of Different Run Sizes" on page 459. Note the following:

- The *X* axis in the plot represents the proportion of the design space, ranging from 0 to 100%.
- The *Y* axis represents relative prediction variance values.
- For a point (x, y) that falls on a given curve, the value x is the proportion of design space with variance less than or equal to y.
- Red dotted crosshairs mark the value that bounds the relative prediction variance for 50% of design space for the reference design.

Figure 16.25 shows that the relative prediction variance for the 24-run design is uniformly smaller than for the other two designs. The 20-run design has uniformly smaller prediction variance than the 16-run design. The red dotted crosshairs indicate that the relative prediction variance for the 20-run design is less than about 0.23 over about 50% of the design space.

You can use the crosshairs tool to find the maximum relative prediction variance that corresponds to any Fraction of Space value. For example, use the crosshairs tool to see that for the 24-run design, 90% of the prediction variance values are below approximately 0.20.

Note: Plots for the same design might vary slightly, since Monte Carlo sampling of the design space is used in constructing the Fraction of Design Space Plot.

Relative Estimation Efficiency

The Relative Estimation Efficiency report compares designs in terms of the standard errors of parameter estimates for parameters in the assumed model. The standard errors control the length of confidence intervals for the parameter estimates. This report provides an efficiency ratio and the relative standard errors.

The relative estimation efficiency values are colored according to a color gradient shown under the table of relative estimation efficiency values. You can control the color legend using the options in the Power Analysis red triangle menu. See "Color Dashboard" on page 470.

Figure 16.26 Relative Estimation Efficiency Comparing Two Split-Plot Designs

Relative Estimation Efficiency

Term	Efficiency of 4 Whole Plots Relative to 8 Whole Plots
Intercept	0.778
X1	0.778
X2	1.004
X3 1	0.924
X3 2	1.154
X1*X2	0.982
X1*X3 1	0.924
X1*X3 2	1.154
X2*X3 1	1.107
X2*X3 2	1.449

Good 1.50 1.25 0.80 0.67 Bad

Relative Std Error of Estimates

Term	4 Whole Plots	8 Whole Plots
Intercept	0.553	0.43
X1	0.553	0.43
X2	0.23	0.231
X3 1	0.25	0.231
X3 2	0.22	0.254
X1*X2	0.24	0.235
X1*X3 1	0.25	0.231
X1*X3 2	0.22	0.254
X2*X3 1	0.256	0.283
X2*X3 2	0.218	0.315

Figure 16.26 shows the Relative Estimation Efficiency outline for the split-plot designs compared in "Split Plot Designs with Different Numbers of Whole Plots" on page 463.

Relative Estimation Efficiency

For a given term, the estimation efficiency of the reference design relative to a comparison design is the relative standard error of the term for the comparison design divided by the relative standard error of the term for the reference design. A value less than one indicates that the reference design is not as efficient as the comparison design. A value greater than one indicates that it is more efficient.

Relative Standard Error of Estimates

The Relative Std Error of Estimates report gives the ratio of the standard deviation of a parameter's estimate to the error standard deviation. These values indicate how large the standard errors of the model's parameter estimates are, relative to the error standard deviation. For the ith parameter estimate, the Relative Std Error of Estimate is defined as follows:

$$SE = \sqrt{(\mathbf{X'X})_{ii}^{-1}}$$

where:

$\mathbf{X}$ is the model matrix defined in "The Alias Matrix" on page 690 in the "Technical Details" appendix, and

$(\mathbf{X'X})_{ii}^{-1}$ is the ith diagonal entry of $(\mathbf{X'X})^{-1}$.

Alias Matrix Summary

The alias matrix addresses the issue of how terms that are not included in the model affect the estimation of the model terms, if they are indeed active. In the Alias Terms outline, you list potentially active effects that are not in your assumed model but that might bias the estimates of model terms. The alias matrix entries represent the degree of bias imparted to model parameters by the Alias Terms effects. See "Alias Terms" on page 469 and "Alias Matrix" on page 479.

The Alias Matrix Summary table lists the terms in the assumed model. These are the terms that correspond to effects listed in the Model outline. Given a design, for each entry in the Term column, the square root of the sum of the squared alias matrix entries for the terms corresponding to effects in the Alias Terms outline is computed. This value is reported in the Root Mean Squared Values column for the given design. For an example, see "Example of Calculation of Alias Matrix Summary Values" on page 479.

Note: The Alias Matrix Summary report appears only if there are effects in the Alias Terms list.

Figure 16.27 Alias Matrix Summary for Two Designs

Alias Matrix Summary

Term	Root Mean Squared Values Plackett-Burman	Root Mean Squared Values Definitive Screening Design
Intercept	0.0000	0.0000
X1	0.2722	0.0000
X2	0.2722	0.0000
X3	0.2722	0.0000
X4	0.2722	0.0000
X5	0.2722	0.0000
X6	0.2722	0.0000
Total	0.2520	0.0000

Good 0.10 0.30 0.50 0.70 Bad

Figure 16.27 shows the Alias Matrix Summary report for the Plackett-Burman and Definitive Screening designs constructed in "Designs of Same Run Size" on page 454, with only main effects in the Model outline. All two-factor interactions are in the Alias Terms list. The table shows that, for the Definitive Screening Design, main effects are uncorrelated with two-factor interactions.

The Root Mean Squares Values are colored according to a color gradient shown under the Alias Matrix Summary table. You can control the color legend using the options in the Alias Matrix Summary red triangle menu. See "Color Dashboard" on page 470.

Alias Matrix

The rows of the Alias Matrix are the terms corresponding to the model effects listed in the Model outline. The columns are terms corresponding to effects listed in the Alias Terms outline. The entry in a given row and column indicates the degree to which the alias term affects the parameter estimate corresponding to the model term.

In evaluating your design, you ideally want one of two situations to occur relative to any entry in the Alias Matrix. Either the entry is small or, if it is not small, the effect of the alias term is small so that the bias is small. If you suspect that the alias term might have a substantial effect, then that term should be included in the model or you should consider an alias-optimal design. In fact, alias-optimality is driven by the squared values of the alias matrix.

For additional background on the Alias Matrix, see "The Alias Matrix" on page 690 in the "Technical Details" appendix. See also Lekivetz, R. (2014).

Example of Calculation of Alias Matrix Summary Values

This example illustrates the calculation of the values that appear in the Alias Matrix Summary outline. In this example, you compare the two designs assuming that only main effects are active.

1. Select **Help > Sample Data**, click **Open the Sample Scripts Directory**, and select Compare Same Run Size.jsl.
2. Right-click in the script window and select **Run Script**.

Two 13-run design tables are constructed:

- Definitive Screening Design
- Plackett-Burman

You are interested only in the Plackett-Burman design. This is the active table.

3. From the Plackett-Burman table, select **DOE > Design Diagnostics > Evaluate Design**.
4. Select X1 through X6 and click **X, Factor**.
5. Click **OK**.
6. Open the Alias Terms outline to confirm that all two-factor interactions are in the Alias Terms list.
7. Open the Alias Matrix outline.

   For each model term listed in the Effect column, the entry in that row for a given column indicates the degree to which the alias term affects the parameter estimate corresponding to the model term.

For example, to obtain the Alias Matrix Summary entry in Figure 16.27 corresponding to X1, square the terms in the row for X1 in the Alias Matrix, average these, and take the square root. You obtain 0.2722.

Absolute Correlations

The Absolute Correlations report summarizes information about correlations between model terms and alias terms.

Figure 16.28 Absolute Correlations Report for Three Designs

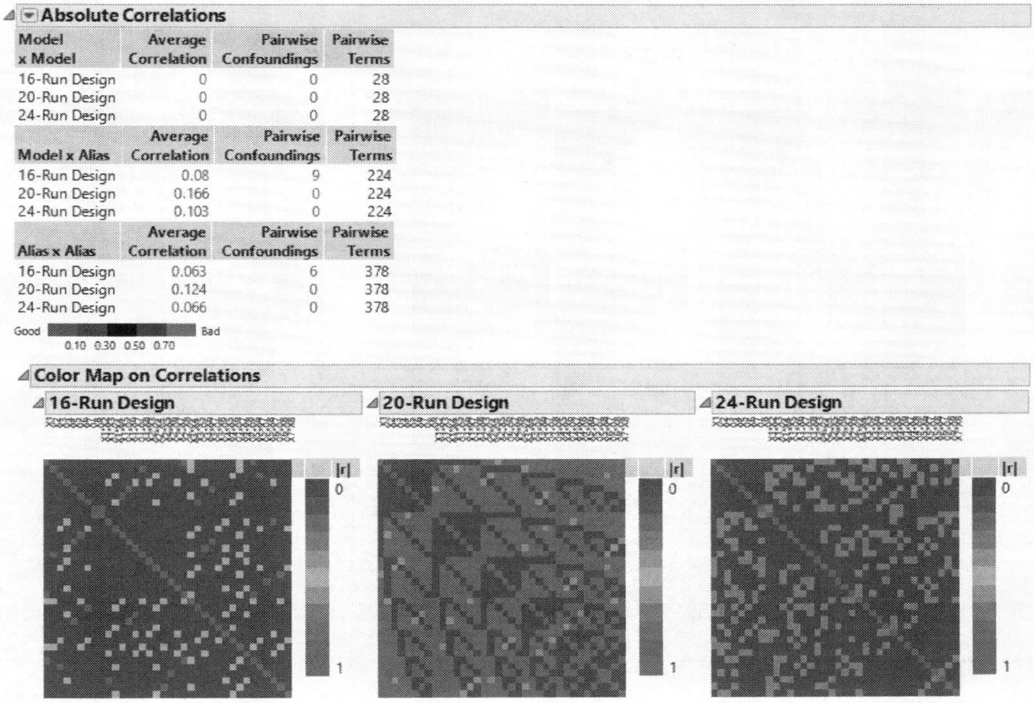

Absolute Correlations

Model x Model	Average Correlation	Pairwise Confoundings	Pairwise Terms
16-Run Design	0	0	28
20-Run Design	0	0	28
24-Run Design	0	0	28

Model x Alias	Average Correlation	Pairwise Confoundings	Pairwise Terms
16-Run Design	0.08	9	224
20-Run Design	0.166	0	224
24-Run Design	0.103	0	224

Alias x Alias	Average Correlation	Pairwise Confoundings	Pairwise Terms
16-Run Design	0.063	6	378
20-Run Design	0.124	0	378
24-Run Design	0.066	0	378

Good 0.10 0.30 0.50 0.70 Bad

Color Map on Correlations

16-Run Design |r| 0 1

20-Run Design |r| 0 1

24-Run Design |r| 0 1

Figure 16.28 shows the Absolute Correlations report for the three designs constructed in "Designs of Different Run Sizes" on page 459, with only main effects in the Model outline.

Absolute Correlations Table

The table in the Absolute Correlations report is divided into three sections:

- Model x Model considers correlations between terms corresponding to effects in the Model list.
- Model x Alias considers correlations between terms corresponding to effects in the Model list and terms corresponding to effects in the Alias list.
- Alias x Alias considers correlations between terms corresponding to effects in the Alias list.

Note: If there are no alias terms, only the Model x Model section appears.

For each section of the report, the following are given:

Average Correlation The average of the correlations for all pairs of terms considered in this section of the report.

Number of Confoundings The number of pairs of terms consisting of confounded terms.

Number of Terms The total number of pairs of terms considered in this section of the report.

The values in the Absolute Correlations table are colored according to a color gradient shown under the table. You can control the color legend using the options in the Absolute Correlations red triangle menu. See "Color Dashboard" on page 470.

Color Map on Correlations

The Color Map on Correlations outline shows plots for each of the designs. The cells of the color map are identified above the map. There are cells for all terms that correspond to effects that appear in either the Model outline or the Alias Terms outline. Each cell is colored according to the absolute value of the correlation between the two terms.

By default, the absolute magnitudes of the correlations are represented by a blue to gray to red intensity color theme. In general terms, the color map for a good design shows a lot of blue off the diagonal, indicating orthogonality or small correlations between distinct terms. Large absolute correlations among effects inflate the standard errors of estimates.

To see the absolute value of the correlation between two effects, place your pointer over the corresponding cell. To change the color theme for the entire plot, right-click in the plot and select Color Theme.

Absolute Correlations and Color Map on Correlations Example

Figure 16.28 shows the Absolute Correlations report for the Plackett-Burman and Definitive Screening designs constructed in "Designs of Different Run Sizes" on page 459. The Model outline contains only main effects, so the Alias Terms outline contains all two-factor interactions. All main effects and two-way interactions are shown in the color maps.

In the Color Map on Correlations for the 16-run design, the red cells off the main diagonal indicate that the corresponding terms have correlation one and therefore are completely confounded. There are nine instances where model terms (main effects) are confounded with alias terms (two factor interactions), and six instances where alias terms are confounded with each other. This is shown in the report under Pairwise Confoundings.

The color maps for the 20- and 24-run designs have no off-diagonal cells that are solid red. It follows that these designs show no instances of confounding between any pair of main or two-way interaction effects. However, it is interesting to note that the 20- and 24-run designs both have a higher Average Correlation for Model x Alias terms than does the 16-run design. Although the 16-run design shows confounding, the average amount of correlation is less than for the 20- and 24-run designs.

Design Diagnostics

The Design Diagnostics outline shows *D*-, *G*-, *A*, and *I*-efficiencies for the reference design relative to the comparison designs. It also shows the Additional Run Size. Given two designs, the one with the higher relative efficiency measure is better.

Figure 16.29 Design Diagnostics for Three Designs

Design Diagnostics

	Efficiency of 16-Run Design Relative to 20-Run Design	Efficiency of 16-Run Design Relative to 24-Run Design
D-efficiency	0.800	0.667
G-efficiency	0.894	0.816
A-efficiency	0.800	0.667
I-efficiency	0.800	0.667
Additional Run Size	-4	-8

Good 1.50 1.25 0.80 0.67 Bad

Figure 16.29 shows the Design Diagnostics report for the three designs constructed in "Designs of Different Run Sizes" on page 459, with only main effects in the Model outline.

The values in the Design Diagnostics table are colored according to a color gradient shown under the table. You can control the color legend using the options in the Design Diagnostics red triangle menu. See "Color Dashboard" on page 470.

Efficiency and Additional Run Size

Relative efficiencies for each of *D*-, *G*-, *A*-, and *I*-efficiency are shown in the Design Diagnostics report. These are obtained by computing each design's efficiency value and then taking the appropriate ratio. The descriptions of the relative efficiency measures are given in "Relative Efficiency Measures" on page 483.

Additional Run Size is the number of runs in the reference design minus the number of runs in the comparison design. If your reference design has more runs than your comparison design, then the Additional Run Size tells you how many additional runs you need to achieve the efficiency of the reference design.

Relative Efficiency Measures

Notation

- **X** is the model matrix
- p is the number of terms, including the intercept, in the model
- $Var(\hat{y}|\underset{\sim}{\mathbf{x}})$ is the relative prediction variance at the point $\underset{\sim}{\mathbf{x}}$. See "Relative Prediction Variance" on page 694 in the "Technical Details" appendix.

Relative Efficiencies

The relative efficiency of the reference design (*Ref*) to the comparison design (*Comp*) is given by the following expressions:

D Efficiency Eff_{Ref} / Eff_{Comp}, where *Eff* for each design is given as follows:

$$Eff = |\mathbf{X'X}|^{1/p}$$

G Efficiency Eff_{Comp} / Eff_{Ref}, where *Eff* for each design is given as follows:

$$Eff = Var(\hat{y}|\underset{\sim}{\mathbf{x}})_{max} = \underset{\underset{\sim}{\mathbf{x}} \text{ in D}}{maximum}[\underset{\sim}{\mathbf{x}}'(\mathbf{X'X})^{-1}\underset{\sim}{\mathbf{x}}]$$

Here, D denotes the design region.

Note: G-Efficiency is calculated using Monte Carlo sampling of the design space. The reported value is based on the larger of $Var(\hat{y}|\underset{\sim}{\mathbf{x}})_{max}$ or the prediction variance from the Monte Carlo sampling. Therefore, calculations for the same design might vary slightly.

A Efficiency Eff_{Comp} / Eff_{Ref}, where *Eff* for each design is given as follows:

$$Eff = \text{Trace}[(\mathbf{X'X})^{-1}]$$

I Efficiency Eff_{Comp} / Eff_{Ref}, where *Eff* for each design is given as follows:

$$Eff = \frac{\int \underset{\sim}{\mathbf{x}}'(\mathbf{X'X})^{-1}\underset{\sim}{\mathbf{x}} d\mathbf{x}}{\int d\mathbf{x}}$$

For details of the calculation, see Section 4.3.5 in Goos and Jones, 2011.

Compare Designs Options

Advanced Options > Split Plot Variance Ratio Specify the ratio of the variance of the random whole plot and the subplot variance (if present) to the error variance. Before setting this value, you must define a hard-to-change factor for your split-plot design, or hard and very-hard-to-change factors for your split-split-plot design. Then you can enter one or two positive numbers for the variance ratios, depending on whether you have specified a split-plot or a split-split-plot design.

Advanced Options > Set Delta for Power Specify the difference in the mean response that you want to detect for model effects. See “Set Delta for Power” on page 112 in the “Custom Designs” chapter.

Chapter 17

Prospective Sample Size and Power

Use the **DOE > Design Diagnostics > Sample Size and Power** command to answer the question "How many runs do I need?" The important quantities are sample size, power, and the magnitude of the effect. These depend on the significance level, alpha, of the hypothesis test for the effect and the standard deviation of the noise in the response. You can supply either one or two of the three values. If you supply only one of these values, the result is a plot of the other two. If you supply two values, the third value is computed.

The **Sample Size and Power** platform can answer the question, "Will I detect the group differences I am looking for, given my proposed sample size, estimate of within-group variance, and alpha level?" In this type of analysis, you must approximate the group means and sample sizes in a data table as well as approximate the within-group standard deviation (σ).

The sample size and power computations determine the sample size necessary for yielding a significant result, given that the true effect size is at least a certain size. It requires that you enter two out of three possible quantities; difference to detect, sample size, and power. The third quantity is computed for the following cases:

- difference between a one sample mean and a hypothesized value
- difference between two sample means
- differences in the means among *k* samples
- difference between a standard deviation and a hypothesized value
- difference between a one sample proportion and a hypothesized value
- difference between two sample proportions
- difference between counts per unit in a Poisson-distributed sample and a hypothesized value.

The calculations assume that there are equal numbers of units in each group. You can apply this platform to more general experimental designs, if they are balanced and an adjustment for the number-of-parameters is specified.

You can also compute the required sample sizes needed for reliability studies and demonstrations.

Launching the Sample Size and Power Platform

The **Sample Size and Power** platform helps you plan your study for a single mean or proportion comparison, a two sample mean or proportion comparison, a one-sample standard deviation comparison, a k sample means comparison, or a counts per unit comparison.

Tip: Formulas for the calculations of many of the sample sizes can be found at http://www.jmp.com/en_us/whitepapers/jmp/power-sample-calculations.html.

Depending upon your experimental situation, you supply one or two quantities to obtain a third quantity. These quantities include:

- required sample size
- expected power
- expected effect size

When you select **DOE > Design Diagnostics > Sample Size and Power**, the panel in Figure 17.1 appears with button selections for experimental situations. The following sections describe each of these selections and explain how to enter the quantities and obtain the desired computation.

Figure 17.1 Sample Size and Power Choices

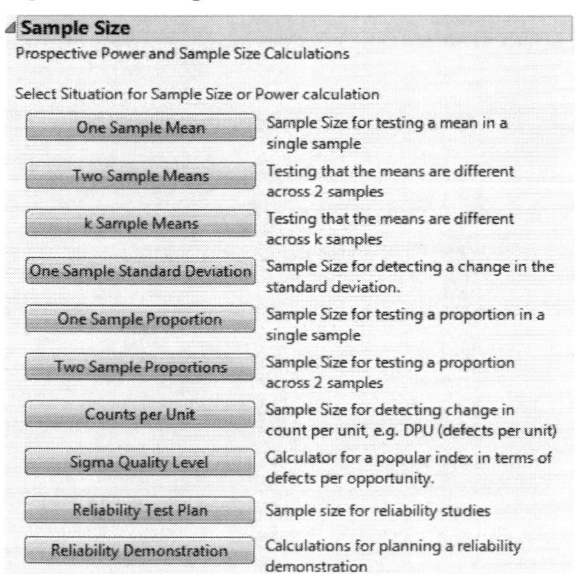

One-Sample and Two-Sample Means

After you click either **One Sample Mean**, or **Two Sample Means** in the initial Sample Size selection list (Figure 17.1), a Sample Size and Power window appears. (See Figure 17.2.)

Figure 17.2 Initial Sample Size and Power Windows for Single Mean (left) and Two Means (right)

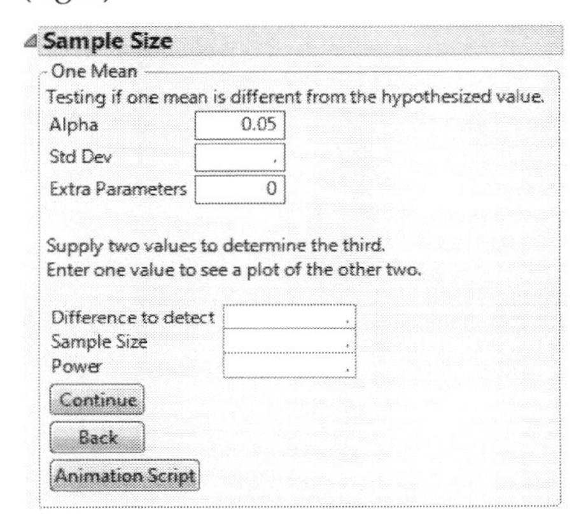

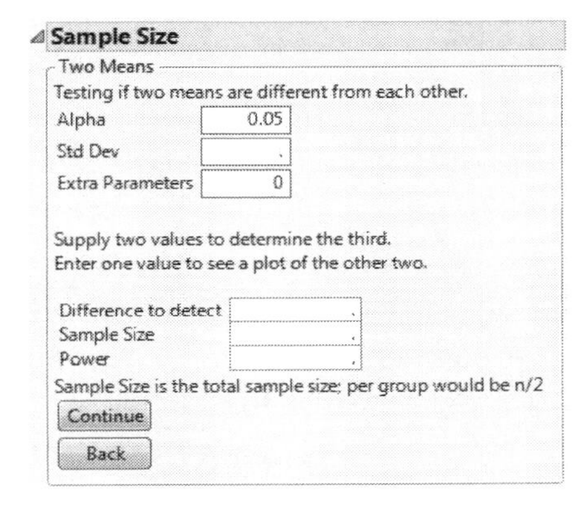

The windows are the same except that the One Mean window has a button at the bottom that accesses an animation script.

The initial Sample Size and Power window requires values for **Alpha, Std Dev** (the error standard deviation), and one or two of the other three values: **Difference to detect, Sample Size**, and **Power**. The Sample Size and Power platform calculates the missing item. If there are two unspecified fields, a plot is constructed, showing the relationship between these two values:

- power as a function of sample size, given specific effect size
- power as a function of effect size, given a sample size
- effect size as a function of sample size, for a given power.

The Sample Size and Power window asks for these values:

Alpha is the probability of a type I error, which is the probability of rejecting the null hypothesis when it is true. It is commonly referred to as the significance level of the test. The default alpha level is 0.05. This implies a willingness to accept (if the true difference between groups is zero) that, 5% (alpha) of the time, a significant difference is incorrectly declared.

Std Dev is the error standard deviation. It is a measure of the unexplained random variation around the mean. Even though the true error is not known, the power calculations are an exercise in probability that calculates what might happen if the true value is the one you specify. An estimate of the error standard deviation could be the root mean square error (RMSE) from a previous model fit.

Extra Parameters is only for multi-factor designs. Leave this field zero in simple cases. In a multi-factor balanced design, in addition to fitting the means described in the situation, there are other factors with extra parameters that can be specified here. For example, in a three-factor two-level design with all three two-factor interactions, the number of extra parameters is five. (This includes two parameters for the extra main effects, and three parameters for the interactions.) In practice, the particular values entered are not that important, unless the experimental range has very few degrees of freedom for error.

Difference to Detect is the smallest detectable difference (how small a difference you want to be able to declare statistically significant) to test against. For single sample problems this is the difference between the hypothesized value and the true value.

Sample Size is the total number of observations (runs, experimental units, or samples) in your experiment. Sample size is not the number per group, but the total over all groups.

Power is the probability of rejecting the null hypothesis when it is false. A large power value is better, but the cost is a higher sample size.

Continue evaluates at the entered values.

Back returns to the previous Sample Size and Power window so that you can either redo an analysis or start a new analysis.

Animation Script runs a JSL script that displays an interactive plot showing power or sample size. See the section, "Sample Size and Power Animation for One Mean" on page 491, for an illustration of the animation script.

Single-Sample Mean

Using the Sample Size and Power window, you can test if one mean is different from the hypothesized value.

For the one sample mean, the hypothesis supported is

$$H_0\colon \mu = \mu_0$$

and the two-sided alternative is

$$H_a\colon \mu \neq \mu_0$$

where μ is the population mean and μ_0 is the null mean to test against or is the difference to detect. It is assumed that the population of interest is normally distributed and the true mean is zero. Note that the power for this setting is the same as for the power when the null hypothesis is H_0: $\mu=0$ and the true mean is μ_0.

Suppose you are interested in testing the flammability of a new fabric being developed by your company. Previous testing indicates that the standard deviation for burn times of this fabric is 2 seconds. The goal is to detect a difference of 1.5 seconds when alpha is equal to 0.05, the sample size is 20, and the standard deviation is 2 seconds. For this example, μ_0 is equal to 1.5. To calculate the power:

1. Select **DOE > Design Diagnostics > Sample Size and Power**.
2. Click the **One Sample Mean** button in the Sample Size and Power Window.
3. Leave **Alpha** as 0.05.
4. Leave **Extra Parameters** as 0.
5. Enter 2 for **Std Dev**.
6. Enter 1.5 as **Difference to detect.**
7. Enter 20 for **Sample Size**.
8. Leave **Power** blank. (See the left window in Figure 17.3.)
9. Click **Continue**.

   The power is calculated as 0.8888478174 and is rounded to 0.89. (See right window in Figure 17.3.) The conclusion is that your experiment has an 89% chance of detecting a significant difference in the burn time, given that your significance level is 0.05, the difference to detect is 1.5 seconds, and the sample size is 20.

Figure 17.3 A One-Sample Example

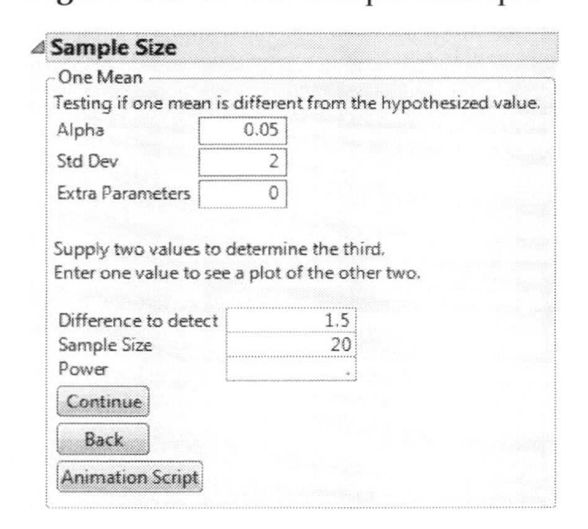

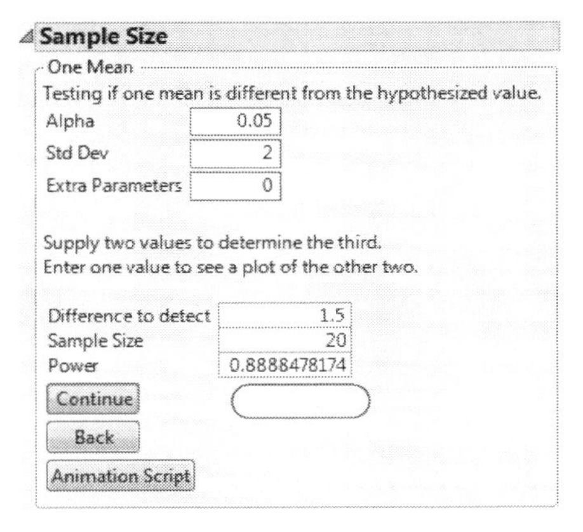

Power versus Sample Size Plot

To see a plot of the relationship of **Sample Size** and **Power**, leave both **Sample Size** and **Power** empty in the window and click **Continue**.

The plots in Figure 17.4, show a range of sample sizes for which the power varies from about 0.1 to about 0.95. The plot on the right in Figure 17.4 shows using the crosshair tool to illustrate the example in Figure 17.3.

Figure 17.4 A One-Sample Example Plot

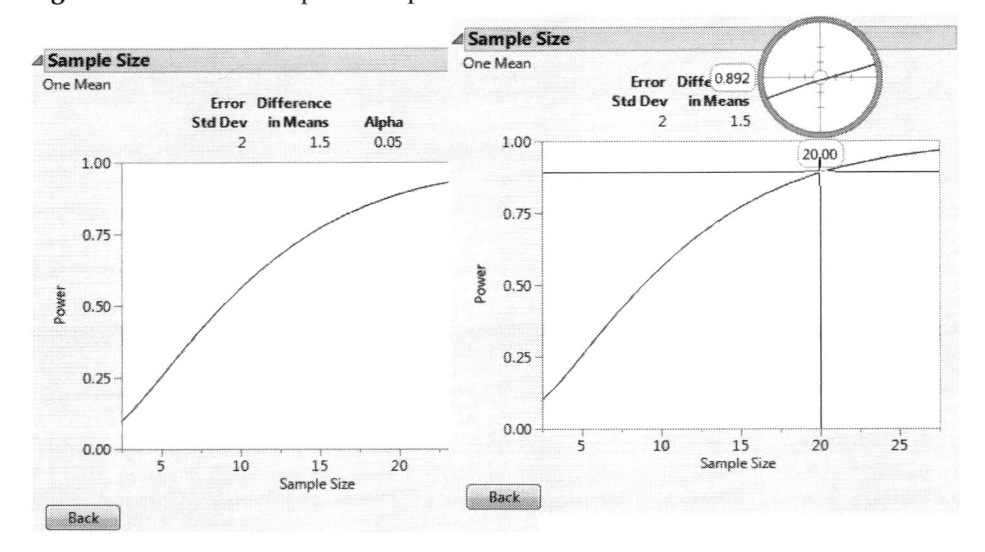

Power versus Difference Plot

When only **Sample Size** is specified (Figure 17.5) and **Difference to Detect** and **Power** are empty, a plot of Power by Difference appears, after clicking **Continue**.

Figure 17.5 Plot of Power by Difference to Detect for a Given Sample Size

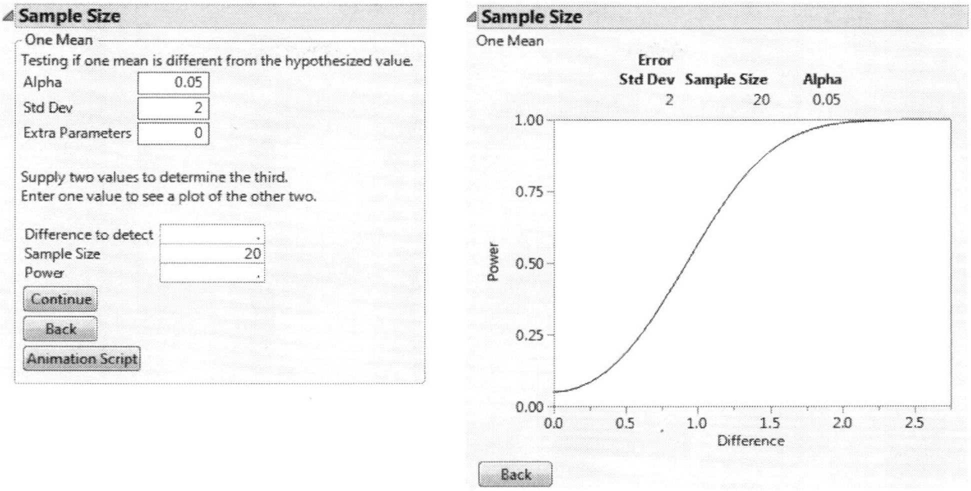

Sample Size and Power Animation for One Mean

Clicking the **Animation Script** button on the Sample Size and Power window for one mean shows an interactive plot. This plot illustrates the effect that changing the sample size has on power. As an example of using the **Animation Script**:

1. Select **DOE > Design Diagnostics > Sample Size and Power**.
2. Click the **One Sample Mean** button in the Sample Size and Power Window.
3. Enter 2 for **Std Dev**.
4. Enter 1.5 as **Difference to detect**.
5. Enter 20 for **Sample Size**.
6. Leave **Power** blank.

   The Sample Size and Power window appears as shown on the left of Figure 17.6.

7. Click **Animation Script**.

   The initial animation plot shows two t-density curves. The red curve shows the t-distribution when the true mean is zero. The blue curve shows the t-distribution when the true mean is 1.5, which is the difference to be detected. The probability of committing a type II error (not detecting a difference when there is a difference) is shaded blue on this plot. (This probability is often represented as β in the literature.) Similarly, the probability of committing a type I error (deciding that the difference to detect is significant when there is no difference) is shaded as the red areas under the red curve. (The red-shaded areas under the curve are represented as α in the literature.)

Select and drag the square handles to see the changes in statistics based on the positions of the curves. To change the values of **Sample Size** and **Alpha**, click on their values beneath the plot.

Figure 17.6 Example of Animation Script to Illustrate Power

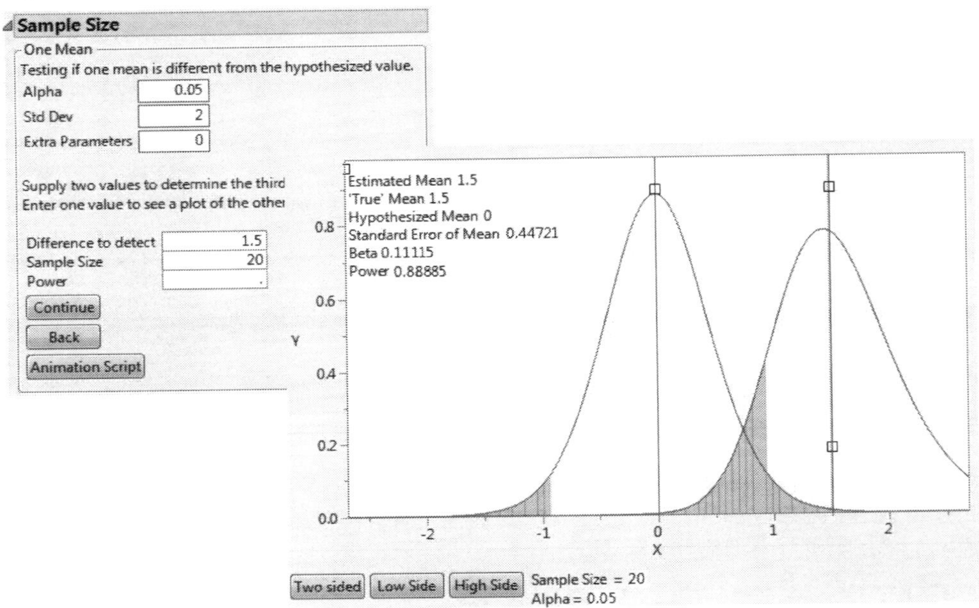

Two-Sample Means

The Sample Size and Power windows work similarly for one and two sample means; the **Difference to Detect** is the difference between two means. The comparison is between two random samples instead of one sample and a hypothesized mean.

For testing the difference between two means, the hypothesis supported is

$$H_0\colon \mu_1 - \mu_2 = D_0$$

and the two-sided alternative is

$$H_a\colon \mu_1 - \mu_2 \neq D_0$$

where μ_1 and μ_2 are the two population means and D_0 is the difference in the two means or the difference to detect. It is assumed that the populations of interest are normally distributed and the true difference is zero. Note that the power for this setting is the same as for the power when the null hypothesis is $H_0\colon \mu_1 - \mu_2 = 0$ and the true difference is D_0.

Suppose the standard deviation is 2 (as before) for both groups, the desired detectable difference between the two means is 1.5, and the sample size is 30 (15 per group). To estimate the power for this example:

1. Select **DOE > Design Diagnostics > Sample Size and Power**.
2. Click the **Two Sample Means** button in the Sample Size and Power Window.
3. Leave **Alpha** as 0.05.
4. Enter 2 for **Std Dev**.
5. Leave **Extra Parameters** as 0.
6. Enter 1.5 as **Difference to detect**.
7. Enter 30 for **Sample Size**.
8. Leave **Power** blank.
9. Click **Continue**.

   The **Power** is calculated as 0.51. (See the left window in Figure 17.7.) This means that you have a 51% chance of detecting a significant difference between the two sample means when your significance level is 0.05, the difference to detect is 1.5, and each sample size is 15.

Plot of Power by Sample Size

To have a greater power requires a larger sample. To find out how large, leave both **Sample Size** and **Power** blank for this same example and click **Continue**. Figure 17.7 shows the resulting plot, with the crosshair tool estimating that a sample size of about 78 is needed to obtain a power of 0.9.

Figure 17.7 Plot of Power by Sample Size to Detect for a Given Difference

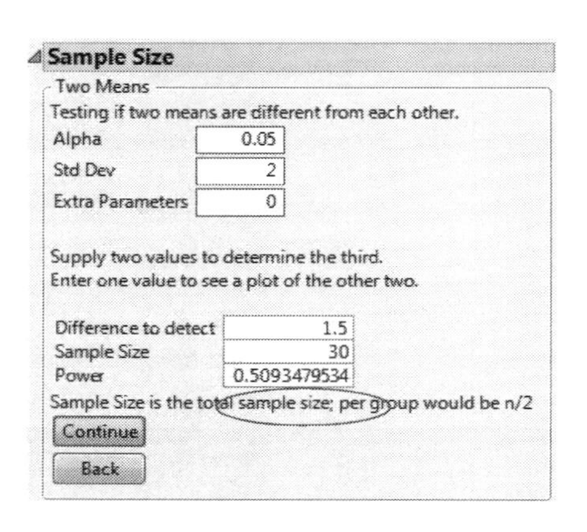

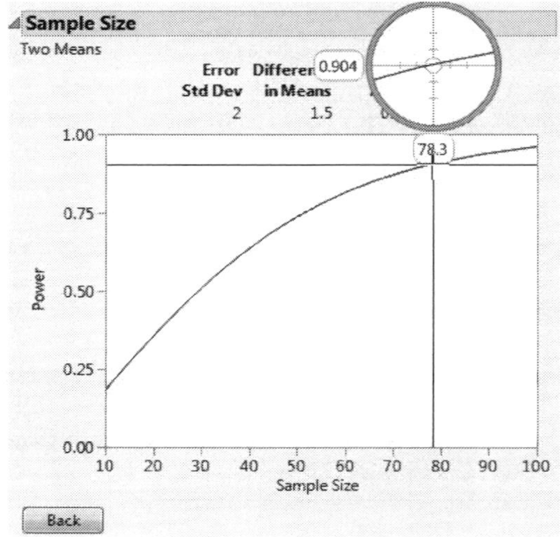

k-Sample Means

Using the **k-Sample Means** option, you can compare up to 10 means. Consider a situation where 4 levels of means are expected to be in the range of 10 to 13, the standard deviation is 0.9, and your sample size is 16.

The hypothesis to be tested is:

H_0: $\mu_1=\mu_2=\mu_3=\mu_4$ versus H_a: at least one mean is different

To determine the power:

1. Select **DOE > Design Diagnostics > Sample Size and Power.**
2. Click the **k Sample Means** button in the Sample Size and Power Window.
3. Leave **Alpha** as 0.05.
4. Enter 0.9 for **Std Dev.**
5. Leave **Extra Parameters** as 0.
6. Enter 10, 11, 12, and 13 as the four levels of means.
7. Enter 16 for **Sample Size.**
8. Leave **Power** blank.
9. Click **Continue.**

The **Power** is calculated as 0.95. (See the left of Figure 17.8.) This means that there is a 95% chance of detecting that at least one of the means is different when the significance level is 0.05, the population means are 10, 11, 12, and 13, and the total sample size is 16.

If both **Sample Size** and **Power** are left blank for this example, the sample size and power calculations produce the **Power** versus **Sample Size** curve. (See the right of Figure 17.8.) This confirms that a sample size of 16 looks acceptable.

Notice that the difference in means is 2.236, calculated as square root of the sum of squared deviations from the grand mean. In this case it is the square root of $(-1.5)^2 + (-0.5)^2 + (0.5)^2 + (1.5)^2$, which is the square root of 5.

Figure 17.8 Prospective Power for k-Means and Plot of Power by Sample Size

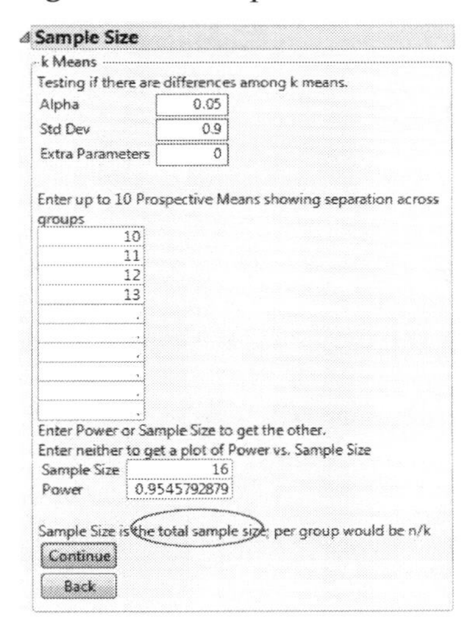

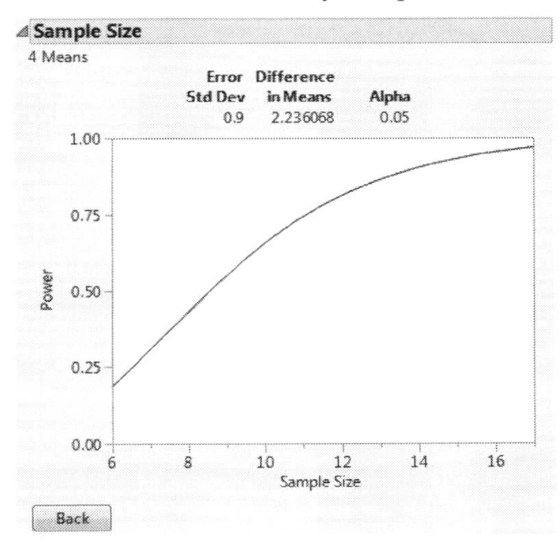

One Sample Standard Deviation

Use the **One-Sample Standard Deviation** option on the Sample Size and Power window (Figure 17.1) to determine the sample size needed for detecting a change in the standard deviation of your data. The usual purpose of this option is to compute a large enough sample size to guarantee that the risk of a type II error, β, is small. (This is the probability of failing to reject the null hypothesis when it is false.)

In the Sample Size and Power window, specify:

Alpha is the significance level, usually 0.05. This implies a willingness to accept (if the true difference between standard deviation and the hypothesized standard deviation is zero) that a significant difference is incorrectly declared 5% of the time.

Hypothesized Standard Deviation is the hypothesized or baseline standard deviation to which the sample standard deviation is compared.

Alternative Standard Deviation can select **Larger** or **Smaller** from the menu to indicate the direction of the change you want to detect.

Difference to Detect is the smallest detectable difference (how small a difference you want to be able to declare statistically significant). For single sample problems this is the difference between the hypothesized value and the true value.

Sample Size is how many experimental units (runs, or samples) are involved in the experiment.

Power is the probability of declaring a significant result. It is the probability of rejecting the null hypothesis when it is false.

In the lower part of the window you enter two of the items and the Sample Size and Power calculation determines the third.

Some examples in this chapter use engineering examples from the online manual of The National Institute of Standards and Technology (NIST). You can access the NIST manual examples at http://www.itl.nist.gov/div898/handbook.

One Sample Standard Deviation Example

One example from the NIST manual states a problem in terms of the variance and difference to detect. The variance for resistivity measurements on a lot of silicon wafers is claimed to be 100 ohm-cm squared. The buyer is unwilling to accept a shipment if the variance is greater than 155 ohm-cm squared for a particular lot (55 ohm-cm squared above the baseline of 100 ohm-cm squared).

In the Sample Size and Power window, the One Sample Standard Deviation computations use the standard deviation instead of the variance. The hypothesis to be tested is:

H_0: $\sigma = \sigma_0$, where σ_0 is the hypothesized standard deviation. The true standard deviation is σ_0 plus the difference to detect.

In this example the hypothesized standard deviation, σ_0, is 10 (the square root of 100) and σ is 12.4499 (the square root of 100 + 55 = 155). The difference to detect is 12.4499 – 10 = 2.4499.

You want to detect an increase in the standard deviation of 2.4499 for a standard deviation of 10, with an alpha of 0.05 and a power of 0.99. To determine the necessary sample size:

1. Select **DOE > Design Diagnostics > Sample Size and Power**.
2. Click the **One Sample Standard Deviation** button in the Sample Size and Power Window.
3. Leave **Alpha** as 0.05.
4. Enter 10 for **Hypothesized Standard Deviation**.
5. Select Larger for **Alternate Standard Deviation**.

6. Enter 2.4499 as **Difference to Detect**.
7. Enter 0.99 for **Power**.
8. Leave **Sample Size** blank. (See the left of Figure 17.9.)
9. Click **Continue**.

   The **Sample Size** is calculated as 171. (See the right of Figure 17.9.) This result is the sample size rounded up to the next whole number.

Note: Sometimes you want to detect a change to a smaller standard deviation. If you select **Smaller** from the **Alternative Standard Deviation** menu, enter a negative amount in the **Difference to Detect** field.

Figure 17.9 Window To Compare Single-Direction One-Sample Standard Deviation

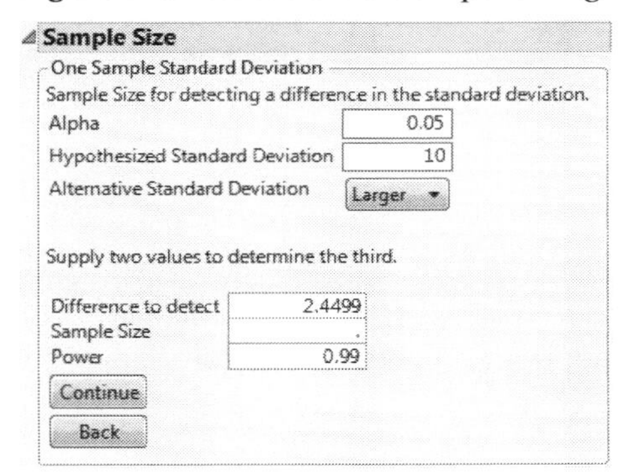

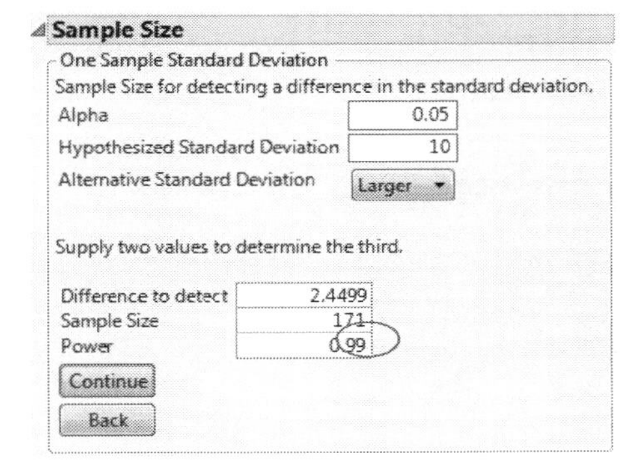

One-Sample and Two-Sample Proportions

The Sample Size windows and computations to test sample sizes and power for proportions are similar to those for testing means. You enter a true **Proportion** and choose an **Alpha** level. Then, for the one-sample proportion case, enter the **Sample Size** and **Null Proportion** to obtain the **Power.** Or, enter the **Power** and **Null Proportion** to obtain the **Sample Size**. Similarly, to obtain a value for **Null Proportion**, enter values for **Sample Size** and **Power**. For the two-sample proportion case, either the two sample sizes or the desired **Power** must be entered. (See Figure 17.10 and Figure 17.11.)

The computations to determine sample size and null proportions use exact methods based on the binomial distribution. Such methods are more reliable than using the normal approximation to the binomial, which can provide erroneous results for small samples or proportions close to 0 or 1.

For a single proportion, results are based on exact power calculations used in conjunction with one of the following:

- The "Add two successes and two failures" adjusted Wald test statistic described in Agresti and Coull (1998). See also Section 3.3 in the white paper by Barker on the JMP website: http://www.jmp.com/en_us/whitepapers/jmp/power-sample-calculations.html
- The Clopper-Pearson exact confidence interval method (Clopper and Pearson, 1934; Agresti and Coull, 1998, Section 1)

For two proportions, results are based on exact power calculations for the adjusted Wald statistic proposed by Agresti and Caffo (2000).

Actual Test Size

The results also show the Actual Test Size. This is the actual Type I error rate for a given situation. Since the binomial distribution is discrete, the actual test size can differ significantly from the stated Alpha level for small samples or proportions near 0 or 1.

One Sample Proportion

Clicking the **One Sample Proportion** option on the Sample Size and Power window yields a One Proportion window. In this window, you can specify the alpha level and the true proportion. The sample size, power, or the hypothesized proportion is calculated. If you supply two of these quantities, the third is computed, or if you enter any one of the quantities, you see a plot of the other two.

For example, if you have a hypothesized proportion of defects, you can use the **One Sample Proportion** window to estimate a large enough sample size to guarantee that the risk of accepting a false hypothesis (β) is small. That is, you want to detect, with reasonable certainty, a difference in the proportion of defects.

For the one sample proportion, the hypothesis supported is

$$H_0: p = p_0$$

and the two-sided alternative is

$$H_a: p \neq p_0$$

where p is the population proportion and p_0 is the null proportion to test against. Note that if you are interested in testing whether the population proportion is greater than or less than the null proportion, you use a one-sided test. The one-sided alternative is either

$$H_a: p < p_0$$

or

$$H_a: p > p_0$$

One-Sample Proportion Window Specifications

In the top portion of the Sample Size window, you can specify or enter values for:

Alpha is the significance level of your test. The default value is 0.05.

Proportion is the true proportion, which could be known or hypothesized. The default value is 0.1.

Method is the method to determine the exact confidence interval. Choices are Exact Agresti-Coull or Exact Clopper-Pearson. For more details, see "One-Sample and Two-Sample Proportions" on page 497.

One-Sided or Two-Sided Specify either a one-sided or a two-sided test. The default setting is the two-sided test.

In the bottom portion of the window, enter two of the following quantities to see the third, or a single quantity to see a plot of the other two.

Null Proportion is the proportion to test against (p_0) or is left blank for computation. The default value is 0.2.

Sample Size is the sample size, or is left blank for computation. If **Sample Size** is left blank, then values for **Proportion** and **Null Proportion** must be different.

Power is the desired power, or is left blank for computation.

One-Sample Proportion Example

As an example, suppose that an assembly line has a historical proportion of defects equal to 0.1, and you want to know the power to detect that the proportion is different from 0.2, given an alpha level of 0.05 and a sample size of 100.

1. Select **DOE** > **Design Diagnostics > Sample Size and Power**.
2. Click **One Sample Proportion**.
3. Leave **Alpha** as 0.05.
4. Leave 0.1 as the value for **Proportion**.
5. Leave the **Method** as Exact Agresti-Coull.
6. Accept the default option of **Two-Sided**. (A one-sided test is selected if you are interested in testing if the proportion is either greater than or less than the **Null Proportion**.)
7. Leave 0.2 as the value for **Null Proportion**.
8. Enter 100 as the **Sample Size**.

9. Click **Continue.**

The **Power** is calculated and is shown as approximately 0.7 (see Figure 17.10). Note the Actual Test Size is 0.0467, which is slightly less than the desired 0.05.

Figure 17.10 Power and Sample Window for One-Sample Proportions

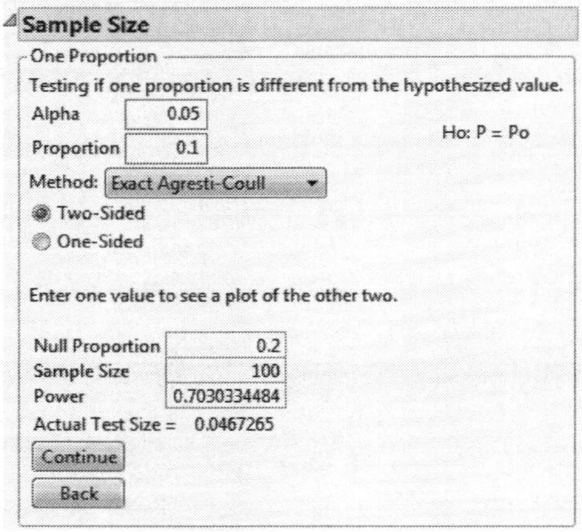

Two Sample Proportions

The **Two Sample Proportions** option computes the power or sample sizes needed to detect the difference between two proportions, p_1 and p_2.

For the two sample proportion, the hypothesis supported is

$$H_0: p_1 - p_2 = D_0$$

and the two-sided alternative is

$$H_a: p_1 - p_2 \neq D_0$$

where *p1* and *p2* are the population proportions from two populations, and D_0 is the hypothesized difference in proportions.

The one-sided alternative is either

$$H_a: (p_1 - p_2) < D_0$$

or

$$H_a: (p_1 - p_2) > D_0$$

Two Sample Proportion Window Specifications

Specifications for the Two Sample Proportions window include:

Alpha is the significance level of your test. The default value is 0.05.

Proportion 1 is the proportion for population 1, which could be known or hypothesized. The default value is 0.5.

Proportion 2 is the proportion for population 2, which could be known or hypothesized. The default value is 0.1.

One-Sided or Two-Sided Specify either a one-sided or a two-sided test. The default setting is the two-sided test.

Null Difference in Proportion is the proportion difference (D_0) to test against, or is left blank for computation. The default value is 0.2.

Sample Size 1 is the sample size for population 1, or is left blank for computation.

Sample Size 2 is the sample size for population 2, or is left blank for computation.

Power is the desired power, or is left blank for computation.

If you enter any two of the following three quantities, the third quantity is computed:

- **Null Difference in Proportion**
- **Sample Size 1** and **Sample Size 2**
- **Power**

Example of Determining Sample Sizes with a Two-Sided Test

As an example, suppose you are responsible for two silicon wafer assembly lines. Based on the knowledge from many runs, one of the assembly lines has a defect rate of 8%; the other line has a defect rate of 6%. You want to know the sample size necessary to have 80% power to reject the null hypothesis of equal proportions of defects for each line.

To estimate the necessary sample sizes for this example:

1. Select **DOE** > **Design Diagnostics > Sample Size and Power.**
2. Click **Two Sample Proportions.**
3. Accept the default value of **Alpha** as 0.05.
4. Enter 0.08 for **Proportion 1**.
5. Enter 0.06 for **Proportion 2.**
6. Accept the default option of **Two-Sided.**
7. Enter 0.0 for **Null Difference in Proportion.**
8. Enter 0.8 for **Power.**

9. Leave **Sample Size 1** and **Sample Size 2** blank.
10. Click **Continue.**

The Sample Size window shows sample sizes of 2554. (see Figure 17.11.) Testing for a one-sided test is conducted similarly. Simply select the **One-Sided** option.

Figure 17.11 Difference Between Two Proportions for a Two-Sided Test

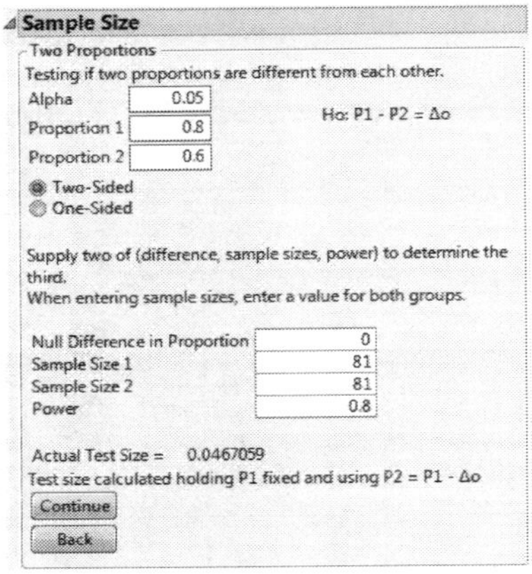

Example of Determining Power with Two Sample Proportions Using a One-Sided Test

Suppose you want to compare the effectiveness of a two chemical additives. The standard additive is known to be 50% effective in preventing cracking in the final product. The new additive is assumed to be 60% effective. You plan on conducting a study, randomly assigning parts to the two groups. You have 800 parts available to participate in the study (400 parts for each additive). Your objective is to determine the power of your test, given a null difference in proportions of 0.01 and an alpha level of 0.05. Because you are interested in testing that the difference in proportions is greater than 0.01, you use a one-sided test.

1. Select **DOE** > **Design Diagnostics** > **Sample Size and Power.**
2. Click **Two Sample Proportions.**
3. Accept the default value of **Alpha** as 0.05.
4. Enter 0.6 for **Proportion 1**.
5. Enter 0.5 for **Proportion 2.**
6. Select **One-Sided.**
7. Enter 0.01 as the **Null Difference in Proportion.**
8. Enter 400 for **Sample Size 1**.
9. Enter 400 for **Sample Size 2.**
10. Leave **Power** blank.

11. Click **Continue**.

Figure 17.12 shows the Two Proportions windows with the estimated **Power** calculation of 0.82.

Figure 17.12 Difference Between Two Proportions for a One-Sided Test

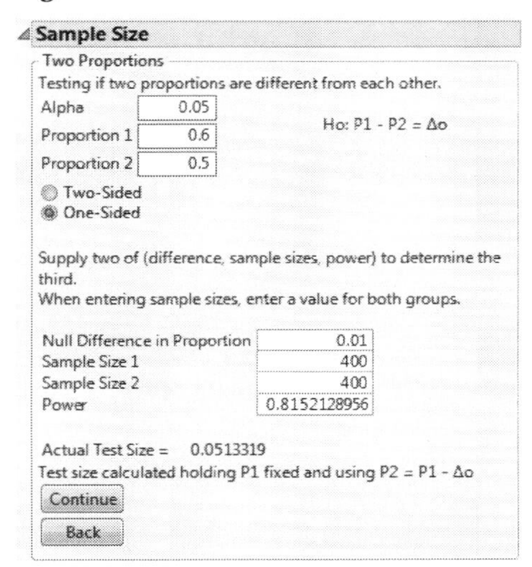

You conclude that there is about an 82% chance of rejecting the null hypothesis at the 0.05 level of significance, given that the sample sizes for the two groups are each 400. Note the Actual Test Size is 0.0513, which is slightly larger than the stated 0.05.

Counts per Unit

You can use the **Counts per Unit** option from the Sample Size and Power window (Figure 17.1) to calculate the sample size needed when you measure more than one defect per unit. A unit can be an area and the counts can be fractions or large numbers.

Although the number of defects observed in an area of a given size is often assumed to have a Poisson distribution, it is understood that the area and count are large enough to support a normal approximation.

Consider the following questions:

- Is the defect density within prescribed limits?
- Is the defect density greater than or less than a prescribed limit?

In the Counts per Unit window, options include:

Alpha is the significance level of your test. The default value is 0.05.

Baseline Count per Unit is the number of targeted defects per unit. The default value is 0.1.

Difference to detect is the smallest detectable difference to test against and is specified in defects per unit, or is left blank for computation.

Sample Size is the sample size, or is left blank for computation.

Power is the desired power, or is left blank for computation.

In the Counts per Unit window, enter **Alpha** and the **Baseline Count per Unit**. Then enter two of the remaining fields to see the calculation of the third. The test is for a one-sided (one-tailed) change. Enter the **Difference to Detect** in terms of the **Baseline Count per Unit** (defects per unit). The computed sample size is expressed as the number of units, rounded to the next whole number.

Counts per Unit Example

As an example, consider a wafer manufacturing process with a target of 4 defects per wafer. You want to verify that a new process meets that target within a difference of 1 defect per wafer with a significance level of 0.05. In the Counts per Unit window:

1. Leave **Alpha** as 0.05 (the chance of failing the test if the new process is as good as the target).
2. Enter 4 as the **Baseline Counts per Unit**, indicating the target of 4 defects per wafer.
3. Enter 1 as the **Difference to detect**.
4. Enter a power of 0.9, which is the chance of detecting a change larger than 1 (5 defects per wafer). In this type of situation, alpha is sometimes called the *producer's risk* and beta is called the *consumer's risk*.
5. Click **Continue** to see the results in Figure 17.13, showing a computed sample size of 38 (rounded to the next whole number).

The process meets the target if there are less than 190 defects (5 defects per wafer in a sample of 38 wafers).

Figure 17.13 Window For Counts Per Unit Example

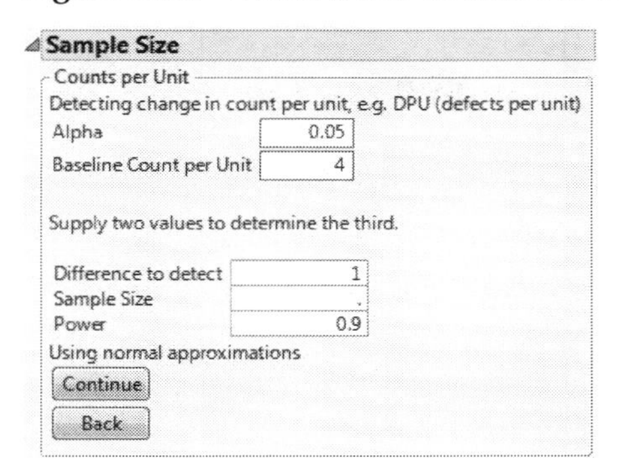

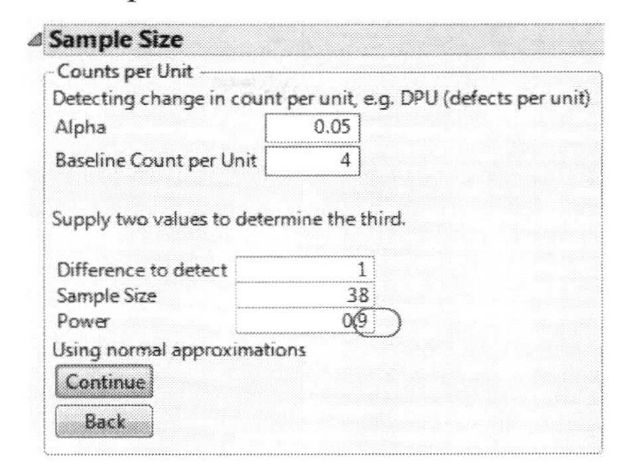

Sigma Quality Level

The Sigma Quality Level feature is a simple statistic that puts a given defect rate on a "six-sigma" scale. For example, on a scale of one million opportunities, 3.397 defects result in a six-sigma process. The computation that gives the Sigma Quality Level statistic is

```
Sigma Quality Level = NormalQuantile(1 - defects/opportunities) + 1.5
```

Two of three quantities can be entered to determine the Sigma Quality Level statistic in the Sample Size and Power window:

- Number of Defects
- Number of Opportunities
- Sigma Quality Level

When you click **Continue**, the sigma quality calculator computes the missing quantity.

Sigma Quality Level Example

As an example, use the Sample Size and Power feature to compute the Sigma Quality Level for 50 defects in 1,000,000 opportunities:

1. Select **DOE > Design Diagnostics > Sample Size and Power**.
2. Click the **Sigma Quality Level** button.
3. Enter 50 for the **Number of Defects.**
4. Enter 1000000 as the **Number of Opportunities**. (See window to the left in Figure 17.14.)
5. Click **Continue**.

   The results are a Sigma Quality Level of 5.39. (See right window in Figure 17.14.)

Figure 17.14 Sigma Quality Level Example 1

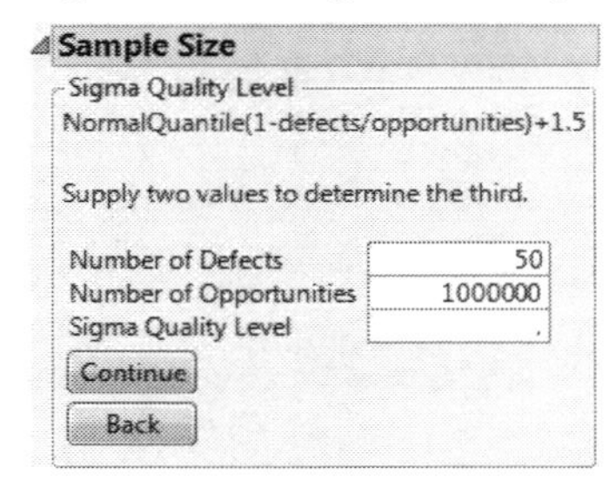

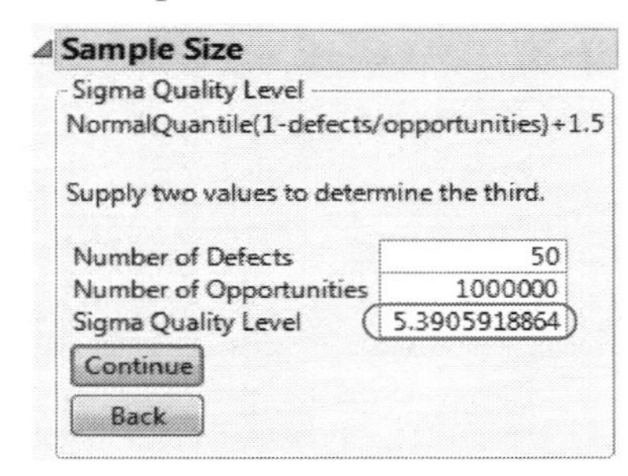

Number of Defects Computation Example

If you want to know how many defects reduce the Sigma Quality Level to "six-sigma" for 1,000,000 opportunities:

1. Select **DOE > Design Diagnostics > Sample Size and Power**.
2. Click the **Sigma Quality Level** button.
3. Enter 6 as **Sigma Quality Level.**
4. Enter 1000000 as the **Number of Opportunities**. (See left window in Figure 17.14.)
5. Leave **Number of Defects** blank.
6. Click **Continue.**

   The computation shows that the **Number of Defects** cannot be more than approximately 3.4. (See right window in Figure 17.15.)

Figure 17.15 Sigma Quality Level Example 2

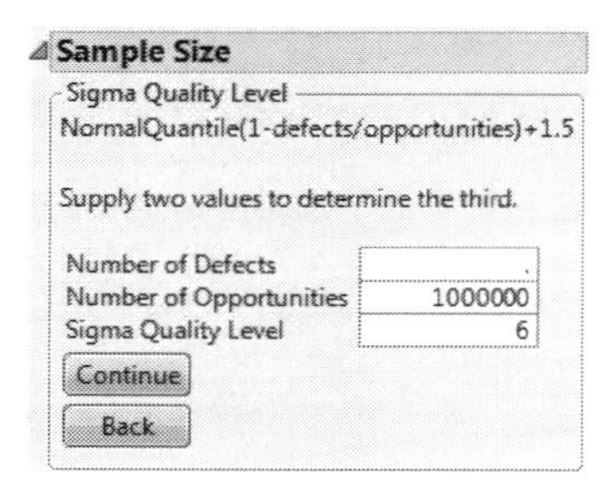

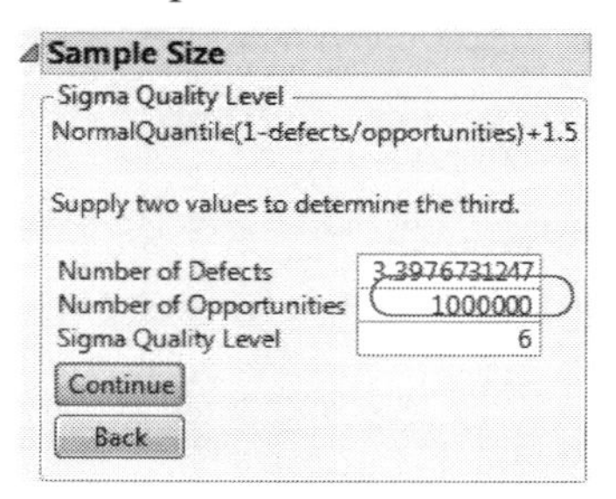

Reliability Test Plan and Demonstration

You can compute required sample sizes for reliability tests and reliability demonstrations using the **Reliability Test Plan** and **Reliability Demonstration** features.

Reliability Test Plan

The **Reliability Test Plan** feature computes required sample sizes, censor times, or precision, for estimating failure times and failure probabilities.

To launch the Reliability Test Plan calculator, select **DOE** > **Design Diagnostics > Sample Size and Power**, and then select **Reliability Test Plan**. Figure 17.16 shows the Reliability Test Plan window.

Figure 17.16 Reliability Test Plan Window

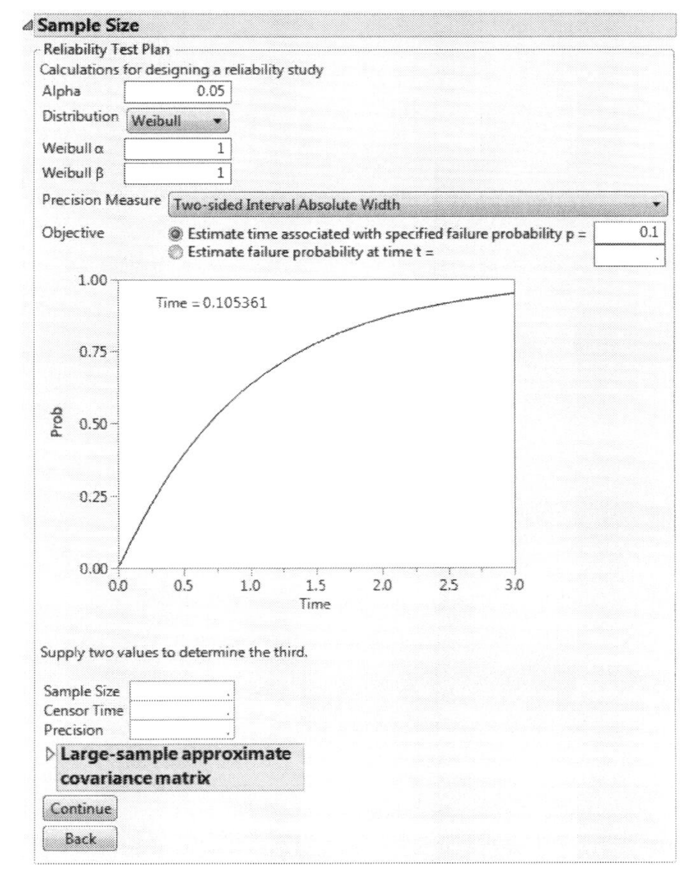

The Reliability Test Plan window has the following options:

Alpha is the significance level. It is also 1 minus the confidence level.

Distribution is the assumed failure distribution, with the associated parameters.

Precision Measure is the precision measure. In the following definitions, U and L correspond to the upper and lower confidence limits of the quantity being estimated (either a time or failure probability), and T corresponds to the true time or probability for the specified distribution.

Interval Ratio is sqrt(U/L), the square root of the ratio of the upper and lower limits.

Two-sided Interval Absolute Width is U-L, the difference of the upper and lower limits.

Lower One-sided Interval Absolute Width is T-L, the true value minus the lower limit.

Two-sided Interval Relative Width is (U-L)/T, the difference between the upper and lower limits, divided by the true value.

Lower One-sided Interval Relative Width is (T-L)/T, the difference between the true value and the lower limit, divided by the true value.

Objective is the objective of the study. The objective can be one of the following two:

- estimate the time associated with a specific probability of failure.
- estimate the probability of failure at a specific time.

CDF Plot is a plot of the CDF of the specified distribution. When estimating a time, the true time associated with the specified probability is written on the plot. When estimating a failure probability, the true probability associated with the specified time is written on the plot.

Sample Size is the required number of units to include in the reliability test.

Censor Time is the amount of time to run the reliability test.

Precision is the level of precision. This value corresponds to the Precision Measure chosen above.

Large-sample approximate covariance matrix gives the approximate variances and covariance for the location and scale parameters of the distribution.

Continue click here to make the calculations.

Back click here to go back to the Power and Sample Size window.

After the **Continue** button is clicked, two additional statistics are shown:

Expected number of failures is the expected number of failures for the specified reliability test.

Probability of fewer than 3 failures is the probability that the specified reliability test will result in fewer than three failures. This is important because a minimum of three failures is required to reliably estimate the parameters of the failure distribution. With only one or two failures, the estimates are unstable. If this probability is large, you risk not being able to achieve enough failures to reliably estimate the distribution parameters, and you should consider changing the test plan. Increasing the sample size or censor time are two ways of lowering the probability of fewer than three failures.

Example

A company has developed a new product and wants to know the required sample size to estimate the time till 20% of units fail, with a two-sided absolute precision of 200 hours. In other words, when a confidence interval is created for the estimated time, the difference between the upper and lower limits needs to be approximately 200 hours. The company can run the experiment for 2500 hours. Additionally, from studies done on similar products, they believe the failure distribution to be approximately Weibull (2000, 3).

To compute the required sample size, do the following steps:

1. Select **DOE** > **Design Diagnostics > Sample Size and Power**.
2. Select **Reliability Test Plan**.
3. Select **Weibull** from the Distribution list.
4. Enter 2000 for the Weibull α parameter.
5. Enter 3 for the Weibull β parameter.
6. Select **Two-sided Interval Absolute Width** from the Precision Measure list.
7. Select **Estimate time associated with specified failure probability**.
8. Enter 0.2 for **p**.
9. Enter 2500 for **Censor Time**.
10. Enter 200 for **Precision**.
11. Click **Continue**. Figure 17.17 shows the results.

Figure 17.17 Reliability Test Plan Results

Sample Size	217
Censor Time	2500
Precision	200
Expected number of failures	186.2229
Probability of fewer than 3 failures	7.4e-179

The required sample size is 217 units if the company wants to estimate the time till 20% failures with a precision of 200 hours. The probability of fewer than 3 failures is small, so the experiment will likely result in enough failures to reliably estimate the distribution parameters.

Reliability Demonstration

A reliability demonstration consists of testing a specified number of units for a specified period of time. If fewer than k units fail, you pass the demonstration, and conclude that the product reliability meets or exceeds a reliability standard.

The **Reliability Demonstration** feature computes required sample sizes and experimental run-times for demonstrating that a product meets or exceeds a specified reliability standard.

To launch the Reliability Demonstration calculator, select **DOE** > **Design Diagnostics > Sample Size and Power**, and then select **Reliability Demonstration**. Figure 17.18 shows the Reliability Demonstration window.

Figure 17.18 Reliability Demonstration Window

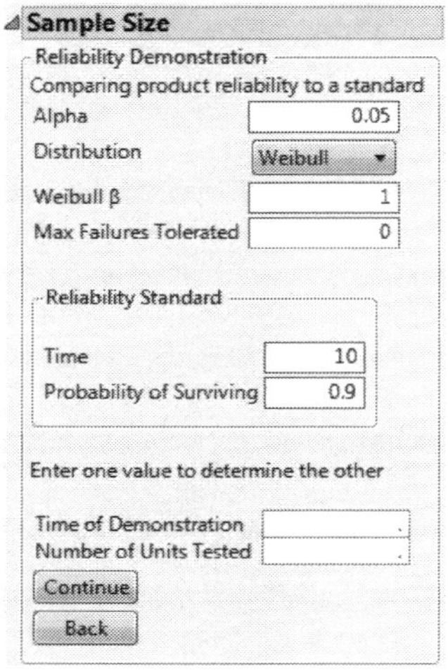

The Reliability Demonstration window has the following options:

Alpha is the alpha level.

Distribution is the assumed failure distribution. After selecting a distribution, specify the associated scale parameter in the text field under the **Distribution** menu.

Max Failures Tolerated is the maximum number of failures you want to allow during the demonstration. If we observe this many failures or fewer, then we say we passed the demonstration.

Time is the time component of the reliability standard you want to meet.

Probability of Surviving is the probability component of the reliability standard you want to meet.

Time of Demonstration is the required time for the demonstration.

Number of Units Tested is the required number of units for the demonstration.

Continue click here to make the calculations.

Back click here to go back to the Power and Sample Size window.

After the Continue button is clicked, a plot appears (see Figure 17.19).

Figure 17.19 Reliability Demonstration Plot

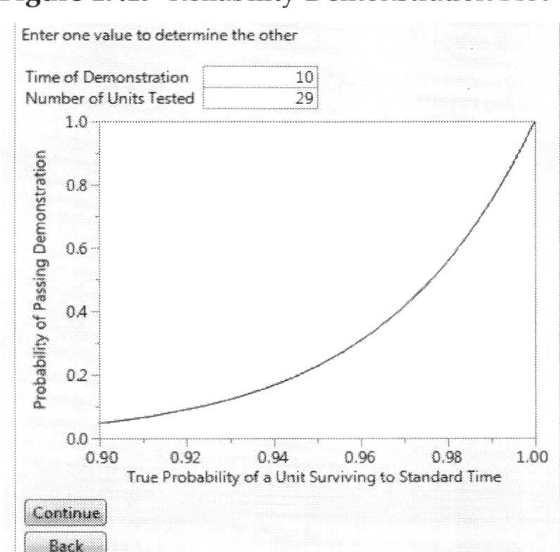

The true probability of a unit surviving to the specified time is unknown. The Y axis of the plot gives the probability of passing the demonstration (concluding the true reliability meets or exceeds the standard) as a function of the true probability of a unit surviving to the standard time. Notice the line is increasing, meaning that the further the truth is above the standard, the more likely you are to detect the difference.

Example

A company wants to get the required sample size for assessing the reliability of a new product against an historical reliability standard of 90% survival after 1000 hours. From prior studies on similar products, it is believed that the failure distribution is Weibull, with a β parameter of 3. The company can afford to run the demonstration for 800 hours, and wants the experiment to result in no more than 2 failures.

To compute the required sample size, do the following steps:

1. Select **DOE > Design Diagnostics > Sample Size and Power.**
2. Select **Reliability Demonstration.**
3. Select **Weibull** from the Distribution list.
4. Enter 3 for the Weibull β.
5. Enter 2 for **Max Failures Tolerated.**
6. Enter 1000 for **Time.**
7. Enter 0.9 for **Probability of Surviving.**
8. Enter 800 for **Time of Demonstration.**

9. Click **Continue**. Figure 17.20 shows the results.

Figure 17.20 Reliability Demonstration Results

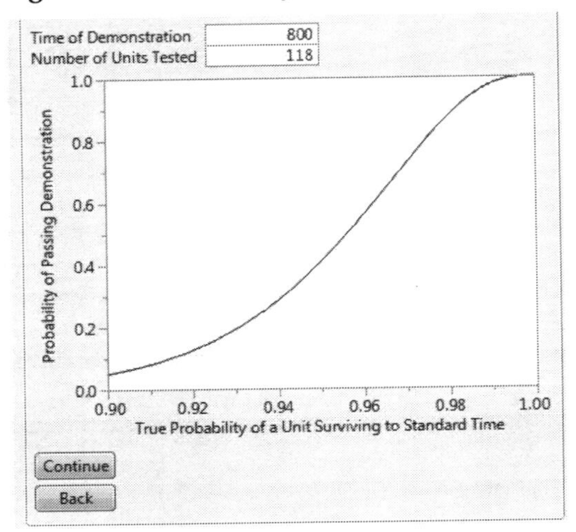

The company needs to run 118 units in the demonstration. Furthermore, if they observe 2 or fewer failures by 800 hours, we can conclude that the new product reliability is at least as reliable as the standard.

Chapter 18

Discrete Choice Designs

Create a Design for Selecting Preferred Product Profiles

A choice (or *discrete choice*) experiment provides data for modeling discrete preferences. Study participants are presented with sets of potential products (or product profiles) with varying attributes. From each set of profiles, a participant selects a preferred profile. For example, in designing a high-end laptop, a computer company might be interested in the relative importance of key features such as: processor speed, hard disk size, screen size, battery life, and price. A choice experiment addresses the relative values of these features to a customer and indicates an optimal set of trade-offs among product features.

The results of a choice experiment are analyzed using conjoint analysis methods. See the Choice Models chapter in the *Consumer Research* book.

Figure 18.1 A Survey with Eight Choice Sets

	Choice Set	Choice ID	Disk Size	Speed	Battery Life	Price
1	1	1	40 GB	2.0 GHz	6 Hrs	$1000
2	1	2	80 GB	1.5 GHz	4 Hrs	$1500
3	2	1	80 GB	1.5 GHz	4 Hrs	$1200
4	2	2	40 GB	1.5 GHz	6 Hrs	$1500
5	3	1	40 GB	1.5 GHz	4 Hrs	$1200
6	3	2	80 GB	2.0 GHz	6 Hrs	$1500
7	4	1	40 GB	2.0 GHz	4 Hrs	$1000
8	4	2	80 GB	1.5 GHz	6 Hrs	$1500
9	5	1	80 GB	2.0 GHz	6 Hrs	$1200
10	5	2	40 GB	1.5 GHz	6 Hrs	$1000
11	6	1	40 GB	1.5 GHz	4 Hrs	$1500
12	6	2	80 GB	1.5 GHz	6 Hrs	$1000
13	7	1	80 GB	1.5 GHz	4 Hrs	$1000
14	7	2	40 GB	2.0 GHz	6 Hrs	$1200
15	8	1	80 GB	1.5 GHz	6 Hrs	$1200
16	8	2	80 GB	2.0 GHz	4 Hrs	$1500

Overview of Choice Designs

Discrete choice experiments support the process of designing a product. They help prioritize product features for a company's market so that the company can design a product that people want to buy. The Choice Design platform creates experiments using factors that are attributes of a product. Selecting the attributes to be studied and their values is of critical importance. You must include all attributes that are likely to influence a consumer's decision to buy the product. For more information and guidelines for designing a choice experiment, see Sall (2008).

Choice Design Terminology

The following terminology is associated with choice designs:

- An *attribute* is a feature of a product.
- A *profile* is a specification of product attributes.
- A *choice set* is a collection of profiles.
- A *survey* is a collection of choice sets.
- A *partial profile* is a profile in a choice design where only a specified number of attributes are varied within each choice set. The remaining attributes are not varied.

In a discrete choice experiment, respondents are presented with a survey containing several choice sets. Choice sets usually contain only a small number of profiles to facilitate the decision process. Within each choice set, each respondent specifies which of the profiles he or she prefers. For example, attributes for a laptop experiment might include speed, storage, screen size, battery life, and price. Different combinations of these attributes comprise product profiles. A choice set might consist of two profiles. From each choice set, a respondent chooses the profile that he or she prefers.

In cases where many attributes are involved, you can construct surveys where each choice set contains *partial profiles*. In a choice set with partial profiles, only a specified number of attributes are varied and the remaining attributes are held constant. This reduces the complexity of the choice task.

Bayesian D-Optimality

Because discrete choice models are nonlinear in their parameters, the efficiency of a choice design depends on the unknown parameters. The Choice Design platform uses a Bayesian approach, optimizing the design over a prior distribution of likely parameter values that you specify. The Bayesian D-optimality criterion is the expected logarithm of the determinant of the information matrix, taken with respect to the prior distribution. The Choice Design platform maximizes this expectation with respect to the prior probability distribution. For details, see "Bayesian D-Optimality and Design Construction" on page 531 and Kessels et al. (2011).

You can also generate the following types of designs:

- Utility-neutral designs - In a utility-neutral design, all choices within a choice set are equally probable. The prior mean is set to 0.
- Local D-optimal designs - A local D-optimal design takes into account the prior on the mean, but does not include any information from a prior covariance matrix.

For more information about utility neutral and local D-optimal designs, see "Utility-Neutral and Local D-Optimal Designs" on page 532.

Example of a Choice Design

About the Experiment

In this example, a coffee shop is interested in making an ideal cup of coffee to satisfy the majority of its customers. The manager has asked you to determine which factors affect customer preferences. Specifically, you need to determine which settings of the following factors (attributes) result in an ideal cup of coffee:

- Grind size (medium or coarse)
- Temperature (195°, 200°, 205°)
- Brewing time (3 minutes, 3.5 minutes, or 4 minutes)
- Charge (1.6 grams/ounce, 2 grams/ounce, or 2.4 grams/ounce)

Each combination of factor levels is a *profile*. Trying to obtain information about preferences by having every respondent sample every possible profile is not practical. However, you can ask a respondent to select a preferred profile from a *choice set* consisting of a small number of profiles.

In this example, you design an experiment where each respondent indicates his or her preference in several choice sets. Your design will have the following structure:

- ten respondents
- twelve choice sets
- two profiles per choice set
- one survey per respondent containing all 12 choice sets

The experiment results in 12 responses per respondent. Analysis of these preferences can be used to draw conclusions about how to make a cup of coffee that pleases most customers.

Create the Design

You can enter factors either manually or automatically using a preexisting table that contains the factors and settings. In this example, for convenience, you use a preexisting table. But, if you are designing a new experiment, you must first enter the factors manually. For details on entering factors manually, see "Attributes" on page 525.

1. Select **DOE** > **Consumer Studies** > **Choice Design**.
2. Select **Help > Sample Data Library** and open Design Experiment/Coffee Choice Factors.jmp.
3. Click the Choice Design red triangle and select **Load Factors**.

Figure 18.2 Choice Design Window with Attributes Defined

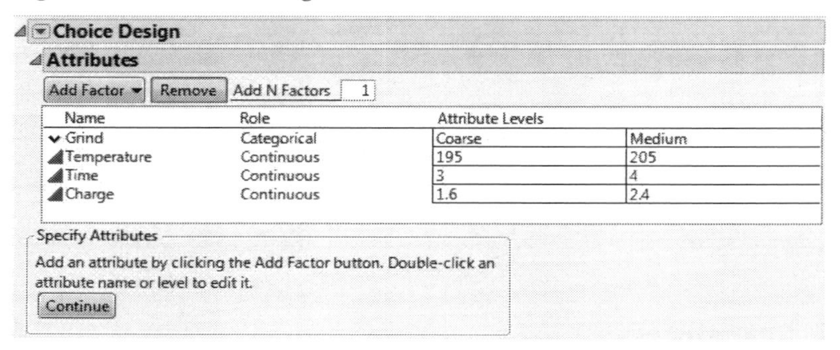

4. Click **Continue**.
5. Open the DOE Model Controls outline.

   In this example, you are only interested in a model that contains the main effects of your four factors. However, if you wanted your design to be capable of estimating additional effects, you would add them in this outline.
6. In the Design Generation panel:
   - Keep the **Number of attributes that can change within a choice set** at 4.
   - Keep the **Number of profiles per choice set** at 2.
   - Type 12 for the **Number of choice sets per survey**.

     In this example, the respondents evaluate 12 choice sets.
   - Keep the **Number of surveys** at 1.
   - Type 10 for the **Expected number of respondents per survey**.

     In this example, there are ten respondents.

Figure 18.3 Completed Design Generation Panel

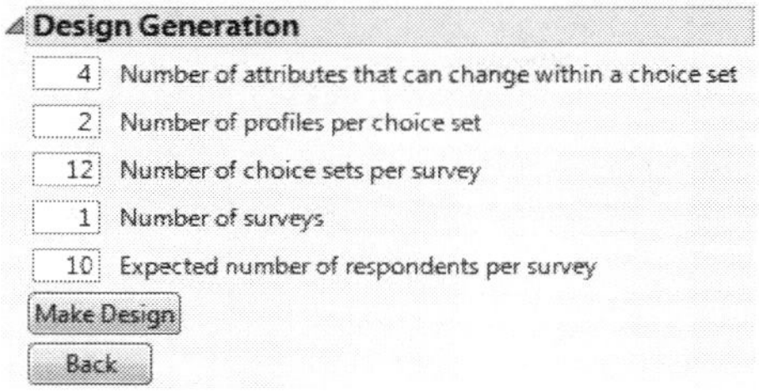

Design Generation
4 Number of attributes that can change within a choice set
2 Number of profiles per choice set
12 Number of choice sets per survey
1 Number of surveys
10 Expected number of respondents per survey
Make Design
Back

Note: Setting the Random Seed in step 7 reproduces the exact results shown in this example. In constructing a design on your own, this step is not necessary.

7. (Optional) Click the Choice Design red triangle and select **Set Random Seed**. Type 12345 and click **OK**.
8. Click **Make Design**.

   There are 12 choice sets, each consisting of two coffee profiles.
9. Select **Output separate tables for profiles and responses**.

   This places the descriptions of the choice sets in a data table called Choice Profiles. A second data table (called Choice Runs) is constructed to facilitate entry of the response information.
10. Click **Make Table**.

    The Choice Profiles table shows the 12 choice sets, each consisting of two profiles. The Choice Runs table enables you to record results in the column Response Indicator. Enter 1 for the preferred profile and 0 for the other profile. Alternatively, if the respondent has no preference, enter 0 for both profiles or leave both missing.

    The **Choice** script in the Choice Profiles table facilitates analysis of experimental results. It opens a completed launch window for a Choice Model. For information about Choice Models, see the Choice Models chapter in the *Consumer Research* book.

Example of a Choice Design with Analysis

In this example, a computer manufacturer is interested in manufacturing a new laptop and wants information about customer preferences before beginning an expensive development process. The manufacturer decides to construct a design consisting of two sets of profiles that will be administered to ten respondents. The goal of the choice design is to understand how potential laptop purchasers view the advantages of a collection of four attributes:

- size of hard drive disk (40 GB or 80 GB)
- speed of processor (1.5 GHz or 2.0 GHz)

- battery life (4 Hrs or 6 Hrs)
- cost of computer ($1000, $1200 or $1500)

To construct the design for the ten respondents, you first conduct a small pilot study using only one respondent. Then you analyze the results and use the parameter estimates as prior information in designing the final study for the ten respondents. Here are the steps:

- "Create a Choice Design for a Pilot Study" on page 518
- "Analyze the Pilot Study Data" on page 519
- "Design the Final Choice Experiment Using Prior Information" on page 521

Create a Choice Design for a Pilot Study

In this section, you construct a choice design for a one-respondent study.

Define Factors and Levels

In this example, you load the factors from an existing table. When designing a new experiment on your own, enter the factors manually.

1. Select **DOE > Consumer Studies > Choice Design**.
2. Select **Help > Sample Data Library** and open Design Experiment/Laptop Factors.jmp.
3. Click the Choice Design red triangle and select **Load Factors**.

Figure 18.4 Choice Design Window with Attributes Defined

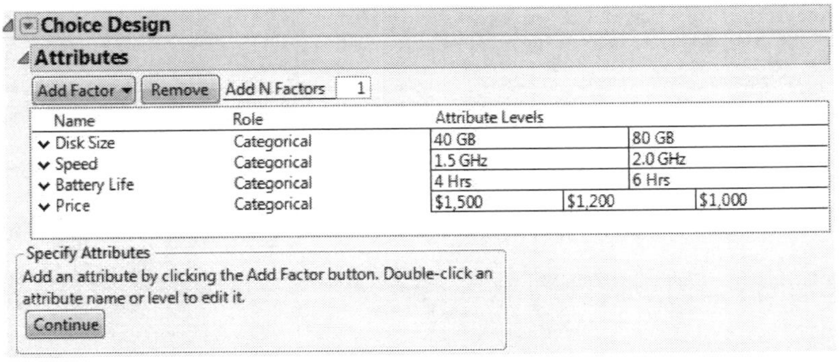

Create the Design

1. Click **Continue**.

   This pilot survey will be given to a single respondent. The default values in the DOE Model Controls, Prior Specification, and Design Generation panels are appropriate as is.

2. Click **Make Design**.

Note: Because choice designs are constructed using a random starting design, your design will likely differ from the one in Figure 18.5.

Figure 18.5 Pilot Design

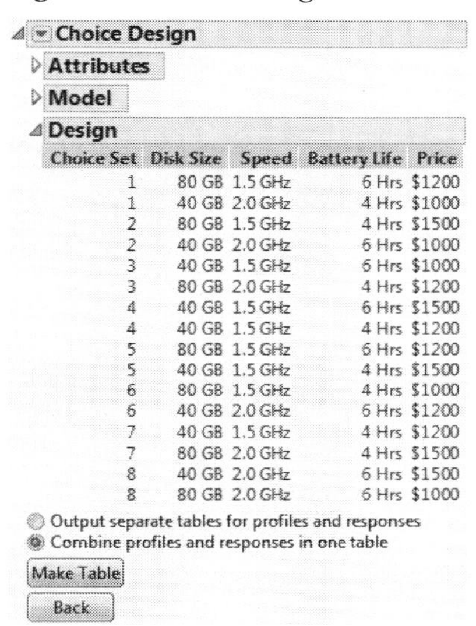

Choice Set	Disk Size	Speed	Battery Life	Price
1	80 GB	1.5 GHz	6 Hrs	$1200
1	40 GB	2.0 GHz	4 Hrs	$1000
2	80 GB	1.5 GHz	4 Hrs	$1500
2	40 GB	2.0 GHz	6 Hrs	$1000
3	40 GB	1.5 GHz	6 Hrs	$1000
3	80 GB	2.0 GHz	4 Hrs	$1200
4	40 GB	1.5 GHz	6 Hrs	$1500
4	40 GB	1.5 GHz	4 Hrs	$1200
5	80 GB	1.5 GHz	6 Hrs	$1200
5	40 GB	1.5 GHz	4 Hrs	$1500
6	80 GB	1.5 GHz	4 Hrs	$1000
6	40 GB	2.0 GHz	6 Hrs	$1200
7	40 GB	1.5 GHz	4 Hrs	$1200
7	80 GB	2.0 GHz	4 Hrs	$1500
8	40 GB	2.0 GHz	6 Hrs	$1500
8	80 GB	2.0 GHz	6 Hrs	$1000

The single survey contains eight choice sets, each consisting of two laptop profiles.

3. Verify that the **Combine profiles and responses in one table** option is selected.

   This places the choice sets and the survey results in the same table.

4. Click **Make Table**.

This survey was designed assuming no prior information. For this reason, some choice sets might not elicit useful information. The plan is to obtain survey results from the single respondent, analyze the results, and then use the results from the pilot survey as prior information in designing the final survey.

Analyze the Pilot Study Data

Now that the pilot survey design is complete, it is administered to single respondent. The respondent chooses one profile from each set, entering 1 for the chosen profile and 0 for the rejected profile. You will analyze the results using the Choice platform.

Note: For details about the Choice platform, see the Choice Models chapter the *Consumer Research* book.

1. Select **Help > Sample Data Library** and open Design Experiment/Laptop Design.jmp.

2. Run the **Choice** script.

Figure 18.6 Choice Model Launch Window

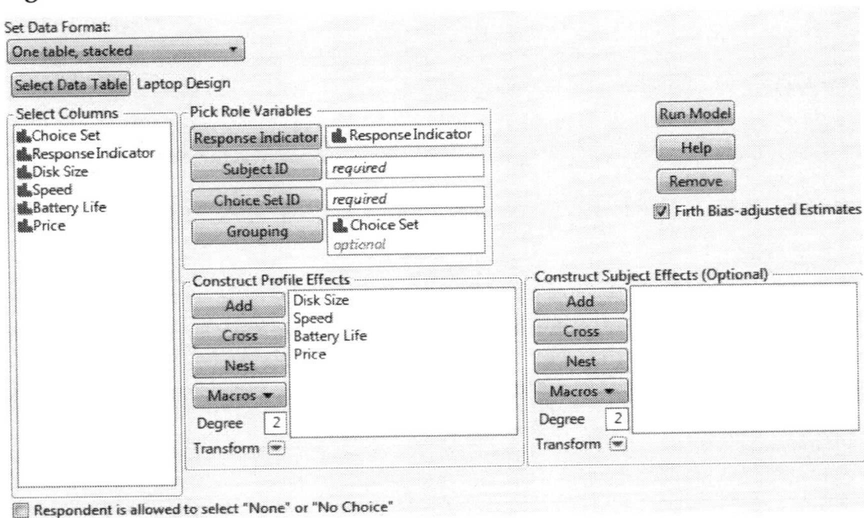

The only grouping variable is **Choice Set** because there is a single survey and a single respondent.

3. Click **Run Model**.

Figure 18.7 Parameter Estimates for Pilot Survey

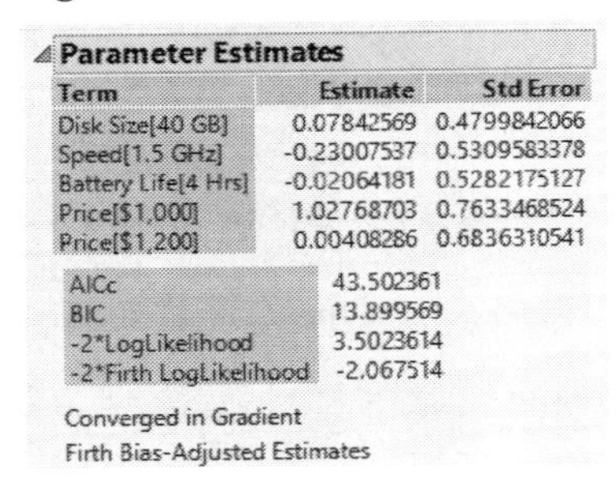

Parameter Estimates

Term	Estimate	Std Error
Disk Size[40 GB]	0.07842569	0.4799842066
Speed[1.5 GHz]	-0.23007537	0.5309583378
Battery Life[4 Hrs]	-0.02064181	0.5282175127
Price[$1,000]	1.02768703	0.7633468524
Price[$1,200]	0.00408286	0.6836310541

AICc	43.502361
BIC	13.899569
-2*LogLikelihood	3.5023614
-2*Firth LogLikelihood	-2.067514

Converged in Gradient
Firth Bias-Adjusted Estimates

To construct the final choice design that you will give to ten respondents, you need prior means and variances for the parameter estimates. The analysis in Figure 18.7 gives estimates of the parameter means (Estimate) and estimates of their standard errors (Std Error). You will treat the standard errors as prior estimates of the standard deviations. Next, to calculate estimates of the variances of the attributes, construct a JMP table and square the standard errors.

4. Right-click in the Parameter Estimates report and select **Make into Data Table**.

5. In the new data table, right-click in the Std Error column header and select **New Formula Column > Transform > Square**.

   A column called Std Error^2 is added to the data table. Its values will serve as your estimates of the prior variance for the choice model parameters.

Figure 18.8 Untitled Data Table with Variance Estimates in Last Column

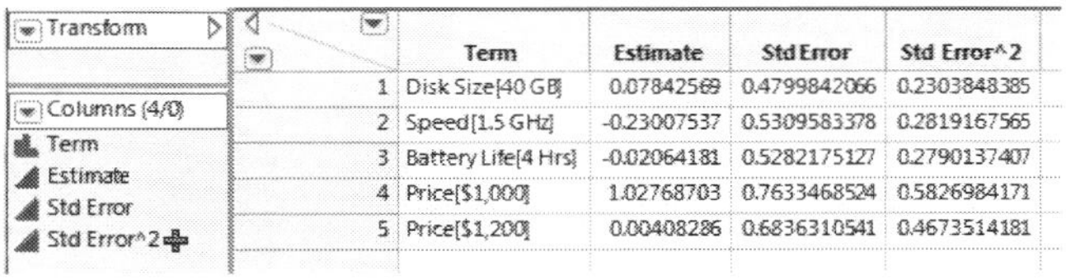

	Term	Estimate	Std Error	Std Error^2
1	Disk Size[40 GB]	0.07842569	0.4799842066	0.2303848385
2	Speed[1.5 GHz]	-0.23007537	0.5309583378	0.2819167565
3	Battery Life[4 Hrs]	-0.02064181	0.5282175127	0.2790137407
4	Price[$1,000]	1.02768703	0.7633468524	0.5826984171
5	Price[$1,200]	0.00408286	0.6836310541	0.4673514181

Transform
Columns (4/0)
Term
Estimate
Std Error
Std Error^2

Note: Do not close the Untitled data table at this point.

Design the Final Choice Experiment Using Prior Information

In this section, you use the prior information obtained from the pilot laptop study to construct a final design. The final design will be administered to a set of ten participants.

1. Select **Help > Sample Data Library** and open Design Experiment/Laptop Factors.jmp.
2. Select **DOE > Consumer Studies > Choice Design**.
3. Click the Choice Design red triangle and select **Load Factors**.
4. Click **Continue**.
5. From the Untitled table, enter the values in the Estimate column into the Prior Mean outline in the Choice Design window, as shown in Figure 18.9.

   You can copy and paste the entire column from the Untitled table, then click in the Disk Size text box under **Prior Mean** in the Prior Mean outline, right-click, and select Paste.
6. From the Untitled table, enter the values in the Std Error^2 column into the diagonal entries in the Prior Variance outline in the Choice Design window, as shown in Figure 18.9. Enter these one-by-one, rounded to three decimal places.

Figure 18.9 Prior Mean and Variance Information from Pilot Study

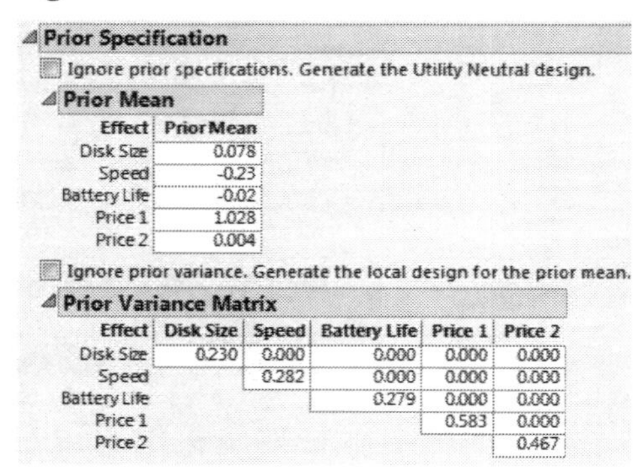

7. In the Design Generation panel, enter 2 for the **Number of surveys** and five for the **Expected number of respondents per survey**.

   This gives instruments for a total of 10 respondents and allows for two different sets of profiles.

8. Click **Make Design**.
9. Click **Make Table**.

   The design table has 160 rows. There are 16 rows for each of the ten study respondents. Each respondent has 8 choice sets, each with 2 profiles. There are two surveys, each given to 5 respondents. The 160 rows result from the following calculation: 2 profiles * 8 choice sets * 2 surveys * 5 respondents = 160 rows.

   The final design is now ready to be administered to the 10 respondents.

Run the Design and Analyze the Results

In this section, you analyze the results obtained when you obtain results from the final design. In particular, you want to know how changing the price or other characteristics of a laptop affects its desirability as perceived by potential buyers. This desirability is called the *utility value* of the laptop attributes.

Determine Significant Attributes

1. Select **Help > Sample Data Library** and open Design Experiment/Laptop Results.jmp.
2. Run the **Choice** script.

Figure 18.10 Choice Model Launch Window

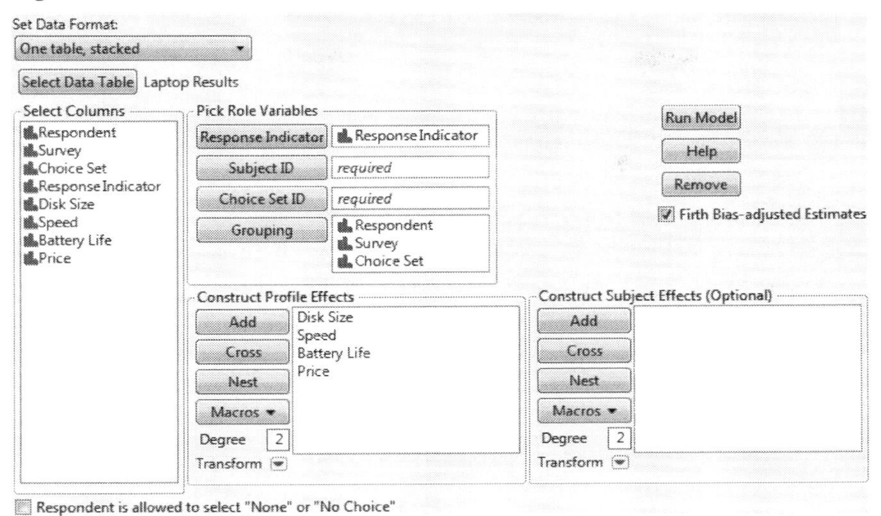

There are three grouping variables, **Respondent**, **Survey**, and **Choice Set**, because there are multiple surveys and respondents.

3. Click **Run Model**.

Figure 18.11 Initial Analysis of the Final Laptop Design

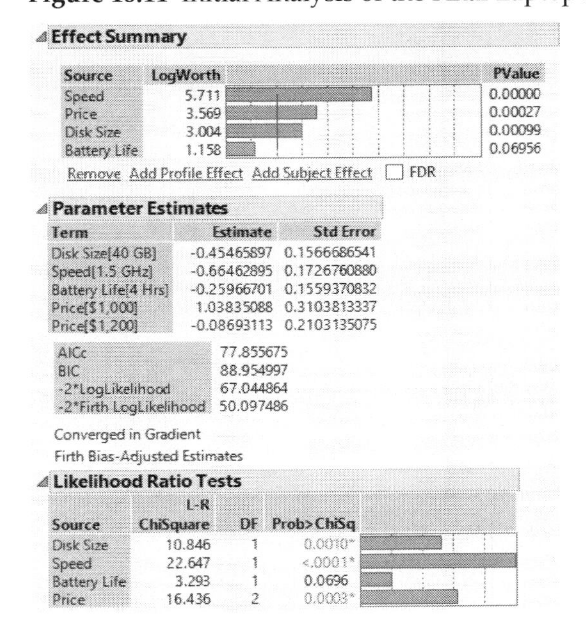

Effect Summary

Source	LogWorth	PValue
Speed	5.711	0.00000
Price	3.569	0.00027
Disk Size	3.004	0.00099
Battery Life	1.158	0.06956

Remove Add Profile Effect Add Subject Effect FDR

Parameter Estimates

Term	Estimate	Std Error
Disk Size[40 GB]	-0.45465897	0.1566686541
Speed[1.5 GHz]	-0.66462895	0.1726760880
Battery Life[4 Hrs]	-0.25966701	0.1559370832
Price[$1,000]	1.03835088	0.3103813337
Price[$1,200]	-0.08693113	0.2103135075

AICc	77.855675
BIC	88.954997
-2*LogLikelihood	67.044864
-2*Firth LogLikelihood	50.097486

Converged in Gradient

Firth Bias-Adjusted Estimates

Likelihood Ratio Tests

Source	L-R ChiSquare	DF	Prob>ChiSq
Disk Size	10.846	1	0.0010*
Speed	22.647	1	<.0001*
Battery Life	3.293	1	0.0696
Price	16.436	2	0.0003*

The Effect Summary and Likelihood Ratio Tests outlines indicate that Disk Size, Speed, and Price are significant at the 0.05 level, and that Battery Life is marginally significant.

Find Unit Cost and Trade Off Costs

Next, use the Profiler to see the utility value and how it changes as the laptop attributes change.

1. Click the Choice Model red triangle and select **Utility Profiler**.

Figure 18.12 Utility Profiler at Price = $1000

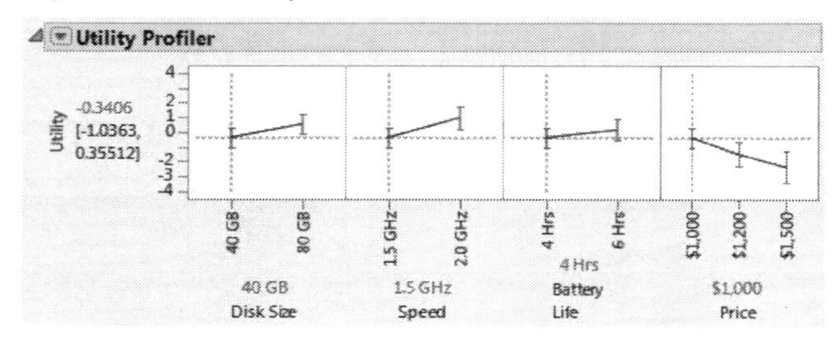

When each attribute value is set to its lowest value, the Utility value is –0.3406. The first thing that you want to do is determine the unit utility cost.

2. Move the slider for Price to $1,500.

Figure 18.13 Utility Profiler at Price = $1500

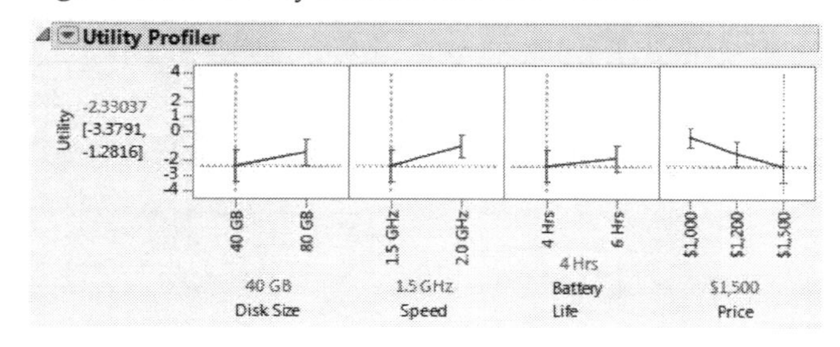

When Price changes from $1,000 to $1,500, the Utility changes from –0.3406 to –2.3303. That is, raising the price of a laptop by $500.00 lowers the utility (or desirability) approximately 2 units. Therefore, you can estimate the unit utility cost to be approximately $250.00.

With this unit utility cost estimate, you can now vary the other attributes, note the change in utility, and find an approximate monetary value associated with attribute changes. For example, the most significant attribute is Speed (see Figure 18.11).

3. In the Utility Profiler, set Price back to $1,000, its lowest value, and change Speed to 2.0 GHz, its higher value.

Figure 18.14 Utility Value of Higher Speed

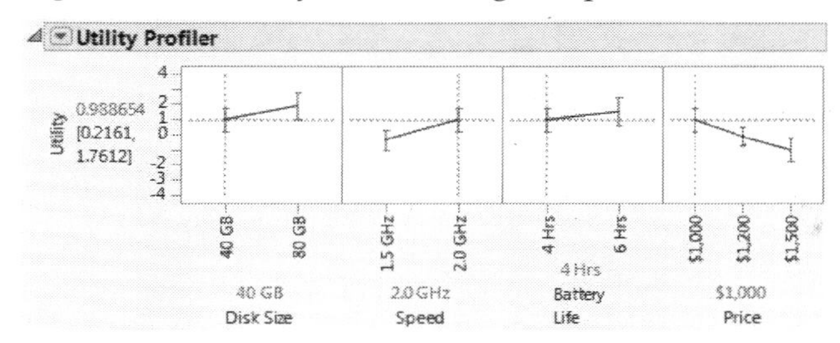

The Utility value changes from the original value shown in Figure 18.12 of –0.3406 to 0.9886, for a total change of 1.3292 units. Using the utility cost estimate or $250.00, the increase in price for a 2.0 GHz laptop over a 1.5 GHz laptop can be computed to be 1.3292*$250.00 = $332.30. This is the dollar value that the Choice study indicates that the manufacturer can use as a basis for pricing this laptop attribute. You can make similar calculations for the other attributes.

Choice Design Window

The Choice Design window walks you through the steps to construct a choice design for modeling attribute preferences. You can specify an assumed model and prior information.

The Choice Design window updates as you work through the design steps. The outlines, separated by buttons that update the outlines, follow the flow in Figure 18.15.

Figure 18.15 Choice Design Flow

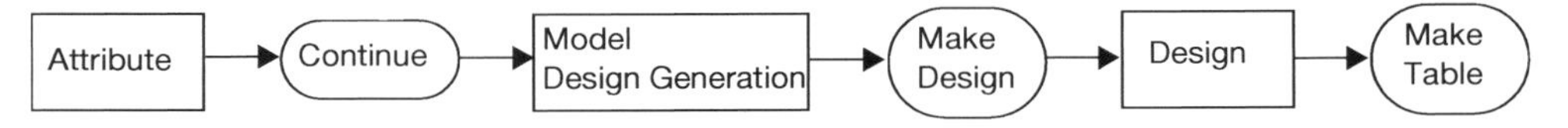

The following sections describe the steps in creating a choice design:

- "Attributes" on page 525
- "Model" on page 527
- "Design Generation" on page 528
- "Design" on page 529
- "Output Options" on page 530

Attributes

Attributes in a choice design can only be categorical.

Tip: When you have completed the Attributes outline, consider selecting **Save Factors** from the red triangle menu. This saves the attribute names, roles, and levels in a data table that you can later reload and reuse.

Figure 18.16 Attributes Outline

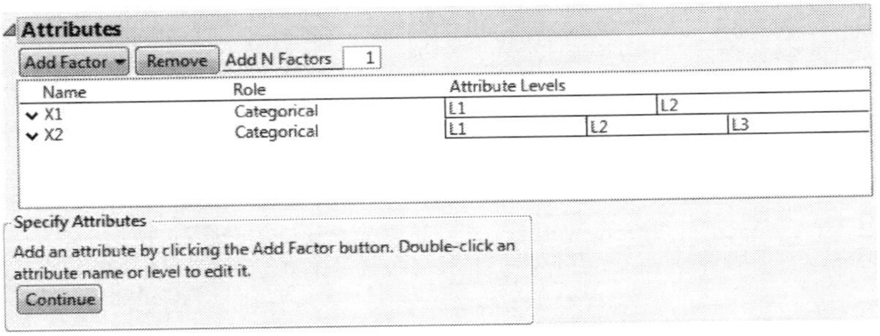

The Attributes outline contains the following buttons:

Add Factor Adds an attribute with the selected number of levels.

Remove Removes the selected attributes.

Add N Factors Adds multiple attributes. Enter the number of attributes to add, click **Add Factor,** and then select the number of levels. Repeat **Add N Factors** to add multiple attributes with different numbers of levels.

The Attributes outline contains the following columns:

Name Name of the attribute. Attributes are given default names of X1, X2, and so on. To change a name, double-click it and type the desired name.

Role Specifies the Design Role of the attribute as Categorical.

Attribute Levels The attribute name or description. To insert Attribute Levels, click on the default levels and type the desired names.

Editing the Attributes Outline

- To edit the Name of an attribute, double-click the attribute name.
- To edit an Attribute Level, click the level.

Attribute Column Properties

For each attribute, JMP saves the Value Ordering column property to the data tables constructed by the Choice platform. The Value Ordering column property specifies that levels appear in reports using the ordering specified in the Attributes outline. For details, see "Value Ordering" on page 679 in the "Column Properties" appendix.

Model

The model outline consists of two parts:

- You can specify your assumed model, which contains all the effects that you want to estimate. See "DOE Model Controls" on page 527.
- You can specify prior knowledge about the attribute levels, which can result in a better design. See "Prior Specification" on page 527.

DOE Model Controls

Specify your assumed model in the DOE Model Controls outline. All main effects are included by default. Click the Interactions button to add all two-way interactions.

Figure 18.17 DOE Model Controls Outline

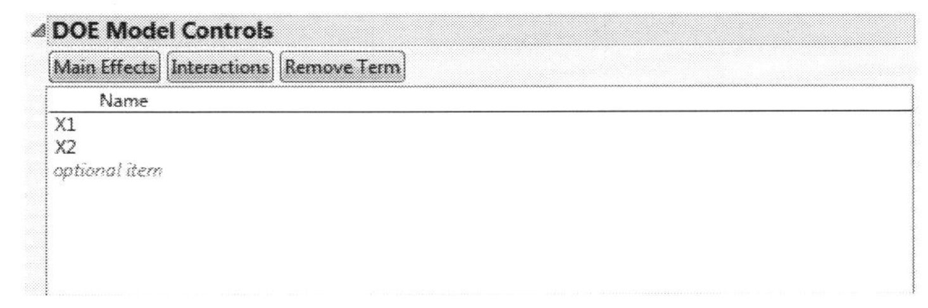

When you construct your design table, JMP saves a Choice script to the data table. The Choice script contains the main effects shown in the DOE Model Controls outline.

The DOE Model Controls outline contains the following buttons:

Main Effects Adds main effects for all attributes in the model.

Interactions Adds all second-order interactions. If you do not want to include all of the interactions, select the interactions that you want to remove and click **Remove Term**.

Remove Term Removes selected effects.

Prior Specification

Enter specifications for a multivariate normal prior distribution on the model parameters. Enter the prior distribution's mean in the Prior Mean outline and its covariance matrix in the Prior Variance outline.

Figure 18.18 Prior Specification Outline

Prior Specification

☐ Ignore prior specifications. Generate the Utility Neutral design.

Prior Mean

Effect	Prior Mean
X1	0.000
X2 1	0.000
X2 2	0.000

☐ Ignore prior variance. Generate the local design for the prior mean.

Prior Variance Matrix

Effect	X1	X2 1	X2 2
X1	1.000	0.000	0.000
X2 1		1.000	0.000
X2 2			1.000

You can ignore the prior specifications using these options:

Ignore prior specifications. Generate the Utility Neutral design. Sets the prior means to 0 and generates a locally D-optimal design. This design is called a *utility-neutral* design. For more information, see Huber and Zwerina (1996).

Ignore prior specifications. Generate the local design for the prior mean. Generates a locally D-optimal design. The local design takes into account the prior on the mean but ignores the covariance matrix. For more information, see Huber and Zwerina (1996).

Design Generation

Enter specifications that define the structure for your design in the Design Generation panel.

Figure 18.19 Design Generation Panel for Coffee Example

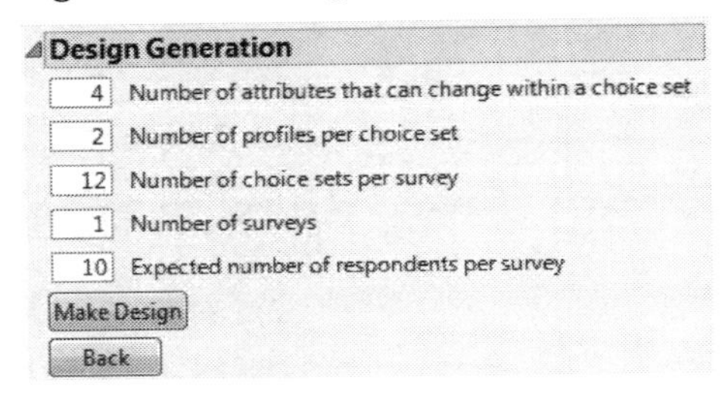

Note: Figure 18.19 is taken from the coffee example. See "Example of a Choice Design" on page 515.

Enter a number for each of the following items:

Number of attributes that can change within a choice set Enter a number less than or equal to the total number of attributes. This is often set to the total number of attributes. However, if you are comparing many attributes and want to simplify the selection process for respondents, set this to a number that is smaller than the total number of attributes.

Number of profiles per choice set Enter the number of profiles that a respondent must choose from in stating a preference.

Number of choice sets per survey Enter the number of preferences that you want to obtain from each respondent.

Number of surveys Enter the number of distinct collections of choice sets. This is useful if you want to administer surveys to multiple respondents.

Expected number of respondents per survey Enter the total number of respondents divided by the number of surveys.

Make Design

Once you have completed the Design Generation outline, click **Make Design** to generate the design. The design appears in the Design outline.

Design

The Design outline shows the runs for a design that is optimal, given your selections. Review the design to ensure that it meets your needs.

Figure 18.20 Design Outline for Coffee Example

Design

Choice Set	Grind	Temperature	Time	Charge
1	Coarse	195	3	2
1	Medium	200	4	1.6
2	Medium	205	3	2
2	Coarse	200	3.5	2.4
3	Coarse	195	3	1.6
3	Medium	200	3.5	2.4
4	Coarse	200	3	2.4
4	Medium	205	3.5	2
5	Coarse	195	3.5	1.6
5	Coarse	200	3	2
6	Medium	195	3.5	2
6	Medium	205	4	2.4
7	Coarse	195	3.5	2.4
7	Medium	200	3	1.6
8	Medium	195	4	2
8	Coarse	205	3	1.6
9	Coarse	200	3.5	1.6
9	Medium	195	3	2.4
10	Medium	200	3	1.6
10	Coarse	205	4	2
11	Coarse	200	3.5	2
11	Medium	195	4	1.6
12	Coarse	200	4	2
12	Medium	205	3.5	2.4

○ Output separate tables for profiles and responses
◉ Combine profiles and responses in one table

Note: Figure 18.20 is taken from the coffee example. See"Example of a Choice Design" on page 515.

Note: The algorithm for finding an optimal design is based on a random starting design. Because of this, the design you obtain is not unique. The design algorithm will generate different designs when you click the **Back** and **Make Design** buttons repeatedly.

Output Options

Select one of the following output options:

Output separate tables for profiles and responses Displays two data tables:

- The Choice Profiles table lists the profiles in each row, identified by Survey and Choice Set columns. Within a choice set, the profiles are identified by Choice ID. This table is useful for constructing the survey instruments.
- The Choice Runs table provides an empty Response column where you can enter respondent preferences. Each row corresponds to a single choice set. The rows are sorted by Respondent, Survey, and Choice Set. The choice set IDs are given in the next columns, followed by the Response column. Enter the choice set ID for the respondent's preference in the Response column.

Combine profiles and responses in one table Provides a single Choice Profiles table with an empty Response Indicator column where you can enter respondent preferences. Each row corresponds to a single profile. The table is sorted by Respondent, Survey, and Choice Set. Enter the value 1 (or another nonzero numerical indicator) for the respondent's preferred profile and a 0 indicator for the other profiles in that choice set.

Note: The values you enter in the Response Indicator column must be numerical.

Make Table

Click **Make Table** to construct the table or tables that you selected in "Output Options" on page 530. In the table panel of the Choice Profiles table, there is a **Choice** script. Run the script and then click **Run Model** to analyze your experimental results.

Caution: Only main effects are added by the **Choice** script. Add interactions manually.

For information about the Choice Model report, see the Choice Models chapter in the *Consumer Research* book.

Choice Design Options

Save Factors Saves the information in the Factors panel to a new data table. Each factor's column contains its levels. Other information is stored as column properties. You can then quickly load the factors and their associated information into most DOE windows.

Note: It is possible to create a factors table by entering data into an empty table, but remember to assign each column an appropriate Design Role. Do this by right-clicking on the column name in the data grid and selecting **Column Properties > Design Role**. In the Design Role area, select the appropriate role.

Load Factors Loads factors that you have saved using the Save Factors option.

Set Random Seed Sets the random seed that JMP uses to control certain actions that have a random component. These actions include:

- simulating responses using the Simulate Responses option
- randomizing Run Order for design construction
- selecting a starting design for designs based on random starts

To reproduce a design or simulated responses, enter the random seed that generated them. For designs using random starts, set the seed before clicking Make Design. To control simulated responses or run order, set the seed before clicking Make Table.

Note: The random seed associated with a design is included in the DOE Dialog script that is saved to the design data table.

Number of Starts Enables you to specify the number of random starts used in constructing the design. See "Bayesian D-Optimality and Design Construction" on page 531.

Advanced Options None available.

Save Script to Script Window Creates the script for the design that you specified in the Choice Design window and saves it in an open script window.

Technical Details

Bayesian D-Optimality and Design Construction

The Bayesian D-optimality criterion is the expected logarithm of the determinant of the information matrix of the maximum likelihood of the parameter estimators in the multinomial logit model, taken with respect to the prior distribution. The Choice Design platform maximizes this expectation with respect to a sample of parameter vectors that represents the prior probability distribution. For details, see Kessels et al. (2011).

For partial profile designs, JMP uses a two-stage design algorithm:

1. The constant attributes in each choice set are determined using an *attribute balance* approach.
2. The levels of the non-constant attributes are determined using Bayesian D-optimality.

Attribute balance means that the algorithm attempts to balance the number of times each attribute is held constant in the entire design. If two or more attributes are held constant, the algorithm attempts to balance the occurrence of pairs of attributes held constant in the design.

The levels of the non-constant attributes are determined to optimize the Bayesian D-optimal criterion. A random starting design is found. Then levels of the non-constant attributes are generated using a coordinate-exchange algorithm and evaluated until the Bayesian D-optimality criterion is optimized. The calculations, which involve integration with respect to a multivariate normal prior, use the quadrature method described in Gotwalt et al. (2009).

Note: The Bayesian D-optimality criterion can result in choice sets where some non-constant attributes have identical levels. This situation occurs when varying the non-constant levels within a profile would result in uninformative choice sets where all profiles have very high or very low probabilities.

Utility-Neutral and Local D-Optimal Designs

You can use the Choice Design platform to generate a utility-neutral design by setting prior means to 0. In a utility-neutral design, all choices within a choice set are equally probable. For more information, see Huber and Zwerina (1996).

You can also generate a local D-optimal design. The local design takes into account the prior of the mean, but does not include any information from a prior covariance matrix. For more information, see Huber and Zwerina (1996).

Chapter 19

MaxDiff Design

Create a Design for Selecting Best and Worst Items

MaxDiff (*maximum difference scaling*) studies are an alternative to studies that use standard preference scales to determine the relative importance of items being rated. In a MaxDiff study, a respondent reports only the most and least preferred options from among a small set of choices. This forces respondents to rank options in terms of preference, which often results in rankings that are more definitive than rankings obtained using standard preference scales.

Use the MaxDiff platform when you need to construct a design consisting of choice sets for one factor that can be presented to respondents as part of a MaxDiff study. Conduct your study and then analyze your data use the MaxDiff analysis platform.

Figure 19.1 A MaxDiff Design Table

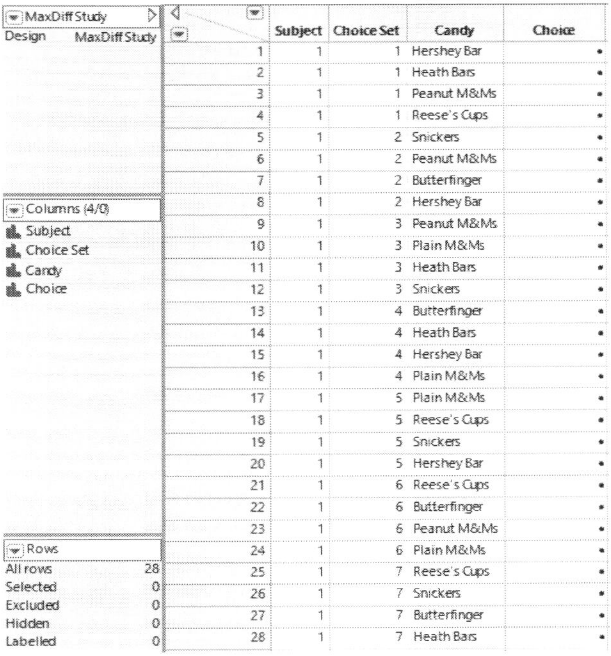

MaxDiff Study
Design MaxDiff Study

Columns (4/0)
Subject
Choice Set
Candy
Choice

Rows
All rows 28
Selected 0
Excluded 0
Hidden 0
Labelled 0

	Subject	Choice Set	Candy	Choice
1	1	1	Hershey Bar	•
2	1	1	Heath Bars	•
3	1	1	Peanut M&Ms	•
4	1	1	Reese's Cups	•
5	1	2	Snickers	•
6	1	2	Peanut M&Ms	•
7	1	2	Butterfinger	•
8	1	2	Hershey Bar	•
9	1	3	Peanut M&Ms	•
10	1	3	Plain M&Ms	•
11	1	3	Heath Bars	•
12	1	3	Snickers	•
13	1	4	Butterfinger	•
14	1	4	Heath Bars	•
15	1	4	Hershey Bar	•
16	1	4	Plain M&Ms	•
17	1	5	Plain M&Ms	•
18	1	5	Reese's Cups	•
19	1	5	Snickers	•
20	1	5	Hershey Bar	•
21	1	6	Reese's Cups	•
22	1	6	Butterfinger	•
23	1	6	Peanut M&Ms	•
24	1	6	Plain M&Ms	•
25	1	7	Reese's Cups	•
26	1	7	Snickers	•
27	1	7	Butterfinger	•
28	1	7	Heath Bars	•

MaxDiff Design Platform Overview

A *choice set* is a collection of items from which a respondent must select an item that is most preferred (*best* item) and one that is least preferred (*worst* item). Use the MaxDiff platform to specify the number of items that appear in a choice set and the number of choice sets to be presented to a respondent.

The MaxDiff platform provides you with a single survey. If you have several respondents, you can administer the same survey to all respondents or construct a survey for each respondent.

The MaxDiff design is constructed to match the incidence matrix of a balanced incomplete block design as closely as possible. This implies that each pair of items occurs equally often in the design. If the number of choice sets allows a balanced incomplete block to be constructed for the specified number of items in a choice set, the design that is constructed is a balanced incomplete block design.

The items or profiles are considered to be the levels of the treatment factor. The choice sets are considered to be blocks. In a MaxDiff study, the block size is smaller than the number of treatments.

Example of a MaxDiff Design

You are the purchaser for your company's office supplies and you need to buy candy for a holiday party. First, you want to determine which candy types people prefer. To figure this out, you conduct a MaxDiff study.

You ask five randomly chosen associates to rate seven types of candies. Based on your experience with previous studies, you realize that it is difficult for raters to rank seven types of items in order of preference. Instead, you create a design that consists of choice sets of size four. To keep the study manageable, you structure the survey as a MaxDiff study: you ask each associate to specify his or her most preferred and least preferred candy in each of the seven choice sets. (These selections result in a balanced incomplete block design. See Cochran and Cox (1957).) You administer the same survey to each associate.

Create the Design

Construct a table that lists the items or profiles for your choice sets. In this example, your table of items, Candy Profiles.jmp, is already constructed.

1. Select **Help > Sample Data Library** and open Design Experiment/Candy Profiles.jmp.

   The table lists the seven candy types of interest.

2. Select **DOE** > **Consumer Studies** > **MaxDiff Design**.
3. From the Select Columns list, select Candy and click **X, Factor**.
4. Click **OK**.
5. In the Design Options outline, do the following:
   - Set the **Number of Profiles per Choice Set** to **4**.
   - Set the **Number of Choice Sets** to **7**.

Note: Setting the Random Seed in step 6 reproduces the exact results shown in this example. In constructing a design on your own, this step is not necessary.

6. (Optional) Click the MaxDiff Study red triangle and select **Set Random Seed**. Type 12345 and click **OK**.
7. Click **Make Design**.

   The Design outline shows 7 choice sets, each consisting of 4 candy types.

8. Click **Make Table**.

Figure 19.2 Design for Candy Preference Survey

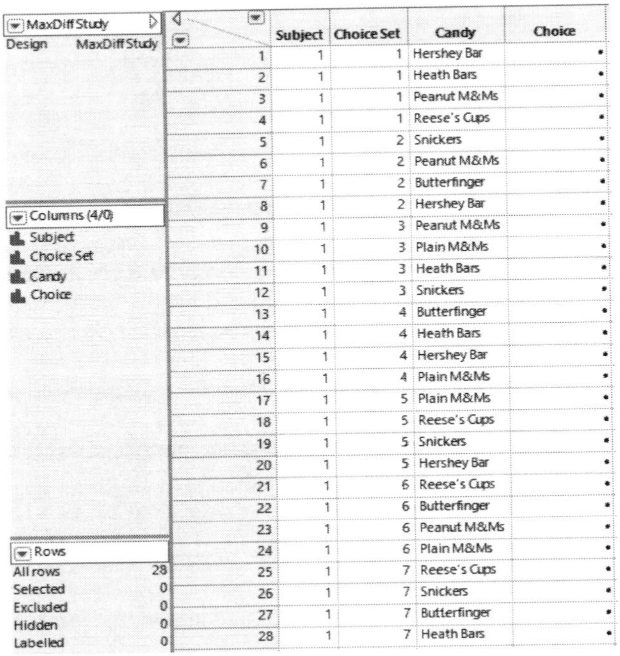

MaxDiff Study
Design MaxDiff Study

Columns (4/0)
Subject
Choice Set
Candy
Choice

Rows
All rows 28
Selected 0
Excluded 0
Hidden 0
Labelled 0

	Subject	Choice Set	Candy	Choice
1	1	1	Hershey Bar	•
2	1	1	Heath Bars	•
3	1	1	Peanut M&Ms	•
4	1	1	Reese's Cups	•
5	1	2	Snickers	•
6	1	2	Peanut M&Ms	•
7	1	2	Butterfinger	•
8	1	2	Hershey Bar	•
9	1	3	Peanut M&Ms	•
10	1	3	Plain M&Ms	•
11	1	3	Heath Bars	•
12	1	3	Snickers	•
13	1	4	Butterfinger	•
14	1	4	Heath Bars	•
15	1	4	Hershey Bar	•
16	1	4	Plain M&Ms	•
17	1	5	Plain M&Ms	•
18	1	5	Reese's Cups	•
19	1	5	Snickers	•
20	1	5	Hershey Bar	•
21	1	6	Reese's Cups	•
22	1	6	Butterfinger	•
23	1	6	Peanut M&Ms	•
24	1	6	Plain M&Ms	•
25	1	7	Reese's Cups	•
26	1	7	Snickers	•
27	1	7	Butterfinger	•
28	1	7	Heath Bars	•

The design table contains a Choice column for recording preferences. For each choice set, record a 1 for the most preferred candy, a -1 for the least preferred, and a 0 for the other two candies.

Analyze the Study Results

You conduct the study and record your data in Candy Survey.jmp.

1. Select **Help > Sample Data Library** and open Design Experiment/Candy Survey.jmp.

   The table shows the results of presenting the survey to each of five respondents, listed in the Subject column.

2. Select **Analyze** > **Consumer Research** > **MaxDiff**.
3. Click Select Data Table, select Candy Survey, and click **OK**.
4. Assign roles to columns as follows:
   – Select Choice and click **Response Indicator**.
   – Select Subject and click **Subject ID**.
   – Select Choice Set and click **Choice Set ID**.
   – Select Candy and click **Add** in the Construct Profile Effects panel.

Figure 19.3 Completed MaxDiff Analysis Launch Window

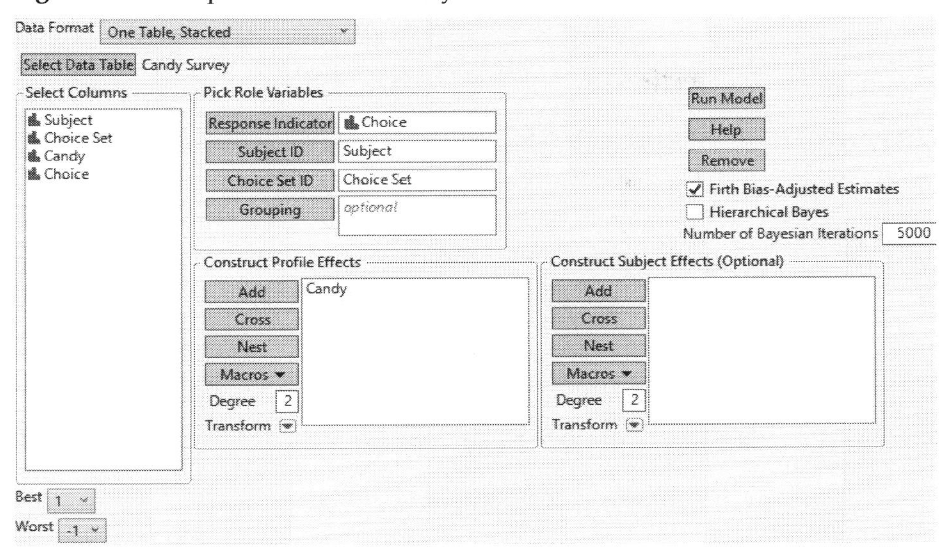

Because you designated the Best choice as 1 and the Worst choice as -1, you make no change to the Best and Worst choice indicators at the bottom left of the launch window.

5. Click **Run Model**.

Figure 19.4 MaxDiff Report

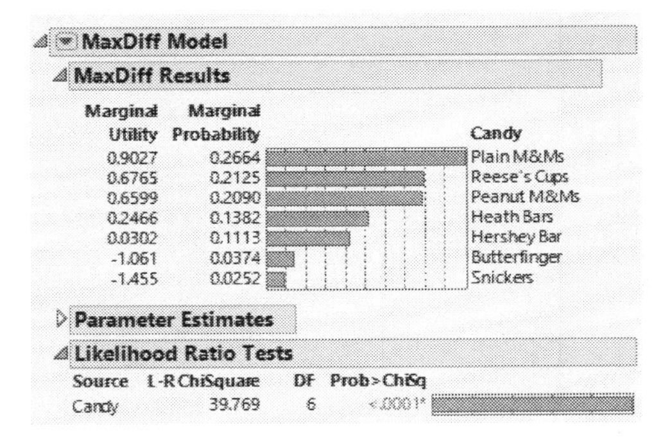

The report indicates that **Candy** is significant. The three candy types with the highest utilities are Plain M&Ms, Reese's Cups, and Peanut M&Ms.

6. Click the MaxDiff Model red triangle and select **All Levels Comparison Report**.

Figure 19.5 All Levels Comparison Report for Candy Types

All Levels Comparison Report

Difference (Row-Column) Standard Error of Difference Wald p-Value	Butterfinger	Heath Bars	Hershey Bar	Peanut M&Ms	Plain M&Ms	Reese's Cups	Snickers
Butterfinger	0	-1.3079	-1.0915	-1.7212	-1.964	-1.7378	0.39343
	0	0.52954	0.52038	0.54692	0.56495	0.54938	0.5059
		0.01964	0.04478	0.0038	0.00162	0.00364	0.44306
Heath Bars	1.30792	0	0.21642	-0.4133	-0.656	-0.4299	1.70135
	0.52954	0	0.47619	0.48741	0.49094	0.46959	0.54758
	0.01964		0.65286	0.40343	0.19185	0.36752	0.0042
Hershey Bar	1.0915	-0.2164	0	-0.6297	-0.8725	-0.6463	1.48493
	0.52038	0.47619	0	0.47406	0.49984	0.49096	0.53808
	0.04478	0.65286		0.19443	0.09149	0.19835	0.00992
Peanut M&Ms	1.72121	0.41329	0.62971	0	-0.2427	-0.0166	2.11463
	0.54692	0.48741	0.47406	0	0.49706	0.49054	0.56901
	0.0038	0.40343	0.19443		0.62896	0.97325	0.00086
Plain M&Ms	1.96396	0.65604	0.87246	0.24275	0	0.22616	2.35738
	0.56495	0.49094	0.49984	0.49706	0	0.49327	0.58562
	0.00162	0.19185	0.09149	0.62896		0.65002	0.00037
Reese's Cups	1.7378	0.42988	0.6463	0.01659	-0.2262	0	2.13123
	0.54938	0.46959	0.49096	0.49054	0.49327	0	0.5714
	0.00364	0.36752	0.19835	0.97325	0.65002		0.00083
Snickers	-0.3934	-1.7013	-1.4849	-2.1146	-2.3574	-2.1312	0
	0.5059	0.54758	0.53808	0.56901	0.58562	0.5714	0
	0.44306	0.0042	0.00992	0.00086	0.00037	0.00083	

The comparison report indicates which pairs of candy types differ significantly in terms of utility. The third entry in each cell is the *p*-value for the difference defined by the row item's utility minus the column item's utility. The intensity of the color for the *p*-value indicates how significant a difference is. The shading, blue or red, indicates whether the difference (row - column) is negative or positive. The *p*-values are not adjusted to control the multiple comparison error rate and should be used only as a guide. For details about the All Level Comparisons Report, see the MaxDiff chapter in the *Consumer Research* book.

MaxDiff Design Launch Window

To use the MaxDiff Design platform, you need a starting data table. Your starting data table must contain a column of character data that lists the items to be presented to respondents for rating.

With your starting data table active in JMP, select **DOE > Consumer Studies > MaxDiff Design**. If you have no active data tables, you are prompted to navigate to your starting data table.

Figure 19.6 MaxDiff Launch Window using Candies.jmp

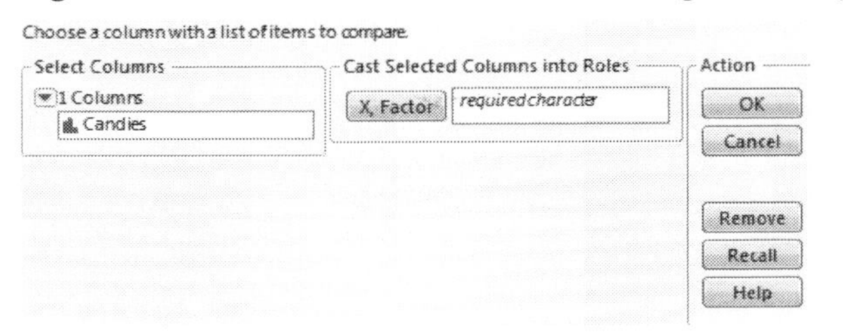

X, Factor The character column that contains the items that respondents will rate.

MaxDiff Window

The MaxDiff Study window updates as you work through the design steps. The outlines, separated by buttons that update the outlines, follow the flow in Figure 19.7

Figure 19.7 MaxDiff Design Flow

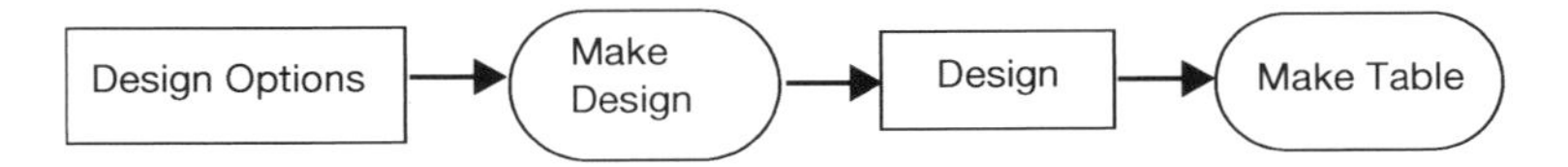

The MaxDiff Study window opens showing the Design Options outline. Once you click Make Design, the Design outline appears. To construct the design table, click Make Table.

Design Options Outline

Specify the following:

Number of Profiles per Choice Set The number of items to be included in each choice set.

Number of Choice Sets The total number of sets of items to be presented to and rated by respondents.

Design Outline

The Design outline identifies the choice sets using consecutive positive integers. The items that comprise each choice set are listed. In the Design outline, you can review the design settings.

Make Table

The Make Table button creates the design table. The design table consists of four columns:

Subject Initially populated by ones. Replace with appropriate identifiers for respondents.

> **Tip:** To easily add respondent identifiers, see the section Fill Columns with Sequential Data in the Enter and Edit chapter of the *Using JMP* book. Use Columns >Recode to change the identifiers to names.

Choice Set A designator for each choice set.

Factor The levels of the factor that you specified. These are the items in the choice set.

Choice A column where you can enter results. Use a numeric value to indicate the best choice, the worst choice, and choices in between. The values 1, -1, and 0 are typically used and are required for analysis by the MaxDiff analysis platform (located at Analyze > Consumer Research > MaxDiff).

MaxDiff Options

Set Random Seed Sets the random seed that JMP uses to control certain actions that have a random component. These actions include:

- simulating responses using the Simulate Responses option
- randomizing Run Order for design construction
- selecting a starting design

To reproduce a MaxDiff design or simulated responses, enter the random seed and Number of Starts that generated them. Do this step before clicking Make Design.

Simulate Responses Select this option to simulate response values. When you click Make Table, a Simulate Choice window opens along with the design table, and Probability and Choice Simulated columns are added to the design table. The Choice Simulated formula column contains random responses and the Probability formula column contains their probabilities.

To change the model used in simulating the responses, enter values in the Marginal Utility column in the Simulate Choice window for all factor levels but the last. Because the marginal utilities must sum to zero, you are not permitted to edit the Marginal Utility for the last level. Once you have specified the remaining Marginal Utility values, the last level of the factor is adjusted accordingly.

Number of Starts The number of random starts used in constructing the design. This value is set to 10 by default.

Advanced Options None applicable.

Chapter 20

JMP PRO Covering Arrays

Detecting Component Interaction Failures

Covering arrays are used in testing deterministic systems where failures occur as a result of interactions among components or subsystems. The design goal is to reveal if any interaction induces a failure in the system. Application areas include software, circuit, and network design.

Since the tests are deterministic, the emphasis driving the design is the need to cover all required interactions. The Covering Arrays platform constructs highly efficient covering arrays. You can also exclude factor level combinations that are not feasible for your testing protocol.

Figure 20.1 Strength 3 Covering Array

Design

Run	X1	X2	X3	X4	X5
1	L1	L1	L1	L1	L1
2	L1	L1	L1	L2	L2
3	L2	L2	L2	L2	L2
4	L2	L1	L2	L2	L1
5	L2	L2	L2	L1	L2
6	L2	L2	L1	L1	L1
7	L2	L1	L1	L1	L2
8	L1	L2	L2	L1	L1
9	L1	L2	L1	L2	L2
10	L1	L2	L2	L2	L1
11	L1	L1	L2	L1	L2
12	L2	L2	L1	L2	L1

JMP PRO Overview of Covering Arrays

You can use covering arrays to test systems where failures occur as a result of interactions among components or subsystems. Covering arrays are often used in areas such as software, circuit, and network design, where the following conditions are likely to be true:

- The cost of testing is usually high.
- Testing focuses on revealing interactions for which failures occur.
- A test run is typically deterministic and results in either success or failure.
- Replicate runs are wasteful because they yield identical results.
- The efficiency of a design is based on how many of the possible conditions are covered without including redundant runs.

Because systems testing is expensive, reducing the amount of testing is critical. Testing all possible interactions is usually prohibitive and often unnecessary. Experience shows that most failures result from the interaction of a small number of components. The size of the largest combination of components likely to drive a failure, called the *strength*, drives the size of the design.

In the Covering Array platform, you specify the required strength of your design. If appropriate, you define factor level combinations that are not permitted. The Covering Array platform constructs a highly efficient design that meets your requirements. It provides metrics that you can use to assess the quality of the design in terms of its coverage. It also provides a script in the data table for the design that enables you to analyze your results.

Covering arrays are often used in situations where certain combinations of factor level settings are not feasible. The Covering Array platform is able to find very efficient covering arrays even when restrictions are placed on factor level combinations.

For background on the structure of covering arrays and algorithms for computing them, see Colbourn (2004), Colbourn et al. (2011), Hartman and Raskin (2004), and Martirosyan (2003). For details about covering arrays with restrictions on factor levels, see Cohen et al. (2007) and Morgan (2009).

JMP PRO Covering Arrays and Strength

A *covering array of strength t* is a design that tests all combinations of *t* factor level settings. Consider an interaction defined by specific settings for *k* factors. If failures occur for all tests involving that interaction, then that interaction *detects* a failure. Using this terminology, a strength *t* design enables you to detect failures associated with any interaction of up to *t* factors.

In the literature, covering arrays are also referred to as *factor covering designs*. For background and details, see Yilmaz et al. (2014), Cohen et al. (2003), and Dalal and Mallows (1998).

To illustrate the nature of covering arrays, consider a situation involving seven categorical factors each with two levels. You want to test all pairwise combinations of factor levels.

A design that might be used in this situation is the 8-run resolution III main effects design given as follows:

Figure 20.2 A Resolution III Design with Strength 2

Run	X1	X2	X3	X4	X5	X6	X7
1	2	2	2	1	1	1	2
2	1	1	2	1	2	2	1
3	2	1	1	2	2	1	2
4	1	1	1	1	1	1	1
5	2	2	2	2	2	2	1
6	1	2	1	2	1	2	2

Note that this design is a strength 2 covering array because all pairwise combinations of levels of any two factors appear. For example, for X1 and X2, the following combinations each appear twice:

- L2 and L1
- L2 and L2
- L1 and L1
- L1 and L2

However, the 6-run design in Figure 20.3 is also a strength 2 covering array:

Figure 20.3 Strength 2 Covering Array

Run	X1	X2	X3	X4	X5	X6	X7
1	L1	L1	L2	L1	L1	L2	L2
2	L2	L2	L2	L2	L1	L1	L2
3	L1	L1	L2	L2	L2	L1	L1
4	L2	L1	L1	L1	L2	L1	L2
5	L1	L2	L1	L1	L1	L1	L1
6	L2	L2	L2	L1	L2	L2	L1
7	L1	L2	L1	L2	L2	L2	L2
8	L2	L1	L1	L2	L1	L2	L1

All pairwise combinations of levels of any two factors appear and this is accomplished in six runs, rather than eight. The Covering Array design is more efficient than the Resolution III design because it achieves strength 2 coverage in fewer runs.

The efficiency of a covering array is measured by the number of runs required to achieve the required coverage. The smaller the number of runs, the more efficient the design.

Example of a Covering Array with No Factor Level Restrictions

The data in this example pertain to interoperability in the area of software testing. There are four factors of interest:

- Web Browser (Safari, IE, Firefox, Chrome, Other)
- Operating System (Windows or Macintosh)
- RAM (4, 8, or 16 MB)
- Connection Speed at three settings (0-1, 1-5, or greater than 5 Mbps)

You are interested in finding out which combinations of these factors are likely to cause failures.

The response is whether the system functions properly for each combination of factor settings.

Testing each combination of settings would require 90 (5x2x3x3) trials. To keep the run size manageable, you decide to require Strength 3 coverage, indicating that all combinations of any three factors are tested.

Create the Design

Create the Strength 3 covering array by following these steps.

1. Select **Help > Sample Data Library** and open Design Experiment/Software Factors.jmp.

   The Software Factors.jmp data table contains the factors and their settings.
2. Select **DOE > Special Purpose > Covering Array**.
3. From the menu next to **Strength: t =** , select 3.
4. From the Covering Array red triangle menu, select **Load Factors** and click **Continue**.

   The Factors outline is populated with the four factors and their levels.

Figure 20.4 Factors Outline for Software Factors

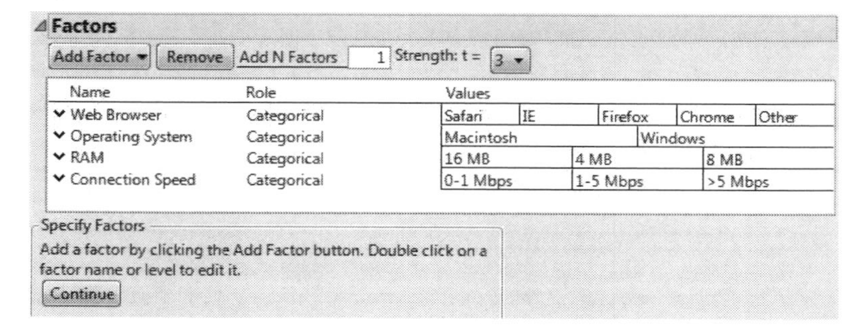

5. Click **Continue**.

   The Restrict Factor Level Combinations outline opens, where you can enter restrictions on the design settings. Because there are no restrictions for this design, do not change the default selection of **None**.

6. Click **Make Design**.

   The Design outline opens to show a 45-run design.

Figure 20.5 Design and Metrics Outlines for Software Factors

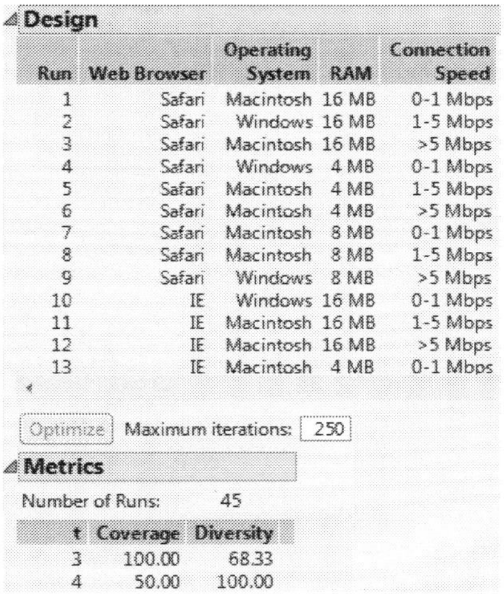

Design

Run	Web Browser	Operating System	RAM	Connection Speed
1	Safari	Macintosh	16 MB	0-1 Mbps
2	Safari	Windows	16 MB	1-5 Mbps
3	Safari	Macintosh	16 MB	>5 Mbps
4	Safari	Windows	4 MB	0-1 Mbps
5	Safari	Macintosh	4 MB	1-5 Mbps
6	Safari	Macintosh	4 MB	>5 Mbps
7	Safari	Macintosh	8 MB	0-1 Mbps
8	Safari	Macintosh	8 MB	1-5 Mbps
9	Safari	Windows	8 MB	>5 Mbps
10	IE	Windows	16 MB	0-1 Mbps
11	IE	Macintosh	16 MB	1-5 Mbps
12	IE	Macintosh	16 MB	>5 Mbps
13	IE	Macintosh	4 MB	0-1 Mbps

Optimize Maximum iterations: 250

Metrics

Number of Runs: 45

t	Coverage	Diversity
3	100.00	68.33
4	50.00	100.00

In the Metrics outline, consider the row that corresponds to $t = 3$. The Coverage is 100%, indicating that the design covers 100% of the three-factor interactions. This is what you want, because you requested a Strength 3 design. For $t = 3$, the Diversity column indicates that 68.33% of the three-factor interactions that appear are distinct. There is some minor repetition of three-factor combinations.

For $t = 4$, the Coverage is 50%, indicating that the design covers half of the four-factor interactions. There are 90 possible distinct combinations of the four factor settings. The 45 runs in the design comprise one-half of these distinct combinations. The Diversity value of 100% reinforces the fact that none of the four-way interactions are repeated.

7. Click **Make Table**.

Figure 20.6 Design Table for Software Factors

Covering Array
Design Covering Array, Stre
DOE Dialog
Analysis

Columns (5/0)
Response
Web Browser *
Operating System *
RAM *
Connection Speed *

Rows
All rows 45
Selected 0
Excluded 0
Hidden 0
Labelled 0

	Response	Web Browser	Operating System	RAM	Connection Speed
1	•	Safari	Macintosh	16 MB	0-1 Mbps
2	•	Safari	Windows	16 MB	1-5 Mbps
3	•	Safari	Macintosh	16 MB	>5 Mbps
4	•	Safari	Windows	4 MB	0-1 Mbps
5	•	Safari	Macintosh	4 MB	1-5 Mbps
6	•	Safari	Macintosh	4 MB	>5 Mbps
7	•	Safari	Macintosh	8 MB	0-1 Mbps
8	•	Safari	Macintosh	8 MB	1-5 Mbps
9	•	Safari	Windows	8 MB	>5 Mbps
10	•	IE	Windows	16 MB	0-1 Mbps
11	•	IE	Macintosh	16 MB	1-5 Mbps
12	•	IE	Macintosh	16 MB	>5 Mbps
13	•	IE	Macintosh	4 MB	0-1 Mbps
14	•	IE	Macintosh	4 MB	1-5 Mbps
15	•	IE	Windows	4 MB	>5 Mbps
16	•	IE	Macintosh	8 MB	0-1 Mbps
17	•	IE	Windows	8 MB	1-5 Mbps
18	•	IE	Macintosh	8 MB	>5 Mbps

The design is presented in a data table. Notice the following in the Table panel at the top left:

- The Design note indicates that this is a strength 3 covering array.
- The DOE Dialog script reproduces the Covering Array window settings.
- The Analysis script analyzes the experimental data.

JMP PRO Analyze the Experimental Data

Now that you have your design table, you can conduct your experiment and record your data in the Response column of the design table (Figure 20.6). Your experimental results are in the Software Data.jmp sample data table.

1. Select **Help > Sample Data Library** and open Design Experiment/Software Data.jmp.
2. In the Table panel, click the green triangle next to the **Analysis** script.

Figure 20.7 Analysis of Software Experimental Data

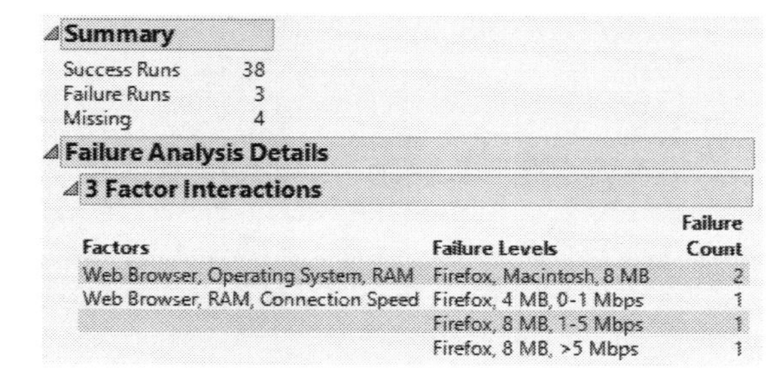
Summary
Success Runs 38
Failure Runs 3
Missing 4

Failure Analysis Details
3 Factor Interactions

Factors	Failure Levels	Failure Count
Web Browser, Operating System, RAM	Firefox, Macintosh, 8 MB	2
Web Browser, RAM, Connection Speed	Firefox, 4 MB, 0-1 Mbps	1
	Firefox, 8 MB, 1-5 Mbps	1
	Firefox, 8 MB, >5 Mbps	1

The Summary outline indicates that three tests failed and four tests did not result in a pass or fail outcome.

The Failure Analysis Details outline gives a breakdown of failures in terms of the associated three-way interactions. The outline lists only combinations of factor levels where all tests resulted in failure. If any test that involves a given three-way combination of settings results in success, then that three-way combination of settings cannot be responsible for system failure.

Two failures were associated with Web Browser set to Firefox, Operating System set to Macintosh, and RAM set to 8 MB. Notice that this combination led to failure regardless of the setting of Connection Speed.

3. Select the first line in the 3 Factor Interactions report.

Figure 20.8 Selection of an Interaction in the Analysis Report

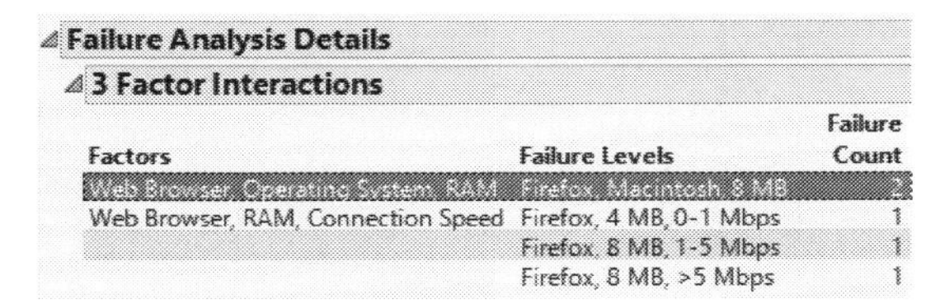

Failure Analysis Details

3 Factor Interactions

Factors	Failure Levels	Failure Count
Web Browser, Operating System, RAM	Firefox, Macintosh, 8 MB	2
Web Browser, RAM, Connection Speed	Firefox, 4 MB, 0-1 Mbps	1
	Firefox, 8 MB, 1-5 Mbps	1
	Firefox, 8 MB, >5 Mbps	1

This action selects the corresponding rows and columns in the data table.

Figure 20.9 Corresponding Selection of Rows and Columns in the Data Table

Software Data
Design Covering Array
Notes Strength covering arr
DOE Dialog
Analysis

Columns (5/3)
Response
Web Browser *
Operating System *
RAM *
Connection Speed *

	Response	Web Browser	Operating System	RAM	Connection Speed
1	1	Safari	Macintosh	16 MB	0-1 Mbps
2	1	Safari	Windows	16 MB	1-5 Mbps
3	1	Safari	Macintosh	16 MB	>5 Mbps
4	1	Safari	Windows	4 MB	0-1 Mbps
5	1	Safari	Macintosh	4 MB	1-5 Mbps
6	1	Safari	Macintosh	4 MB	>5 Mbps
7	1	Safari	Macintosh	8 MB	0-1 Mbps
8	1	Safari	Macintosh	8 MB	1-5 Mbps
9	1	Safari	Windows	8 MB	>5 Mbps
10	1	IE	Windows	16 MB	0-1 Mbps
11	1	IE	Macintosh	16 MB	1-5 Mbps
12	1	IE	Macintosh	16 MB	>5 Mbps
13	1	IE	Macintosh	4 MB	0-1 Mbps

Three failures were associated with combinations of Web Browser, RAM, and Connection Speed. Note that two of these failures, Firefox, 8 MB, 1-5 Mbps and Firefox, 8 MB, >5 Mbps, are among the two failures for the Web Browser, Operating System, and RAM interaction. Selecting any of these rows in the report selects the corresponding rows and columns in the data table.

JMP PRO Example of a Covering Array with Factor Level Restrictions

The following example is patterned after an example described in Dalal and Mallows (1998). An originating phone (Near Phone) calls a receiving phone (Far Phone). Each phone call goes through an interface of type A or B. Five factors are of interest:

- Market: USA, UK, Canada, France, Mexico
- Near Phone: ISDN, Bus (Business), Coin, Res (Residential)
- Near Interface: A or B
- Far Phone: ISDN, Bus (Business), Coin, Res (Residential)
- Far Interface: A or B

You are interested in which combinations of pairs of these factors are likely to cause failures. However, certain combinations are not possible:

- An ISDN line on either phone (Near or Far) cannot use interface A.
- Business and Residential lines on the originating phone (Near) cannot use interface B.

JMP PRO Create the Design

The factors and their settings are given in the data table Phone Factors.jmp. Create a Strength 2 covering array by following these steps.

JMP PRO Load Factors

1. Select **Help > Sample Data Library** and open Design Experiment/Phone Factors.jmp.

   The Phone Factors.jmp data table contains the factors and their settings.

2. Select **DOE > Special Purpose > Covering Array**.

   Notice that the menu next to **Strength: t =** , is set to 2 by default.

3. From the Covering Array red triangle menu, select **Load Factors**.

   The Factors outline is populated with the five factors and their levels.

Figure 20.10 Factors Outline for Phone Factors

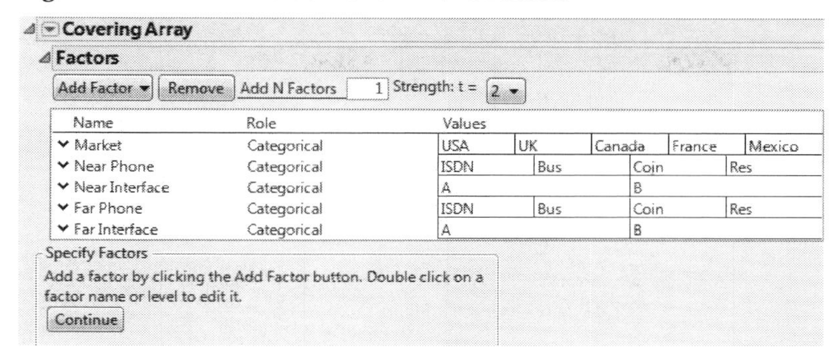

4. Click **Continue**.

   The Restrict Factor Level Combinations outline opens.

JMP PRO Restrict Factor Level Combinations

You can specify disallowed combinations in two ways:

- Use Disallowed Combinations Filter
- Use Disallowed Combinations Script

The filter gives an intuitive way to specify disallowed combinations. The script provides a quick and easy way to specify disallowed combinations, but requires that you have written or saved a script. In this example, if you do not want to specify combinations using the filter, skip to "Specify Disallowed Combinations Using a Script" on page 551.

Recall that the restrictions are the following:

- An ISDN line on either phone (Near or Far) cannot use interface A.
- Business and Residential phones on the originating phone (Near) cannot use interface B.

JMP PRO Specify Disallowed Combinations Using the Filter

Use this approach to enter disallowed combinations using the filter interface. Alternatively, you can paste a script as shown in "Specify Disallowed Combinations Using a Script" on page 551.

1. Select **Use Disallowed Combinations Filter**.
2. From the Add Filter Factors list, select Near Phone and Near Interface and click **Add**.
3. Hold down the Ctrl key and click ISDN under Near Phone and A under Near Interface.

Figure 20.11 Disallowed Combinations Panel Showing First Constraints

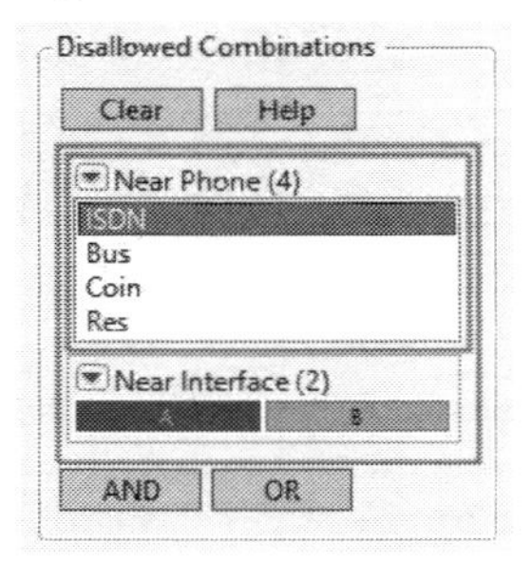

Both blocks should turn dark. You have added the constraint that an ISDN line on the originating phone (Near) cannot use interface A.

4. Click **OR**.
5. From the Add Filter Factors list, select Far Phone and Far Interface and click **Add**.
6. Hold down the Ctrl key and click ISDN under Far Phone and A under Far Interface.

   You have added the constraint that an ISDN line on the receiving phone (Far) cannot use interface A.
7. Click **OR**.
8. From the Add Filter Factors list, select Near Phone and Near Interface and click **Add**.
9. Hold down the Ctrl key and click Bus and Res under Near Phone and B under Near Interface.

   You have added the restriction that Business and Residential lines on the originating phone (Near) cannot use interface B.

Figure 20.12 Completed Disallowed Combinations Filter

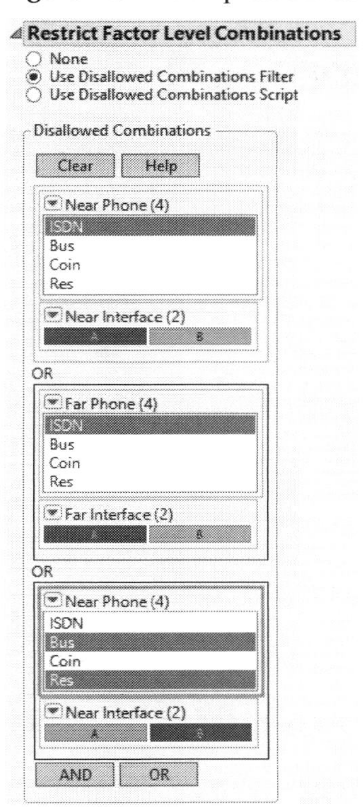

JMP PRO Specify Disallowed Combinations Using a Script

Alternatively, you can specify disallowed combinations by constructing a script. After loading your factors ("Load Factors" on page 548), do the following:

1. Click **Continue**.
2. Select **Use Disallowed Combinations Script**.
3. Copy the following script and paste it in the Disallowed Combinations Expression script box:

```
(Near Phone == "ISDN" & Near Interface == "A") |
(Far Phone == "ISDN" & Far Interface == "A")  |
(Near Phone == "Bus" & Near Interface == "B") |
(Near Phone == "Res" & Near Interface == "B")
```

Figure 20.13 Completed Disallowed Combinations Script Window

Restrict Factor Level Combinations

- None
- Use Disallowed Combinations Filter
- Use Disallowed Combinations Script

Disallowed Combinations Expression

```
1   (Near Phone == "ISDN" & Near Interface == "A") |
2   (Far Phone == "ISDN" & Far Interface == "A")  |
3   (Near Phone == "Bus" & Near Interface == "B") |
4   (Near Phone == "Res" & Near Interface == "B")
```

Note: In DOE platforms other than Covering Arrays, a script for disallowed combinations must specify the level number rather than the level name.

JMP PRO Construct the Design Table

Note: Setting the Random Seed in the next two steps reproduces the exact results shown in this example. When constructing a design on your own, these steps are not necessary.

1. From the red triangle menu, select **Set Random Seed**.
2. Enter 632 and click **OK**.
3. Click **Make Design**.

   The Design outline opens to show a 20-run design. A Metrics outline is added to the window.

Figure 20.14 Metrics Outline for Phone Design

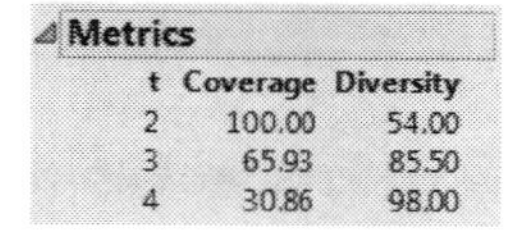

Metrics

t	Coverage	Diversity
2	100.00	54.00
3	65.93	85.50
4	30.86	98.00

   The Metrics outline indicates that Strength 2 coverage is 100%. This means that all permissible two-factor combinations are represented in the design. The design also covers 65% of all three-factor combinations.

4. Click **Make Table**.

   The design is placed in a design table. A column for the response is provided, as well as various scripts.

Figure 20.15 Covering Array Design Table

Covering Array
Design Covering Array, Streng
DOE Dialog
Disallowed Combinations
Analysis

Columns (6/0)
Response
Market *
Near Phone *
Near Interface *
Far Phone *
Far Interface *

Rows
All rows 20
Selected 0
Excluded 0
Hidden 0
Labelled 0

	Response	Market	Near Phone	Near Interface	Far Phone	Far Interface
1	•	France	Bus	A	Coin	A
2	•	USA	Res	A	Res	A
3	•	USA	Coin	A	Coin	B
4	•	Mexico	Res	A	Res	A
5	•	Canada	Res	A	Bus	B
6	•	Canada	Coin	B	ISDN	B
7	•	UK	Bus	A	ISDN	B
8	•	USA	ISDN	B	ISDN	B
9	•	France	ISDN	B	Res	B
10	•	Mexico	Bus	A	Bus	A
11	•	Mexico	Coin	B	Coin	B
12	•	Canada	Bus	A	Res	A
13	•	France	Coin	B	Bus	A
14	•	USA	Bus	A	Bus	B
15	•	UK	Res	A	Coin	A
16	•	UK	ISDN	B	Bus	A
17	•	UK	Coin	B	Res	A
18	•	Mexico	ISDN	B	ISDN	B
19	•	Canada	ISDN	B	Coin	B
20	•	France	Res	A	ISDN	B

JMP PRO Analyze the Experimental Data

Now that you have your design table, you can conduct your experiment and record your data in the Response column of the design table. Your experimental results are in the Phone Data.jmp sample data table.

1. Select **Help > Sample Data Library** and open Design Experiment/Phone Data.jmp.
2. In the Table panel, click the green triangle next to the **Analysis** script.

Figure 20.16 Analysis of Phone Experimental Data

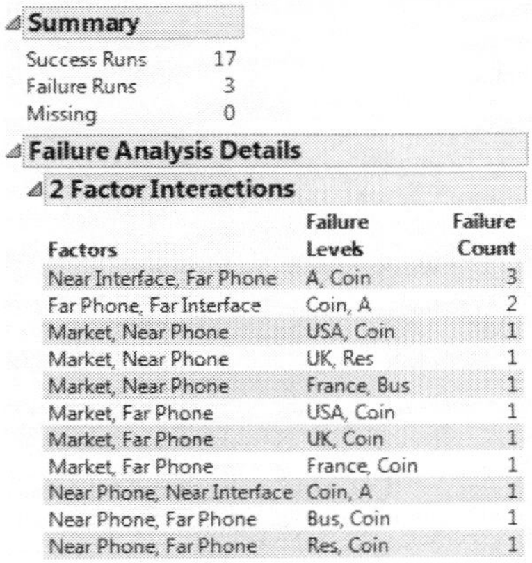

Summary

Success Runs	17
Failure Runs	3
Missing	0

Failure Analysis Details

2 Factor Interactions

Factors	Failure Levels	Failure Count
Near Interface, Far Phone	A, Coin	3
Far Phone, Far Interface	Coin, A	2
Market, Near Phone	USA, Coin	1
Market, Near Phone	UK, Res	1
Market, Near Phone	France, Bus	1
Market, Far Phone	USA, Coin	1
Market, Far Phone	UK, Coin	1
Market, Far Phone	France, Coin	1
Near Phone, Near Interface	Coin, A	1
Near Phone, Far Phone	Bus, Coin	1
Near Phone, Far Phone	Res, Coin	1

The Summary outline indicates that three tests failed.

The Failure Analysis Details outline contains a 2 Factor Interactions report, because a two-way interaction is the lowest level interaction that detects a failure.

The 2 Factor Interactions report shows the combinations that might have caused the three failures. It is possible that one combination, Near Interface set to A and Far Phone set to Coin, is responsible for all three failures. Or it is possible that two or three other combinations caused the three failures.

3. Select the first line in the 2 Factor Interactions report.

   In the data table, rows 1, 3, and 15 are selected. Failures occur for these combinations, regardless of the settings for Market, Near Phone, and Far Interface. But note that other combinations of factor settings could account for these failures as well.

JMP PRO Covering Array Window

The covering array window updates as you work through the design steps. For more information about the general flow of DOE windows, see "The DOE Workflow: Describe, Specify, Design" on page 54. The outlines, separated by buttons that update the outlines, follow the flow in Figure 20.17.

Figure 20.17 Covering Array Flow

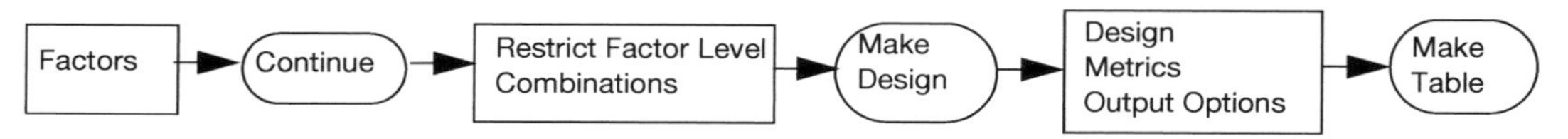

JMP PRO Factors

Add factors in the Factors outline.

Figure 20.18 Factors Outline

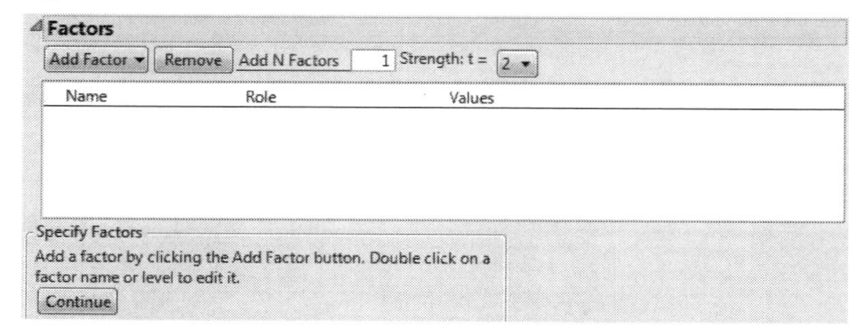

Add Factor Enters the number of factors specified in **Add N Factors**. All factors are categorical. Select or specify the number of levels.

Remove Removes the selected factors.

Add N Factors Adds multiple factors with a specific number of levels. Enter the number of factors to add, click **Add Factor**, and select or specify the number of levels. Repeat **Add N Factors** to add multiple factors with different numbers of levels.

Strength t = Select a value to specify the strength of the array.

Tip: When you have completed your Factors panel, select **Save Factors** from the red triangle menu. This saves the factor names and values in a data table that you can later reload. See "Covering Array Options" on page 563.

JMP PRO Factors Table

The Factors table contains the following columns:

Name The name of the factor. When a factor is added, it is given a default name of X1, X2, and so on. To change this name, double-click it and enter the desired name.

Role Specifies the Design Role of the factor. The Design Role for all covering array factors is Categorical. The Design Role column property for the factor is saved to the data table. This property ensures that the factor is modeled appropriately.

Values The settings for the factors. To insert Values, click on the default values and enter the desired values. The value ordering in the design table is the order of the values as entered from left to right.

JMP PRO Editing the Factors Table

In the Factors outline, notice the following:

- To edit a factor name, double-click the factor name.
- Categorical factors have a down arrow to the left of the factor name. Click the arrow to add a level.
- To remove a factor level, click the value, click **Delete**, and click outside the text box.
- To edit a value, click the value in the Values column.

JMP PRO Factor Column Properties

For each factor, the Value Labels column property is saved to the design table. The Value Labels column property represents values in a column with specified labels. These labels are shown in the data table and are used in plots and reports. For details, see "Value Labels" on page 683 in the "Column Properties" appendix.

JMP PRO Restrict Factor Level Combinations

When you complete the Factors outline and click Continue, the Restrict Factor Level Combinations outline appears. This outline enables you to specify factor level combinations that are prohibited. Unless you have loaded a constraint or included one as part of a script, the **None** option is selected. To specify constraints, select one of the other options:

Use Disallowed Combinations Filter Defines sets of constraints based on restricting values of individual factors. You can define both AND and OR constraints. See "Use Disallowed Combinations Filter" on page 556.

Use Disallowed Combinations Script Defines disallowed combinations and other constraints as Boolean JSL expressions in a script editor box. See "Use Disallowed Combinations Script" on page 558.

JMP PRO Use Disallowed Combinations Filter

This option uses an adaptation of the Data Filter to facilitate specifying disallowed combinations. For detailed information about using the Data Filter, see the JMP Reports chapter in the *Using JMP* book.

To add disallowed combinations:

1. Select factors from the Add Filter Factors list and click **Add**.
2. Specify the disallowed combinations by selecting levels.

Note: The red triangle options in the Add Filter Factors menu are the same as those found in the Select Columns panel of many platform launch windows. See "Column Filter Menu" the Get Started chapter in the *Using JMP* book for additional details.

When you click Add, the initial panel is updated. The Disallowed Combinations control panel shows the selected factors and provides options for further control.

The Covering Array platform allows only categorical factors. For categorical factors, the possible levels are shown either as labeled blocks or, when the number of levels is large, as list entries. Select a level to disallow it. To select multiple levels, hold down the Ctrl key. The block or list entries are highlighted to indicate the levels that have been disallowed. When you add a factor to the Disallowed Combinations panel, the number of levels of the categorical factor is given in parentheses following the factor name.

Disallowed Combinations Options

Clear Clears all disallowed factor level settings that you have specified. This does not clear the selected factors.

Start Over Removes all selected factors and returns you to the initial list of factors.

AND Opens the Add Filter Factors list. Selected factors become an AND group. Any combination of factor levels specified within an AND group is disallowed.

To add a factor to an AND group later on, click the group's outline to see a highlighted rectangle. Select AND and add the factor.

To remove a single factor, select **Delete** from its red triangle menu.

OR Opens the Add Filter Factors list. Selected factors become a separate AND group. For AND groups separated by OR, a combination is disallowed if it is specified in at least one AND group.

Red Triangle Options for Factors

A factor can appear in several OR groups. An occurrence of the factor in a specific OR group is referred to as an *instance* of the factor.

Delete Removes the selected instance of the factor from the Disallowed Combinations panel.

Clear Selection Clears any selection for that instance of the factor.

Invert Selection Deselects the selected values and selects the values not previously selected for that instance of the factor.

Display Options Changes the appearance of the display. Options include the following:

- **Blocks Display** shows each level as a block.
- **List Display** shows each level as a member of a list.
- **Single Category Display** shows each level.
- **Check Box Display** adds a check box next to each value.

Find Provides a text box beneath the factor name where you can enter a search string for levels of the factor. Press the Enter key or click outside the text box to perform the search. Once **Find** is selected, the following Find options appear in the red triangle menu:

- **Clear Find** clears the results of the Find operation and returns the panel to its original state.
- **Match Case** uses the case of the search string to return the correct results.
- **Contains** searches for values that include the search string.
- **Does not contain** searches for values that do not include the search string.
- **Starts with** searches for values that start with the search string.
- **Ends with** searches for values that end with the search string.

Use Disallowed Combinations Script

This option opens a script window where you insert a script that identifies the combinations that you want to disallow. The script must evaluate as a Boolean expression. When the expression evaluates as true, the specified combination is disallowed.

When creating the expression, use the name of the level in quotation marks. Do not use the ordinal value of the level. For example, Figure 20.19 shows the script that you entered in the phone interface example, "Specify Disallowed Combinations Using a Script" on page 551.

Note: In DOE platforms other than Covering Arrays, a script for disallowed combinations must specify the level number rather than the level name.

Figure 20.19 Script Window Showing Names of Levels in Quotes

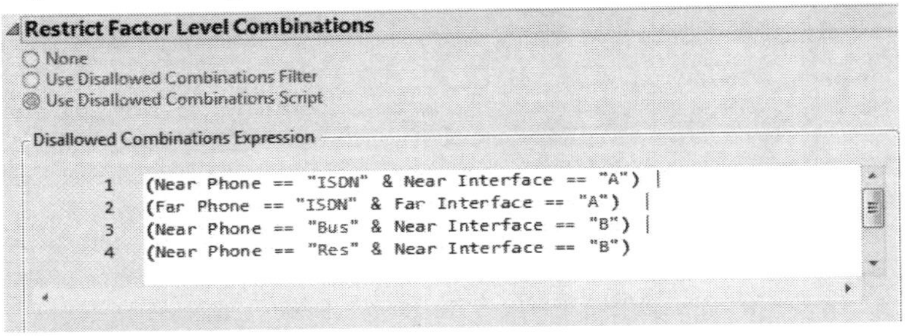

Restrict Factor Level Combinations

- None
- Use Disallowed Combinations Filter
- Use Disallowed Combinations Script

Disallowed Combinations Expression

```
1  (Near Phone == "ISDN" & Near Interface == "A") |
2  (Far Phone == "ISDN" & Far Interface == "A")  |
3  (Near Phone == "Bus" & Near Interface == "B") |
4  (Near Phone == "Res" & Near Interface == "B")
```

Design

When you click Make Design, the Design and Metrics outlines appear. For designs that require extensive computation, a progress bar appears.

The Design outline shows the design that you have constructed. The first column lists a Run order. You might need to use the scroll bar to view all the runs. The remaining columns show factor settings for each run.

Optimize

Select Optimize to reduce the size of a design that was constructed by the Covering Array platform or that you have loaded using the Load Design red triangle option. Optimize is not available for designs constructed by the Covering Array platform that are known to be optimal. In particular, all unconstrained strength 2 designs for two-level factors constructed by the platform are optimal. Also, any unconstrained strength t design for $t+1$ factors is optimal for any t.

For details about the algorithm, see "Algorithm for Optimize" on page 564.

Note: Optimize is time intensive, but can be run repeatedly to yield incrementally better designs.

Use the **Maximum iterations** option to specify a maximum number of iterations to be used in optimizing the design.

JMP PRO Unsatisfiable Constraints

If a set of constraints prohibits the construction of a covering array where all required factor levels are represented, it is said to be *unsatisfiable*.

Example of Unsatisfiable Constraints

Consider a strength 2 design for three factors, each at three levels.

1. Select **DOE > Special Purpose > Covering Array**.
2. Next to **Add N Factors**, type 3.
3. From the **Add Factor** menu, select **3 Level**.
4. Click **Continue**.
5. Select **Use Disallowed Combinations Filter**.
6. From the **Add Filter Factors** list, select all three factors and click **Add**.
7. Hold down the Ctrl key and select the following levels:
   – For X1, select L1.
   – For X2, select L1, L2, and L3.
   – For X3, select L3.

Figure 20.20 Completed Restrict Factor Level Combinations Panel

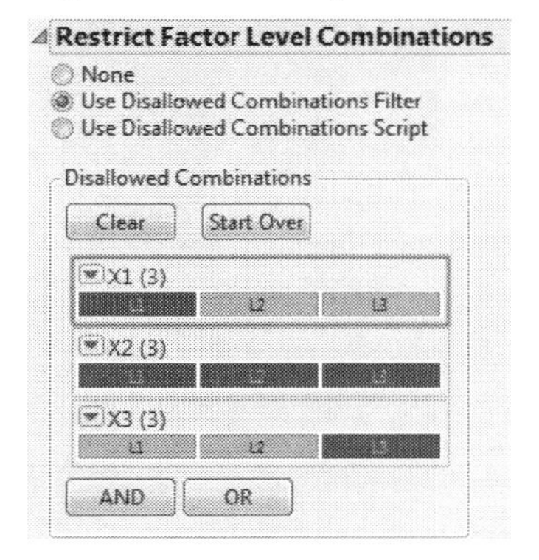

Note: Setting the Random Seed in the next two steps reproduces the exact results shown in this example. When constructing a design on your own, these steps are not necessary.

8. From the Covering Array red triangle menu, select **Set Random Seed**.
9. Enter 12345 and click **OK**.
10. Click **Make Design**.

Figure 20.21 Design and Metrics Outlines

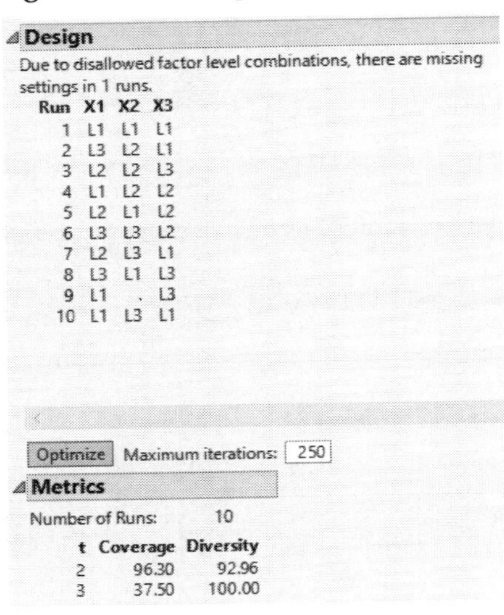

A note beneath the Design outline indicates that one run has a missing setting due to the constraints. That run is run 9. To ensure that the covering array has strength 2, the combination of X1 set to L1 and X3 set to L3 is required. But for these settings, the constraints prohibit all settings for X2.

JMP PRO Metrics

The Metrics outline gives you information about how well the design meets the strength requirements. See Dalal and Mallows (1998) for background on these metrics for unconstrained designs.

t The number of factors.

Coverage The ratio of the number of distinct *t*-factor settings that appear in the design to the total possible number of *t*-factor settings, expressed as a percentage. A *t*-coverage of 100%

indicates that all possible t-factor settings are covered by the design. Note that each t-factor setting can appear multiple times.

For constrained and unsatisfiable designs, the definition of Coverage is adjusted for the number of t-factor settings that are possible once the constraints have been applied to all t-factor combinations. See "Formulas for Metrics" on page 564.

Diversity The ratio of the number of distinct t-factor settings in the design to the total number of occurrences of t-factor settings in the design, expressed as a percentage. The t-diversity measures how well the design avoids replication. A t-diversity of 100% indicates that no t-factor settings are repeated. A t-diversity of 50% indicates that the average number of times that distinct t-factor settings appear is two.

For constrained and unsatisfiable designs, the definition of Diversity is adjusted for the number of runs with missing settings. See "Formulas for Metrics" on page 564.

JMP PRO Output Options

Make Table Constructs the Covering Array data table.

Back Takes you back to the Factors outline. You can make changes to the previous outlines and regenerate the design.

Note: If you have defined Disallowed Combinations in the Restrict Factor Level Combinations outline, these are retained as a script. The script is shown in the Use Disallowed Combinations Script panel when you click Continue.

JMP PRO The Covering Array Data Table

The Covering Array data table contains a first column where you can enter the response. The remaining columns give the factor settings.

The Table panel in the upper left contains a Design note indicating that the design is a Covering Array and giving the Strength of the design. The Table panel also contains the following scripts.

DOE Dialog Re-creates the Covering Array window that you used to generate the design table.

Disallowed Combinations Shows factor level restrictions that you entered in the Restrict Factor Level Combinations outline.

Analysis Provides an analysis of your experimental data. For details, see "Analysis Script" on page 562. For background, see Zhang and Zhang (2011).

Tip: To run a script, click the green triangle next to the script name.

Figure 20.22 Partial View of Covering Array Table for Software Data.jmp Showing Scripts

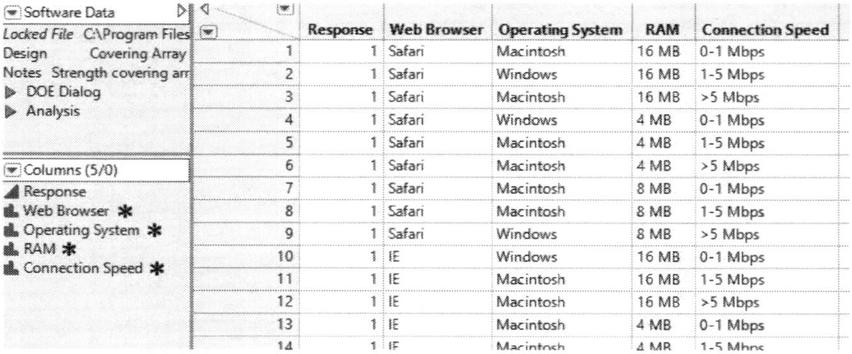

Software Data
Locked File C:\Program Files
Design Covering Array
Notes Strength covering arr
DOE Dialog
Analysis

Columns (5/0)
Response
Web Browser
Operating System
RAM
Connection Speed

	Response	Web Browser	Operating System	RAM	Connection Speed
1	1	Safari	Macintosh	16 MB	0-1 Mbps
2	1	Safari	Windows	16 MB	1-5 Mbps
3	1	Safari	Macintosh	16 MB	>5 Mbps
4	1	Safari	Windows	4 MB	0-1 Mbps
5	1	Safari	Macintosh	4 MB	1-5 Mbps
6	1	Safari	Macintosh	4 MB	>5 Mbps
7	1	Safari	Macintosh	8 MB	0-1 Mbps
8	1	Safari	Macintosh	8 MB	1-5 Mbps
9	1	Safari	Windows	8 MB	>5 Mbps
10	1	IE	Windows	16 MB	0-1 Mbps
11	1	IE	Macintosh	16 MB	1-5 Mbps
12	1	IE	Macintosh	16 MB	>5 Mbps
13	1	IE	Macintosh	4 MB	0-1 Mbps
14	1	IE	Macintosh	4 MB	1-5 Mbps

JMP PRO Analysis Script

The Analysis script assumes the following about the **Response** column:

- The responses are recorded as 0 for failure and 1 for success.
- Missing values are permitted.
- The **Response** column is continuous.
- You can rename or move the **Response** column.

The Analysis script produces a report with two outlines:

- The Summary outline gives the number of runs resulting in Success, Failure, and the number of runs for which the response is Missing.
- The Failure Analysis Details report contains a <k> Factor Interactions report. The value of *k* is the smallest number of interactions that detect a failure. (For a definition of *detect*, see "Covering Arrays and Strength" on page 542.) The three columns contain the following:
  - The **Factors** column lists all *k*-factor combinations that detect failures.
  - The **Failure Levels** column lists the values of the *k* factors in the **Factors** column that detect failures.
  - The **Failure Count** column gives the number of failures corresponding to the *k*-factor combination of **Failure Levels**.

Note: A failure observation can appear in more than one of the *k*-factor combinations listed in the **Failure Levels** column.

The rows in the <k> Factor Interactions report are dynamically linked to the data table. If you select one or more rows in the report, the corresponding rows are selected in the data table.

Covering Array Options

The red triangle menu in the Covering Array platform contains these options:

Save Factors Creates a data table containing a column for each factor that contains its factor levels. Each factor's column contains these column properties: Design Role, Value Ordering, and Factor Changes. Saving factors enables you to quickly load them into a DOE window.

Note: You can create a factors table for a Covering Array by entering data into an empty table, but remember to assign each column an appropriate Design Role of Categorical. Right-click on the column name in the data grid and select **Column Properties > Design Role.** In the Design Role area, select Categorical.

Load Factors Loads factors that you have saved using the Save Factors option into the Factors outline.

Load Design Loads a design from the active data table. If no data table is active, you are prompted to open one. When you select Load Design, a menu appears that enables you to select the columns that you want to specify as factors in the design. All columns are imported as categorical. Columns and their values are listed in the Factors outline. The Design outline shows a Run for each row in the data table and gives the values of the factors for each run.

The Load Design options enables you to obtain metrics, modify, or construct an Analysis script for an existing design:

– The Metrics outline shows *t*-Coverage and *t*-Diversity for the specified design.

– You can click **Back** to impose factor level restrictions and then construct a new design.

– Clicking **Make Table** constructs a design table where you can enter responses. The table contains an Analysis script for the design.

Set Random Seed Sets the random seed that JMP uses to control certain actions that have a random component. For Covering Arrays, the seed selects a starting design and an iteration count. For most designs, the random seed guarantees reproducibility of the design, but not of the run order.

Note: Upper limits on time as well as iteration count are used to limit design construction time. For some large and high strength designs, depending on the machine, the time limit might override the iteration limit. For such designs, the random seed will not guarantee reproducibility.

To reproduce a design, enter the random seed used to generate it before clicking Make Design.

Note: The random seed associated with a design is included in the DOE Dialog script that is saved to the design data table.

Advanced Options Not available for Covering Arrays.

Save Script to Script Window Creates the script for the design that you specified in the Covering Array window and places it in an open script window.

JMP PRO Technical Details

JMP PRO Algorithm for Optimize

The Optimize button invokes an algorithm that is conceptually similar to a class of covering array optimizers sometimes referred to as *post-construction randomized optimizers* (Nayeri et al., 2013). However, JMP's algorithm differs from most in that it also addresses designs with constraints. In particular, it optimizes constrained covering arrays as well as unsatisfiable, constrained covering arrays.

The algorithm assumes that the design to be optimized is a covering array of the specified strength. For a K factor design of strength t, the algorithm iteratively examines all ${}_K\mathrm{C}_t$ factor projections to determine whether runs can be eliminated or merged. Consequently, as K or t increases, the run time of the algorithm quickly escalates. To improve performance, the JMP implementation is threaded to use as many CPU cores as are available on your workstation.

JMP PRO Formulas for Metrics

The formulas for Coverage and Diversity depend on whether there are constraints. The following notation is used:

${}_u\mathrm{C}_v$ is the number of combinations of u things taken v at a time

t is the strength of the design

K is the number of factors

$M = {}_K\mathrm{C}_t$

$i = 1, 2, ..., M$ is an index that orders all combinations, or *projections*, of t factors

v_{ik} is the number of levels for the kth factor

n_i is the number of distinct t tuples in the design for the ith projection

p_i is the product of the v_{ik} for the factors in the ith projection

r is the number of runs in the design

JMP PRO Unconstrained Design

Coverage and Diversity are given by the following:

$$Coverage = \frac{1}{M}\sum_{i=1}^{M} n_i/p_i$$

$$Diversity = \frac{1}{M}\sum_{i=1}^{M} n_i/r$$

JMP PRO Constrained Design

In a constrained design, certain *t* tuples are not allowed. This can result in missing values for some *t* tuples. For some combinations of *t* factors, there might be no valid *t* tuples whatsoever. Coverage and diversity must be defined in terms of the possible valid combinations. For this reason, the formulas for constrained designs require additional notation:

a_i is the number of invalid *t* tuples arising from factors in the ith projection

m is the number of projections where there are no valid *t* tuples

q_i is the number of runs in the design with missing values for any factor in the ith projection

$r_i = r - q_i$

$M' = M - m$

Coverage and Diversity are given by the following:

$$Coverage = \frac{1}{M'}\sum_{i=1}^{M'} n_i/(p_i - a_i)$$

$$Diversity = \frac{1}{M'}\sum_{i=1}^{M'} n_i/r_i$$

If there are no invalid *t* tuples (M′ = M) and if there are no missing values ($r_i = r$, for all i), then the definitions for coverage and diversity for constrained designs reduce to the definitions for unconstrained designs. For details, see Morgan (2014).

Chapter 21

Space-Filling Designs

Space-filling designs are useful in situations where run-to-run variability is of far less concern than the form of the model. Consider a sensitivity study of a computer simulation model. In this situation, and for any mechanistic or deterministic modeling problem, any variability is small enough to be ignored. For systems with no variability, replication, randomization, and blocking are irrelevant.

The Space Filling platform provides designs for situations with both continuous and categorical factors. For continuous factors, space-filling designs have two objectives:

- maximize the distance between any two design points
- space the points uniformly

Figure 21.1 Space-Filling Design

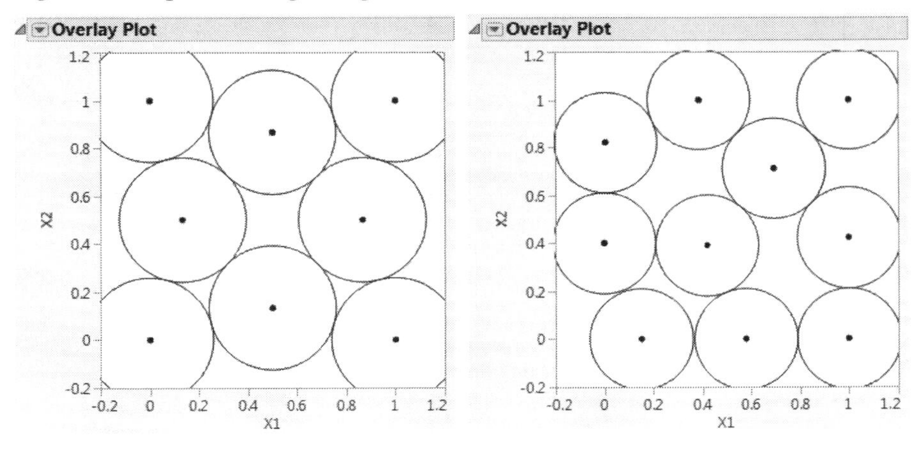

Overview of Space-Filling Designs

Space-filling designs are useful for modeling systems that are deterministic or near-deterministic. One example of a deterministic system is a computer simulation. Such simulations can be very complex involving many variables with complicated interrelationships. A goal of designed experiments on these systems is to find a simpler empirical model that adequately predicts the behavior of the system over limited ranges of the factors.

In experiments on systems where there is substantial random noise, the goal is to minimize the variance of prediction. In experiments on deterministic systems, there is no variance but there is *bias*. Bias is the difference between the approximation model and the true mathematical function. The goal of space-filling designs is to bound the bias.

There are two schools of thought on how to bound the bias. One approach is to spread the design points out as far from each other as possible consistent with staying inside the experimental boundaries. The other approach is to space the points out evenly over the region of interest.

The **Space Filling** designer supports the following design methods:

Note: If the number of runs is 500 or less, a Gaussian Process model is saved to the data table. If the number of runs exceeds 500, a Neural model is saved to the data table.

Sphere Packing maximizes the minimum distance between pairs of design points. See "Sphere-Packing Designs" on page 580 and "Create the Sphere-Packing Design for the Borehole Data" on page 599.

Latin Hypercube maximizes the minimum distance between design points but requires even spacing of the levels of each factor. This method produces designs that mimic the uniform distribution. The Latin Hypercube method is a compromise between the Sphere-Packing method and the Uniform design method. See "Latin Hypercube Designs" on page 583.

Uniform minimizes the discrepancy between the design points (which have an empirical uniform distribution) and a theoretical uniform distribution. See "Uniform Designs" on page 586.

Minimum Potential spreads points out inside a sphere around the center. See "Minimum Potential Designs" on page 589.

Maximum Entropy measures the amount of information contained in the distribution of a set of data. See "Maximum Entropy Designs" on page 592.

Gaussian Process IMSE Optimal creates a design that minimizes the integrated mean squared error of the Gaussian process over the experimental region. See "Gaussian Process IMSE Optimal Designs" on page 593.

Fast Flexible Filling The Fast Flexible Filling method forms clusters from random points in the design space. These clusters are used to choose design points according to an optimization criterion. This is the only method that can accommodate categorical factors and constraints on the design space. You can specify linear constraints and disallowed combinations. See "Fast Flexible Filling Designs" on page 594 and "Creating and Viewing a Constrained Fast Flexible Filling Design" on page 597.

Space Filling Design Window

The Space Filling Design window updates as you work through the design steps. The outlines that appear, separated by buttons that update the window, follow the flow in Figure 21.2.

Figure 21.2 Space Filling Design Flow

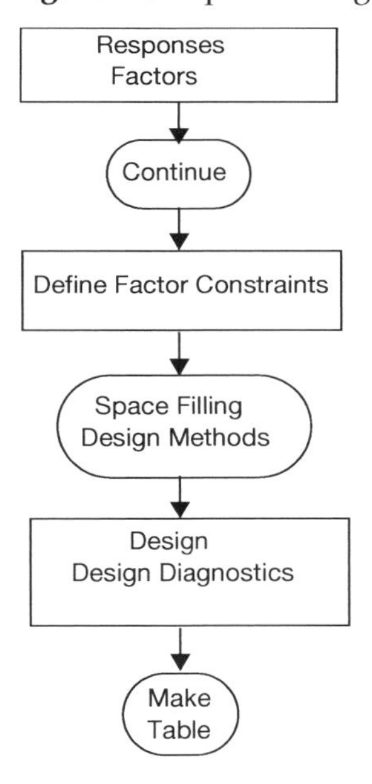

This section describes the outlines in the Space Filling Design window.

Responses

Use the Responses outline to specify one or more responses.

Tip: When you have completed the Responses outline, consider selecting **Save Responses** from the red triangle menu. This option saves the response names, goals, limits, and importance values in a data table that you can later reload in DOE platforms.

Figure 21.3 Responses Outline

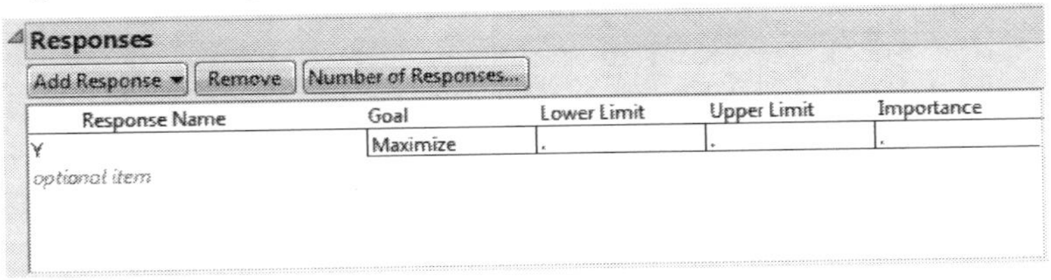

Add Response Enters a single response with a goal type of Maximize, Match Target, Minimize, or None. If you select Match Target, enter limits for your target value. If you select Maximize or Minimize, entering limits is not required but can be useful if you intend to use desirability functions.

Remove Removes the selected responses.

Number of Responses Enters additional responses so that the number that you enter is the total number of responses. If you have entered a response other than the default Y, the Goal for each of the additional responses is the Goal associated with the last response entered. Otherwise, the Goal defaults to Match Target. Click the Goal type in the table to change it.

The Responses outline contains the following columns:

Response Name The name of the response. When added, a response is given a default name of Y, Y2, and so on. To change this name, double-click it and enter the desired name.

Goal, Lower Limit, Upper Limit The Goal tells JMP whether you want to maximize your response, minimize your response, match a target, or that you have no response goal. JMP assigns a Response Limits column property, based on these specifications, to each response column in the design table. It uses this information to define a desirability function for each response. The Profiler and Contour Profiler use these desirability functions to find optimal factor settings. For further details, see the Profiler chapter in the *Profilers* book and "Response Limits" on page 653 in the "Column Properties" appendix.

– A Goal of Maximize indicates that the best value is the largest possible. If there are natural lower or upper bounds, you can specify these as the Lower Limit or Upper Limit.

– A Goal of Minimize indicates that the best value is the smallest possible. If there are natural lower or upper bounds, you can specify these as the Lower Limit or Upper Limit.

– A Goal of Match Target indicates that the best value is a specific target value. The default target value is assumed to be midway between the Lower Limit and Upper Limit.

– A Goal of None indicates that there is no goal in terms of optimization. No desirability function is constructed.

Note: If your target response is not midway between the Lower Limit and the Upper Limit, you can change the target after you generate your design table. In the data table, open the Column Info window for the response column (**Cols > Column Info**) and enter the desired target value.

Importance When you have several responses, the Importance values that you specify are used to compute an overall desirability function. These values are treated as weights for the responses. If there is only one response, then specifying the Importance is unnecessary because it is set to 1 by default.

Editing the Responses Outline

In the Responses outline, note the following:

- Double-click a response to edit the response name.
- Click the goal to change it.
- Click on a limit or importance weight to change it.
- For multiple responses, you might want to enter values for the importance weights.

Response Limits Column Property

The Goal, Lower Limit, Upper Limit, and Importance that you specify when you enter a response are used in finding optimal factor settings. For each response, the information is saved in the generated design data table as a Response Limits column property. JMP uses this information to define the desirability function. The desirability function is used in the Prediction Profiler to find optimal factor settings. For further details about the Response Limits column property and examples of its use, see "Response Limits" on page 653 in the "Column Properties" appendix.

If you do not specify a Lower Limit and Upper Limit, JMP uses the range of the observed data for the response to define the limits for the desirability function. Specifying the Lower Limit and Upper Limit gives you control over the specification of the desirability function. For more details about the construction of the desirability function, see the Profiler chapter in the *Profilers* book.

Factors

Add factors in the Factors outline.

Figure 21.4 Factors Outline

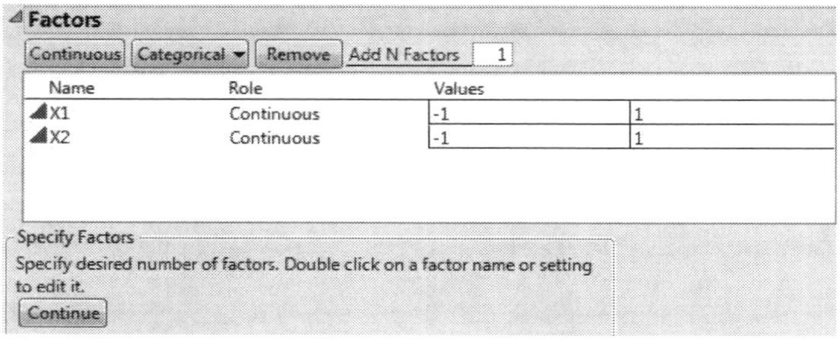

The Factors outline contains these options:

Continuous Enters the number of continuous factors specified in **Add N Factors**.

Categorical Enters the number of nominal factors specified in **Add N Factors.**

Remove Removes the selected factors.

Add N Factors Adds multiple factors of a given type. Enter the number of factors to add and click Continuous or Categorical. Repeat **Add N Factors** to add multiple factors of different types.

Tip: When you have completed your Factors panel, select **Save Factors** from the red triangle menu. This saves the factor names and values in a data table that you can later reload. See "Space Filling Design Options" on page 578.

Factors Outline

The Factors outline contains the following columns:

Name The name of the factor. When added, a factor is given a default name of X1, X2, and so on. To change this name, double-click it and enter the desired name.

Role Specifies the Design Role of the factor. The Design Role column property for the factor is saved to the data table. This property ensures that the factor type is modeled appropriately.

Values The experimental settings for the factors. To insert Values, click on the default values and enter the desired values.

Editing the Factors Outline

In the Factors outline, note the following:

- To edit a factor name, double-click the factor name.

- Categorical factors have a down arrow to the left of the factor name. Click the arrow to add a level.
- To remove a factor level, click the value, click **Delete**, and click outside the text box.
- To edit a value, click the value in the Values column.

Factor Types

Continuous Numeric data types only. A continuous factor is a factor that you can conceptually set to any value between the lower and upper limits you supply, given the limitations of your process and measurement system.

Categorical Either numeric or character data types. For a categorical factor, the value ordering is the order of the values as entered from left to right. This ordering is saved in a Value Ordering column property after the design data table is created.

Factor Column Properties

For each factor, various column properties are saved to the data table.

Design Role Each factor is assigned the Design Role column property. The Role that you specify in defining the factor determines the value of its Design Role column property. The Design Role property reflects how the factor is intended to be used in modeling the experimental data. Design Role values are used in the Augment Design platform.

Factor Changes Each factor is assigned the Factor Changes column property with a setting of Easy. In space-filling designs, it is assumed that factor levels can be changed for each experimental run. Factor Changes values are used in the Evaluate Design and Augment Design platforms.

Coding If the Role is Continuous, the Coding column property for the factor is saved. This property transforms the factor values so that the low and high values correspond to –1 and +1, respectively. The estimates and tests in the Fit Least Squares report are based on the transformed values.

Value Ordering If the Role is Categorical or Blocking, the Value Ordering column property for the factor is saved. This property determines the order in which levels of the factor appear.

Define Factor Constraints

Note: Constraints can be specified only for designs constructed using the Fast Flexible Filling method.

Use Define Factor Constraints to restrict the design space. Unless you have loaded a constraint or included one as part of a script, the **None** option is selected. To specify constraints, select one of the other options:

Specify Linear Constraints Specifies inequality constraints on linear combinations of factors. Only available for factors with a Role of Continuous or Mixture. See "Specify Linear Constraints".

Note: When you save a script for a design that involves a linear constraint, the script expresses the linear constraint as a *less than or equal to* inequality (≤).

Use Disallowed Combinations Filter Defines sets of constraints based on restricting values of individual factors. You can define both AND and OR constraints. See "Use Disallowed Combinations Filter".

Use Disallowed Combinations Script Defines disallowed combinations and other constraints as Boolean JSL expressions in a script editor box. See "Use Disallowed Combinations Script".

Specify Linear Constraints

In cases where it is impossible to vary continuous factors independently over the design space, you can specify linear inequality constraints. Linear inequalities describe factor level settings that are allowed.

Click **Add** to enter one or more linear inequality constraints.

Add Adds a template for a linear expression involving all the continuous factors in your design. Enter coefficient values for the factors and select the direction of the inequality to reflect your linear constraint. Specify the constraining value in the box to the right of the inequality. To add more constraints, click **Add** again.

Note: The Add option is disabled if you have already constrained the design region by specifying a Sphere Radius.

Remove Last Constraint Removes the last constraint.

Check Constraints Checks the constraints for consistency. This option removes redundant constraints and conducts feasibility checks. A JMP alert appears if there is a problem. If constraints are equivalent to bounds on the factors, a JMP alert indicates that the bounds in the Factors outline have been updated.

Use Disallowed Combinations Filter

This option uses an adaptation of the Data Filter to facilitate specifying disallowed combinations. For detailed information about using the Data Filter, see the JMP Reports chapter in the *Using JMP* book.

Select factors from the Add Filter Factors list and click **Add**. Then specify the disallowed combinations by using the slider (for continuous factors) or by selecting levels (for categorical factors).

The red triangle options for the Add Filter Factors menu are those found in the Select Columns panel of many platform launch windows. See the Get Started chapter in the *Using JMP* book for additional details about the column selection menu.

When you click Add, the Disallowed Combinations control panel shows the selected factors and provides options for further control. Factors are represented as follows, based on their modeling types:

Continuous Factors For a continuous factor, a double-arrow slider that spans the range of factor settings appears. An expression that describes the range using an inequality appears above the slider. You can specify disallowed settings by dragging the slider arrows or by clicking on the inequality bounds in the expression and entering your desired constraints. In the slider, a solid blue highlight represents the disallowed values.

Categorical Factor For a categorical factor, the possible levels are displayed either as labeled blocks or, when the number of levels is large, as list entries. Select a level to disallow it. To select multiple levels, hold the Control key. The block or list entries are highlighted to indicate the levels that have been disallowed. When you add a categorical factor to the Disallowed Combinations panel, the number of levels of the categorical factor is given in parentheses following the factor name.

Disallowed Combinations Options

The control panel has the following controls:

Clear Clears all disallowed factor level settings that you have specified. This does not clear the selected factors.

Start Over Removes all selected factors and returns you to the initial list of factors.

AND Opens the Add Filter Factors list. Selected factors become an AND group. Any combination of factor levels specified within an AND group is disallowed.

To add a factor to an AND group later on, click the group's outline to see a highlighted rectangle. Select AND and add the factor.

To remove a single factor, select **Delete** from its red triangle menu.

OR Opens the Add Filter Factors list. Selected factors become a separate AND group. For AND groups separated by OR, a combination is disallowed if it is specified in at least one AND group.

Red Triangle Options for Factors

A factor can appear in several OR groups. An occurrence of the factor in a specific OR group is referred to as an *instance* of the factor.

Delete Removes the selected instance of the factor from the Disallowed Combinations panel.

Clear Selection Clears any selection for that instance of the factor.

Invert Selection Deselects the selected values and selects the values not previously selected for that instance of the factor.

Display Options Available only for categorical factors. Changes the appearance of the display. Options include:

– **Blocks Display** shows each level as a block.

– **List Display** shows each level as a member of a list.

– **Single Category Display** shows each level.

– **Check Box Display** adds a check box next to each value.

Find Available only for categorical factors. Provides a text box beneath the factor name where you can enter a search string for levels of the factor. Press the Enter key or click outside the text box to perform the search. Once **Find** is selected, the following Find options appear in the red triangle menu:

– **Clear Find** clears the results of the Find operation and returns the panel to its original state.

– **Match Case** uses the case of the search string to return the correct results.

– **Contains** searches for values that include the search string.

– **Does not contain** searches for values that do not include the search string.

– **Starts with** searches for values that start with the search string.

– **Ends with** searches for values that end with the search string.

Use Disallowed Combinations Script

Use this option to disallow particular combinations of factor levels using a JSL script. This option can be used with continuous factors or mixed continuous and categorical factors.

This option opens a script window where you insert a script that identifies the combinations that you want to disallow. The script must evaluate as a Boolean expression. When the expression evaluates as true, the specified combination is disallowed.

When forming the expression for a categorical factor, use the ordinal value of the level instead of the name of the level. If a factor's levels are high, medium, and low, specified in that order in the Factors outline, their associated ordinal values are 1, 2, and 3. For example, suppose that you have two continuous factors, X1 and X2, and a categorical factor X3 with three levels: L1, L2, and L3, in order. You want to disallow levels where the following holds:

$$e^{X_1} + 2X_2 < 0 \text{ and } X_3 = L2$$

Enter the expression `(Exp(X1) + 2*X2 < 0) & (X3 == 2)` into the script window.

Figure 21.5 Expression in Script Editor

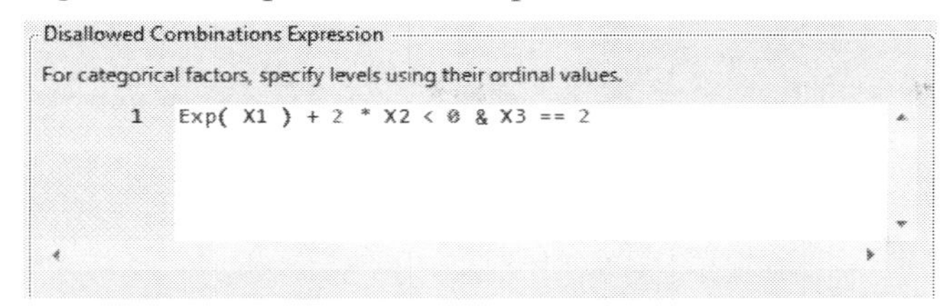

(In the figure, unnecessary parentheses were removed by parsing.) Notice that functions can be entered as part of the Boolean expression.

Space Filling Design Methods

The following methods for constructing space-filling designs are available:

- "Sphere-Packing Designs" on page 580
- "Latin Hypercube Designs" on page 583
- "Uniform Designs" on page 586
- "Minimum Potential Designs" on page 589
- "Maximum Entropy Designs" on page 592
- "Gaussian Process IMSE Optimal Designs" on page 593
- "Fast Flexible Filling Designs" on page 594

Design

The Design outline shows the runs for the space-filling screening design.

Design Diagnostics

The Design Diagnostics outline shows the values for the factors scaled from zero to one. The Minimum Distance is based on these scaled values and is the minimum distance from each point to its nearest neighbor. The row number for the nearest neighbor is given in the Nearest Point column. The discrepancy value shown below the table is the integrated difference between the design points based and a uniform distribution.

Design Table

Make Table Constructs the Space Filling Design data table.

Back Takes you back to where you were before clicking **Make Design**. You can make changes to the previous outlines and regenerate the design.

Space Filling Design Options

The red triangle menu in the Space Filling Design platform contains these options:

Save Responses Saves the information in the Responses panel to a new data table. You can then quickly load the responses and their associated information into most DOE windows. This option is helpful if you anticipate re-using the responses.

Load Responses Loads responses that you saved using the Save Responses option.

Save Factors Saves the information in the Factors panel to a new data table. Each factor's column contains its levels. Other information is stored as column properties. You can then quickly load the factors and their associated information into most DOE windows.

Note: It is possible to create a factors table by entering data into an empty table, but remember to assign each column an appropriate Design Role. Do this by right-clicking on the column name in the data grid and selecting **Column Properties > Design Role**. In the Design Role area, select the appropriate role.

Load Factors Loads factors that you saved using the Save Factors option.

Save Constraints (Unavailable for some platforms) Saves factor constraints that you defined in the Define Factor Constraints or Linear Constraints outline into a data table, with a column for each constraint. You can then quickly load the constraints into most DOE windows.

In the constraint table, the first rows contain the coefficients for each factor. The last row contains the inequality bound. Each constraint's column contains a column property called ConstraintState that identifies the constraint as a "less than" or a "greater than" constraint. See "ConstraintState" on page 686 in the "Column Properties" appendix.

Load Constraints (Unavailable for some platforms) Loads factor constraints that you saved using the Save Constraints option.

Set Random Seed Sets the random seed that JMP uses to control certain actions that have a random component. These actions include the following:

- simulating responses using the Simulate Responses option
- randomizing Run Order for design construction
- selecting a starting design for designs based on random starts

To reproduce a design or simulated responses, enter the random seed that generated them. For designs using random starts, set the seed before clicking Make Design. To control simulated responses or run order, set the seed before clicking Make Table.

Note: The random seed associated with a design is included in the DOE Dialog script that is saved to the design data table.

Simulate Responses Adds response values and a column containing a simulation formula to the design table. Select this option before you click Make Table.

When you click Make Table, the following occur:

- A set of simulated response values is added to each response column.
- For each response, a new a column that contains a simulation model formula is added to the design table. The formula and values are based on the model that is specified in the design window.
- A Model window appears where you can set the values of coefficients for model effects and specify one of three distributions: Normal, Binomial, or Poisson.
- A script called **DOE Simulate** is saved to the design table. This script re-opens the Model window, enabling you to re-simulate values or to make changes to the simulated response distribution.

Make selections in the Model window to control the distribution of simulated response values. When you click Apply, a formula for the simulated response values is saved in a new column called <Y> Simulated, where Y is the name of the response. Clicking Apply again updates the formula and values in <Y> Simulated.

For additional details, see "Simulate Responses" on page 106 in the "Custom Designs" chapter.

Note: JMP PRO You can use Simulate Responses to conduct simulation analyses using the JMP Pro Simulate feature. For information about Simulate and some DOE examples, see the Simulate chapter in the *Basic Analysis* book.

FFF Optimality Criterion For the Fast Flexible Filling design method, enables you to select between the MaxPro criterion (the default) and the Centroid criterion. See "FFF Optimality Criterion" on page 594.

Number of Starts Specifies the number of times that the algorithm for the chosen design type initiates to construct a new design. The best design, based on the criterion for the given design type, is returned. Set to 1 by default for all design types. Not used for Fast Flexible Filling Designs.

Advanced Options > Set Average Cluster Size For the Fast Flexible Filling design method, enables you to specify the average number of randomly generated points used to define each cluster or, equivalently, each design point.

Save Script to Script Window Creates the script for the design that you specified in the Custom Design platform and saves it in an open script window.

Sphere-Packing Designs

The Sphere-Packing design method maximizes the minimum distance between pairs of design points. The effect of this maximization is to spread the points out as much as possible inside the design region.

Creating a Sphere-Packing Design

1. Select **DOE > Special Purpose > Space Filling Design**.
2. Enter responses and factors.

   See “Responses” on page 569.
3. Alter the factor level values, if necessary. For example, Figure 21.6 shows the two existing factors, X1 and X2, with values that range from 0 to 1 (instead of the default –1 to 1).

Figure 21.6 Space-Filling Dialog for Two Factors

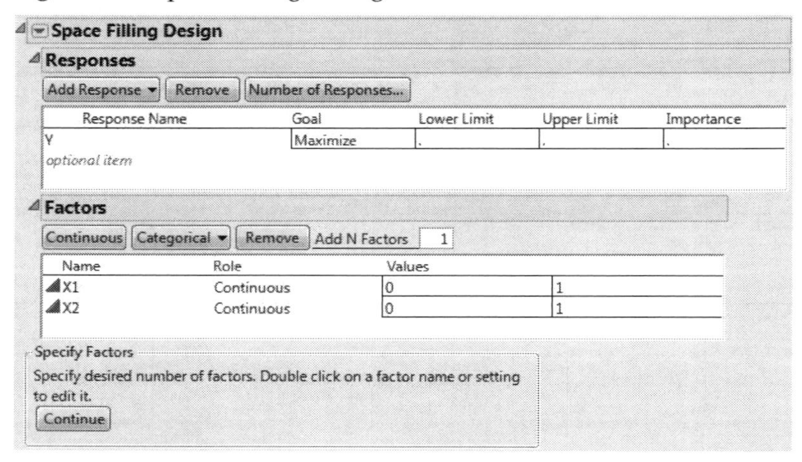

4. Click **Continue**.
5. In the design specification dialog, specify a sample size (**Number of Runs**). Figure 21.7 shows a sample size of eight.

Figure 21.7 Space-Filling Design Dialog

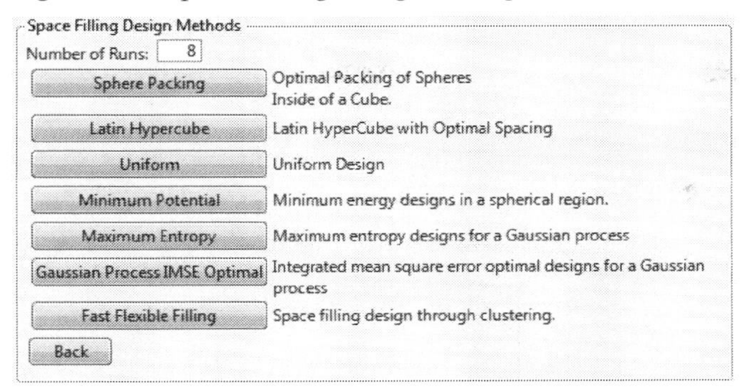

6. Click **Sphere Packing.**

   JMP creates the design and displays the design runs and the design diagnostics. Figure 21.8 shows the Design Diagnostics panel open with 0.518 as the **Minimum Distance**. Your results might differ slightly from the ones below, but the minimum distance is the same.

Figure 21.8 Sphere-Packing Design Diagnostics

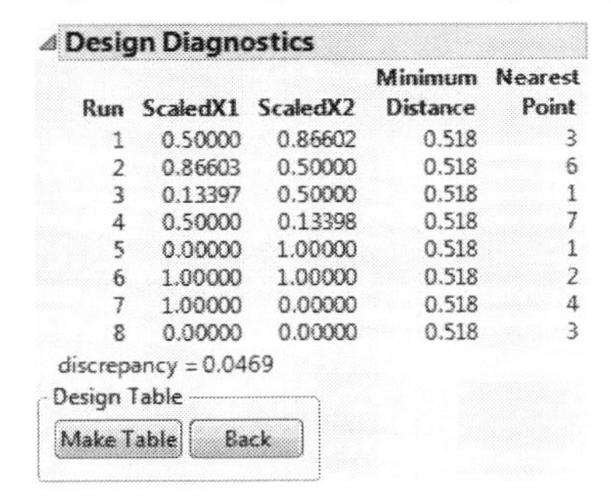

Design Diagnostics

Run	ScaledX1	ScaledX2	Minimum Distance	Nearest Point
1	0.50000	0.86602	0.518	3
2	0.86603	0.50000	0.518	6
3	0.13397	0.50000	0.518	1
4	0.50000	0.13398	0.518	7
5	0.00000	1.00000	0.518	1
6	1.00000	1.00000	0.518	2
7	1.00000	0.00000	0.518	4
8	0.00000	0.00000	0.518	3

discrepancy = 0.0469

Design Table

Make Table Back

7. Click **Make Table**. Use this table to complete the visualization example, described next.

Visualizing the Sphere-Packing Design

To visualize the nature of the Sphere-Packing technique:

- Create an overlay plot.
- Adjust the plot's frame size.
- Add circles using the minimum distance from the diagnostic report shown in Figure 21.8 as the radius for the circles.

Example

Using the table you just created, proceed as follows:

1. Select **Graph > Overlay Plot**.
2. Specify X1 as **X** and X2 as **Y**, and then click **OK**.
3. Adjust the frame size so that the frame is square by right-clicking the plot and selecting **Size/Scale > Size to Isometric**.
4. Right-click the plot and select **Customize**. When the Customize panel appears, click the plus sign to see a text edit area and enter the following script:
   `For Each Row(Circle({:X1, :X2}, 0.518/2))`
   where `0.518` is the minimum distance number that you noted in the Design Diagnostics panel. This script draws a circle centered at each design point with radius 0.259 (half the diameter, 0.518), as shown on the left in Figure 21.9. This plot shows the efficient way JMP packs the design points.
5. Now repeat the procedure exactly as described in the previous section, but with a sample size of 10 instead of eight.

   Remember to change `0.518` in the graphics script to the minimum distance produced by 10 runs. When the plot appears, again set the frame size and create a graphics script using the minimum distance from the diagnostic report as the diameter for the circle. You should see a graph similar to the one on the right in Figure 21.9. Note the irregular nature of the sphere packing. In fact, you can repeat the process a third time to get a slightly different picture because the arrangement is dependent on the random starting point.

Figure 21.9 Sphere-Packing Example with Eight Runs (left) and 10 Runs (right)

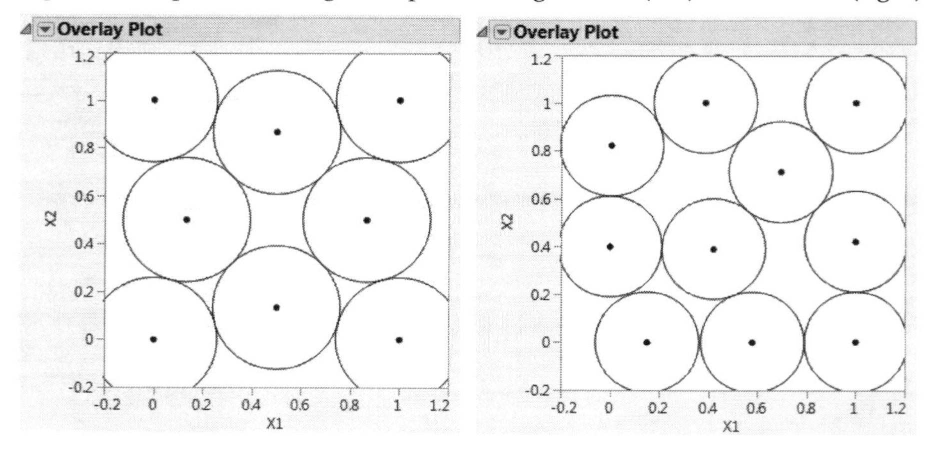

Latin Hypercube Designs

In a Latin Hypercube, each factor has as many levels as there are runs in the design. The levels are spaced evenly from the lower bound to the upper bound of the factor. Like the sphere-packing method, the Latin Hypercube method chooses points to maximize the minimum distance between design points, but with a constraint. The constraint maintains the even spacing between factor levels.

Creating a Latin Hypercube Design

To use the Latin Hypercube method:

1. Select **DOE > Special Purpose > Space Filling Design.**
2. Enter responses, if necessary, and factors.

   See "Responses" on page 569.
3. Alter the factor level values, if necessary. Figure 21.10 shows adding two factors to the two existing factors and changing their values to 1 and 8 instead of the default –1 and 1.

Figure 21.10 Space-Filling Dialog for Four Factors

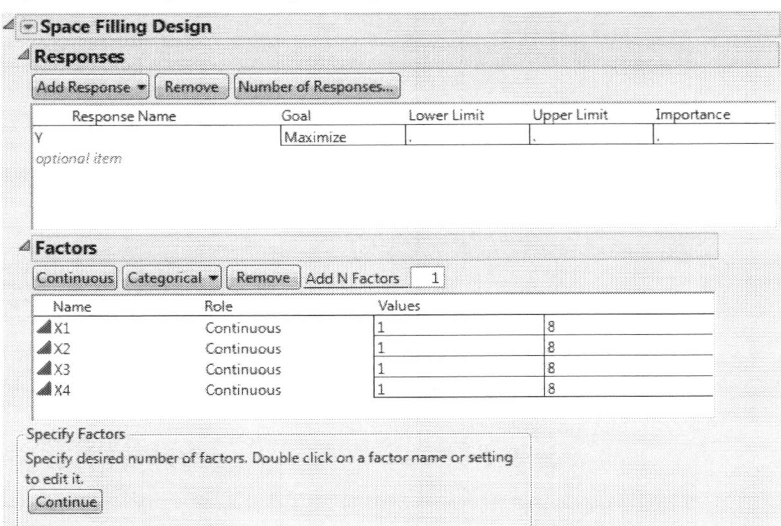

4. Click **Continue.**
5. In the design specification dialog, specify a sample size (**Number of Runs**). This example uses a sample size of eight.

6. Click **Latin Hypercube** (see Figure 21.7). Factor settings and design diagnostics results appear similar to those in Figure 21.11, which shows the Latin Hypercube design with four factors and eight runs.

Note: The purpose of this example is to show that each column (factor) is assigned each level only once, and each column is a different permutation of the levels.

Figure 21.11 Latin Hypercube Design for Four Factors and Eight Runs with Eight Levels

Space Filling Latin Hypercube

Design

Run	X1	X2	X3	X4
1	4.00000	7.00000	1.00000	4.00000
2	2.00000	6.00000	5.00000	7.00000
3	1.00000	4.00000	6.00000	1.00000
4	7.00000	5.00000	3.00000	8.00000
5	5.00000	2.00000	8.00000	6.00000
6	6.00000	8.00000	7.00000	3.00000
7	8.00000	3.00000	4.00000	2.00000
8	3.00000	1.00000	2.00000	5.00000

Design Diagnostics

Run	ScaledX1	ScaledX2	ScaledX3	ScaledX4	Minimum Distance	Nearest Point
1	0.42857	0.85714	0.00000	0.42857	0.782	2
2	0.14286	0.71429	0.57143	0.85714	0.782	1
3	0.00000	0.42857	0.71429	0.00000	0.926	2
4	0.85714	0.57143	0.28571	1.00000	0.795	2
5	0.57143	0.14286	1.00000	0.71429	0.845	2
6	0.71429	1.00000	0.85714	0.28571	0.892	7
7	1.00000	0.28571	0.42857	0.14286	0.892	6
8	0.28571	0.00000	0.14286	0.57143	0.892	1

discrepancy = 0.0393

Visualizing the Latin Hypercube Design

To visualize the nature of the Latin Hypercube technique:

- Create an overlay plot
- Adjust the plot's frame size
- Add circles using the minimum distance from the diagnostic report as the radius for the circle

Example

1. Create another Latin Hypercube design using the default X1 and X2 factors.
2. Be sure to change the factor values so that they are 0 and 1 instead of the default –1 and 1.
3. Click **Continue.**
4. Specify a sample size of eight (**Number of Runs**).
5. Click **Latin Hypercube**. Factor settings and design diagnostics are shown in Figure 21.12.

Figure 21.12 Latin Hypercube Design with Two Factors and Eight Runs

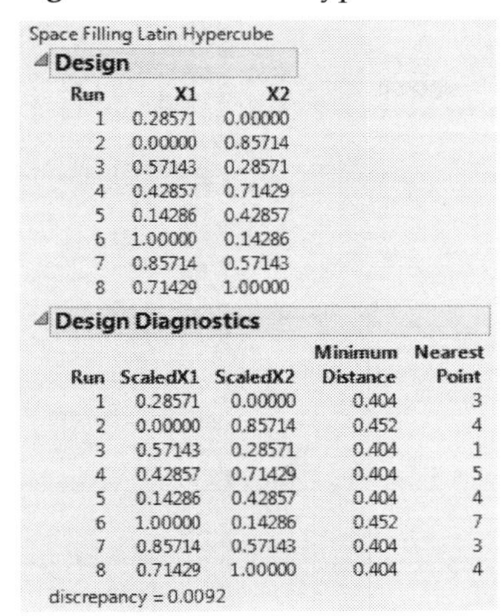
Space Filling Latin Hypercube

Design

Run	X1	X2
1	0.28571	0.00000
2	0.00000	0.85714
3	0.57143	0.28571
4	0.42857	0.71429
5	0.14286	0.42857
6	1.00000	0.14286
7	0.85714	0.57143
8	0.71429	1.00000

Design Diagnostics

Run	ScaledX1	ScaledX2	Minimum Distance	Nearest Point
1	0.28571	0.00000	0.404	3
2	0.00000	0.85714	0.452	4
3	0.57143	0.28571	0.404	1
4	0.42857	0.71429	0.404	5
5	0.14286	0.42857	0.404	4
6	1.00000	0.14286	0.452	7
7	0.85714	0.57143	0.404	3
8	0.71429	1.00000	0.404	4

discrepancy = 0.0092

6. Click **Make Table**.
7. Select **Graph > Overlay Plot**.
8. Specify X1 as **X** and X2 as **Y**, and then click **OK**.
9. Right-click the plot and select **Size/Scale > Size to Isometric** to adjust the frame size so that the frame is square.
10. Right-click the plot, select **Customize** from the menu. In the Customize panel, click the large plus sign to see a text edit area, and enter the following script:

    ```
    For Each Row(Circle({:X1, :X2}, 0.404/2))
    ```

    where `0.404` is the minimum distance number that you noted in the Design Diagnostics panel (Figure 21.12). This script draws a circle centered at each design point with radius 0.202 (half the diameter, 0.404), as shown on the left in Figure 21.13. This plot shows the efficient way JMP packs the design points.
11. Repeat the above procedure exactly, but with 10 runs instead of eight (step 5). Remember to change `0.404` in the graphics script to the minimum distance produced by 10 runs.

You should see a graph similar to the one on the right in Figure 21.13. Note the irregular nature of the sphere packing. In fact, you can repeat the process to get a slightly different picture because the arrangement is dependent on the random starting point.

Figure 21.13 Comparison of Latin Hypercube Designs with Eight Runs (left) and 10 Runs (right)

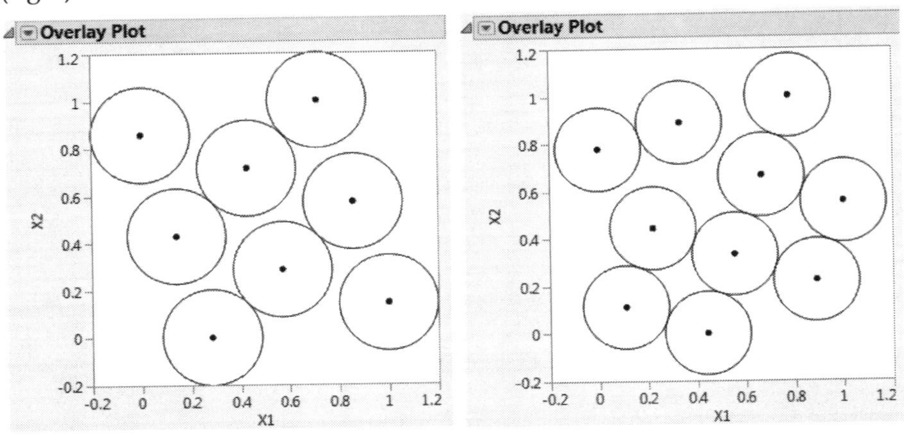

Note that the minimum distance between each pair of points in the Latin Hypercube design is smaller than that for the Sphere-Packing design. This is because the Latin Hypercube design constrains the levels of each factor to be evenly spaced. The Sphere-Packing design maximizes the minimum distance without any constraints.

Uniform Designs

The Uniform design minimizes the discrepancy between the design points (empirical uniform distribution) and a theoretical uniform distribution.

Note: These designs are most useful for getting a simple and precise estimate of the integral of an unknown function. The estimate is the average of the observed responses from the experiment.

1. Select **DOE > Special Purpose > Space Filling Design**.
2. Enter responses, if necessary, and factors.

   See “Responses” on page 569.
3. Alter the factor level values to 0 and 1.
4. Click **Continue.**
5. In the design specification dialog, specify a sample size. This example uses a sample size of eight (**Number of Runs**).
6. Click the **Uniform** button. JMP creates this design and displays the design runs and the design diagnostics as shown in Figure 21.14.

Note: The emphasis of the Uniform design method is not to spread out the points. The minimum distances in Figure 21.14 vary substantially.

Figure 21.14 Factor Settings and Diagnostics for Uniform Space-Filling Designs with Eight Runs

Space Filling Uniform Design

Design

Run	X1	X2
1	0.93092	0.69041
2	0.69276	0.18514
3	0.06860	0.32218
4	0.18696	0.81304
5	0.56496	0.93540
6	0.81410	0.43443
7	0.32218	0.06860
8	0.43643	0.56357

Design Diagnostics

Run	ScaledX1	ScaledX2	Minimum Distance	Nearest Point
1	0.93092	0.69041	0.281	6
2	0.69276	0.18514	0.277	6
3	0.06860	0.32218	0.359	7
4	0.18696	0.81304	0.353	8
5	0.56496	0.93540	0.393	8
6	0.81410	0.43443	0.277	2
7	0.32218	0.06860	0.359	3
8	0.43643	0.56357	0.353	4

discrepancy = 0.0046

7. Click **Make Table**.

A Uniform design does not guarantee even spacing of the factor levels. However, increasing the number of runs and running a distribution on each factor (use **Analyze** > **Distribution**) shows flat histograms.

Figure 21.15 Histograms Are Flat for Each Factor When Number of Runs Is Increased to 20

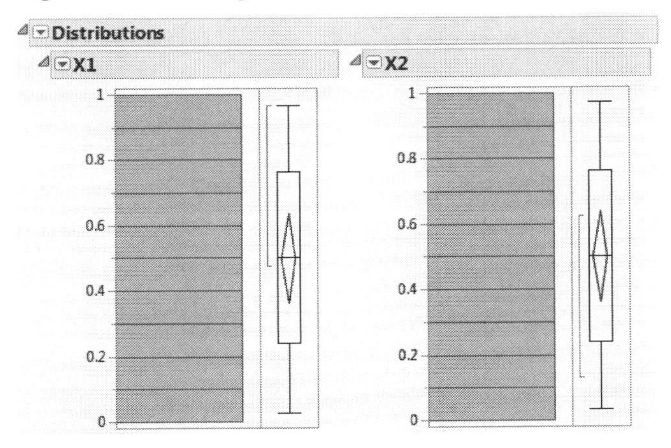

Comparing Sphere-Packing, Latin Hypercube, and Uniform Methods

To compare space-filling design methods, create the **Sphere Packing**, **Latin Hypercube**, and **Uniform** designs, as shown in the previous examples. The Design Diagnostics tables show the values for the factors scaled from zero to one. The minimum distance is based on these scaled values and is the minimum distance from each point to its closest neighbor. The discrepancy value is the integrated difference between the design points and the uniform distribution.

Figure 21.16 shows a comparison of the design diagnostics for three eight-run space-filling designs. Note that the discrepancy for the Uniform design is the smallest (best). The discrepancy for the Sphere-Packing design is the largest (worst). The discrepancy for the Latin Hypercube takes an intermediate value that is closer to the optimal value.

Also note that the minimum distance between pairs of points is largest (best) for the Sphere-Packing method. The Uniform design has pairs of points that are only about half as far apart. The Latin Hypercube design behaves more like the Sphere-Packing design in spreading the points out.

For both spread and discrepancy, the Latin Hypercube design represents a healthy compromise solution.

Figure 21.16 Comparison of Diagnostics for Three Eight-Run Space-Filling Methods

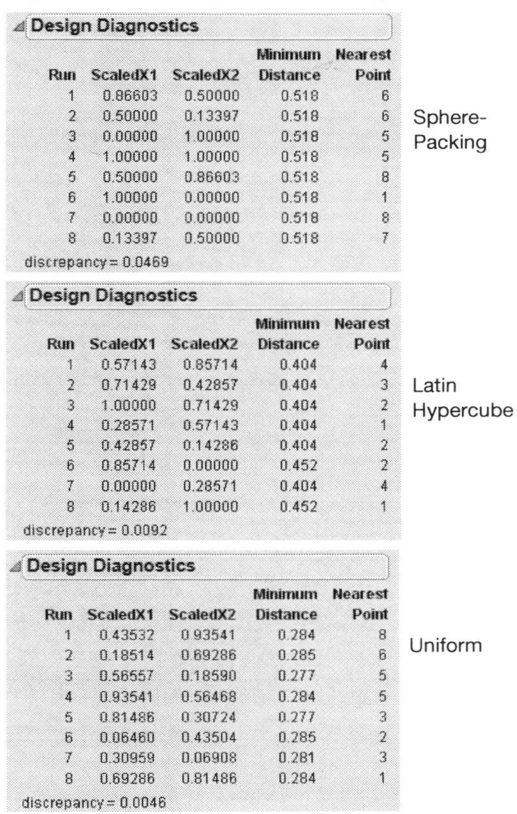

Design Diagnostics (Sphere-Packing)

Run	ScaledX1	ScaledX2	Minimum Distance	Nearest Point
1	0.86603	0.50000	0.518	6
2	0.50000	0.13397	0.518	6
3	0.00000	1.00000	0.518	5
4	1.00000	1.00000	0.518	5
5	0.50000	0.86603	0.518	8
6	1.00000	0.00000	0.518	1
7	0.00000	0.00000	0.518	8
8	0.13397	0.50000	0.518	7

discrepancy = 0.0469

Design Diagnostics (Latin Hypercube)

Run	ScaledX1	ScaledX2	Minimum Distance	Nearest Point
1	0.57143	0.85714	0.404	4
2	0.71429	0.42857	0.404	3
3	1.00000	0.71429	0.404	2
4	0.28571	0.57143	0.404	1
5	0.42857	0.14286	0.404	2
6	0.85714	0.00000	0.452	2
7	0.00000	0.28571	0.404	4
8	0.14286	1.00000	0.452	1

discrepancy = 0.0092

Design Diagnostics (Uniform)

Run	ScaledX1	ScaledX2	Minimum Distance	Nearest Point
1	0.43532	0.93541	0.284	8
2	0.18514	0.69286	0.285	6
3	0.56557	0.18590	0.277	5
4	0.93541	0.56468	0.284	5
5	0.81486	0.30724	0.277	3
6	0.06460	0.43504	0.285	2
7	0.30959	0.06908	0.281	3
8	0.69286	0.81486	0.284	1

discrepancy = 0.0046

Another point of comparison is the time it takes to compute a design. The Uniform design method requires the most time to compute. Also, the time to compute the design increases rapidly with the number of runs. For comparable problems, all the space-filling design methods take longer to compute than the *D*-optimal designs in the Custom Designer.

Minimum Potential Designs

The Minimum Potential design spreads points out inside a sphere. To understand how this design is created, imagine the points as electrons with springs attached to every other point, as illustrated to the right. The coulomb force pushes the points apart, but the springs pull them together. The design is the spacing of points that minimizes the potential energy of the system.

Figure 21.17 Minimum Potential Design

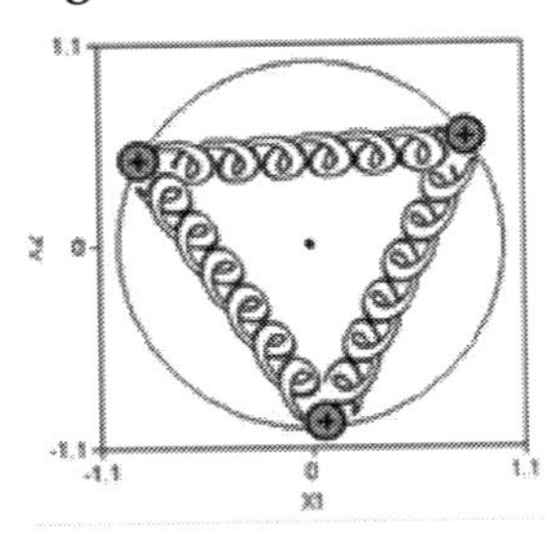

Minimum Potential designs:

- have spherical symmetry
- are nearly orthogonal
- have uniform spacing

To see a Minimum Potential example:

1. Select **DOE > Special Purpose > Space Filling Design.**
2. Add 1 continuous factor.

   See "Factors" on page 571.
3. Alter the factor level values to 0 and 1, if necessary.
4. Click **Continue.**
5. In the design specification dialog (shown on the left in Figure 21.18), enter a sample size (**Number of Runs**). This example uses a sample size of 12.
6. Click the **Minimum Potential** button. JMP creates this design and displays the design runs and the design diagnostics (shown on the right in Figure 21.18).

Figure 21.18 Space-Filling Methods and Design Diagnostics for Minimum Potential Design

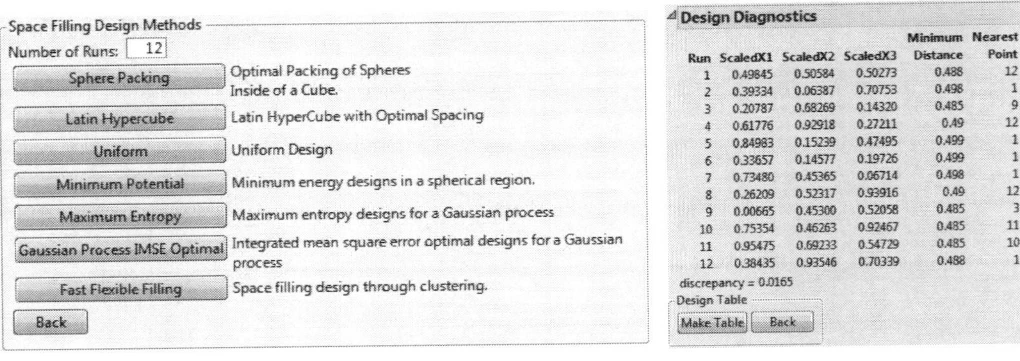

Run	ScaledX1	ScaledX2	ScaledX3	Minimum Distance	Nearest Point
1	0.49845	0.50584	0.50273	0.488	12
2	0.39334	0.06387	0.70753	0.498	1
3	0.20787	0.68269	0.14320	0.485	9
4	0.61776	0.92918	0.27211	0.49	12
5	0.84983	0.15239	0.47495	0.499	1
6	0.33657	0.14577	0.19726	0.499	1
7	0.73480	0.45365	0.06714	0.498	1
8	0.26209	0.52317	0.93916	0.49	12
9	0.00665	0.45300	0.52058	0.485	3
10	0.75354	0.46263	0.92467	0.485	11
11	0.95475	0.69233	0.54729	0.485	10
12	0.38435	0.93546	0.70339	0.488	1

7. Click **Make Table**.

You can see the spherical symmetry of the Minimum Potential design using the Scatterplot 3D graphics platform.

1. After you make the JMP design table, choose the **Graph** > **Scatterplot 3D** command.
2. In the Scatterplot 3D launch dialog, select X1, X2, and X3 as **Y, Columns** and click **OK** to see the initial three-dimensional scatterplot of the design points.
3. To see the results similar to those in Figure 21.19:
   – Select the **Normal Contour Ellipsoids** option from the menu in the Scatterplot 3D title bar.
   – Make the points larger. Right-click on the plot and select **Settings**, and then increase the **Marker Size** slider.

Now it is easy to see the points spread evenly on the surface of the ellipsoid.

Figure 21.19 Minimum Potential Design Points on Sphere

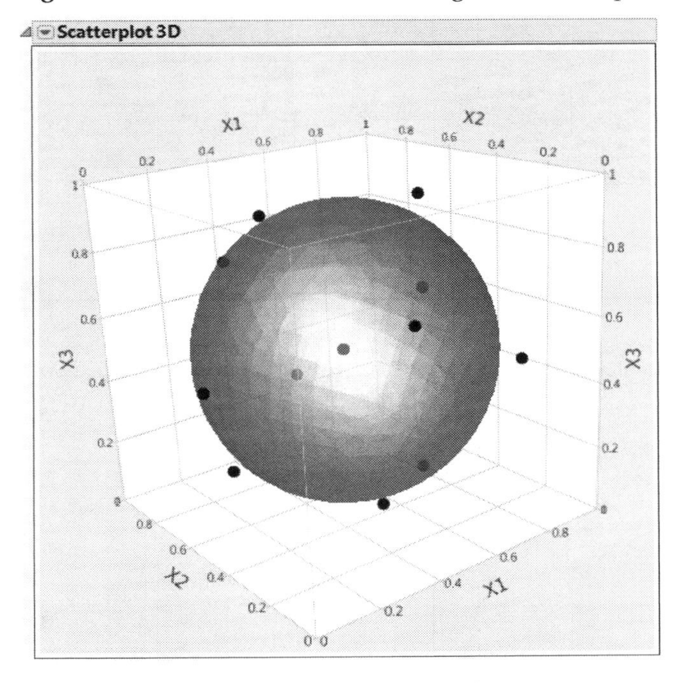

Maximum Entropy Designs

The Latin Hypercube design is currently the most popular design assuming you are going to analyze the data using a Gaussian-Process model. Computer simulation experts like to use the Latin Hypercube design because all projections onto the coordinate axes are uniform.

However, as the example in Figure 21.20 shows, the Latin Hypercube design does not necessarily do a great job of space filling. This is a two-factor Latin Hypercube with 16 runs and with the factor level settings set between -1 and 1. Note that this design seams to leave a hole in the bottom right of the overlay plot.

Figure 21.20 Two-factor Latin Hypercube Design

Space Filling Latin ...
Design Space Filling Latin
▶ Screening
▶ Model
▶ DOE Dialog

Columns (3/0)
X1 ✱
X2 ✱
Y ✱

	X1	X2
1	-0.733333333	0.7333333333
2	-0.6	-0.6
3	0.2	-0.066666667
4	0.4666666667	-0.466666667
5	-0.066666667	0.3333333333
6	-1	-0.866666667
7	-0.866666667	-0.2
8	0.0666666667	1
9	0.7333333333	0.2
10	0.8666666667	-0.733333333
11	-0.466666667	0.0666666667
12	-0.333333333	-1
13	0.3333333333	0.4666666667
14	1	0.6
15	0.6	0.8666666667
16	-0.2	-0.333333333

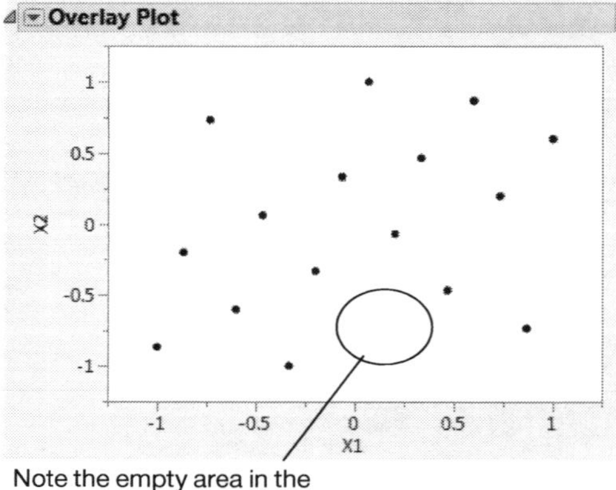

The Maximum Entropy design is a competitor to the Latin Hypercube design for computer experiments because it optimizes a measure of the amount of information contained in an experiment. See the technical note below. With the factor levels set between -1 and 1, the two-factor Maximum Entropy design shown in Figure 21.21 covers the region better than the Latin hypercube design in Figure 21.20. The space-filling property generally improves as the number of runs increases without bound.

Figure 21.21 Two-Factor Maximum Entropy Design

Maximum Entropy
Design Maximum Entropy
Screening
Model
DOE Dialog

Columns (3/0)
X1
X2
Y

Rows

All rows	16
Selected	0
Excluded	0
Hidden	0
Labelled	0

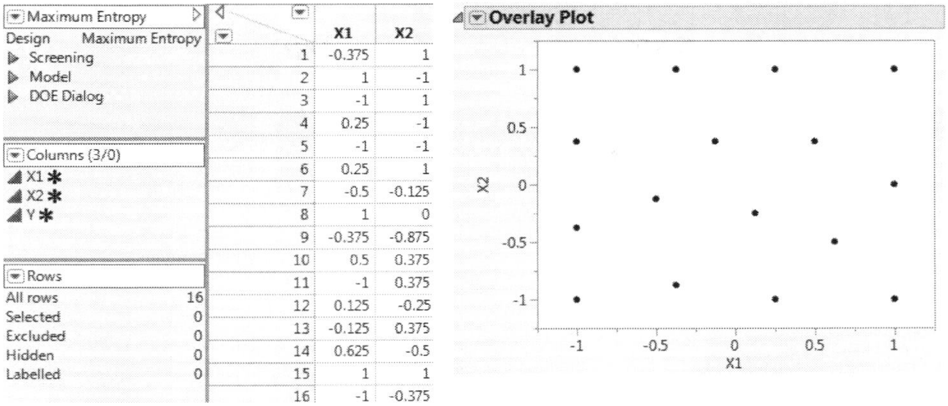

	X1	X2
1	-0.375	1
2	1	-1
3	-1	1
4	0.25	-1
5	-1	-1
6	0.25	1
7	-0.5	-0.125
8	1	0
9	-0.375	-0.875
10	0.5	0.375
11	-1	0.375
12	0.125	-0.25
13	-0.125	0.375
14	0.625	-0.5
15	1	1
16	-1	-0.375

Overlay Plot

Technical Maximum Entropy designs maximize the Shannon information (Shewry and Wynn (1987)) of an experiment, assuming that the data come from a normal (m, s^2 R) distribution, where

$$R_{ij} = \exp\left(-\sum_k \theta_k (x_{ik} - x_{jk})^2\right)$$

is the correlation of response values at two different design points, x_i and x_j. Computationally, these designs maximize $|\mathbf{R}|$, the determinant of the correlation matrix of the sample. If x_i and x_j are far apart, then $\mathbf{R}_{ij}$ approaches zero. If x_i and x_j are close together, then $\mathbf{R}_{ij}$ is near one.

Gaussian Process IMSE Optimal Designs

The Gaussian process IMSE optimal design method constructs designs that are suitable for Gaussian process models. Gaussian process models fit a wide variety of surfaces. Gaussian process IMSE optimal designs minimize the integrated mean squared error of the Gaussian process model over the experimental region. The Gaussian process IMSE optimal design method uses a correlation structure similar to that of the kriging model. See Jones and Johnson (2009).

Covariance Parameter Vector

In a Gaussian Process IMSE Optimal Design formulation of the Gaussian process model, the covariance parameter vector determines the correlation structure. There is a Theta for each factor. A theta equal to 0 corresponds to a correlation of 1, causing the fitted surface to be flat in the corresponding factor's direction. As theta increases, the correlation decreases, allowing the surface to be flexible in the factor's direction.

In the Covariance Parameter Vector outline, in the list of values under Thetas, you can enter values that reflect your prior knowledge of the surface.

Comparison of Gaussian Process IMSE Optional Design with Latin Hypercube Design

Gaussian process IMSE optimal designs are competitors to the Latin Hypercube design. You can compare the IMSE optimal design to the Latin Hypercube (shown previously in Figure 21.20). The table and overlay plot in Figure 21.22 show a Gaussian IMSE optimal design. You can see that the design provides uniform coverage of the factor region.

Figure 21.22 Comparison of Two-factor Latin Hypercube and Gaussian IMSE Optimal Designs

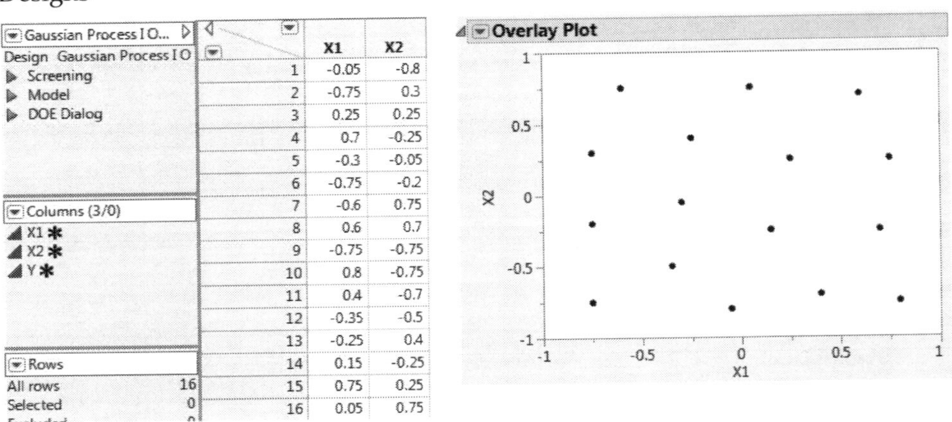

	X1	X2
1	-0.05	-0.8
2	-0.75	0.3
3	0.25	0.25
4	0.7	-0.25
5	-0.3	-0.05
6	-0.75	-0.2
7	-0.6	0.75
8	0.6	0.7
9	-0.75	-0.75
10	0.8	-0.75
11	0.4	-0.7
12	-0.35	-0.5
13	-0.25	0.4
14	0.15	-0.25
15	0.75	0.25
16	0.05	0.75

Note: Both the Maximum Entropy design and the Gaussian Process IMSE Optimal design were created using 100 random starts.

Fast Flexible Filling Designs

Note: If you have Categorical factors or factor constraints, then Fast Flexible Filling is the only Method available.

FFF Optimality Criterion

The algorithms for Fast Flexible Filling designs begin by generating a large number of random points within the specified design region. These points are then clustered using a Fast Ward algorithm into a number of clusters that equals the Number of Runs that you specified.

The final design points can be obtained by using the default MaxPro (*maximum projection*) optimality criterion or by selecting the Centroid criterion. You can find these options under FFF Optimality Criterion in the report's red triangle menu.

MaxPro For p factors and n equal to the specified Number of Runs, the MaxPro criterion strives to find points in the clusters that minimize the following criterion:

$$C_{MaxPro} = \sum_{i}^{n-1} \sum_{j=i+1}^{n} \left[1 / \prod_{k=1}^{p} (x_{ik} - x_{jk})^2 \right]$$

The MaxPro criterion maximizes the product of the distances between potential design points in a way that involves all factors. This supports the goal of providing good space-filling properties on projections of factors. See Joseph et al. (2015). The Max Pro option is the default.

Centroid This method places a design point at the centroid of each cluster. It has the property that the average distance from an arbitrary point in the design space to its closest neighboring design point is smaller than for other designs.

Note: You can set a preference to always use a given optimality criterion. Select File > Preferences > Platforms > DOE. Check FFF Optimality Criterion and select your preferred criterion.

Categorical Factors

When you have categorical factors, the algorithm proceeds as follows:

- The total number of design points is balanced across the total number of combinations of levels of the categorical factors. Suppose that there are m combinations of levels and that k design points are allocated to each of these.
- A large number of points within the design space defined by the continuous variables is generated. These are grouped into k primary clusters.
- Each of the k primary clusters of points is further clustered into m sub-clusters.
- Within each primary cluster, a design point is calculated for each of the m sub-clusters using the specified FFF optimality criterion.
- For each of the k primary clusters, one of the m combinations of levels is randomly assigned to each of the m sub-cluster design points. This yields a total of km design points.

Set Average Cluster Size

The Set Average Cluster Size option is found under Advanced Options in the Space Filling Design red triangle menu. This option enables you to specify the average number of randomly- generated points used to define each cluster or, equivalently, each design point.

By default, if the Number of Runs is set to 200 or less, a total of 10,000 randomly generated points are used as the basis for the clustering algorithm. When the number of Runs exceeds 200, a default value of 50 is used. Increasing this value can be particularly useful in designs with a large number of factors or where disallowed combinations restrict the distribution of points used in the clustering algorithm.

Note: Depending on the number of factors and the specified value for Number of Runs, you might want to increase the average number of initial points per design point by selecting **Advanced Options > Set Average Cluster Size.**

Constraints

Once you complete the Factors outline, click **Continue**. The Define Factor Constraints outline appears. Use this outline to restrict the design region. For details about the outline, see "Define Factor Constraints" on page 573.

You can use the Use Disallowed Combinations Filter and Use Disallowed Combinations Script options to specify disallowed factor level combinations. Or, you can use the Specify Linear Constraints option to specify bounds in terms of linear inequalities. However, the design is generated differently for these two methods.

Use Disallowed Combinations Filter and Use Disallowed Combinations Script

When disallowed combinations are specified, the random points that form the basis for the clustering algorithm are randomly distributed within the unconstrained design region. Then disallowed points are removed and clustering proceeds with the remaining points.

Note: Depending on the nature of the constraints and the specified Number of Runs, the default coverage of the unconstrained design space by the initial randomly generated points might not be sufficient to produce the required Number of Runs. In this case, you might obtain a JMP Alert indicating that the algorithm "Could not find sufficient number of points." To increase the initial number of points that form the basis for the clustering algorithm, specify a larger average number of initial points per design point by selecting **Advanced Options > Set Average Cluster Size.** (See "Set Average Cluster Size" on page 596).

Specify Linear Constraints

When you use the **Specify Linear Constraints** option, the random points that form the basis for the clustering algorithm are randomly distributed within the constrained design region. The clustering algorithm uses these points.

Creating and Viewing a Constrained Fast Flexible Filling Design

Constructing the Design

1. Select **DOE > Special Purpose > Space Filling Design**.
2. Enter Values of 0 and 1 for both X1 and X2.
3. Click **Continue.**
4. In the Define Linear Constraints outline, select **Specify Linear Constraints.**

   Notice that Fast Flexible Filling is the only available Space Filling Design Method.
5. Select **Add.**
6. Enter the following coefficients and bound:

   1 for X1

   1 for X2

   0.8 for the bound

Figure 21.23 Linear Constraint

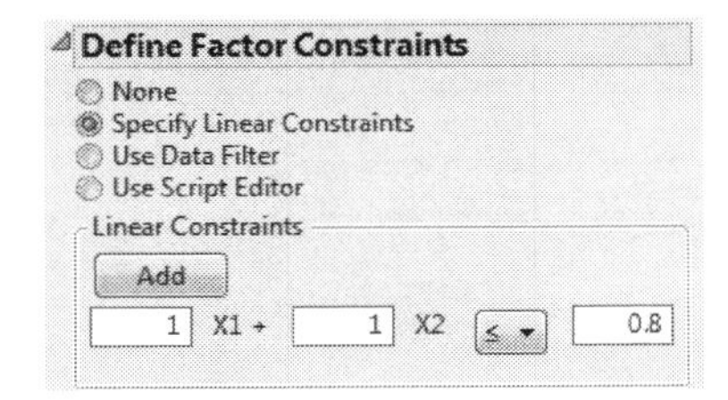

7. Type 200 next to **Number of Runs.**
8. Select **Fast Flexible Filling.**

   JMP creates a design that satisfies the constraints. Open the Design outline to view the design.
9. Select **Make Table** to construct the data table.

Constructing the Plot

1. With the data table active, select **Graph > Graph Builder**.
2. Drag X1 to the drop zone labeled **X**.

3. Drag **X2** to the drop zone labeled **Y**.
4. Remove the **Smoother** by clicking the smoother icon.
5. In the Graph Builder red triangle menu, click **Show Control Panel** to deselect it.

You should see a graph similar to the one in Figure 21.24. Note that the points satisfy the linear constraint $X1 + X2 \le 0.8$.

Figure 21.24 Fast Flexible Filling Design with One Linear Constraint

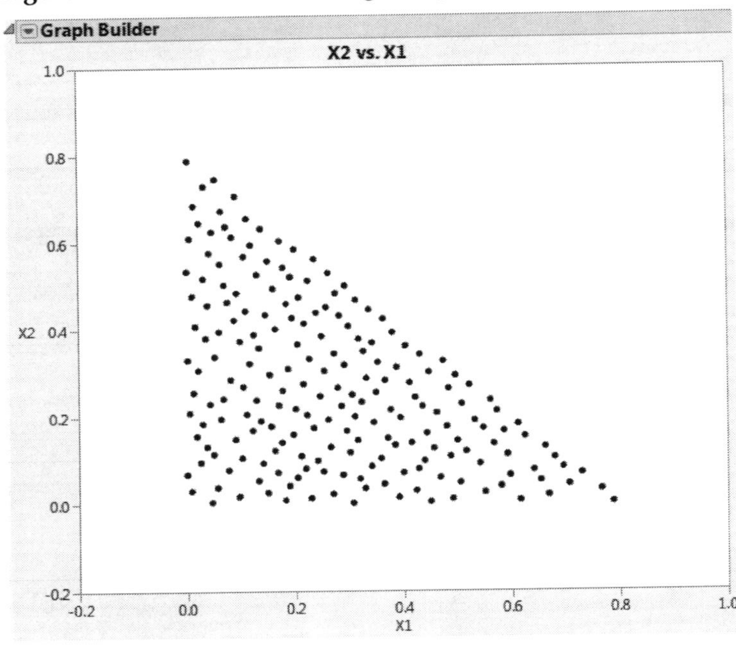

Borehole Model: A Sphere-Packing Example

Worley (1987) presented a model of the flow of water through a borehole that is drilled from the ground surface through two aquifers. The response variable y is the flow rate through the borehole in m^3/year and is determined by the following equation:

$$y = \frac{2\pi T_u(H_u - H_l)}{\ln(r/r_w)\left[1 + \frac{2LT_u}{\ln(r/r_w)r_w^{\,2}K_w} + \frac{T_u}{T_l}\right]}$$

There are eight inputs to this model:

r_w = radius of borehole, 0.05 to 0.15 m

r = radius of influence, 100 to 50,000 m

T_u = transmissivity of upper aquifer, 63,070 to 115,600 m^2/year

H_u = potentiometric head of upper aquifer, 990 to 1100 m

T_l = transmissivity of lower aquifer, 63.1 to 116 m^2/year

H_l = potentiometric head of lower aquifer, 700 to 820 m

L = length of borehole, 1120 to 1680 m

K_w = hydraulic conductivity of borehole, 9855 to 12,045 m/year

This example is atypical of most computer experiments because the response can be expressed as a simple, explicit function of the input variables. However, this simplicity is useful for explaining the design methods.

Create the Sphere-Packing Design for the Borehole Data

To create a Sphere-Packing design for the borehole problem:

1. Select **DOE > Special Purpose > Space Filling Design**.
2. Click the red triangle icon on the Space Filling Design title bar and select **Load Factors**.
3. Select **Help > Sample Data Library** and open Design Experiment/Borehole Factors.jmp (Figure 21.25).

Figure 21.25 Factors Panel with Factor Values Loaded for Borehole Example

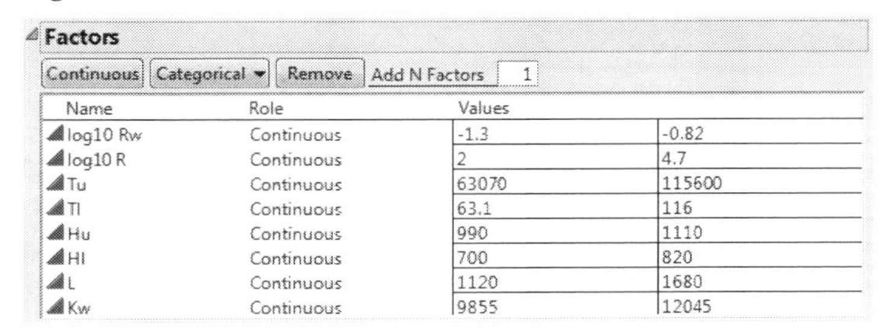

Name	Role	Values	
log10 Rw	Continuous	-1.3	-0.82
log10 R	Continuous	2	4.7
Tu	Continuous	63070	115600
Tl	Continuous	63.1	116
Hu	Continuous	990	1110
Hl	Continuous	700	820
L	Continuous	1120	1680
Kw	Continuous	9855	12045

Note: The logarithm of r and r_w are used in the following discussion.

4. Click **Continue**.
5. Specify a sample size (Number of Runs) of 32 as shown in Figure 21.26.

Figure 21.26 Space-Filling Design Method Panel Showing 32 Runs

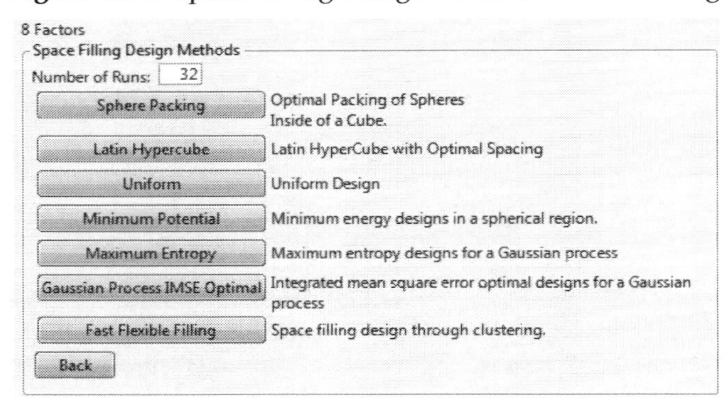

6. Click the **Sphere Packing** button to produce the design.
7. Click **Make Table** to make a table showing the design settings for the experiment.

To see a completed data table for this example, select **Help > Sample Data Library** and open Design Experiment/ Borehole Sphere Packing.jmp. Because the designs are generated from a random seed, the settings that you obtain will differ from those shown in the completed table.

The Borehole Sphere Packing.jmp data table contains a Fit Model script that you can use to analyze the data. Columns containing the true model, the prediction formula, and the prediction bias are included in the data table.

Guidelines for the Analysis of Deterministic Data

It is important to remember that deterministic data have no random component. As a result, p-values from fitted statistical models do not have their usual meanings. A large F statistic (low p-value) is an indication of an effect due to a model term. However, you cannot make valid confidence intervals about the size of the effects or about predictions made using the model.

Residuals from any model fit to deterministic data are not a measure of noise. Instead, a residual shows the model bias for the current model at the current point. Distinct patterns in the residuals indicate new terms to add to the model to reduce model bias.

Results of the Borehole Experiment

The example described in the previous sections produced the following results:

- A stepwise regression of the response, log y, versus the full quadratic model in the eight factors, led to the prediction formula column.
- The prediction bias column is the difference between the true model column and the prediction formula column.

- The prediction bias is relatively small for each of the experimental points. This indicates that the model fits the data well.

In real world examples, the true model is generally not available in a simple analytical form. As a result, it is impossible to know the prediction bias at points other than the observed data without doing additional runs.

In this case, the true model column contains a formula that allows profiling the prediction bias to find its value anywhere in the region of the data. To understand the prediction bias in this example:

1. Select **Graph > Profiler**.
2. Highlight the prediction bias column and click the **Y, Prediction Formula** button.
3. Check the **Expand Intermediate Formulas** box, shown at the bottom on the Profiler dialog in Figure 21.27.

   The prediction bias formula is a function of columns that are also created by formulas.

Figure 21.27 Profiler Dialog for Borehole Sphere-Packing Data

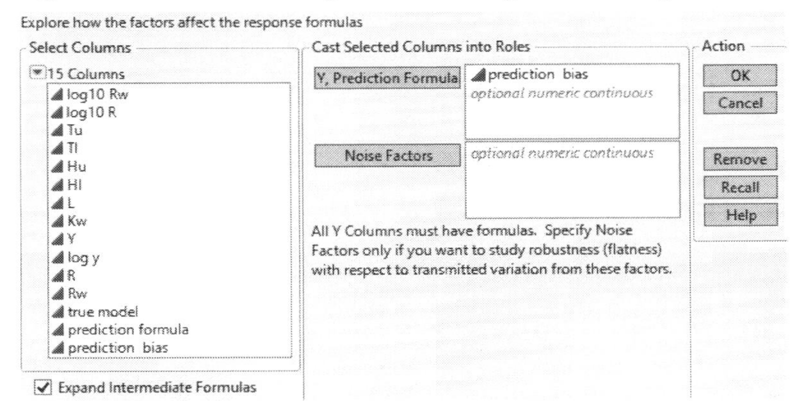

4. Click **OK**.

   The profile plots in Figure 21.28 show the prediction bias at the center of the design region. If there were no bias, the profile traces would be constant between the value ranges of each factor. In this example, the variables Hu and Hl show nonlinear effects.

Figure 21.28 Profiler for Prediction Bias for Borehole Sphere-Packing Data

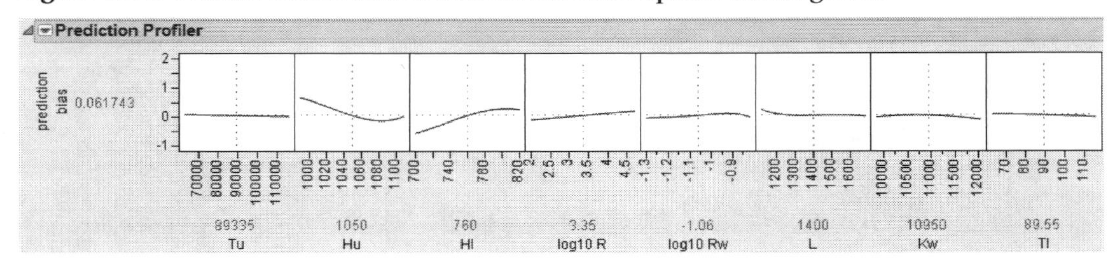

The range of the prediction bias on the data is smaller than the range of the prediction bias over the entire domain of interest. To see this, look at the distribution analysis (**Analyze > Distribution**) of the prediction bias in Figure 21.29. Note that the maximum bias is 1.826 and the minimum is –0.684 (the range is 2.51).

Figure 21.29 Distribution of the Prediction Bias

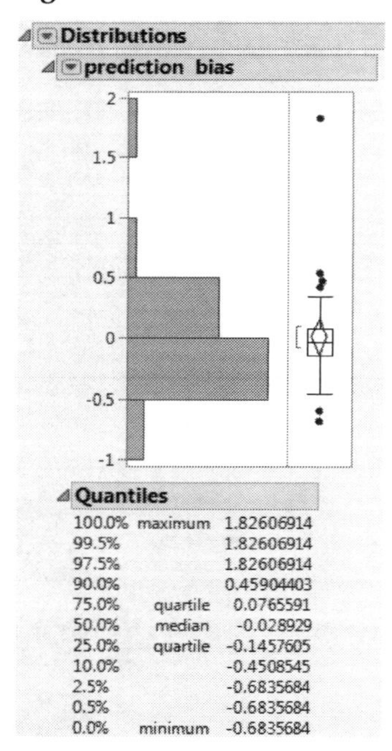

100.0%	maximum	1.82606914
99.5%		1.82606914
97.5%		1.82606914
90.0%		0.45904403
75.0%	quartile	0.0765591
50.0%	median	-0.028929
25.0%	quartile	-0.1457605
10.0%		-0.4508545
2.5%		-0.6835684
0.5%		-0.6835684
0.0%	minimum	-0.6835684

The top plot in Figure 21.30 shows the maximum bias (2.91) over the entire domain of the factors. The plot at the bottom shows the comparable minimum bias (–4.84). This gives a range of 7.75. This is more than three times the size of the range over the observed data.

Figure 21.30 Prediction Plots Showing Maximum and Minimum Bias over Factor Domains

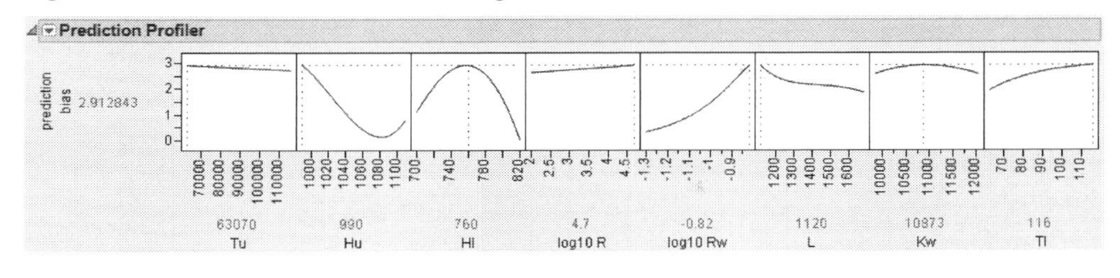

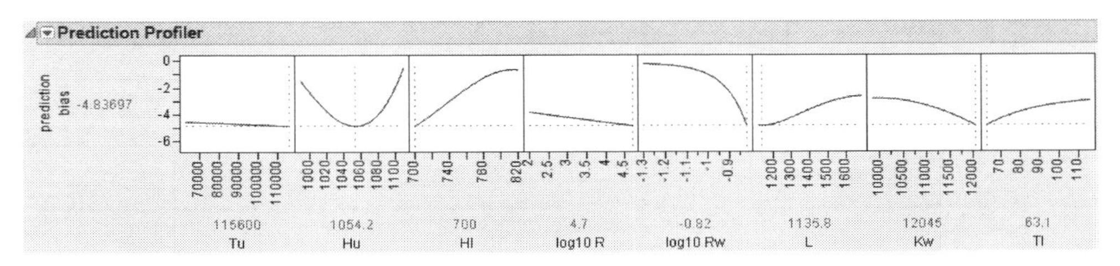

Keep in mind that, in this example, the true model is known. In any meaningful application, the response at any factor setting is unknown. The prediction bias over the experimental data underestimates the bias throughout the design domain.

There are two ways to assess the extent of this underestimation:

- Cross validation refits the data to the model while holding back a subset of the points and looks at the error in estimating those points.
- Verification runs (new runs performed) at different settings to assess the lack of fit of the empirical model.

Chapter 22

Accelerated Life Test Designs

Designing Experiments for Accelerated Life Tests

Product reliability at normal use conditions is often so high that the time required to test the product until it fails is prohibitive. Rather than test the product at normal use conditions, you can test the product under conditions that are more severe than normal use conditions. These severe conditions cause the product to degrade faster and fail more quickly. You can then use this accelerated failure data to predict product reliability at normal use conditions.

Use the Accelerated Life Test (ALT) Design platform to design plans for accelerated life testing experiments. You can design initial experiments or augment existing experiments.

Figure 22.1 Profiler Showing Failure Probabilities for ALT Experiment

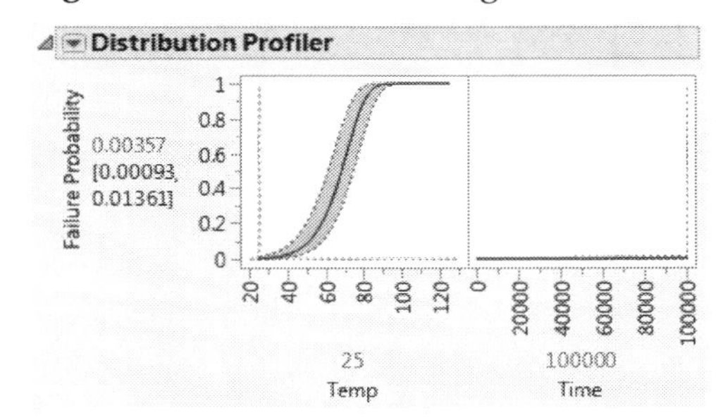

Overview of Accelerated Life Test Designs

When product reliability at normal use conditions is high, you can test the product under conditions that are more severe than normal so that failures occur more quickly. You can use this accelerated failure data to predict product reliability at normal use conditions.

Such a test is called an *accelerated life test* (ALT).The factors that are set at higher than normal use levels are called *acceleration factors*. The models for accelerated life tests are typically nonlinear models. For details about nonlinear models, see “Nonlinear Models” on page 648 in the “Nonlinear Designs” chapter.

The ALT Design platform creates and evaluates designs for situations involving one or two acceleration factors (and their interaction). You can optimize your design using D-optimality or two types of I-optimality criteria.

Creating a design requires initial estimates of the acceleration model parameters. Since those parameters are usually unknown, you can specify a multivariate normal prior distribution to describe their uncertainty. To decrease the variance of estimates, you can augment existing experiments in the Accelerated Life Test (ALT) Design platform.

Example of an Accelerated Life Test Design

In this example, suppose that you need to design an accelerated life test for a mechanical component. The single acceleration factor is torque, and the normal use stress is 35 Nm (newton meters). You want to estimate the B10 life, which is the life at which 10% of the units fail at the normal use stress.

Your test plan has the following characteristics:

- A total of 100 units are available for testing.
- The life distribution is assumed to be Weibull.
- The life-stress relationship is given by the logarithmic transformation.
- You will test at torque three stress levels: 50, 75, and 100 Nm.
- You have some prior knowledge to help you guess failure times at the test levels. More details appear in "Obtain Prior Estimates" on page 607.
- The test will continue for 5000 cycles.
- You will monitor the process for failures on a continuous basis.

Obtain Prior Estimates

To create an accelerated life test design, you need to provide prior estimates of the parameters. Here is a convenient approach to obtaining prior estimates:

1. Use your process knowledge to create a table of hypothetical but likely, failure times at a small number of stress levels.
2. Use the Fit Life by X platform to fit a model and obtain estimates of the model parameters.
3. Use these estimates as your prior values for creating a design using the ALT Design platform.

Following the approach outlined above, you create a data table containing estimates of the number of failure cycles for a balanced design. Your table consists of five units at each of the three stress levels that you will use in your design. Your data table is Torque Prior.jmp, found in the Design Experiment sub folder.

1. Select **Help > Sample Data Library** and open Design Experiment/Torque Prior.jmp.
2. Select **Analyze > Reliability and Survival > Fit Life by X**.
3. Select Cycles and click **Y, Time to Event**.
4. Select Torque and click **X**.
5. From the **Relationship** list, select **Log**.
6. From the **Distribution** list, select **Weibull**.

Figure 22.2 Fit Life by X Launch Window

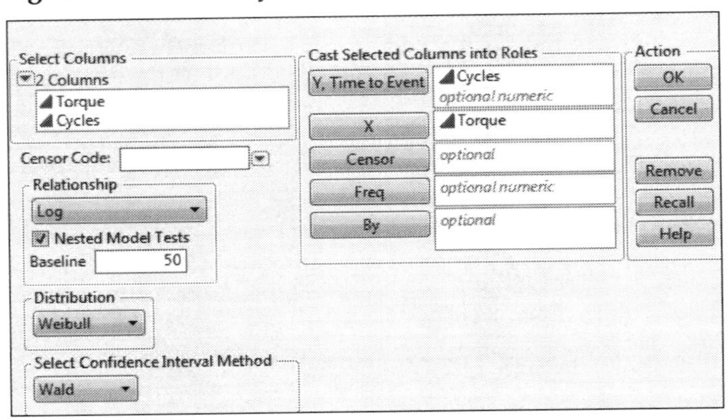

7. Click **OK**.
8. Scroll down to the Weibull Results report and open the Covariance Matrix report.

Figure 22.3 Fit Life by X Model for Prior Data

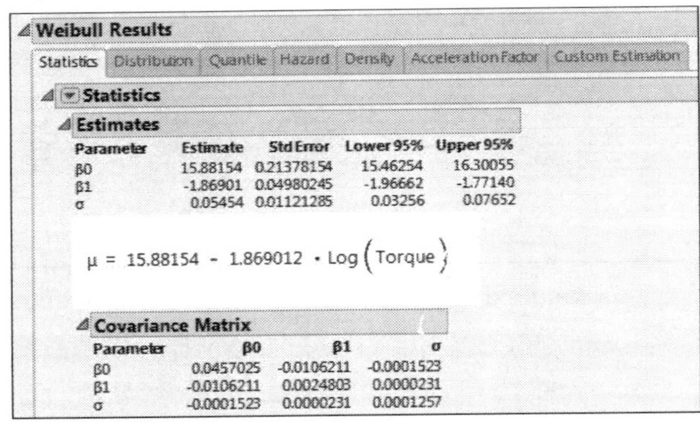

Parameter	Estimate	Std Error	Lower 95%	Upper 95%
β0	15.88154	0.21378154	15.46254	16.30055
β1	-1.86901	0.04980245	-1.96662	-1.77140
σ	0.05454	0.01121285	0.03256	0.07652

$$\mu = 15.88154 - 1.869012 \cdot \text{Log}(\text{Torque})$$

Parameter	β0	β1	σ
β0	0.0457025	-0.0106211	-0.0001523
β1	-0.0106211	0.0024803	0.0000231
σ	-0.0001523	0.0000231	0.0001257

The model for the mean is given in the Estimates report. The Estimate column contains the parameter estimates for the intercept (β0), the linear coefficient (β1), and the scale (σ). The estimated variances and covariances for the parameter estimates are given in the Covariance Matrix report. You will use these parameter estimates and their covariances as your prior values in constructing your ALT design.

Enter Basic Specifications

1. Select **DOE > Special Purpose > Accelerated Life Test Design.**

   Notice that **Design for one accelerating factor** and **Continuous Monitoring** are selected by default.

2. Click **Continue**.
3. Under **Factor Name**, click X1 and type Torque.

   Notice that the **Number of Levels** is set to 3 by default.
4. Select **Log** under Factor Transformation.
5. Enter 35 first for the **High Usage Condition** and then for the **Low Usage Condition**.

   Setting both the high and low usage conditions at the common value of 35 indicates that 35 represents the normal usage condition.

Figure 22.4 Completed ALT Specification Window

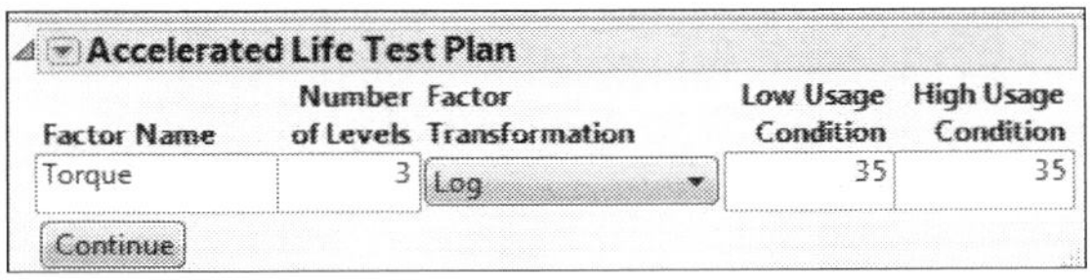

Accelerated Life Test Plan

Factor Name	Number of Levels	Factor Transformation	Low Usage Condition	High Usage Condition
Torque	3	Log	35	35

Continue

6. Click **Continue**.
7. Enter 50, 75, and 100 for the **Torque Level Values**.
8. Ensure that **Weibull** is selected as the **Distribution Choice**.

Enter Prior Information and Remaining Specifications

1. Under **Prior Mean**, enter your prior acceleration model parameter estimates, rounding the values shown in the Estimates report in Figure 22.3 to two decimal places:
   – Next to **Intercept**, type 15.88.
   – Next to **Torque**, type -1.87.
   – Next to **scale**, type 0.05.
2. Open the Prior Variance Matrix outline.
3. De-select the option **Ignore prior variance. Generate the local design for the prior mean.**
4. Enter the estimated covariances from the prior acceleration model, rounding the values shown in the Covariance Matrix report in Figure 22.3 to five decimal places:
   – In the row and column for Intercept, click on 0.10000 and type 0.04570.
   – In the row for Intercept and column for Torque, click on 0.00000 and type -0.01062.
   – In the row for Intercept and column for scale, click on 0.00000 and type -0.00015.
   – In the row and column for Torque, click on 0.10000 and type 0.00248.
   – In the row for Torque and column for scale, click on 0.00000 and type 0.00002.
   – In the row and column for scale, click on 0.10000 and type 0.00013.

Figure 22.5 Completed Prior Specification Outline

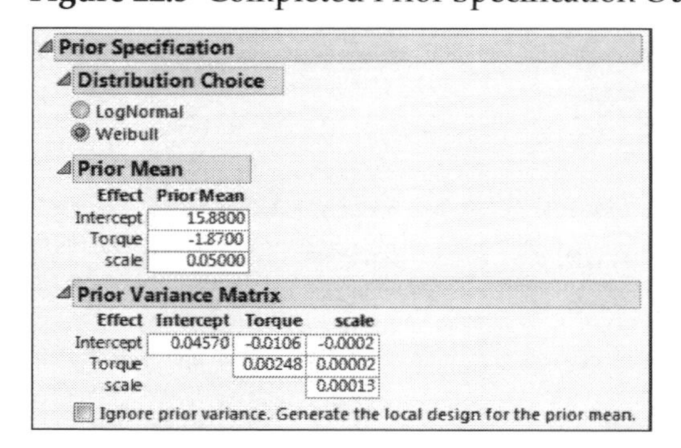

The test will be conducted over 5000 cycles. However, you are interested in predictions for as many as 10,000 cycles.

5. Under Diagnostic Choices, enter 10,000 for both boxes for **Time range of interest**. Leave the **Probability of interest** value set to 0.1.

   This indicates that you are interested in estimating the time by which 10% of the units fail (B10 life).

6. Under Design Choices, enter 5000 for **Length of test.**
7. Enter 100 for **Number of units under test.**

   Figure 22.11 shows the completed Accelerated Life Test Plan outline.

Figure 22.6 Completed Design Details Window

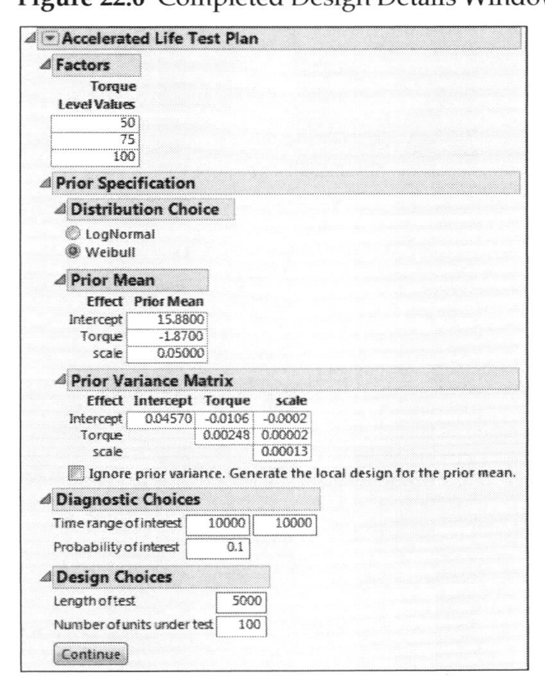

Create the Design

1. Click **Continue**.

Figure 22.7 Balanced Design Diagnostics

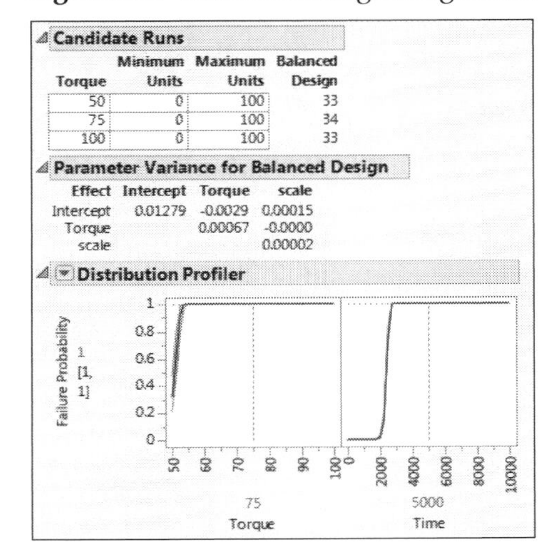

The number of runs in a balanced design appears in the Candidate Runs outline. The Parameter Variance for Balanced Design outline shows the covariance matrix for the parameters for this design. The Distribution Profiler also appears. When you obtain the optimal design, you can compare the Parameter Variance and Distribution Profiler results for the optimal design to those for the balanced design to see the reduction in uncertainty.

2. From the ALT Plan red triangle menu, select **ALT Optimality Criterion** > **Make Probability I-Optimal Design**.

   This tells JMP to use the Probability optimality criterion when creating the design in the next step. For details about this criterion, see "Make Probability I-Optimal Design" on page 624.

3. Click **Make Design**.

   The optimal experimental design is shown, along with other results.

Figure 22.8 Optimal Design

Design

Torque	N Units	Expected Failures	All Censored Probability
50	72	23.2	0.0
75	8	8.0	0.0
100	20	20.0	0.0

The optimal design is computed based on the levels of the test runs, the total number of units to be tested, and the prior information that you specified. The optimal design consists of testing the following number of units at each torque level:

– 72 units at 50 Nm

– 8 units at 75 Nm

– 20 units at 100 Nm

Example of Augmenting an Accelerated Life Test Design

This example shows how to use the Accelerated Life Test Design platform to augment an existing design.

In this example, 50 capacitor units are tested at three temperatures (85°, 105°, and 125° Celsius) for 1500 hours. The results are recorded in the Capacitor ALT.jmp sample data table. The resulting model is used to predict the fraction of the population that is failing at 100,000 hours at normal use temperature of 25° Celsius.

Review Current Predictions

1. Select **Help > Sample Data Library** and open Design Experiment/Capacitor ALT.jmp.
2. Click the green triangle to run the **Fit Life by X** table script.

3. In the Distribution Profiler, found in the Comparisons report on the Distribution tab, do the following:
   – Click 105 above **Temp** and change it to 25.
   – Click 750.5 above **Hours** and change it to 100,000.

Figure 22.9 Distribution Profiler for Capacitor Model

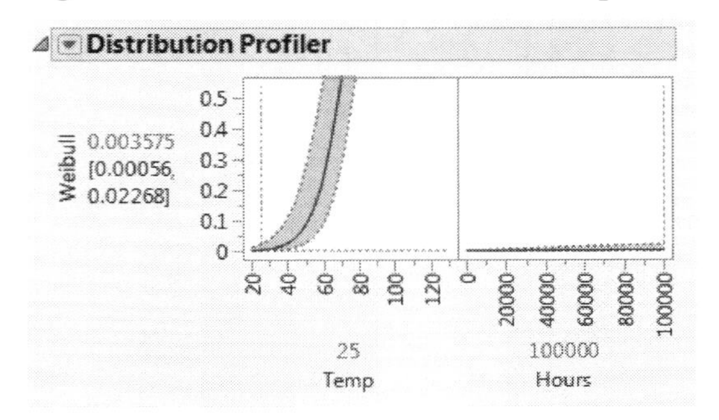

Based on your current study, the predicted fraction of the population that fails at 25° at 100,000 hours is 0.00358, with a confidence interval of 0.00056 to 0.02268. You want to estimate the failure fraction more precisely. To decrease the width of the confidence interval, augment your study with additional tests.

Augment the Design

You want to augment your design to obtain more precise estimates of the predicted failure fractions. Your original design used temperature settings of 85, 105, and 125. In your augmented design, you want to test at temperature values of 90, 110, and 125. Note that two of these settings are new. To augment the design with optimally selected runs, follow these steps:

1. Select **DOE** > **Special Purpose > Accelerated Life Test Design**.
2. Select **Design for one accelerating factor** and click **Continue**.
3. Enter Temp for **Factor Name**.
4. Enter 5 for **Number of Levels**.

   Even though your augmented runs span only three levels (90, 110, and 125), you must specify the levels used in the original experiment as well, making a total of five levels. The Factor Transformation is set to Arrhenius Celsius by default.
5. Enter 25 for both **Low Usage Condition** and **High Usage Condition**.
6. Click **Continue**.
7. Enter 85, 90, 105, 110, and 125 for the **Temp Level Values**.

There are three levels from the original experiment (85, 105, and 125). The augmented design will have two new levels (90 and 110) and one of the levels from the first experiment (125). All levels must be listed.

8. Ensure that **Weibull** is selected as the **Distribution Choice**.
9. Under **Prior Mean**, enter the acceleration model parameter estimates from the Fit Life by X Estimates report, found in the Weibull Results report on the Statistics tab.

Figure 22.10 Parameter Estimates and Fitted Model from Weibull Results Report

Estimates

Parameter	Estimate	Std Error	Lower 95%	Upper 95%
β0	-35.19979	4.6912686	-44.39451	-26.00508
β1	1.38891	0.1566297	1.08192	1.69589
σ	1.30471	0.1119128	1.08536	1.52405

$$\mu = -35.19979 + \frac{(1.388906 \cdot 11605)}{(\text{Temp} + 273.15)}$$

– Enter -35.200 for **Intercept**.

– Enter 1.389 for **Temp**.

  This is an estimate of the activation energy and is the coefficient of the inverse temperature, measured in degrees Kelvin, multiplied by Boltsmann's constant.

– Enter 1.305 for **scale**.

  For the Weibull distribution, JMP uses a parameterization that depends on a location parameter μ and scale parameter σ. In terms of the usual α and β parameterization, the scale parameter is $\sigma = 1/\beta$. See "JMP's Weibull Parameterization" on page 626.

In the Accelerated Life Test Plan window, you could specify uncertainty about your prior means in the Prior Variance Matrix outline. In this example, do not make any changes to the Prior Variance Matrix outline. Your design is created assuming that the Prior Means are the true parameter values.

10. Under Diagnostic Choices, enter 100,000 for both boxes for **Time range of interest**. Leave the **Probability of interest** value set to 0.1.
11. Under Design Choices, enter 1500 for **Length of test**.

    The test will be conducted over 1500 hours, which was the length of the original design.

12. Enter 250 for **Number of units under test**.

    The previous experiment tested 150 units, and the augmented experiment will test an additional 100 units, for a total of 250.

    Figure 22.11 shows the completed Accelerated Life Test Plan outline.

Figure 22.11 Completed Design Details Window

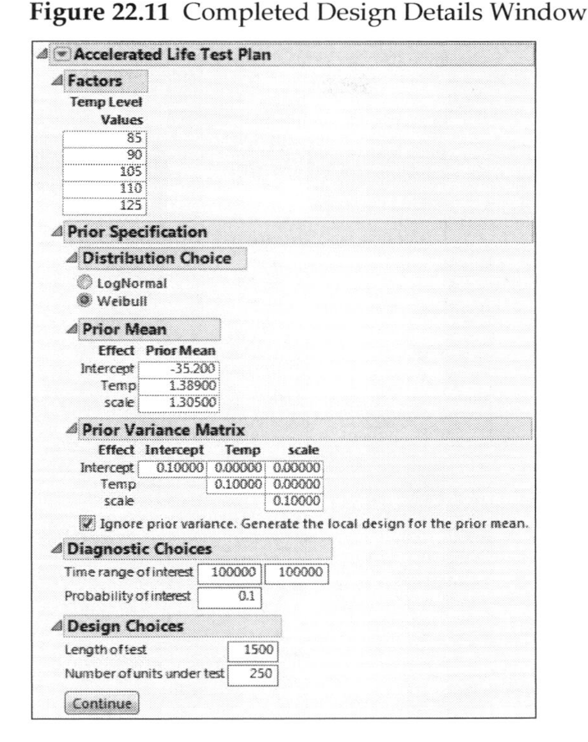

13. Click **Continue**.
14. To account for the units at each setting of Temp in the previous experiment, enter the following under **Candidate Runs**.
    - For the Temperatures of 85, 105, and 125, enter Minimum Units of 50 for each.
15. From the ALT Plan red triangle menu, select **ALT Optimality Criterion** > **Make Probability I-Optimal Design**.

    This tells JMP to use the Probability optimality criterion when creating the design in step 16. For details about this criterion, see "Make Probability I-Optimal Design" on page 624.
16. Click **Make Design**.

    The optimal experimental design is shown, along with other results.

Figure 22.12 Optimal Design

Design

Temp	N Units	Expected Failures	All Censored Probability
85	50	6.9	0.0
90	89	18.8	0.0
105	50	30.0	0.0
110	0	0.0	0.0
125	61	60.5	0.0

The optimal design is computed based on the levels of the test runs, the minimum number of units under test, the total number of units to be tested (this is the information in the Candidate Runs outline), and other information that you specified earlier. The optimal design consists of testing the following number of units at each temperature level:

- 50 units at 85º. Since the previous experiment already tested 50 units at 85º, no additional units are needed.
- 89 units at 90º. The next experiment will test 89 units at 90º.
- 50 units at 105º. Since the previous experiment test 50 units at 105º, no additional units are needed.
- 0 units at 110º. The next experiment will not test any units at this level.
- 61 units at 125º. Since the previous experiment test 50 units at 125º, 11 additional units are needed.

Compare the Augmented Design to the Original Study

1. In the Distribution Profiler, enter 25 for Temperature and 100,000 for Time. The estimate of the fraction of the population that is failing is 0.00357, with a 95% confidence interval of 0.00093 to 0.01361. This interval is narrower than the one from the initial experiment, which ranges from 0.003575 to 0.02268 (Figure 22.9).

Figure 22.13 Distribution Profiler for Temp = 25 and Time = 100000

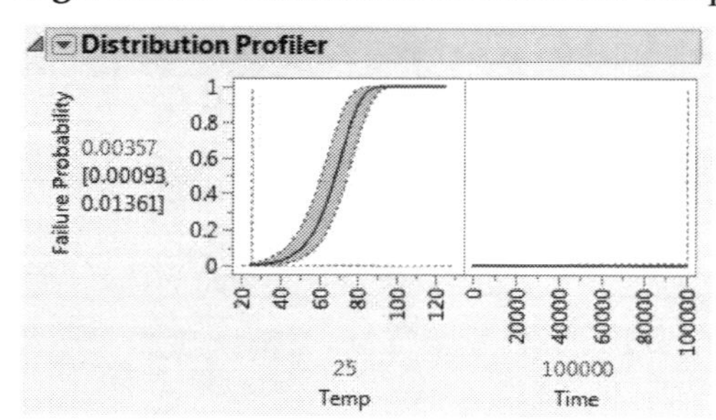

Accelerated Life Test Plan Window

The Accelerated Life Test Plan window is updated as you work through the design steps. The outlines that appear, separated by buttons that update the window, follow the flow in Figure 22.14.

Figure 22.14 Accelerated Life Test Plan Flow

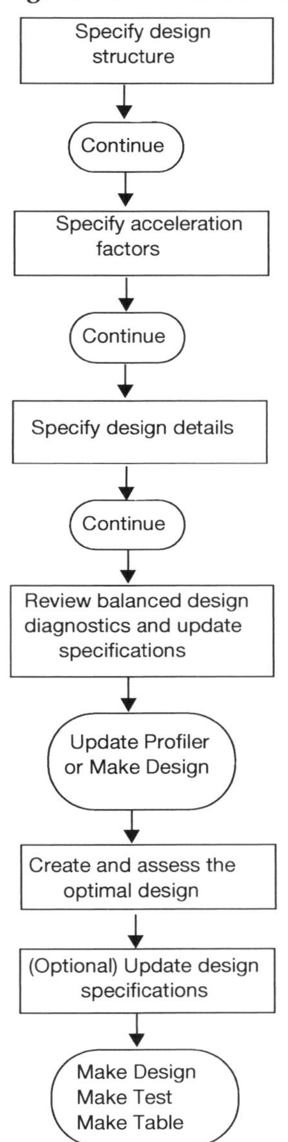

This section describes the outlines in the Accelerated Life Test Plan flow.

Specify the Design Structure

Select the model structure and the type of inspection that describe your design.

Figure 22.15 Initial ALT Design Window

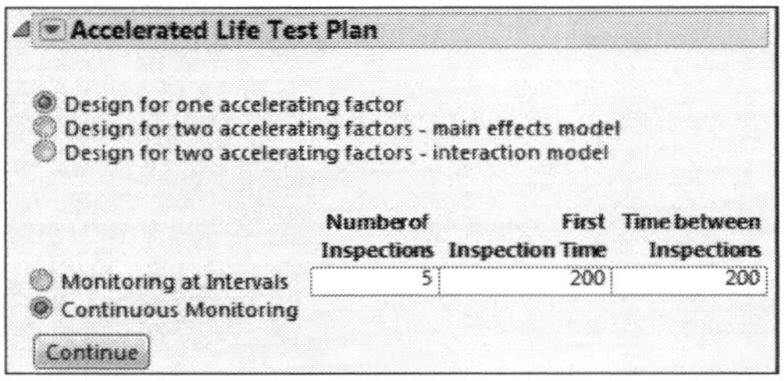

Select one of the first three options to indicate the number of acceleration factors in your design. If you have two factors, indicate whether you want to fit a main effects model or a model that contains an interaction term for the two factors.

Monitoring at Intervals Assumes that units will be inspected for failures at intervals. Failure times are interval censored. Enter the number of inspections, the time of the first inspection, and a time between inspections. For inspection intervals that are irregular, you can change the inspection times later in the Design Choices outline.

Continuous Monitoring Assumes that exact failure times are recorded. Failure times beyond the length of the test are right censored.

Specify Acceleration Factors

Specify details about the acceleration factor or factors.

Figure 22.16 ALT Specification Window

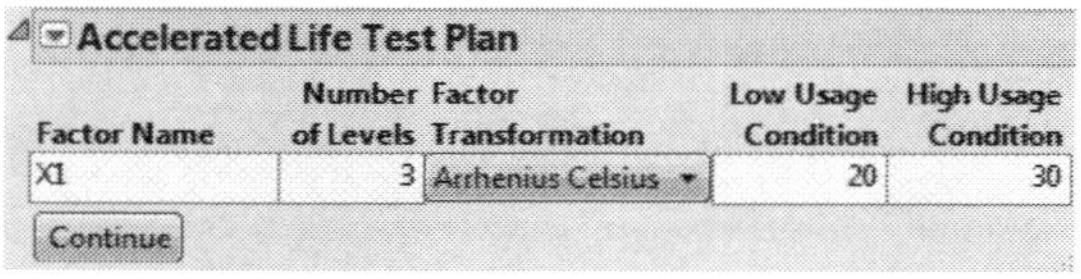

Factor Name Enter a name for each acceleration factor.

Number of Levels For each acceleration factor, enter the number of proposed levels that you want to include in the experiment.

Factor Transformation Select a transformation for each acceleration factor. This transformation describes the life-stress relationship, which is the manner in which the life distribution changes across stress levels. The transformations are Arrhenius Celsius, Reciprocal, Log, Square Root, and Linear.

Low Usage Condition For each acceleration factor, enter a lower bound for its value in typical usage conditions.

Note: The Low Usage Condition and High Usage Condition values can be identical.

High Usage Condition For each acceleration factor, enter an upper bound for its value in typical usage conditions.

Specify Design Details

Specify the factor levels, details of the prior distribution, the time range and probability of interest, and the length of the test and number of units to be tested.

Figure 22.17 Distribution Details

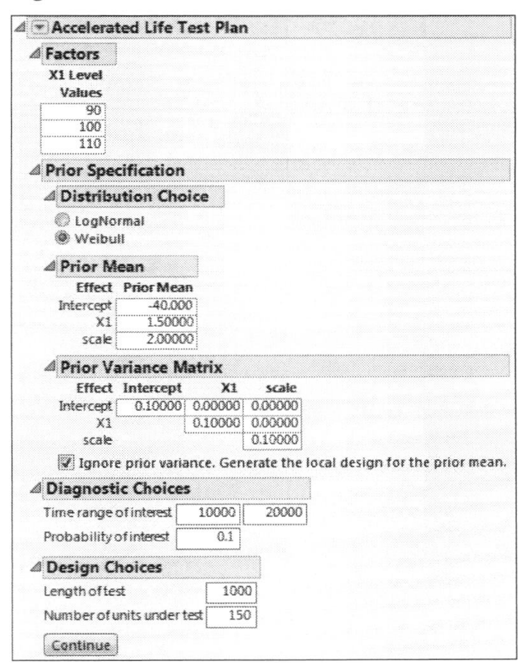

Factors Enter the settings for the acceleration factors.

Distribution Choice Select a life distribution (Weibull or LogNormal) for each acceleration factor. For more information, see "Statistical Details" on page 624.

Prior Mean Enter prior estimates of the acceleration model parameters. The prior estimates are hyperparameters in a Bayesian prior distribution. The Prior Mean values can be a best guess based on subject matter knowledge or they can be based on a previous study.

Prior Variance Matrix Enter values for the variances and covariances of the prior distribution for the acceleration model parameters. These variances and covariances reflect uncertainty relative to the prior estimates of the acceleration model parameters. Large variances indicate greater uncertainty.

Ignore prior variance Select this option to ignore the prior variances and covariances for the prior distribution. When these variances and covariances are ignored, the design is created by treating the values entered under Prior Mean as the true parameter values. This design is close to optimal if the prior mean parameters are close to the true values. However, this design is not robust to misspecification of the parameter estimates. If you are unsure about your prior estimates, use the Prior Variance Matrix to reflect your uncertainty.

Diagnostic Choices Specify values used to construct Time I-Optimal and Probability I-Optimal designs. See "ALT Optimality Criterion" on page 624. Enter values for the following:

Time range of interest Specify the time interval over which you want to estimate the fraction of the population that is failing. Enter the lower value in the left box and the upper value in the right box. If you are interested in a specific time point, enter that value in both boxes.

Probability of interest Specify the failure fraction for which you want an estimate of time. For example, if you want to estimate the time at which 10% of the units fail, then enter 0.10.

Design Choices Specify values relating to the length of the test, inspection intervals, and the number of units being tested. Enter values for the following:

Length of test (Available only for Continuous Monitoring.) Length of time during which units will be on test. When you make the design table, record each unit's failure time or whether it was right censored.

Inspection Times (Available only for Monitoring at Intervals.) Times at which inspections are conducted. When you make the design table, these times are used to construct Start Time and End Time columns. The number of units failing in each interval is recorded.

Number of units under test The number of units in the experiment.

- If you are designing an initial experiment, enter the number of units that you plan to test.
- If you are augmenting a previous experiment, enter the number of units tested in the previous experiment plus the number of units for the next experiment.

Review Balanced Design Diagnostics and Update Specifications

Three new outlines are added to the window. Two of these give results for a balanced design, enabling you to compare results for a balanced design to those for an optimal design:

- "Candidate Runs" on page 621

- "Parameter Variance for Balanced Design" on page 621
- "Distribution Profiler" on page 621

Figure 22.18 Additional Outline Nodes

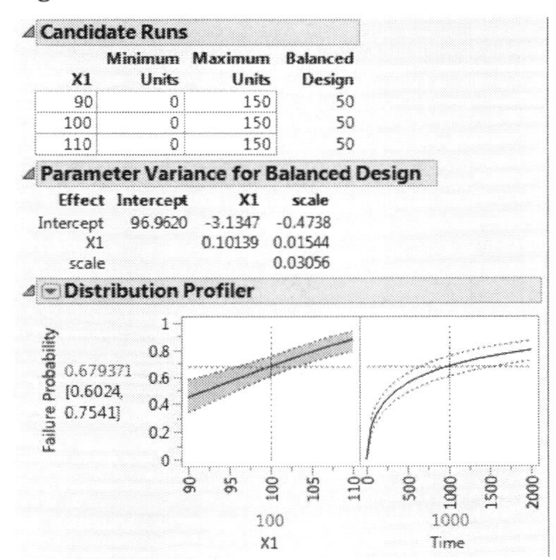

Candidate Runs Enter the minimum and maximum number of units allowed at each setting of the acceleration factors. If you are augmenting a previous experiment, for each setting, include the number of units already run at that setting in the Minimum Units.

Parameter Variance for Balanced Design Gives a matrix proportional to the covariance matrix for the estimates of the acceleration model parameters for the balanced design.

Denote the matrix of first partial derivatives of the model with respect to the parameters, $\boldsymbol{\theta}$, by $\mathbf{X}$. Denote the error variance by σ^2. Under general conditions, the least squares estimator of $\boldsymbol{\theta}$ is asymptotically unbiased with asymptotic covariance matrix given as follows:

$$Cov(\hat{\boldsymbol{\theta}}) = \sigma^2(\mathbf{X'X})^{-1}$$

The Parameter Variance for Balanced Design outline gives $(\mathbf{X'X})^{-1}$, where derivatives are calculated numerically. The calculation assumes that the values specified as Prior Mean are the true parameter values.

For more information, see "Nonlinear Models" on page 648 in the "Nonlinear Designs" chapter.

Distribution Profiler Provides a way to explore failure probabilities, based on the balanced design, as you vary the acceleration factors and time. The probabilities are based on the assumption that the values specified as Prior Mean are the true parameter values.

Click **Update Profiler** to update the profiler if changes are made to Distribution Choice, Prior Mean, or Design Choices.

Create and Assess the Optimal Design

Click **Make Design** to create the optimal design and see results that address the quality of the design. By default, the optimal design is D-optimal. You can change the optimality criterion by selecting the red triangle option ALT Optimality Criterion.

Three new outlines are added to the window: Design, Parameter Variance for Optimal Design, and Optimality Criteria.

Figure 22.19 Design Outlines

Design

X1	N Units	Expected Failures	All Censored Probability
90	61	27.5	0.0
100	0	0.0	0.0
110	89	78.0	0.0

Parameter Variance for Optimal Design

Effect	Intercept	X1	scale
Intercept	73.7806	-2.4008	-0.4421
X1		0.07816	0.01442
scale			0.02812

Optimality Criteria

D Criterion	-9.492
Quantile Criterion	5.3434
Probability Criterion	1.2332

Design Gives the number of units to be tested at each design setting. For a single factor, the first column gives the levels of the factor. For two factors, the first two columns give the design settings.

N Units Number of units to be tested at each design setting.

Expected Failures Expected number of failures for the design setting. The expected number is computed using the prior model specification.

All Censored Probability Probability that none of the units tested at the design setting will fail. The probability is computed using the prior model specification.

Note: The expected failures and censoring probabilities enable you to judge whether your prior specifications are reasonable.

Parameter Variance for Optimal Design Gives a matrix proportional to the covariance matrix for the estimates of the acceleration model parameters for the optimal design. The calculation assumes that the values specified as Prior Mean are the true parameter values. For more information, see "Parameter Variance for Balanced Design" on page 621.

Note: Compare the values in the Parameter Variance for Optimal Design matrix to those in the Parameter Variance for Balanced Design matrix to determine the extent to which the optimal design reduces the variance of estimates.

Optimality Criteria Values for three optimality criteria appear:

D Criterion D-optimality of the design. See "Make D-Optimal Design" on page 624.

Quantile Criterion Time I-optimality of the design. See "Make Time I-Optimal Design" on page 624.

Probability Criterion Probability I-optimality of the design. See "Make Probability I-Optimal Design" on page 624.

Update the Design and Create Design Tables

You can view the design, create a data table that summarizes the design, or create a data table where you can record your experimental results.

Make Design Updates the optimal design if any changes are made to the Distribution Choice, Prior Mean, Prior Variance Matrix, or Design Choices.

Make Test Plan Creates a data table where each row corresponds to a distinct design setting. The table shows the acceleration factor design settings and the number of units to include at those design settings.

Make Table Creates a table that you can use for recording your failure-time data.

- For Continuous Monitoring, the table contains a row for each unit to be tested and the design settings for that unit. Record each unit's failure time in the Failure Time column or whether the observation was right censored in the Censored column.
- For Monitoring at Intervals, the table contains a row for each design setting and time interval combination. The time intervals are defined by the Start Time and End Time columns, which are based on the Inspection Times entered in the Design Choices outline. For each setting and time interval, record the number failing in the Number Failing column.

Platform Options

The Accelerated Life Test Plan red triangle menu contains the following options:

Simulate Responses Adds simulated responses to the table when you click **Make Table**. The simulated responses are created by taking random draws from the chosen distribution at the parameter values specified under Prior Mean. If a simulated response exceeds the specified test length, the observation is censored at the test length value.

ALT Optimality Criterion Gives three choices for design optimality:

Make D-Optimal Design Creates a Bayesian D-optimal design if the number of Monte Carlo spheres is greater than 0. The optimality criterion is the expectation of the logarithm of the determinant of the information matrix with respect to the prior distribution. If the number of Monte Carlo spheres is 0, then the design is a locally D-optimal design. It follows that D-optimality focuses on precise estimates of the coefficients.

Make Time I-Optimal Design Creates a design that minimizes the average prediction variance with respect to the prior distribution when predicting the time to failure over the Time range of interest at the failure probability specified in the Diagnostic Choices outline. See "Diagnostic Choices" on page 620.

Make Probability I-Optimal Design Creates a design that minimizes the average prediction variance with respect to the prior distribution when predicting the failure probability over the Time range of interest specified in the Diagnostic Choices outline. See "Diagnostic Choices" on page 620.

Advanced Options Gives the N Monte Carlo Spheres option, which enables you to set the number of nonzero radius values used in the integration. To find a nonlinear design that optimizes a given optimality criterion, JMP must minimize the integral of the log of the determinant of the Fisher information matrix with respect to the prior distribution of the parameters. Such an integral must be calculated numerically. For details about how the integration is performed, see "Nonlinear Design Options" on page 647 in the "Nonlinear Designs" chapter.

Tip: By default N Monte Carlo Spheres is set to four. Higher values result in better numerical accuracy but with more computation time.

Save Script to Script Window Creates the script for the design that you specified in the Accelerated Life Test Plan window and places it in an open script window.

Statistical Details

In the ALT design platform, you can select either a lognormal or Weibull failure distribution. The parameterizations for the probability density function (pdf) and cumulative distribution function (cdf) for each distribution are given in this section. For additional detail on the Weibull distribution, see the Life Distribution chapter in the *Reliability and Survival Methods* book.

Lognormal

Lognormal distributions are used commonly for failure times when the range of the data is several powers of 10. This distribution is often conceptualized as the multiplicative product of many small positive independently and identically distributed random variables. This distribution is appropriate when the logarithms of the data values appear normally distributed. The pdf is usually characterized by strong right-skewness.

The lognormal family is parameterized by a location parameter, μ, and a shape parameter, σ. The lognormal pdf and cdf are given as follows, where the logarithm is to the base *e*:

$$f(x;\mu,\sigma) = \frac{1}{x\sigma}\phi_{nor}\left[\frac{\log(x)-\mu}{\sigma}\right], \quad x > 0$$

$$F(x;\mu,\sigma) = \Phi_{nor}\left[\frac{\log(x)-\mu}{\sigma}\right],$$

The functions

$$\phi_{nor}(z) = \frac{1}{\sqrt{2\pi}}\exp\left(-\frac{z^2}{2}\right)$$

and

$$\Phi_{nor}(z) = \int_{-\infty}^{z}\phi_{nor}(w)dw$$

are the pdf and cdf, respectively, for the standard normal distribution (N(0,1)).

Weibull

The Weibull distribution can be used to model failure time data with either an increasing or a decreasing hazard rate. It is used frequently in reliability analysis because of its tremendous flexibility in modeling many different types of data, based on the values of the shape parameter.

The Weibull pdf and cdf are commonly represented as follows:

$$f(x;\alpha,\beta) = \frac{\beta}{\alpha^\beta}x^{(\beta-1)}\exp\left[-\left(\frac{x}{\alpha}\right)^\beta\right]; \qquad x > 0, \alpha > 0, \beta > 0$$

$$F(x;\alpha,\beta) = 1-\exp\left[-\left(\frac{x}{\alpha}\right)^\beta\right]$$

where α is a scale parameter, and β is a shape parameter. The Weibull distribution reduces to an exponential distribution when $\beta = 1$.

JMP's Weibull Parameterization

An alternative parameterization is commonly used in the literature and in JMP. In the JMP parameterization, σ is the scale parameter and μ is the location parameter. These are related to the α and β parameterization as follows:

$$\alpha = \exp(\mu)$$

and

$$\beta = \frac{1}{\sigma}$$

With these parameters, the pdf and the cdf of the Weibull distribution are expressed as a log-transformed smallest extreme value distribution (SEV) using a location-scale parameterization with $\mu = \log(\alpha)$ and $\sigma = 1/\beta$:

$$f(x;\mu,\sigma) = \frac{1}{x\sigma}\phi_{sev}\left[\frac{\log(x)-\mu}{\sigma}\right], \qquad x > 0, \sigma > 0$$

$$F(x;\mu,\sigma) = \Phi_{sev}\left[\frac{\log(x)-\mu}{\sigma}\right]$$

where

$$\phi_{sev}(z) = \exp[z - \exp(z)]$$

and

$$\Phi_{sev}(z) = 1 - \exp[-\exp(z)]$$

are the pdf and cdf, respectively, for the standardized smallest extreme value ($\mu = 0$, $\sigma = 1$) distribution.

Chapter 23

Nonlinear Designs

When the goal of your experiment is to fit a model that is nonlinear in the unknown parameters, use a nonlinear design to place design points in areas that are key to fitting the nonlinear model. Although you could use an orthogonal design that is optimal for a linear model, such designs, in general, do not place design points in locations that minimize the uncertainty (or maximize the precision) of the estimates of the fitted parameters.

The efficiency of a nonlinear design depends on the values of the unknown parameters. This creates a circular problem in that to find the best design, you need to know the parameters in advance. JMP uses a Bayesian approach to construct a nonlinear design that maximizes the average efficiency over specified ranges of the values of the parameter. To properly specify these ranges, you must have some insight about the system of interest.

Nonlinear designs offer these advantages compared to designs for linear models:

- Predictions using a well-chosen model are likely to be good over a wider range of factor settings.
- It is possible to model response surfaces with complex curvature and with asymptotic behavior.

Figure 23.1 Design Points for a Nonlinear Model

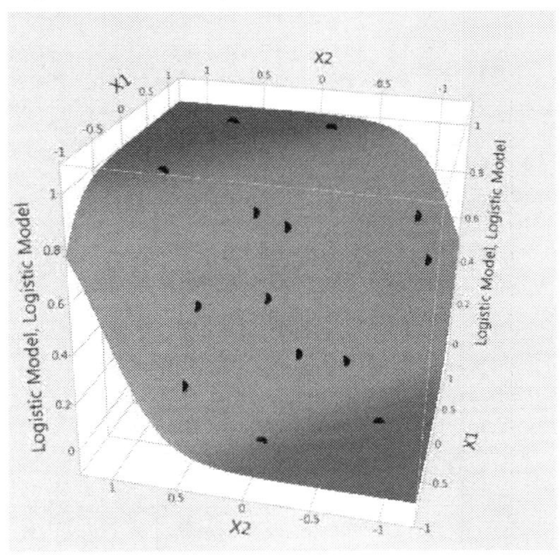

Overview of Nonlinear Designs

Construct designs to fit models that are nonlinear in their parameters using the Nonlinear Design platform. You can construct optimal designs or optimally augment existing data for nonlinear models. Nonlinear designs based on information that is descriptive of the underlying process can yield more accurate estimates of model parameters and prediction of process behavior than is possible with standard designs for polynomial models. For background on nonlinear models, see "Nonlinear Models" on page 648.

The efficiency of a design for a nonlinear model depends on the unknown values of the parameters that the design is intended to estimate. For this reason, JMP uses a Bayesian approach to construct designs that are efficient over a wide range of likely parameter values. You can specify a range of values for the unknown parameters and a distribution for the prior. The prior distribution choices include Uniform, Normal, Lognormal, and Exponential.

The Nonlinear Design platform uses a Bayesian approach, optimizing the design over a prior distribution of likely parameter values that you specify. The Bayesian D-optimality criterion is the expectation of the logarithm of the determinant of the information matrix with respect to a sample of parameter vectors that represents this prior probability distribution. The information matrix entries depend on the prediction variances at the design points. Little information is contributed by observations with low variance, where the response is almost certain. It follows that an optimal design places some design settings at high-variance points. For more information, see Gotwalt et al. (2009).

A principal of optimal design is that, over the feasible region of experimentation, the optimal design places points in locations with the highest variance of prediction. Though this may seem counter-intuitive, if an alternative design put points at other locations, the prediction variance at the design points of the optimal design would be even higher. For models that are linear in the parameters, the high-variance points tend to be at the vertices of the experimental region. But this is not necessarily true for models that are nonlinear in the parameters.

Note: Nonlinear designs are computed using a random starting design. For this reason, nonlinear designs that you obtain for identical specifications usually differ.

To use the Nonlinear Design platform, you must have an existing data table. That data table must contain the following:

- A column for the response.
- A column for each factor.
- A column that contains a formula showing the relationship between the factors and the response. This formula must include the unknown parameters.

Note: This is the same format as is required for a data table used in the Nonlinear platform for modeling.

Your table can come in one of two forms:

- It might be a template, containing only column information and no rows. See "Create a Nonlinear Design with No Prior Data" on page 629.
- It might contain rows with predictor information. In this case, the predictor values are included in the nonlinear design. See "Augment a Design Using Prior Data" on page 634.

Examples of Nonlinear Designs

This section contains the following examples:

- "Create a Nonlinear Design with No Prior Data" on page 629
- "Augment a Design Using Prior Data" on page 634
- "Create a Design for a Binomial Response" on page 638

Create a Nonlinear Design with No Prior Data

This example shows how to create a design when you have not yet collected data, but have a guess for the unknown parameters. In this example, you model the fractional yield (Observed Yield) of an intermediate product in a chemical reaction. The fractional yield is a function of reaction time and temperature. See Box and Draper (1987).

Create the Design

1. Select **Help > Sample Data Library** and open Design Experiment/Reaction Kinetics Start.jmp.

   Notice the following:

   – The data table contains no rows because no data have been collected.

   – The columns for the predictors, Reaction Temperature and Reaction Time, have the Coding, Design Role, and Factor Changes properties. To see these properties, click * in the Columns panel. They tell JMP how to treat these predictors when constructing a design. For information about how to save these column properties, see the "Column Properties" chapter on page 651.

   – The Observed Yield column will contain response data obtained by running the experiment.

   – The Yield Model column contains the formula that relates the predictors to the response, Observed Yield. Click + in the Columns panel to see the formula. The formula is nonlinear in the parameters t1 and t3.

2. Select **DOE > Special Purpose > Nonlinear Design**.
3. Select Observed Yield and click **Y, Response**.
4. Select Yield Model and click **X, Predictor Formula**.
5. Click **OK**.

   In this example, the values 510 and 540 for Reaction Temperature and 0.1 and 0.3 for Reaction Time were specified using the Coding column property. Alternatively, you can specify a reasonable range of values directly in the Factors outline.
6. Change the values of the parameter t1 to 25 and 50, and t3 to 30 and 35.

   These new values represent a reasonable range of parameter values for the experimental situation. The default values were constructed based on the initial parameter values that were specified in the definition of the prediction formula. For information about constructing formulas, see the Formula Editor chapter in the *Using JMP* book.

   Notice that the prior distribution, shown under Distribution, for each of t1 and t3 is set to Normal by default.
7. Change the number of runs to 12 in the Design Generation panel.

Figure 23.2 Completed Outlines for Reaction Kinetics Experiment

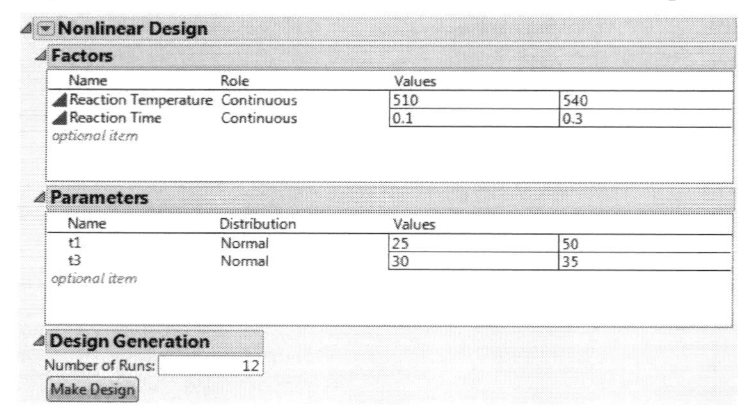

8. Click **Make Design**.
9. Click **Make Table**.

Figure 23.3 Design Table

Nonlinear Design
Design Nonlinear Design
Model

Columns (4/0)
Reaction Temperature
Reaction Time
Observed Yield
Yield Model

	Reaction Temperature	Reaction Time	Observed Yield	Yield Model
1	540	0.3	•	58.093235
2	540	0.11	•	57.3016426
3	540	0.3	•	58.093235
4	530	0.2	•	54.9142081
5	510	0.1	•	11.2590804
6	540	0.3	•	58.093235
7	540	0.3	•	58.093235
8	510	0.1	•	11.2590804
9	517	0.1	•	17.8546023
10	530	0.21	•	55.9838913
11	540	0.11	•	57.3016426
12	540	0.3	•	58.093235

Your design should be similar to the one shown in Figure 23.3. The runs might be in a different order, and the values for Reaction Temperature and Reaction Time, and consequently those computed for Yield Model, can be slightly different. Notice that values appear in the Yield Model column because the column contains the formula for the model. Also notice that the table contains a **Model** script that you can use to fit a nonlinear model to your observations.

Now that you have created your design table, run your experiment, and record the responses in the Observed Yield column. From this point on, work with the data table Reaction Kinetics.jmp, found in the Design Experiment folder.

Explore the Design

Before analyzing your results, construct a plot to see the design settings.

1. Select **Help > Sample Data Library** and open Design Experiment/Reaction Kinetics.jmp.
1. Select **Graph > Graph Builder**.
2. Drag and drop Reaction Temperature into the **Y** zone.
3. Drag and drop Reaction Time into the **X** zone.
4. Click the second icon above the graph to deselect the Smoother.

Figure 23.4 Design Settings

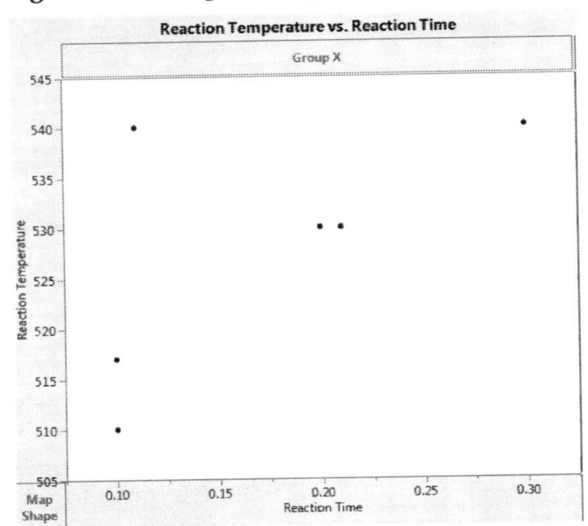

Notice that the points are located in three areas. There are no points at low temperature and high time (the lower right corner of the graph). Unlike orthogonal designs, nonlinear designs do not necessarily place design points at the corners of the design region. In this example, design points at low temperature and high time would be inefficient.

To see the density of design points in the remaining three corners, use the Contour tool.

5. Click [icon] to turn on the Contour tool.
6. Click **Done**.

Figure 23.5 Design Settings with Density Contours

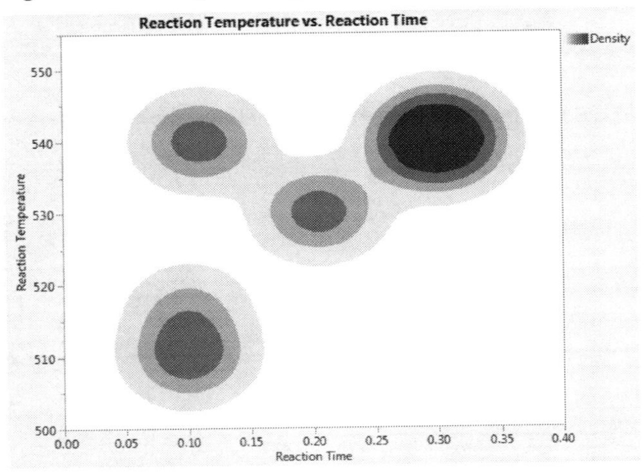

Notice that there are comparatively few points at low time and high temperature. From the design table, you can see that there are only three such points. Because of the model and the parameter specifications, the optimal design places more design points at high time and high temperature.

Analyze the Results

Now that you visually explored your design, analyze your results.

Note: Rather than conduct step 1 through step 4, you can run the **Model** script.

1. Select **Analyze > Specialized Modeling > Nonlinear**.
2. Select Observed Yield and click **Y, Response.**
3. Select Yield Model and click **X, Predictor Formula**.

   Notice that the model appears in the Options for fitting custom formulas panel.
4. Click **OK**.
5. Click **Go** in the Control Panel.

   The iterative search for a solution proceeds until one of the Stop Limit values is reached. Then, the Solution and Correlation of Estimates reports appear.
6. Click the Nonlinear Fit red triangle menu and select **Profilers > Profiler.**
7. To maximize the yield, click the Prediction Profiler red triangle menu and select **Optimization and Desirability > Maximize Desirability**.

Figure 23.6 Time and Temperature Settings for Maximum Yield

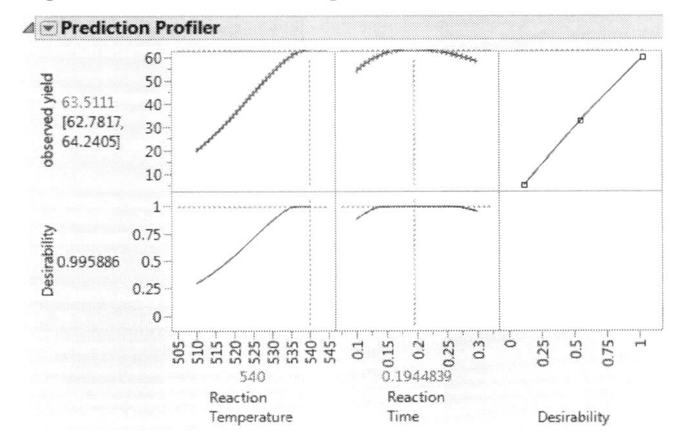

The estimated maximum yield is approximately 63.5% at a reaction temperature of 540 degrees Kelvin and a reaction time of 0.1945 minutes.

Augment a Design Using Prior Data

This example shows how to create a nonlinear design when you have prior data. In this example, the data pertain to a chemical reaction. You want to model the rate of uptake (velocity) of available organic substrate as a function of the concentration of that substrate. See Meyers (1986). You have already run an experiment, but you want to leverage your results to obtain more precise estimates of the parameters.

Obtain Prior Parameter Estimates

Use your existing experimental data to obtain better estimates of the parameter values.

1. Select **Help > Sample Data Library** and open Nonlinear Examples/Chemical Kinetics.jmp.
2. Click the plus sign next to Model (x) in the Columns panel.
3. Click **Table Columns** in the top left of the formula editor window and select **Parameters**.

   The parameter values in the formula element panel (VMax = 1 and k = 1) are your initial guesses. They are used to compute the Model (x) values in the data table. For your next experiment, you want to replace these with better estimates.
4. Click **Cancel** to close the formula editor window.
5. Select **Analyze > Specialized Modeling > Nonlinear**.
6. Select Velocity (y) and click **Y, Response**.
7. Select Model (x) and click **X, Predictor Formula**.

   Notice that the formula given by Model (x) appears in the Options for fitting custom formulas panel.

Figure 23.7 Nonlinear Analysis Launch Window

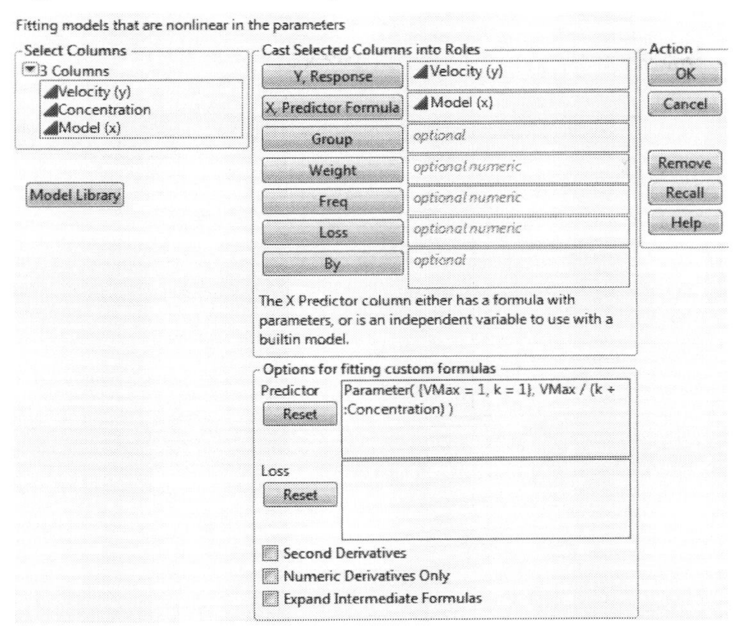

8. Click **OK**.
9. In the Control Panel, click **Go**.

   The iterative search for a solution proceeds until one of the Stop Limit values is reached. Then, the Solution and Correlation of Estimates reports appear. Also, an option appears in the Control Panel enabling you to add confidence limits to the Solution report.
10. In the Control Panel, click **Confidence Limits**.

    Confidence intervals for the parameters VMax and k appear in the Solution report.

Figure 23.8 Nonlinear Fit Results

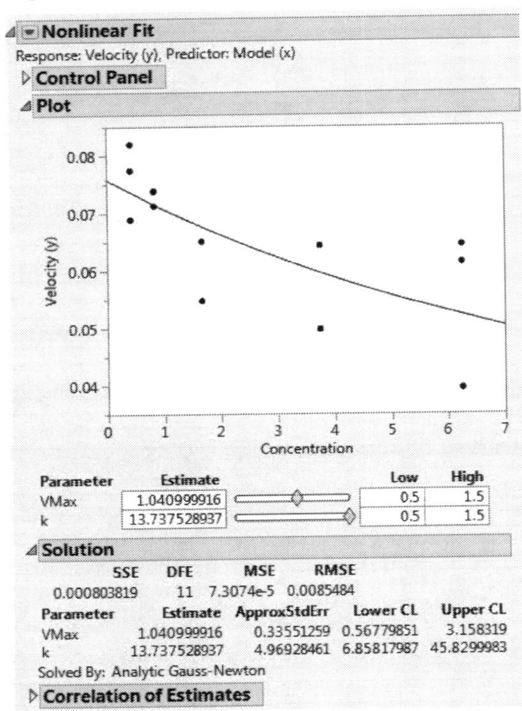

The Lower CL and Upper CL values for VMax and k define ranges of values for these parameters. Next, use these intervals to define a range for the prior values in your augmented nonlinear design.

Note: Do not close the Nonlinear Fit report because these results are needed in the next steps.

Augment the Design

Now, create a design to estimate the nonlinear parameters more precisely.

1. With the Chemical Kinetics.jmp data table active, select **DOE > Special Purpose > Nonlinear Design**.
2. Select Velocity (y) and click **Y, Response**.
3. Select Model (x) and click **X, Predictor Formula**.
4. Click **OK**.

Figure 23.9 Nonlinear Design Outlines for Factors and Parameters

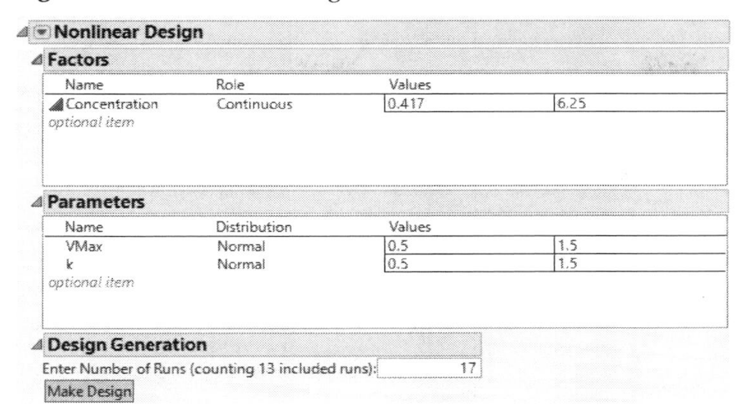

In the Chemical Kinetics.jmp data, the values for Concentration range from 0.417 to 6.25. Therefore, these values initially appear as the low and high values in the Factors outline. You want to change these values to encompass a broader interval.

5. Click 0.417 and type 0.1.
6. Press Tab over to 6.25 and type 7.

   Leave the prior Distribution for each parameter set to Normal.

   The range of Values for the parameters reflects the uncertainty of your knowledge about them. You should specify a range that you think covers 95% of possible parameter values. The confidence limits from the Nonlinear Fit report shown in Figure 23.8 provide such a range. Replace the Values for the parameters in the Parameters outline with the confidence limits, rounding to three decimal places.

7. In the DOE Nonlinear Design window, enter these values into the Parameters for VMax and k:
   – VMax: 0.568 and 3.158
   – k: 6.858 and 45.830

Figure 23.10 Updated Values for Factor and Parameters

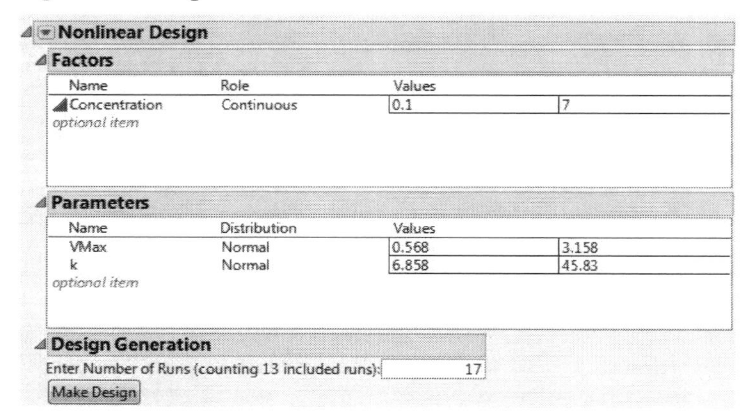

8. Enter 40 for the **Number of Runs** in the Design Generation panel.
9. Click **Make Design.**

   The Design outline opens, showing the Concentration and Velocity (y) values for the original 13 runs and new Concentration settings for the additional 27 runs.
10. Click **Make Table.**

    This creates a new JMP design table that contains the settings and results for the original 13-run design and settings for 27 new runs. Instead of creating a new data table, you can add the new runs to your existing data table by clicking Augment Table instead of Make Table.

The new runs reflect the broader interval of Concentration values and the range of values for VMax and k obtained from the original experiment, which are used to define the prior distribution. Both should lead to more precise estimates of k and Vmax.

Create a Design for a Binomial Response

In some applications, the only measurement type available is a pass/fail (binomial) measurement. In this example, two factors are of interest, X1 and X2, which you will vary between -1 and 1. You will construct a nonlinear design for the binomial response and then view it in the context of your proposed nonlinear model.

Logistic Model

Model the probability of success for your binomial response (Y) using a logistic model:

$$\pi(X_1, X_2) = \frac{1}{1 + e^{-(\beta_0 + \beta_1 X_1 + \beta_2 X_2)}}$$

This model is nonlinear in the unknown parameters β_0, β_1, and β_2. Your goal is to estimate these parameters using an experimental design.

Prior Knowledge

To construct a design using Nonlinear Design, you need to specify your prior knowledge (or uncertainty) about each parameter value using a distribution. You can specify a best guess for each parameter value, but you have a lot of uncertainty relative to these values. Your best guess about the values of the parameters and 95% ranges for them are as follows:

- β_0 is 0, but might range from -2 to 2
- β_1 is 5, but might range from 0 to 10
- β_2 is 5, but might range from 0 to 10

For parameter values in these ranges, the logistic function is nonlinear. So you expect that a design constructed using Nonlinear Design will differ from an orthogonal design. In particular, you expect that the nonlinear design will place factor settings at points where the predicted response has high variance.

Data Table for Launch Window

To construct a nonlinear design, you must first have a data table containing columns for the predictors and a column containing a formula that represents the nonlinear model that you are fitting. The Binomial Optimal Start.jmp data table, found in the Design Experiment folder, contains the following:

- Columns X1 and X2 for the two predictors. The Coding property defined for each of these columns causes the initial factor settings to be -1 and 1.
- A column for the response, Y.
- A column called Logistic Model that contains a formula relating the predictors to the response. To view the formula, click on the plus sign to the right of Logistic Model in the Columns panel. See Figure 23.11.
- Your initial guesses for the parameters **b0**, **b1**, and **b2**. When you defined these parameters, you were asked to specify a value. You set this value to your initial guess. These values are shown in the formula element panel at the top left of the formula editor window. See Figure 23.11.
- A column called Variance that contains the formula for the variance of the predicted value based on the assumed logistic model. When you construct your design, this column indicates which design points have comparatively high variances.

Figure 23.11 Formula Relating Predictors to Binomial Probability

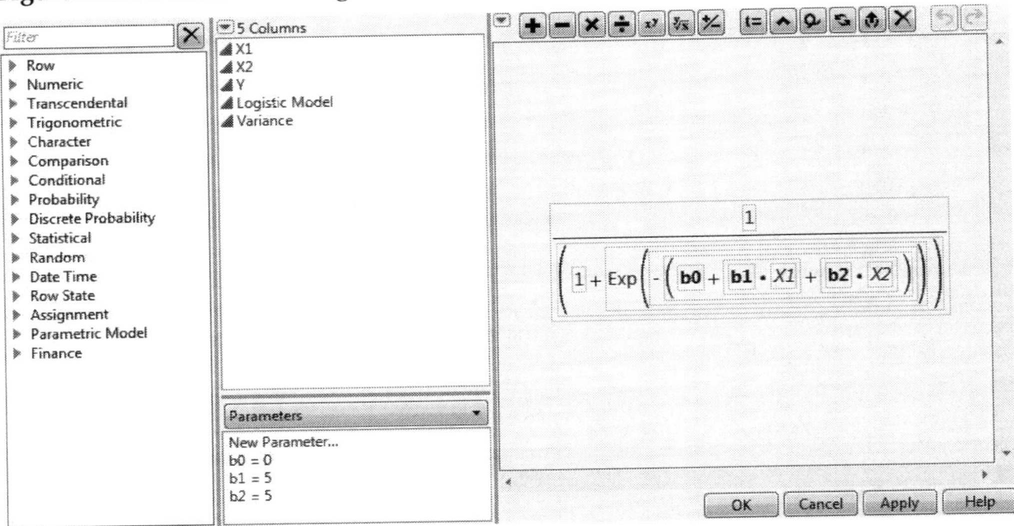

Create the Design

1. Select **Help > Sample Data Library** and open Design Experiment/Binomial Optimal Start.jmp.
2. Select **DOE > Special Purpose > Nonlinear Design**.
3. Select Y and click **Y, Response**.
4. Select Logistic Model and click **X, Predictor Formula**.
5. Click **OK**.

Figure 23.12 Nonlinear Design Window

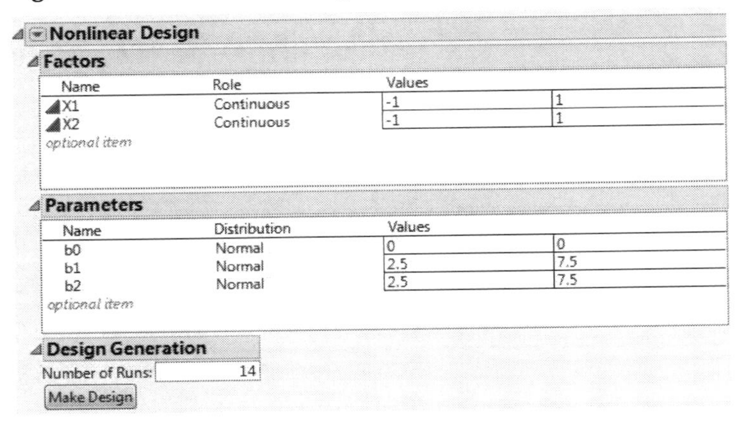

The Factors outline shows the two factors with the appropriate factors settings specified as values. The Parameters outline shows the three parameters with each prior distribution set to a normal distribution. JMP computes default Values based on your initial guesses for the parameter values. You are comfortable assuming a Normal prior, but your uncertainty about the parameters requires that you specify a wider range of values.

6. Enter the following under **Values** for the three parameters:
   – b0: -2 and 2
   – b1: 0 and 10
   – b2: 0 and 10

Figure 23.13 Nonlinear Design Window with Parameter Values

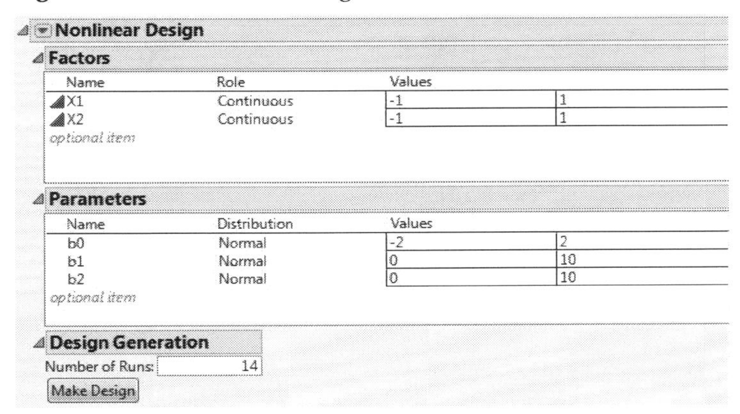

The default number of runs, which you accept, is 14.

7. Click **Make Design**.
8. Click **Augment Table**.

   This adds the 14 runs to Binomial Optimal Start.jmp. Your design table will be different because the optimization algorithm has a random component.

Figure 23.14 Augmentation of Binomial Optimal Start.jmp

BinomialOptimalStart
- Model

Columns (5/0)
- X1
- X2
- Y
- Logistic Model
- Variance

Rows
All rows 14

	X1	X2	Y	Logistic Model	Variance
1	0.9664414322	-1	•	0.4581	0.2482
2	0.1634983639	0.0184514596	•	0.7129	0.2047
3	-0.071872536	-0.154946789	•	0.2434	0.1842
4	1	-0.884473402	•	0.6405	0.2303
5	-0.215542941	0.8466637325	•	0.9591	0.0392
6	0.168538651	-0.490084808	•	0.1669	0.1390
7	-0.652939915	-0.043100755	•	0.0299	0.0290
8	-0.746371597	0.5324145785	•	0.2554	0.1902
9	0.7962333596	0.6329208108	•	0.9992	0.0008
10	-0.108648662	0.091663742	•	0.4788	0.2495
11	-0.140162195	-0.843657249	•	0.0073	0.0072
12	-0.498234843	0.5112146738	•	0.5162	0.2497
13	0.9564547528	-0.150140359	•	0.9826	0.0171
14	0.0294611563	0.2320326696	•	0.7871	0.1676

Now that you have constructed your design, proceed to examine where the design points are located relative to the proposed logistic model. The Variance column gives the prediction variance at each design point, based on the logistic model.

View the Design

1. With Binomial Optimal Start.jmp active, select **Graph > Graph Builder**.
2. Select X1 and drag it to the **X** zone.
3. Select X2 and drag it to the **Y** zone.
4. De-select the Smoother, which is the second icon above the graph.
5. If you need to, drag each axis so that the -1.0 and 1.0 axis labels appear.
6. Click **Done**.

   Because your design differs from the one in Figure 23.14, your plot will differ from the one in Figure 23.15.

Figure 23.15 Design Settings

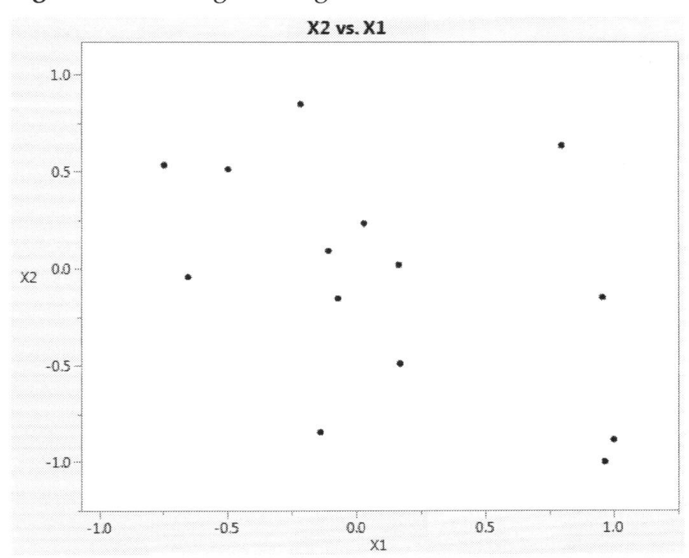

Notice that there are no points at X1 = -1. The only point on a corner of the design region corresponds to X1 = 1 (more precisely, 0.996) and X2 = -1. There are several points in the central part of the design region.

To better see these points in relation to the model, construct a surface plot.

7. Select **Graph > Surface Plot.**
8. Select X1, X2, and Logistic Model and click **Columns.**
9. Click **OK.**
10. In the Dependent Variables outline, locate Logistic Model under Formula. In the Point Response Column Style list, click on **none** and select **Logistic Model.**

    This adds points to the Surface Plot.
11. Right-click in the plot and select **Settings.**
12. Drag the **Marker Size** indicator to the right.
13. Click **Done.**
14. Rotate the plot to view the design points.

    Because your design differs from the one in Figure 23.14, your plot will differ from the one in Figure 23.16.

Figure 23.16 Prediction Model with Design Points

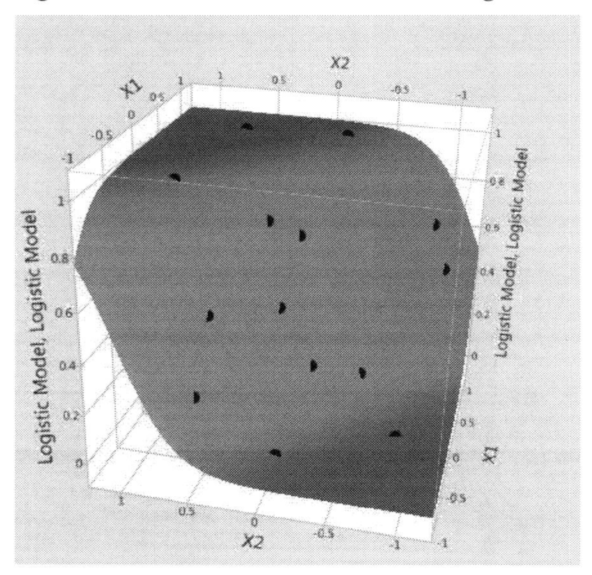

Notice that many of the design points are in areas where the prediction model has a steep slope. These are high-variance points. Experimental results at these design points provide information to fit the model in areas where it is unstable. By way of contrast, an orthogonal design would place all design points at the corners of the design region. This would provide no information about nonlinear behavior within the design region.

Nonlinear Design Launch Window

To use the Nonlinear Design platform, you must have an existing data table that contains the following:

- A column for the response.
- A column for each factor.
- A column that contains a formula showing the relationship between the factors and the response. This formula must include the unknown parameters.

For information about formulas, see the Formula Editor chapter in *Using JMP*.

The table can contain values for the predictors and response. If it does, the design that you construct augments the design that is implicit in the table. There can be no row containing missing predictor values.

With your starting data table active, select **DOE > Special Purpose > Nonlinear Design**.

Figure 23.17 Nonlinear Launch Window

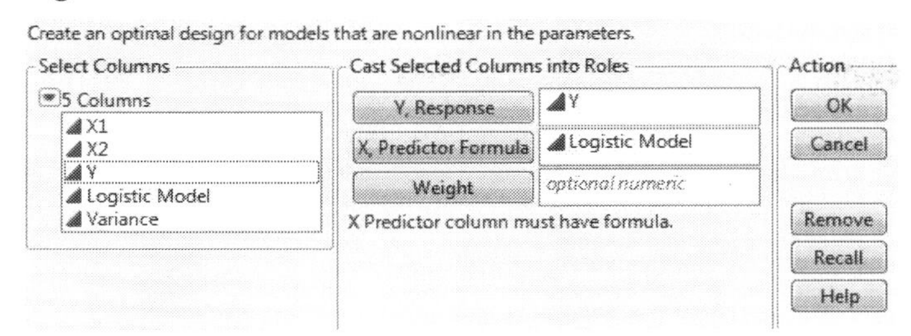

Y, Response The numeric column for response values.

X, Predictor Formula The numeric column that contains the formula for the nonlinear model. This formula must contain parameters.

Weight (Optional) A numeric column that assigns weights to the observations.

Nonlinear Design Window

The Nonlinear Design window updates as you work through the design steps. The outlines, separated by buttons that update the outlines, follow the flow in Figure 23.18.

Figure 23.18 Nonlinear Design Flow

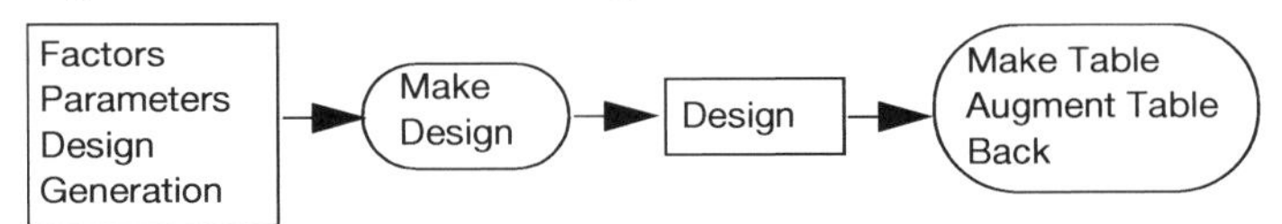

The initial design window shows the Factors, Parameters, and Design Generation outlines.

Figure 23.19 Initial Design Window

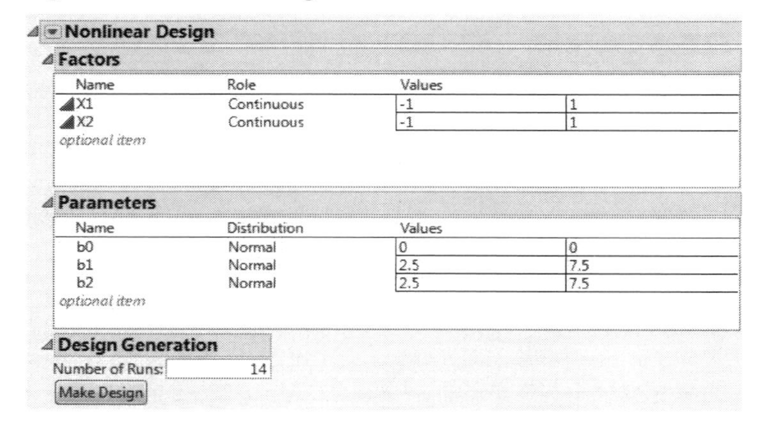

The next sections cover the following:

- "Factors" on page 646
- "Parameters" on page 646
- "Design Generation" on page 647
- "Make Table or Augment Table" on page 647

Factors

The column names used in the model formula are automatically inserted in the Name column of the Factors outline. Each factor's role is set to Continuous.

For each factor, the Values are initially set to -1 and 1. Or, if you have defined a value using the Coding column property, those values are used instead. You can change these values in the factors outline.

Parameters

The parameter names used in the model formula are automatically inserted in the Name column of the Parameters outline.

For each parameter, the Values are initially set to a symmetric interval around the initial value specified in the parameter definition. This interval is obtained by taking the initial value's distance to 0, and constructing an interval of this width around the initial value. These values are used in defining prior distributions for the model parameters.

Note: Adjust the Values for the parameters in conjunction with your choice of distribution to reflect your uncertainty about the model parameters.

Four families of prior distributions are listed under Distribution. The Values that you specify for the parameters determine which member of the family of prior distributions that is used. Denote the low value by *low* and the high value by *high*. Then the distributions are determined as follows:

- Uniform: The distribution is uniform on the interval (*low*, *high*).
- Normal, Lognormal, Exponential: The distribution is the one where *low* is the 0.025 quantile and where *high* is the 0.975 quantile.

Design Generation

JMP provides a suggested number of runs, determined as follows:

- If you are not augmenting a data table, the number of runs is four times the number of parameters plus two.
- If you are augmenting a data table, the number of runs is the number of runs in the data table plus two times the number of parameters.

Note: If you are augmenting a design, the number of runs that JMP suggests or that you specify includes those runs corresponding to observations in your data table. Adjust the number of runs appropriately.

Design

When you click Make Design, JMP constructs the design and adds a Design outline to the Nonlinear Design window. In the Design outline, you can review the factor level settings.

Make Table or Augment Table

The Make Table button creates a new design table. If your original table included existing runs, the new table also includes the existing runs. The **Augment Table** button adds the new runs to your existing table.

Nonlinear Design Options

The red triangle menu in the Nonlinear Design platform contains these options:

Save Responses Saves the information in the Responses panel to a new data table. You can then quickly load the responses and their associated information into most DOE windows. This option is helpful if you anticipate reusing the responses.

Load Responses Not available.

Save Factors Saves the information in the Factors panel to a new data table. Each factor's column contains its levels. Other information is stored as column properties. You can then quickly load the factors and their associated information into most DOE windows.

Load Factors Not available.

Save Constraints Not available.

Load Constraints Not available.

Simulate Responses Adds response values to the design table. Select this option before you click Make Table. Then, in the resulting design table, the response columns contain simulated values.

Note: To set a preference to always simulate responses, select **File > Preferences > Platforms > DOE** and select **Simulate Responses**.

Number of Starts Sets the number of times that a nonlinear design is created using the quadrature method. Among the designs created, the platform selects the design that maximizes the optimality criterion.

Advanced Options > Number of Monte Carlo Samples Sets the number of octahedra per sphere used in computing the optimality criterion. The default value is one octahedron. See "Radial-Spherical Integration of the Optimality Criterion" on page 649.

Advanced Options > N Monte Carlo Spheres Sets the number of nonzero radius values used in computing the optimality criterion. The default is two. See "Radial-Spherical Integration of the Optimality Criterion" on page 649.

Note: If N Monte Carlo Spheres (the number of radii) is set to zero, then only the center point is used in the calculations. This gives a local design that is optimal for the initial values of the parameters. For some situations, this is adequate.

Statistical Details

This section contains the following information:

Nonlinear Models

Denote the vector of n responses by $\mathbf{Y} = (Y_1, Y_2, \ldots, Y_n)'$. A nonlinear model is defined by the following properties:

- The Y_i are independent and identically distributed with an exponential family distribution.
- The expected value of each Y_i given a vector of predictor values $\mathbf{x}_i$ is a nonlinear function of parameters, $\boldsymbol{\theta}$. Denote this function as follows:

  $$E(Y_i(\mathbf{x})) = f(\boldsymbol{\theta}, \mathbf{x}_i)$$

- Each Y_i is expressed as follows:

  $$Y_i = f(\boldsymbol{\theta}, \mathbf{x}_i) + \varepsilon_i$$

- The vector of errors, $\boldsymbol{\varepsilon} = (\varepsilon_1, \varepsilon_2, \ldots, \varepsilon_n)'$ has mean $\mathbf{0}$ and covariance matrix $\sigma^2\mathbf{I}$, where $\mathbf{I}$ is the n x n identity matrix.

Denote the matrix of first partial derivatives of the function f with respect to the parameters $\boldsymbol{\theta}$ by X. Under general conditions, the least squares estimator of $\boldsymbol{\theta}$ is asymptotically unbiased, with asymptotic covariance matrix given as follows:

$$Cov(\hat{\boldsymbol{\theta}}) = \sigma^2(\mathbf{X}'\mathbf{X})^{-1}$$

For the proof of this result, see Wu (1981) and Jennrich (1969).

Radial-Spherical Integration of the Optimality Criterion

The optimality criterion is the expectation of the logarithm of the determinant of the information matrix with respect to the prior distribution. Consequently, finding an optional nonlinear design requires minimizing the integral of the log of the determinant of the Fisher information matrix with respect to the prior distribution of the parameters. This integral must be calculated numerically. The approach used in the Nonlinear Design platform is based on Gotwalt et al. (2009).

For normal distribution priors, the integral is reparameterized into a radial direction and a number of angular directions equal to the number of parameters minus one. The radial part of the integral is computed using Radau-Gauss-Laguerre quadrature with an evaluation at radius = 0. This is done by constructing a certain number of hyperoctahedra and randomly rotating each of them.

If the prior distribution is not normal, then the integral is reparameterized so that the new parameters have a normal distribution. Then the radial-spherical integration method is applied.

Note: If the prior distribution for the parameters does not lend itself to a solution and the process fails, a message is added to the window that the Fisher information matrix is singular in a region of the parameter space. When this occurs, consider changing the prior distribution or the ranges of the parameters.

Finding the Optimal Design

The method used to find an optimal design is similar to the coordinate exchange algorithm described in Meyer and Nachtsheim (1995). For details about how the nonlinear optimal design is obtained, see Gotwalt et al. (2009). The general approach proceeds as follows:

- Random designs are tested until a nonsingular starting design is found.
- Iterations are conducted, where each iteration consists of a pass through all the runs.
- For each run, factors are optimized one at a time.
- The objective function is the Bayesian D-optimality criterion. This is the expectation of the logarithm of the determinant of the information matrix with respect to the prior distribution.
- Iterations terminate once the change in the objective function is small.

Column Properties

Understanding Column Properties Assigned by DOE

When you construct a design using the DOE platforms, column properties are saved to the data table that contains the resulting design. This appendix provides detail about only those column properties that are saved to designs that the DOE platform constructs. Examples illustrate how each column property is assigned and used by the DOE platforms. Descriptions of column properties not assigned by the DOE platforms are provided in The Column Info chapter in the *Using JMP* book.

Some of the column properties described in this appendix are useful in general modeling situations. To use the properties more generally, you can specify them yourself. This ability is particularly useful when your design has not been created by a DOE platform. Some of the examples in this appendix illustrate situations where you add a column property on your own.

Figure A.1 Column Property Asterisks and Column Info Window

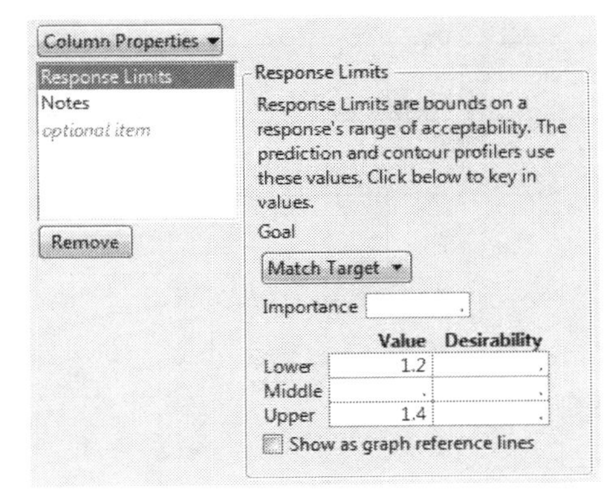

Adding and Viewing Column Properties

The DOE platforms automatically save certain column properties to the design tables that they construct. However, some of the column properties associated with designed experiments are useful in general modeling situations. To use these column properties with data tables that have not been created using DOE platforms, you can add them yourself.

Adding a Column Property

To assign a column property to one or more columns, do the following:

1. Select the column or columns to which you want to assign a property.
2. Do one of the following:
   – Right-click the header area, select **Column Properties**, and select the property.
   – Right-click the header area, select **Column Info**, and select the property from the Column Properties menu.
   – Select **Cols > Column Info** and select the property from the Column Properties menu.
3. In the column property panel that appears, specify values and select options as appropriate.
4. Click **Apply** to add the column property or click **OK** to add the column property and close the column properties window.

Tip: A column might already contain a property that you want to apply to other columns. Use the Standardize Attributes command to apply that property to other columns. For details, see The Column Info Window chapter in the *Using JMP* book.

Viewing a Column Property

To view the properties assigned to a specific column, in the columns panel, click the column property asterisk icon *. Click a property to see its settings or to edit it. Figure A.2 shows the column properties assigned to Stretch in the Bounce Data.jmp sample data table, located in the Design Experiment folder.

Figure A.2 Asterisk Icon for Stretch Revealing Two Column Properties

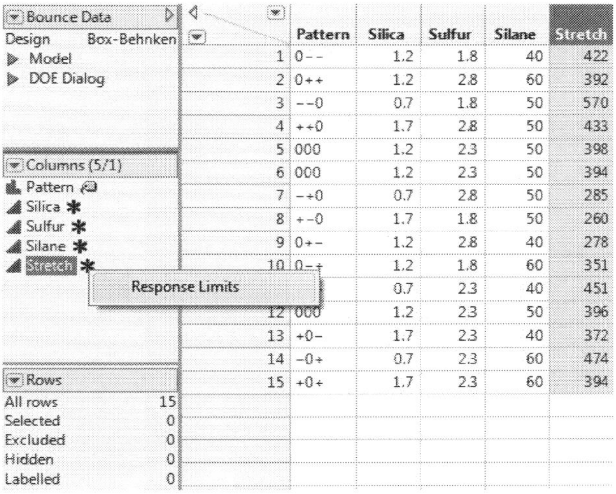

	Pattern	Silica	Sulfur	Silane	Stretch
1	0--	1.2	1.8	40	422
2	0++	1.2	2.8	60	392
3	--0	0.7	1.8	50	570
4	++0	1.7	2.8	50	433
5	000	1.2	2.3	50	398
6	000	1.2	2.3	50	394
7	-+0	0.7	2.8	50	285
8	+-0	1.7	1.8	50	260
9	0+-	1.2	2.8	40	278
10	0-+	1.2	1.8	60	351
		0.7	2.3	40	451
12	000	1.2	2.3	50	396
13	+0-	1.7	2.3	40	372
14	-0+	0.7	2.3	60	474
15	+0+	1.7	2.3	60	394

You can also view column properties by accessing the Column Properties list in the Column Info menu. Select the column or columns whose column properties you want to view and do one of the following:

- Right-click the header area, select **Column Info**, and select the property from the Column Properties list.
- Select **Cols > Column Info** and select the property from the Column Properties list.

Response Limits

Using the Response Limits column property, you can specify the following:

- bounds on the range of variation for a response
- a desirability goal
- a measure of the importance of the response
- desirability values

The Response Limits column property defines a desirability function for the response. The Profiler and Contour Profiler use desirability functions to find optimal factor settings. See the Profiler chapter in the *Profilers* book.

Figure A.3 shows the Response Limits panel in the Column Info window for the response Stretch in the Bounce Data.jmp sample data table, found in the Design Experiment folder.

Figure A.3 Example of the Response Limits Panel

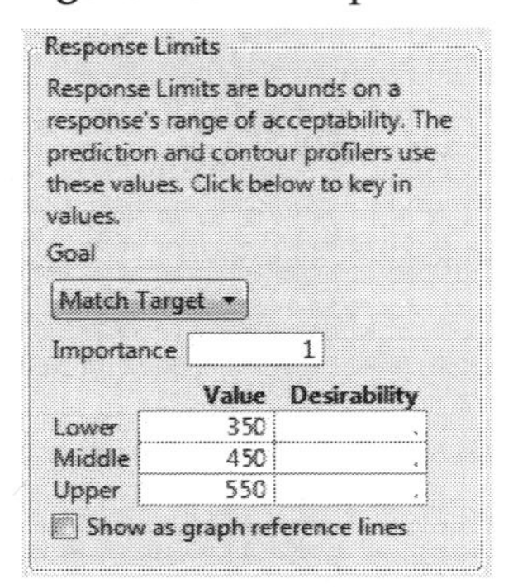

The Response Limits panel consists of the following areas:

Goal Select your response goal from the menu. Available goals are Maximize, Match Target, Minimize, and None. JMP defines a desirability function for the response to match the selected goal. If you specify limits, the desirability function is defined using these limits. If you do not specify limits, JMP bases the desirability function on conservative limit values derived from the distribution of the response. If None is selected as the goal, then all response values are considered equally desirable. For further details, see "Responses" on page 83 in the "Custom Designs" chapter.

Importance Enter a relative weighting for each of several responses in computing the overall desirability function. The Importance value can be any positive number. When no Importance value is specified, JMP treats all responses in a given analysis as having equal Importance values. If there is only one response, it receives Importance 1.

Value Specify Lower and Upper limits and a Middle value for your response. JMP uses these values to construct a desirability function for the response. If you do not specify limits, JMP bases the desirability function on conservative limit values. If your goal is Match Target and no Middle value is specified, then the target is defined to be the midpoint of the Lower and Upper limits.

Desirability Specify values that reflect the desirability of your Lower, Middle, and Upper values. Desirability values should be between 0 and 1. If you do not specify Desirability values, JMP assigns values in accordance with the selected Goal.

Show as graph reference lines Shows horizontal reference lines for the Lower, Middle, and Upper values in the Actual by Predicted Plot and the Prediction Profiler. This option applies only if limits are specified.

Response Limits Example

The Coffee Data.jmp sample data table (located in the Design Experiment folder) contains the results of an experiment that was performed to optimize the Strength of coffee. For a complete description of the experimental design and analysis, see "The Coffee Strength Experiment" on page 39 in the "Starting Out with DOE" chapter.

Your goal is to find factor settings that enable you to brew coffee with a target strength of 1.3, which is considered to be the most desirable value. Values less than 1.2 and greater than 1.4 are completely undesirable. The desirability of values between 1.2 and 1.4 decreases as their distance from 1.3 increases.

1. Select **Help > Sample Data Library** and open Design Experiment/Coffee Data.jmp.
2. In the Table panel, click the green triangle next to the **DOE Dialog** script.

   The DOE Dialog script re-creates the Custom Design dialog that was used to create the experimental design in Coffee Data.jmp.
3. Open the Responses outline.

Figure A.4 Responses Outline in Custom Design Window

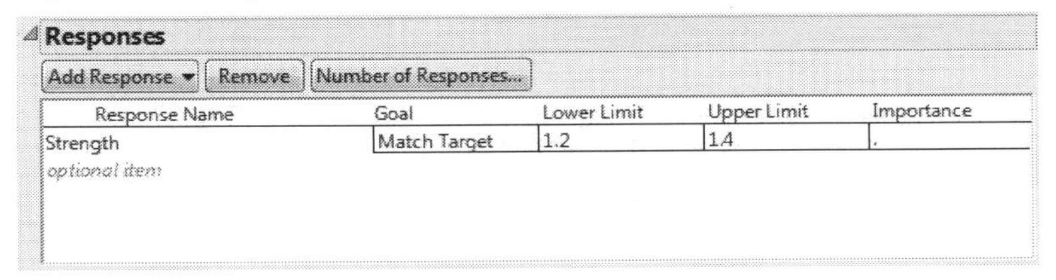

   When you designed this experiment, you specified a response Goal of Match Target with a Lower Limit of 1.2 and an Upper Limit of 1.4. Since there is only one response, you did not specify a value for Importance, because it is 1 by default. When you constructed the design table, JMP assigned the Response Limits column property to Strength.
4. Close the Custom Design window.
5. In the Coffee Data.jmp sample data table, select the Strength column and select **Cols > Column Info**.
6. Select Response Limits in the Column Properties list.

Figure A.5 Response Limits Column Property for Strength

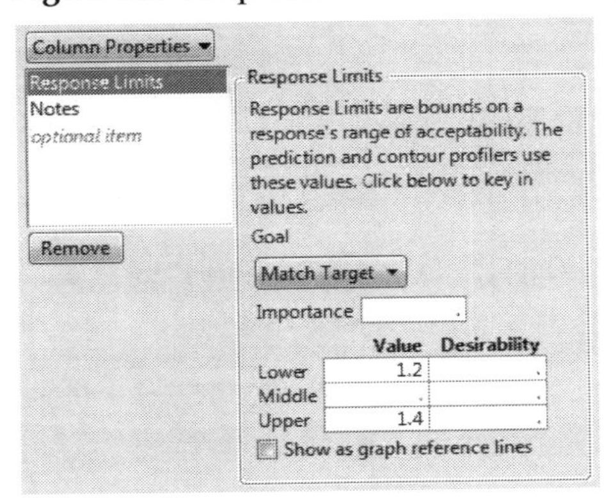

Notice the following:

– The Goal is set to **Match Target**.

– Importance is missing. When Importance is missing, JMP treats all responses in a given analysis as having equal Importance values. So JMP assigns Strength an Importance value of one.

– The Lower limit is 1.2.

– The Upper limit is 1.4.

– No Middle value is specified.

  Because no Middle value is specified, the target is defined to be the midpoint of the Lower and Upper limits, which is 1.3.

– No Desirability values are specified.

7. Select the **Show as graph reference lines** option.

   This option shows horizontal reference lines for the Lower, Middle, and Upper values in the Actual by Predicted Plot and the Prediction Profiler.

8. Click **OK**.

9. In the Coffee Data.jmp data table, click the green triangle next to the **Reduced Model** script.

10. Click **Run**.

    The Prediction Profiler appears at the bottom of the report.

Figure A.6 Profiler Showing Desirability Function for Strength

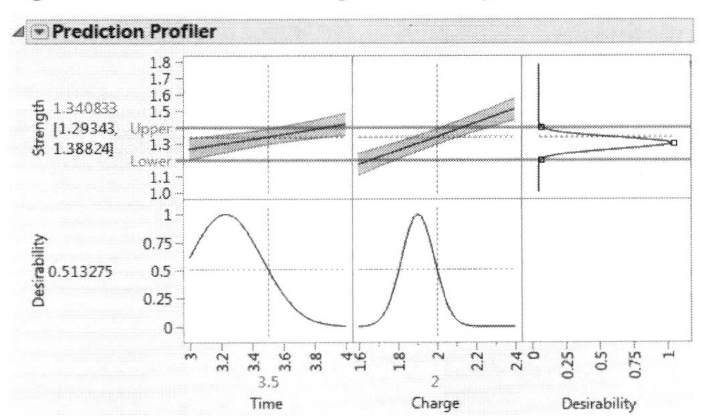

The desirability function for Strength appears in the plot at the right above Desirability. This plot appears because the data table contains a Response Limits column property for Strength. The Prediction Profiler also shows reference lines for the Lower and Upper limits for Strength.

11. Hold down the Ctrl key and click the Strength plot for Desirability.

Figure A.7 Response Goal Window for Strength

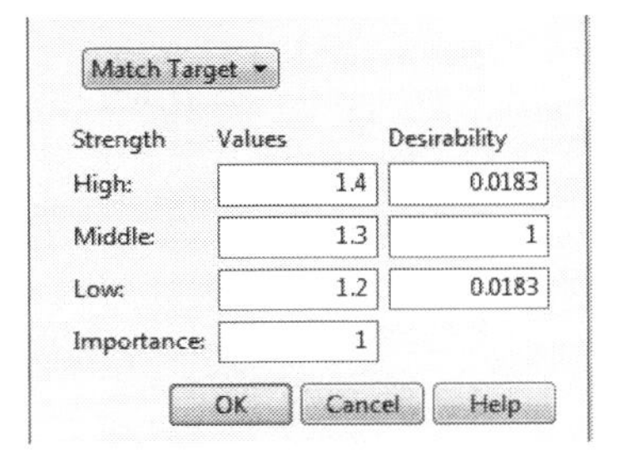

Notice the following:

– JMP determines the Middle value to be the midpoint of the High and Low limits that you specified in the Response Limits column property.

– Because the Goal is set to Match Target, JMP sets the Desirability for the Middle value to 1.

– JMP sets the Desirability for the High and Low values to very small numbers, 0.0183.

– The Desirability plot in Figure A.6 shows how JMP uses the Desirability values shown in Figure A.7. The Desirability function for Strength is essentially 0 beyond the Low and High values and it increases to 1 gradually as Strength approaches the target of 1.3. The Importance value is set to 1 since there is only one response in the model.

12. Click **Cancel** to exit the window.
13. Select **Optimization and Desirability > Maximize Desirability** from the Prediction Profiler red triangle menu.

    The settings for Time and Charge are updated to show settings for the factors that maximize the desirability function for Strength. However, many other settings also maximize the desirability function. See the Contour Profiler chapter in the *Profilers* book for information about how to identify other settings that maximize the desirability function.

Editing Response Limits

In the Vinyl Data.jmp sample data table, a Response Limits column property is already assigned to the response thickness. The property has a goal of maximizing thickness. Suppose that instead of maximizing thickness, you want the sheets of vinyl to have a thickness between 6 and 10, with a target thickness of 8.5.

1. Select **Help > Sample Data Library** and open Design Experiment/Vinyl Data.jmp.
2. Select the thickness column and select **Cols > Column Info.**

   The Response Limits property appears in the Column Properties list as the only property assigned to thickness. The Response Limits panel appears to the right of the list.
3. Click **Maximize** and select **Match Target**.
4. Type 1 for the **Importance** value.
5. Under Value, type 6 for the Lower value, 8.5 for the Middle value, and 10 for the Upper value.

   This is an example of asymmetric response limits. Values of thickness as small as 6 or as large as 10 are acceptable. However, the target for thickness is 8.5.
6. Select **Show as graph reference lines.**

   This option shows horizontal reference lines for the Lower, Middle, and Upper values in the Actual by Predicted Plot and the Prediction Profiler.

Figure A.8 Completed Response Limits Panel

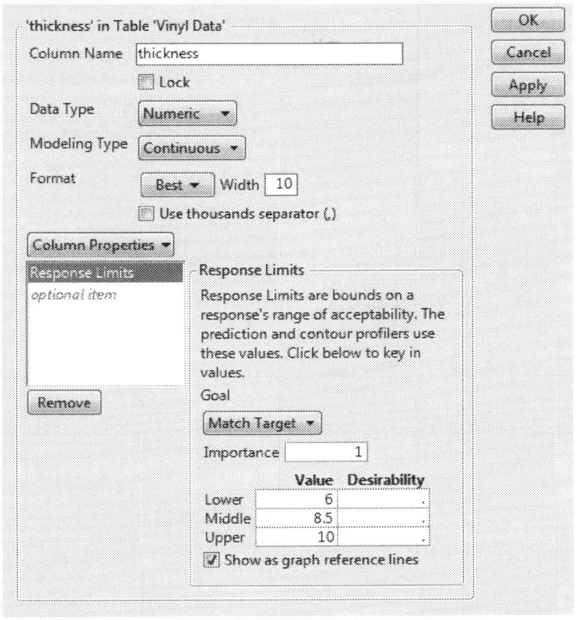

7. Click **OK**.
8. In the Vinyl Data.jmp data table, click the green triangle next to the **Model** script.

   Note that m1, m2, and m3 are mixture factors. Also, the design involves a random Whole Plots factor. Because of this, the default Method is REML (Recommended).
9. Click **Run**.
10. From the red triangle next to Response thickness, select **Row Diagnostics > Plot Actual by Predicted**.

    The reference lines for the Lower, Middle, and Upper limits appear on the Actual by Predicted Plot.
11. From the red triangle next to Response thickness, select **Factor Profiling > Profiler**.

Figure A.9 Prediction Profiler Showing Asymmetric Desirability Function

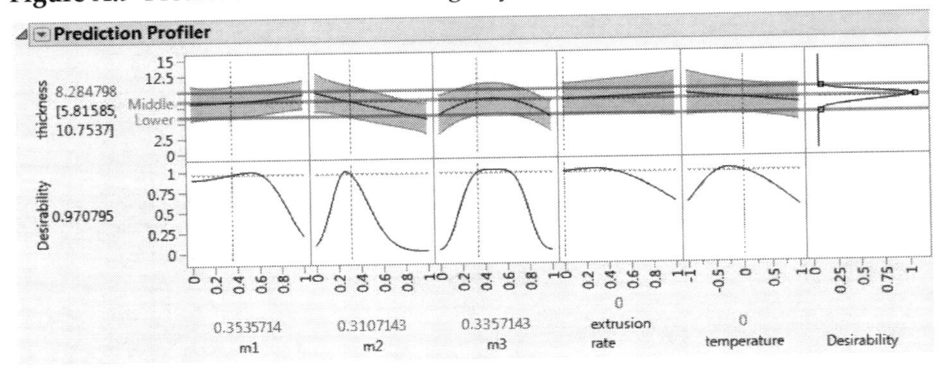

The plot at the right above Desirability shows the desirability function that JMP has constructed for thickness. The desirability is 1 at the Middle limit of 8.5. The desirability is essentially 0 for thickness values below 6 and above 10.

12. Hold down the Ctrl key and click the thickness plot for Desirability.

Figure A.10 Response Goal Window for Thickness

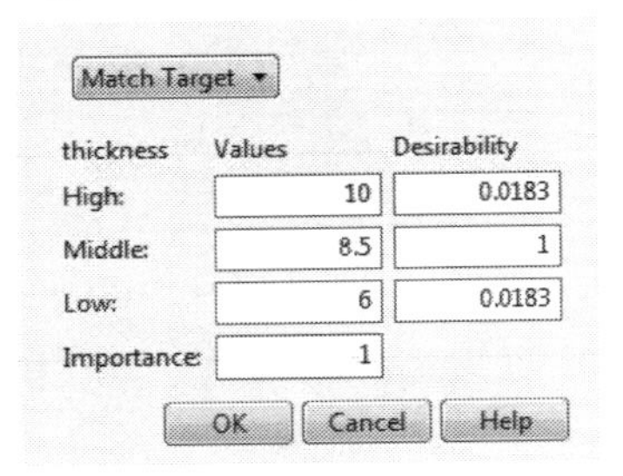

This window shows your settings for the High, Middle, and Low Values. It also shows the Desirability values that JMP assigns, based on your goal of Match Target. The Desirability function shown in Figure A.9 is a continuous curve that matches the Desirability settings in Figure A.10 at the High, Middle, and Low Values. At other values, the Desirability function assigns desirabilities that are consistent with the selected goal.

13. Click **Cancel**.
14. Select **Optimization and Desirability > Maximize Desirability** from the Prediction Profiler red triangle menu.

    The settings for the factors are updated to show values that maximize the desirability function for thickness. Keep in mind that many other settings also maximize the desirability function. The predicted response at these optimal settings is 8.5. Recall that you set 8.5 as the target setting, with limits of 6 and 10.

15. Close the Vinyl Data.jmp sample data table without saving the changes.

Design Role

Factors in designed experiments, as well as in more general models, can behave in various ways. JMP uses the design role column property to describe these behaviors. These are the possible design roles:

- Continuous
- Discrete Numeric
- Categorical
- Blocking
- Covariate
- Mixture
- Constant
- Uncontrolled
- Random Block
- Signal
- Noise

In many of the JMP DOE platforms, you can specify factors with different design roles. In some platforms, your design requirements cause JMP to define factors. For example, Whole Plots and Subplots are factors that JMP creates when you specify very-hard-to-change and hard-to-change factors. In platforms where various design roles can occur, when JMP creates the design table for your design, each factor is assigned the Design Role column property.

- For descriptions of the design roles other than Random Block, Signal, and Noise, see "Factor Types" on page 87 in the "Custom Designs" chapter.
- For a description of the Random Block design role, see "Changes and Random Blocks" on page 89 in the "Custom Designs" chapter.
- For descriptions of the Signal and Noise design roles, see "Factors" on page 415 in the "Taguchi Designs" chapter.

Design Role Example

The experiment in the Odor Control Original.jmp sample data table studies the effect of three factors on odor. You designed the 15-run experiment and it was conducted. However, when the results were reported to you, you learned that the experiment was conducted over three days. The first 5 runs were conducted on Day 1, the second 5 runs on Day 2, and the remaining 5 runs on Day 3.

Since variations in temperature and humidity might have an effect on the response, you want to include Day as a random blocking factor. It is easy to add a column for Day to the design table. But you want to use the Evaluate Design platform to compare the design with the unexpected block to your original design. You also want the ability to use the Augment Design platform in case you need to augment the design. To use the Evaluate Design and Augment Design platforms, you need to add the Design Role column property to your new Day column.

1. Select **Help > Sample Data Library** and open Odor Control Original.jmp.
2. Select the first column, Run.
3. Select **Cols > New Columns.**
4. Next to **Column Name**, type Day.
5. From the **Initialize Data** list, select **Sequence Data**.
6. Enter the following:
   – 1 for **From**
   – 3 for **To**
   – 1 for **Step**
   – 5 for **Repeat each value N times**
7. Next to **Number of columns to add**, type 1.
   – Click **OK**.

Figure A.11 Completed New Columns Window

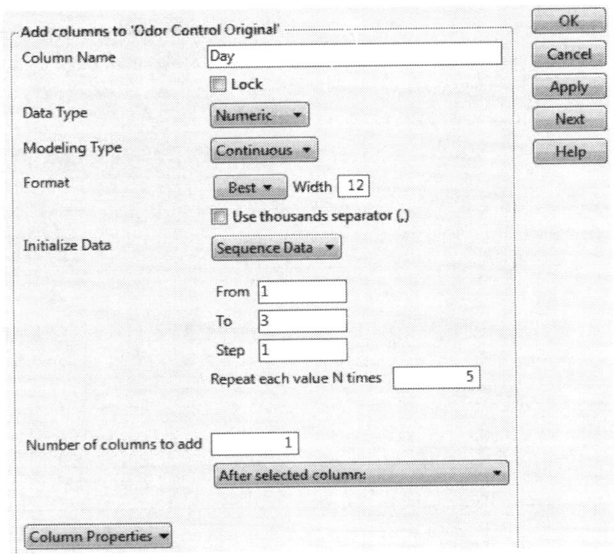

8. Click **OK**.

   The Day column is added as the second column in the data table.

9. Select the Day column.
10. Select **Cols > Column Info**.
11. From the Column Properties list, select **Design Role**.
12. In the Design Role panel, click **Continuous** and select **Random Block**.
13. Click **OK**.

    In the columns panel, an asterisk appears next to Day.

14. Click the asterisk next to Day to verify that the Design Role column property has been assigned.
15. Close the Odor Control Original.jmp sample data table without saving the changes.

Coding

The Coding column property applies only to columns with a numeric data type. It applies a linear transformation to the data in the column. In the Coding column property window, you specify a Low Value and a High Value. The Low Value and High Value in your original data are transformed to –1 and +1. JMP uses the transformed data values whenever the column is entered as a model effect in the Fit Model platform.

The coding property is useful for the following reasons:

- Coded predictors lead to parameter estimates that are more easily interpreted and compared.
- Coded predictors help reduce multicollinearity in models with interaction and higher-order terms.

When any DOE platform other than Accelerated Life Test Design creates a design, JMP defines a Coding column property for each non-mixture factor with a numeric data type. Figure A.12 shows the Coding column property panel for the column Feed Rate in the Reactor 20 Custom.jmp sample data table, found in the Design Experiment folder.

Figure A.12 Coding Property Panel for Feed Rate

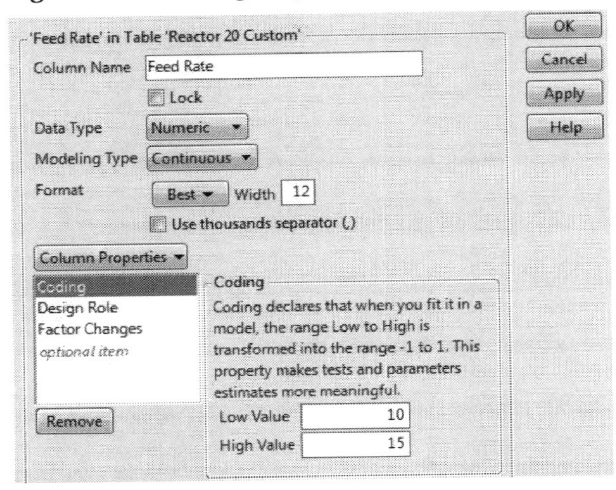

Low and High Values

When the Coding property is applied as part of design construction in a DOE platform, the Low Value is initially set to the minimum setting of the factor and the High Value is initially set to the maximum setting.

When you apply the Coding property to a column that does not contain that property, JMP inserts the minimum value as the Low Value and the maximum value as the High Value. You can change these values as needed.

Caution: After the Coding column property is assigned to a column, JMP does not automatically update it when you make changes to the values in the column. If you change the values in a column that has a Coding column property, review the High Value and Low Value to ensure that they are still appropriate.

The Coding column property centers each value in a column by subtracting the midpoint of the High Value and Low Value. It then divides by half the range. Suppose that H is the High Value and L is the Low Value. Then every X in the column is transformed to the following:

$$\frac{X-(H+L)/2}{(H-L)/2}$$

For each factor, the transformed values have a midpoint equal to 0 and range from -1 to +1.

Coding Column Property and Center Polynomials

The Center Polynomials option is located in the Fit Model launch window, within the Model Specification red triangle menu. Center Polynomials centers a continuous column involved in a polynomial term by subtracting the mean of each value in the column. For more details, see the Model Specification chapter in the *Fitting Linear Models* book.

If the Coding column property is assigned to a column, then the Center Polynomials option has no effect on that column. In a polynomial term involving that column, the values are centered and scaled as specified by their Coding property. Suppose that another column in the model does not have the Coding property and that you select Center Polynomials. Then that column is centered by its mean in any polynomial term where it appears.

Coding Example

The Reactor 20 Custom.jmp sample data table contains data from a 20-run design that was constructed using the Custom Design platform. The experiment investigates the effects of five factors on a yield response (Percent Reacted) for a chemical process.

1. Select **Help > Sample Data Library** and open Design Experiment/Reactor 20 Custom.jmp.
2. In the Table panel, click the green triangle next to the **DOE Dialog** script.
3. Open the **Factors** outline.

Figure A.13 Factors Outline for Design Used in Reactor 20 Custom.jmp

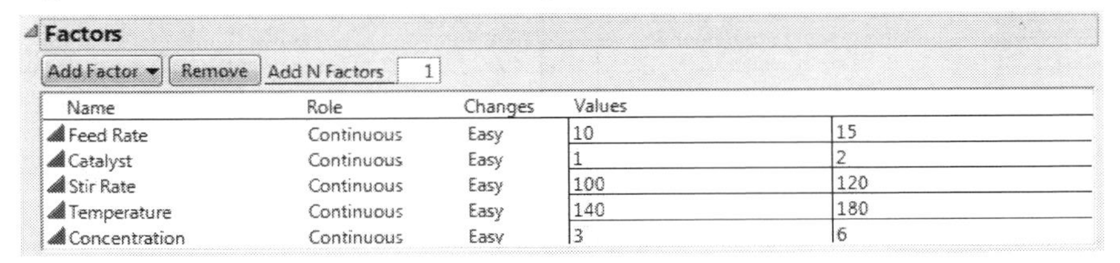

Notice that the settings for Temperature range from 140 to 180. When the design was generated, the Coding column property was assigned to Temperature. The Low Value is set to 140 and the High Value is set to 180.

4. Close the Custom Design window.
5. In the Reactor 20 Custom.jmp sample data table, click the asterisk next to Temperature in the columns panel and select **Coding**.

   The Column Info window appears and shows the Coding column property panel. You can see that JMP added the column property, specifying the Low Value and High Value, when it constructed the design table. In fact, by repeating this step, you can verify that JMP added the Coding property for all five factors.

Figure A.14 Coding Panel for Temperature

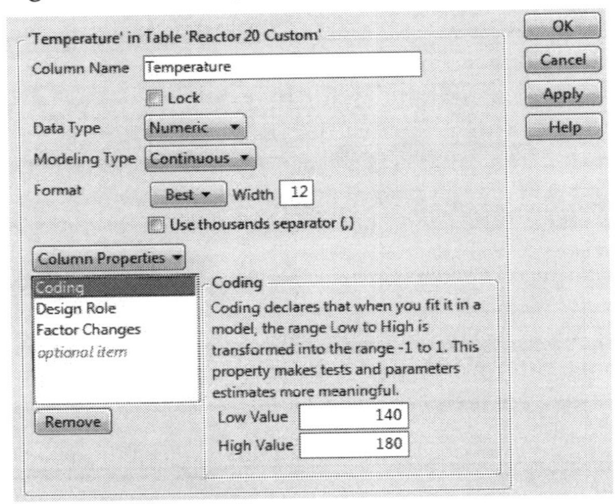

6. Click **Cancel** to close the Column Info window.
7. In the **Reactor 20 Custom.jmp** sample data table, click the green triangle next to the **Reduced Model** script.

   This script fits a model that contains only the five effects determined to be significant based on an analysis of the full model.
8. Click **Run**.

Figure A.15 Effect Summary Report for Reduced Model

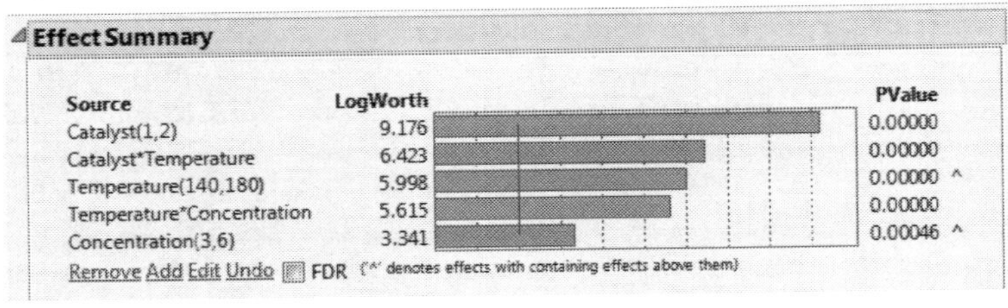

Source	LogWorth	PValue
Catalyst(1,2)	9.176	0.00000
Catalyst*Temperature	6.423	0.00000
Temperature(140,180)	5.998	0.00000 ^
Temperature*Concentration	5.615	0.00000
Concentration(3,6)	3.341	0.00046 ^

In the Source list, the High and Low values used in the Coding column property appear in parentheses to the right of the main effects, **Catalyst**, **Temperature**, and **Concentration**. The ranges imposed by the Coding property are not shown for the interaction effects.

Tip: Notice the "^" symbols to the right of the PValues for **Temperature** and **Concentration**. These symbols indicate that these main effects are components of interaction effects with smaller *p*-values. If an interaction effect is included in the model, then the principle of effect heredity requires that all component effects are also in the model. See "Effect Heredity" on page 59 in the "Starting Out with DOE" chapter.

---

9. Select **Estimates > Show Prediction Expression** from the red triangle next to Response Percent Reacted.

   Look at the Prediction Expression outline to see how coding affects the prediction formula.

Figure A.16 Prediction Expression for Reduced Model

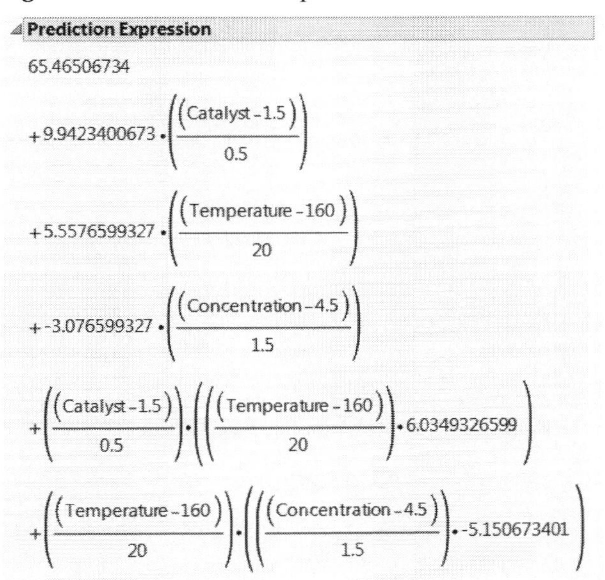

Each factor is transformed as specified by the Coding column property. For example, for Temperature, notice the following:

– The Low Value in the Coding property was set to 140. The Temperature value of 140 is transformed to -1.

– The High Value in the Coding property was set to 180. The Temperature value of 180 is transformed to +1.

– The midpoint of the Low and High values is 160. The Temperature value of 160 is transformed to 0.

The transformed values help you compare the effects. The estimated coefficient for Catalyst is 9.942 and the estimated coefficient for Concentration is -3.077. It follows that the predicted effect of Catalyst on Percent Reacted is more than three times as large as the effect of Concentration on Percent Reacted. Also, the coefficients indicate that predicted Percent Reacted increases as Catalyst increases and decreases as Concentration increases.

The transformed values help you interpret the coefficients:

- When all factors are at their midpoints, their transformed values are 0. The predicted Percent Reacted is the intercept, which is 65.465.

- When Catalyst and Concentration are at their midpoints, a 20 unit increase in Temperature increases the Percent Reacted by 5.558 units.
- Suppose that Concentration is at its midpoint, so that its transformed value is 0:
  - When Catalyst is at its midpoint, a 20 unit increase in Temperature increases the Percent Reacted by 5.558 units.
  - When Catalyst is at its high setting, a 20 unit increase in Temperature increases the Percent Reacted by 5.558 + 6.035 = 11.593 units.

It follows that the coefficient of the interaction term, 6.035, is the increase in the slope of the model for predicted Percent Reacted for a 0.5 unit change in Catalyst.

Assigning Coding

The experimental data in the Tiretread.jmp sample data table results from an experiment to study the effects of SILICA, SILANE, and SULFUR on four measures of tire tread performance. In this example, you will consider only one of the responses, ABRASION.

You will first fit a model using the uncoded factors. Then you will assign the coding property to the factors and rerun the model to obtain meaningful parameter estimates.

1. Select **Help > Sample Data Library** and open Tiretread.jmp.
2. Select **Analyze > Fit Model**.
3. Select ABRASION and click **Y**.
4. Select SILICA, SILANE, and SULFUR and click **Macros > Response Surface**.
5. Check **Keep dialog open**.
6. Click **Run**.
7. Select **Estimates > Show Prediction Expression** from the Response ABRASION red triangle menu.

Figure A.17 Prediction Expression for Model with Uncoded Factors

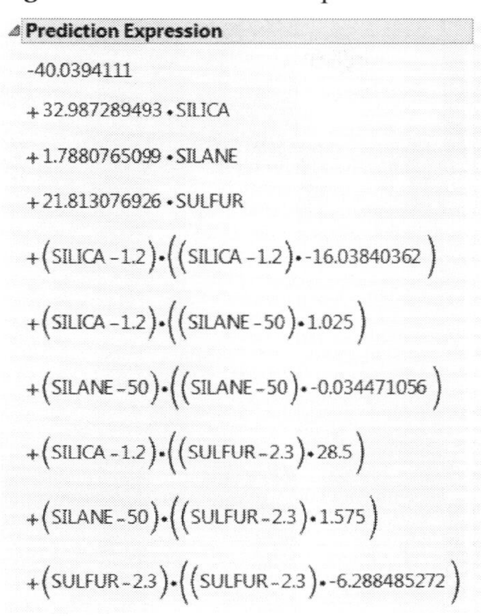

The coefficients do not help you compare effect sizes. The sizes of the coefficients do not reflect the impact of the effects on ABRASION over the range of their settings. Also, the coefficients are not easily interpreted. For example, the coefficients do not facilitate your understanding of the predicted response when SILICA is at the midpoint of its range.

Apply the Coding column property to the three factors to see how coding makes the coefficients more meaningful.

8. In the Tiretread.jmp data table, select SILICA, SILANE, and SULFUR in the Columns panel. Right-click the highlighted column titles and select **Standardize Attributes**.
9. Select **Column Properties > Coding** in the Standardize Properties panel.
10. Click **OK**.

    An asterisk appears in the Columns panel next to SILICA, SILANE, and SULFUR indicating that these have been assigned a column property.
11. In the Fit Model window, click **Run**.
12. Select **Estimates > Show Prediction Expression** from the Response ABRASION red triangle menu.

Figure A.18 Prediction Expression for Model with Coded Factors

Prediction Expression

$$
\begin{aligned}
&139.11923872\\
&+26.934121871\cdot\left(\frac{(SILICA-1.2)}{0.8165}\right)\\
&+29.199289407\cdot\left(\frac{(SILANE-50)}{16.33}\right)\\
&+17.81037731\cdot\left(\frac{(SULFUR-2.3)}{0.8165}\right)\\
&+\left(\frac{(SILICA-1.2)}{0.8165}\right)\cdot\left(\left(\frac{(SILICA-1.2)}{0.8165}\right)\cdot -10.69235863\right)\\
&+\left(\frac{(SILICA-1.2)}{0.8165}\right)\cdot\left(\left(\frac{(SILANE-50)}{16.33}\right)\cdot 13.666781125\right)\\
&+\left(\frac{(SILANE-50)}{16.33}\right)\cdot\left(\left(\frac{(SILANE-50)}{16.33}\right)\cdot -9.192358626\right)\\
&+\left(\frac{(SILICA-1.2)}{0.8165}\right)\cdot\left(\left(\frac{(SULFUR-2.3)}{0.8165}\right)\cdot 19.000159125\right)\\
&+\left(\frac{(SILANE-50)}{16.33}\right)\cdot\left(\left(\frac{(SULFUR-2.3)}{0.8165}\right)\cdot 21.000175875\right)\\
&+\left(\frac{(SULFUR-2.3)}{0.8165}\right)\cdot\left(\left(\frac{(SULFUR-2.3)}{0.8165}\right)\cdot -4.192358626\right)
\end{aligned}
$$

The coefficients for the coded factors enable you to compare effect sizes. SILANE has the largest effect on ABRASION over the range of design settings. The effects of SILICA and the SILANE*SULFUR interaction are large as well.

The coefficients for the coded factors are also more easily interpreted. For example, when all factors are at the center of their ranges, the predicted value of ABRASION is the intercept, 139.12.

13. Close the Tiretread.jmp sample data table without saving the changes.

Mixture

The Mixture column property is useful when a column in a data table represents a component of a mixture. The components of a mixture are constrained to sum to a constant. Because of this, they differ from non-mixture factors. The Mixture column property serves two purposes:

- It identifies a column as a mixture component.

  If you add a column with the Mixture column property as a model effect in the Analyze > Fit Model window, JMP automatically generates a no-intercept model.

- It defines the coding for a mixture component.

  Coding for mixture components differs from that for non-mixture factors. However, as with non-mixture factors, a benefit of coding for mixture factors is that it helps you interpret parameter estimates. See "PseudoComponent Coding" on page 672.

Figure A.19 shows the Mixture column property panel for the factor m1 in the Vinyl Data.jmp sample data table, found in the Design Experiment folder.

Figure A.19 Mixture Column Property Panel

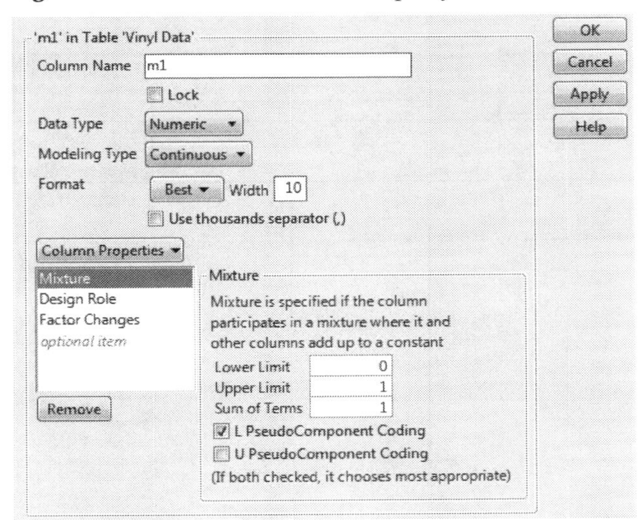

In the Mixture column property panel, you can specify the following:

Lower Limit Specifies the low value used in PseudoComponent Coding. When the Mixture property is applied as part of design construction in a DOE platform, the Lower Limit is set to the minimum setting of the factor. When you apply the Mixture property to a column that does not contain that property, JMP inserts the minimum value as the Lower Limit. You can change this value as needed.

Upper Limit Specifies the high value used in PseudoComponent Coding. When the Mixture property is applied as part of design construction in a DOE platform, the Upper Limit is set to the maximum setting of the factor. When you apply the Mixture property to a column that does not contain that property, JMP inserts the maximum value as the Upper Limit. You can change this value as needed.

Sum of Terms Specifies the sum of the mixture components. When you apply the Mixture property to a column that does not contain that property, JMP inserts a default value of 1 for the Sum of Terms.

L PseudoComponent Coding Transforms data values so that the Lower Limit corresponds to 0.

U PseudoComponent Coding Transforms data values so that the Upper Limit corresponds to 0.

PseudoComponent Coding

A pseudo-component is a linear transformation. Let S denote the sum of the mixture components. Suppose that *i* columns $X_1, X_2, \ldots, X_q$ have been assigned the Mixture column property. Suppose that the columns and effects constructed from these columns are entered as effects in the Fit Model window.

Define the following:

$$L = \sum_{i=1}^{q} L_i, \text{ where } L_i \text{ is the Lower Limit for } X_i$$

$$U = \sum_{i=1}^{q} U_i, \text{ where } U_i \text{ is the Upper Limit for } X_i$$

Let x_i denote a value of the column X_i. The L PseudoComponent at x_i is defined as follows:

$$x_i^L = (X_i - L_i)/(S - L)$$

The U PseudoComponent at x_i is defined as follows:

$$x_i^U = (U_i - X_i)/(U - S)$$

If you select both **L PseudoComponent Coding** and **U PseudoComponent Coding**, the Fit Model platform uses the *L* coding if $(S - L) < (U - S)$. Otherwise, the *U* coding is used.

In Fit Model, mixture factors are transformed using pseudo-components before computing parameter estimates. This helps make parameter estimates more meaningful. In reports dealing with parameter estimates, the mixture main effects are given by the coding transformation. Other reports, such as the profilers, are based on the uncoded values.

Mixture Example

The data in the Donev Mixture Data.jmp sample data table, found in the Design Experiment folder, are based on an example from Atkinson and Donev (1992). The design includes three mixture factors and one non-mixture factor. The response and factors are as follows:

- The response is the electromagnetic Damping of an acrylonitrile powder.
- The three mixture ingredients are copper sulphate (CuSO4), sodium thiosulphate (Na2S2O3), and Glyoxal.
- The non-mixture environmental factor of interest is the Wavelength of light.

Though Wavelength is theoretically continuous, the researchers were interested only in predictions at three discrete wavelengths. As a result, Wavelength is treated as a categorical factor with three levels.

For details about using Custom Design to construct a design for this situation, see "Mixture Experiments" on page 170 in the "Examples of Custom Designs" chapter.

1. Select **Help > Sample Data Library** and open Design Experiment/Donev Mixture Data.jmp.
2. Click the asterisk next to CuSO4 in the columns panel and select **Mixture**.

Figure A.20 Mixture Column Property Panel for CuSO4

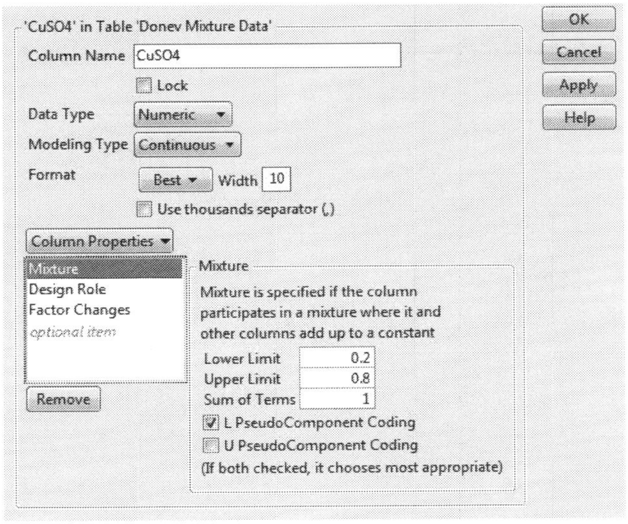

Notice the following:

- The Lower Limit is 0.2, the minimum design setting for CuSO4.
- The Upper Limit is 0.8, the maximum design setting for CuSO4.
- The Sum of Terms is set to 1. This is the sum of the three mixture factors.
- The **L PseudoComponent Coding** option is selected. See "PseudoComponent Coding" on page 672.

3. Click **Cancel**.
4. Click the asterisk next to Glyoxal in the columns panel and select **Mixture**.

   For this factor, note the following:

   - The Lower Limit is 0, the minimum design setting for Glyoxal.
   - The Upper Limit is 0.6, the maximum design setting for Glyoxal.

5. Click **Cancel**.
6. In the Donev Mixture Data.jmp data table, click the green triangle next to the **Model** script.

   The model contains the main effects of the mixture factors and two-way interactions for all four factors.

7. Click **Run**.

   In the Parameter Estimates report, the mixture factors appear in their pseudo-component coded form. When the mixture factors appear in interactions, they are not denoted in coded form. Nevertheless, the model fitting is based on the pseudo-components. The first three terms in the Parameter Estimates report (Figure A.21), show the coded form for the mixture factors.

Figure A.21 Parameter Estimates Report

Parameter Estimates

Term	Estimate	Std Error	t Ratio	Prob>|t|
(CuSO4-0.2)/0.6	6.1910821	0.918805	6.74	0.0005*
(Na2S2O3-0.2)/0.6	4.0089179	0.918805	4.36	0.0048*
Glyoxal/0.6	8.1666667	0.921638	8.86	0.0001*
CuSO4*Na2S2O3	11.293949	4.728922	2.39	0.0542
CuSO4*Glyoxal	4.3511692	4.512775	0.96	0.3722
CuSO4*Wavelength[L1]	-3.847343	1.113399	-3.46	0.0135*
CuSO4*Wavelength[L2]	1.8781509	1.079113	1.74	0.1324
Na2S2O3*Glyoxal	-18.4845	4.512775	-4.10	0.0064*
Na2S2O3*Wavelength[L1]	-0.275689	1.064004	-0.26	0.8042
Na2S2O3*Wavelength[L2]	0.4452853	1.103681	0.40	0.7006
Glyoxal*Wavelength[L1]	0.198283	1.090627	0.18	0.8617
Glyoxal*Wavelength[L2]	0.1905384	1.090647	0.17	0.8671

8. Select **Estimates > Show Prediction Expression** from the Response Damping red triangle menu.

   The Prediction Expression report shows the model that was fit. Note that the mixture factors are transformed using the L PseudoComponent coding.

Figure A.22 Prediction Expression for Damping Model

Prediction Expression

$$
\begin{aligned}
&6.1910820765 \cdot \left(\frac{(\text{CuSO4} - 0.2)}{0.6}\right) \\
&+ 4.0089179235 \cdot \left(\frac{(\text{Na2S2O3} - 0.2)}{0.6}\right) \\
&+ 8.1666666667 \cdot \left(\frac{\text{Glyoxal}}{0.6}\right) \\
&+ \left(\frac{(\text{CuSO4} - 0.2)}{0.6}\right) \cdot \left(\left(\frac{(\text{Na2S2O3} - 0.2)}{0.6}\right) \cdot 11.293949004\right) \\
&+ \left(\frac{(\text{CuSO4} - 0.2)}{0.6}\right) \cdot \left(\left(\frac{\text{Glyoxal}}{0.6}\right) \cdot 4.3511691804\right) \\
&+ \left(\frac{(\text{CuSO4} - 0.2)}{0.6}\right) \cdot \text{Match}(\text{Wavelength})\begin{pmatrix} \text{"L1"} \Rightarrow -3.847342876 \\ \text{"L2"} \Rightarrow 1.8781509412 \\ \text{"L3"} \Rightarrow 1.9691919345 \\ \text{else} \Rightarrow . \end{pmatrix} \\
&+ \left(\frac{(\text{Na2S2O3} - 0.2)}{0.6}\right) \cdot \left(\left(\frac{\text{Glyoxal}}{0.6}\right) \cdot -18.48450251\right) \\
&+ \left(\frac{(\text{Na2S2O3} - 0.2)}{0.6}\right) \cdot \text{Match}(\text{Wavelength})\begin{pmatrix} \text{"L1"} \Rightarrow -0.275688526 \\ \text{"L2"} \Rightarrow 0.44528528 \\ \text{"L3"} \Rightarrow -0.169596754 \\ \text{else} \Rightarrow . \end{pmatrix} \\
&+ \left(\frac{\text{Glyoxal}}{0.6}\right) \cdot \text{Match}(\text{Wavelength})\begin{pmatrix} \text{"L1"} \Rightarrow 0.1982830114 \\ \text{"L2"} \Rightarrow 0.1905384076 \\ \text{"L3"} \Rightarrow -0.388821419 \\ \text{else} \Rightarrow . \end{pmatrix}
\end{aligned}
$$

Suppose that you are interested in predictions at Wavelength L2. Suppose also that Na2S2O3 and Glyoxal are set to their low values, 0.2 and 0 respectively, and that CuSO4 is set to its high value, 0.8. In this case, the predicted Damping equals the parameter estimate for CuSO4 (6.191) plus the parameter estimate for CuSO4*Wavelength[L2] (1.878). You can verify this in the Prediction Profiler.

9. Select **Save Columns > Save Coding Table** from the Response Damping red triangle menu.

Figure A.23 First Three Columns of Coding Table Showing Coded Mixture Factors

	(CuSO4-0.2)/0.6	(Na2S2O3-0.2)/0.6	Glyoxal/0.6
1	1	0	0
2	0.4	0.6	0
3	0	1	0
4	0.5	0	0.5
5	0	0.5	0.5
6	0	0	1
7	1	0	0
8	0.6	0.4	0
9	0	1	0
10	0.5	0	0.5
11	0	0.5	0.5
12	0	0	1
13	1	0	0
14	0.4	0.6	0
15	0	1	0
16	0.5	0	0.5
17	0	0.5	0.5
18	0	0	1

For this particular design, the L PseudoComponent coding transforms the mixture factors to range between 0 and 1. Note that this does not happen in general.

Factor Changes

The Factor Changes column property indicates how difficult it is to change factor settings in a designed experiment. The possible specifications for Factor Changes are Easy, Hard, and Very Hard. For example, Figure A.24 shows the Factor Changes column property panel for the factor A1 in the Battery Data.jmp sample data table, located in the Design Experiment folder.

Figure A.24 Factor Changes Column Property Panel

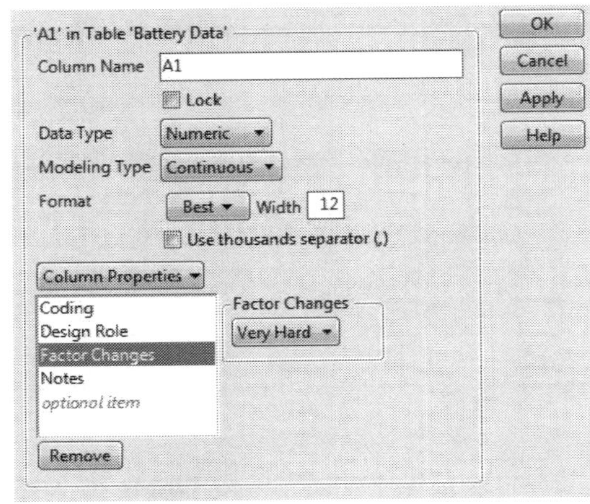

- When a design contains factors that are hard-to-change and very-hard-to-change, it must also include a subplot and a whole plot factor:
  - The levels of the whole plot factor define the groups of runs for which the levels of the very-hard-to-change factors are held constant.
  - The levels of the subplot factor define the groups of runs for which the levels of the hard-to-change factors are held constant.
- When a design contains only factors that are hard-to-change, but no factors that are very-hard-to-change, it should include a whole plot factor. The levels of the whole plot factor define the groups of runs for which the levels of the hard-to-change factors are held constant. For more details, see "Changes and Random Blocks" on page 89 in the "Custom Designs" chapter.

Augment and Evaluate Design

For the Evaluate Design and Augment Design platforms, the Factor Changes column property identifies factors with Changes specified as Hard or Very Hard. However, these platforms also require that the whole plot and subplot factors be entered as model effects in the launch windows. This is because the whole plot and subplot factors are part of the design structure.

Custom Design

The Custom Design platform enables you to create designs where all factor changes are Easy. You can also construct split-plot, split-split plot, or two-way split-plot (strip-plot) designs. When constructing these designs, you need to identify the factors whose values are hard-to-change or very-hard-to-change. In the Factors outline, you can identify factors as having Changes that are Easy, Hard, or Very Hard. When the Custom Design platform constructs the design table, the Factor Changes property is assigned to every factor that appears in the Factors outline.

The Custom Design platform is the only platform that constructs designs for factors with Changes that are Hard or Very Hard. Other DOE platforms also assign the Factor Changes column property to factors that they construct, but the value of the column property is set to Easy for their factors.

If you Load Factors in the Custom Design window using a table of factors, you can assign the Factor Changes column property to columns in that table. When you Load Factors using that table, your Factor Changes specifications appear in the Factors outline.

Factor Changes Example

The Battery Data.jmp sample data table, found in the Design Experiment folder, contains data from an experiment that studies the open current voltage of batteries (OCV). The design is a two-way split-plot design. For further background, see “Examples of Custom Designs” chapter on page 193.

1. Select **Help > Sample Data Library** and open Design Experiment/Battery Data.jmp.
2. Click the asterisk to the right of the factor C1 in the columns panel.
3. Select **Factor Changes**.

Figure A.25 Factor Changes Panel for C1

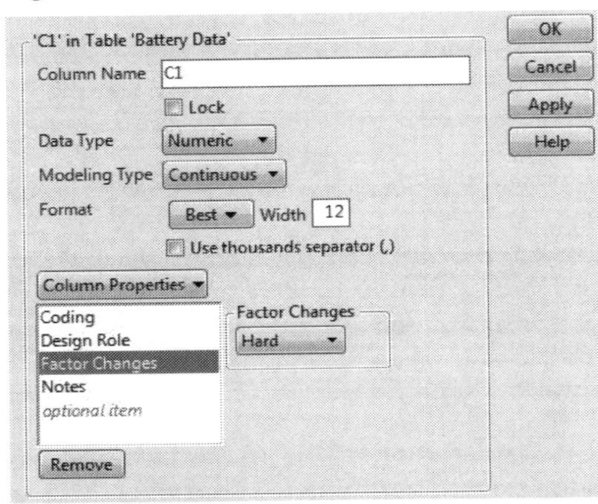

The value of Factor Changes for C1 is Hard. Figure A.24 shows that the value of Factor Changes for A1 is Very Hard.

4. Click **OK**.
5. In the data table, click the green triangle next to the **DOE Dialog** script.
6. Open the **Factors** outline.

Figure A.26 Factors Outline for Battery Experiment

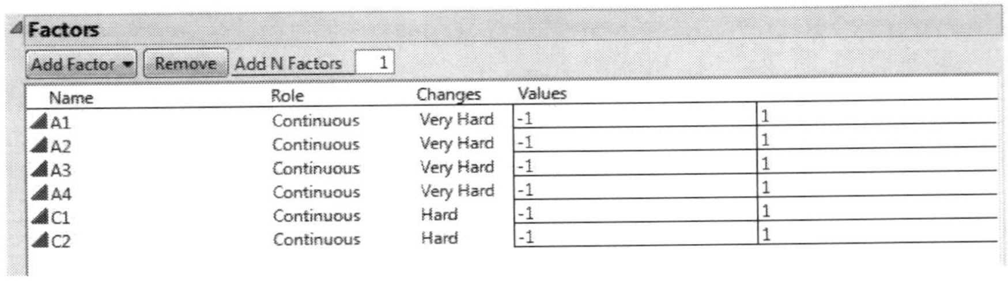

The factors A1, A2, A3, and A4 have Changes set to Very Hard, and the factors C1 and C2 have Changes set to Hard. When the Custom Design platform constructs the design table, it saves these specifications to the appropriate columns as Factor Changes column properties.

In the Design outline, notice the Whole Plots and Subplots factors.

Figure A.27 Design Outline Partial View

Design

Run	Whole Plots	Subplots	A1	A2	A3	A4	C1	C2
1	1	1	-1	-1	1	-1	-1	-1
2	1	2	-1	-1	1	-1	1	-1
3	1	3	-1	-1	1	-1	-1	1
4	2	4	-1	1	1	1	1	1
5	2	5	-1	1	1	1	-1	1
6	2	6	-1	1	1	1	1	-1
7	3	7	1	1	1	-1	-1	1
8	3	8	1	1	1	-1	1	1
9	3	9	1	1	1	-1	-1	-1
10	4	10	1	-1	1	1	1	1
11	4	11	1	-1	1	1	1	-1
12	4	12	1	-1	1	1	-1	-1

To account for the factor changes that are Hard and Very Hard, two factors are created by the Custom Design platform. The Whole Plots factor groups the runs where the Very Hard factor levels are constant and the Subplots factor groups the runs where the Hard factors levels are constant. These factors need to be included as model effects when you enter columns with the Factor Changes column property in the Evaluate Design and Augment Design platforms.

Value Ordering

The Value Ordering column property assigns an order to the values in a column. That order is then used in plots and analyses. You can specify the order in which you want values to appear in reports.

Note: For certain values that have a natural ordering, such as days of the week, JMP automatically orders these in the appropriate way in reports. See The Column Info Window chapter in the *Using JMP* book.

Figure A.28 shows the Value Ordering panel for the Type column in the Car Physical Data.jmp sample data table. Reports that involve the values of Type place these levels in the order Sporty, Small, Compact, Medium, and Large. Use the buttons to the right of the Value Ordering list to specify your desired ordering for the values.

Figure A.28 Value Ordering Column Property for Type

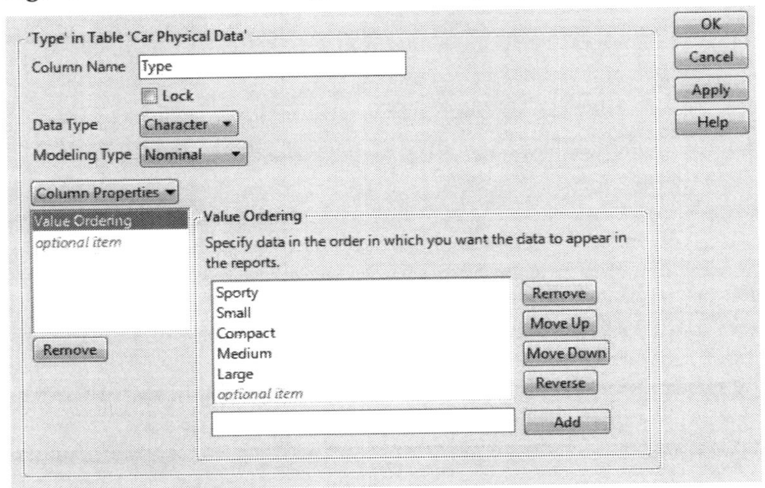

In designs created using most DOE platforms, categorical factors, including the constructed factors Whole Plots and Subplots, and blocking factors are assigned the Value Ordering property. This property orders the levels according to the order in which they appear in the Factors outline. The levels of constructed factors are consecutive integers and the Value Ordering property specifies this natural ordering. You can modify the Value Ordering specification for any factor to meet your needs.

The Value Ordering property is not assigned by the Covering Array or Taguchi Arrays platforms. The Covering Array platform assigns the Value Labels column property. See "Value Labels" on page 683.

Value Ordering Example

Suppose that you want the values for a factor to appear in a different order in the Prediction Profiler. Consider an example of a wine tasting experiment, constructed using Custom Design. Wine is rated by five experts, each listed as a Rater in the Wine Data.jmp sample data table. Rater is a fixed blocking factor. Nine factors are studied. Rating is the response.

1. Select **Help > Sample Data Library** and open Design Experiment/Wine Data.jmp.
2. In the Table panel, click the green triangle next to the **Reduced Model** script.
3. Click **Run**.

   The Prediction Profiler appears at the bottom of the report.

Figure A.29 Profiler with Original Value Ordering

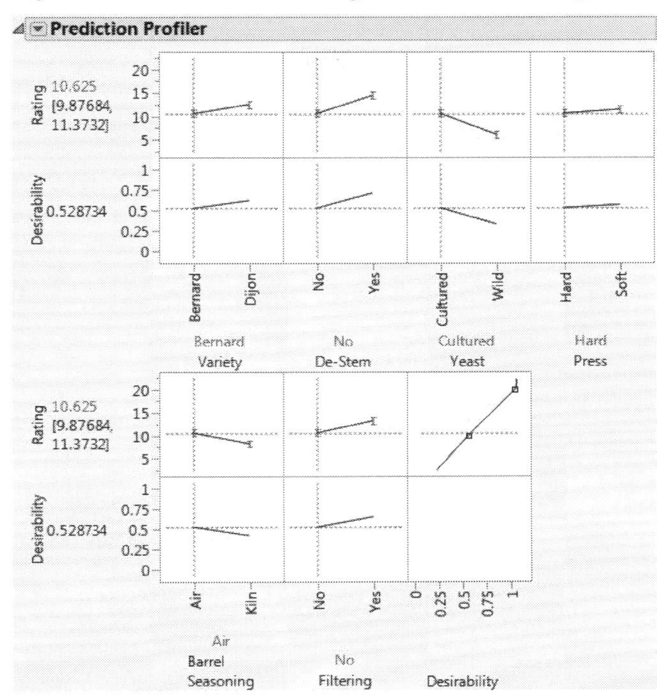

Notice that the values for De-Stem and Filtering appear in the order No followed by Yes. You want to reverse these, so that the Yes level appears first.

4. Close the Response Rating report.
5. In the data table, click the asterisk next to De-Stem in the columns panel and select **Value Ordering**.
6. Click **Reverse**.
7. Click **OK**.
8. Click the asterisk next to Filtering in the columns panel and select **Value Ordering**.
9. Click **Reverse**.
10. Click **OK**.
11. Again, click the green triangle next to the **Reduced Model** script.
12. Click **Run**.

Figure A.30 Profiler with New Value Ordering

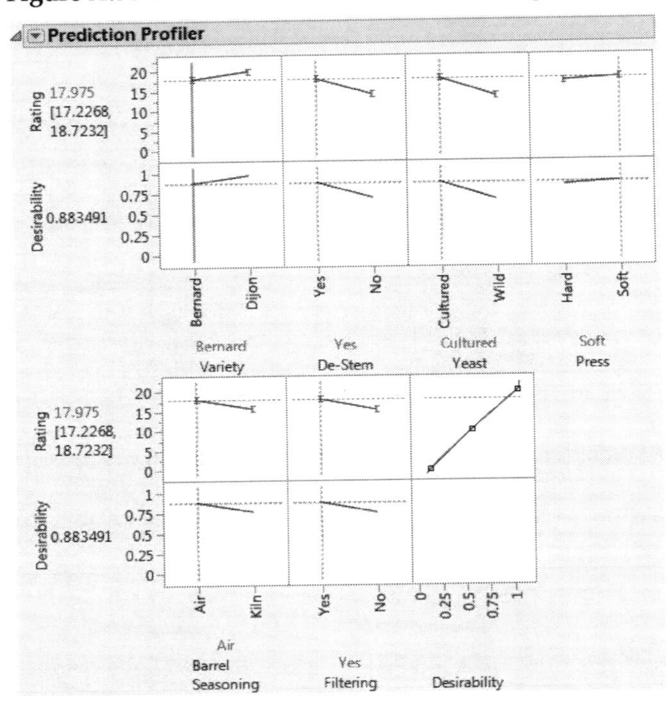

The levels for De-Stem and Filtering now appear in the order Yes followed by No.

13. Close the Wine Data.jmp sample data table without saving the changes.

Assigning Value Ordering

Consider the Candy Bars.jmp sample data table. Of the 18 brands lists under Brand, Hershey and M&M/Mars have the largest numbers of types of candy as listed in the Name column. You want these two brands to appear first in reports.

1. Select **Help > Sample Data Library** and open Candy Bars.jmp.
2. Select the Brand column.
3. Select **Cols > Column Info.**
4. Under Column Properties, select **Value Ordering**.
5. In the Value Ordering list, select Hershey.
6. Click **Move Up** five times.

   Hershey is now first in the Value Ordering list.
7. In the Value Ordering list, select M&M/Mars.

8. Click **Move Up** seven times.

   M&M/Mars is now second in the Value Ordering list.

9. Click **OK**.

   An asterisk indicating the Value Ordering column property appears next to Brand in the columns panel. JMP now lists Hershey and M&M/Mars first in reports involving Brand.

10. Select **Analyze > Distribution**.
11. Select Calories and click **Y, Columns**.
12. Select Brand and click **By**.
13. Click **OK**.
14. While holding down the Ctrl key, from the red triangle next to Calories select **Display Options > Horizontal Layout**.

    Note that the Distribution reports for Hershey and M&M/Mars appear first among the 18 brands.

15. Close the Candy Bars.jmp sample data table without saving the changes.

Value Labels

The Value Labels column property represents values in a column with specified labels. These labels are displayed in the data table and are used in plots and reports. In the data table, you can view the original values by double-clicking within a cell. For details about how to assign and work with the Value Labels column property, see The Column Info Window chapter in the *Using JMP* book.

The Covering Arrays platform is the only DOE platform that assigns the Value Labels column property. The Covering Arrays platform saves factors to the data table with a Nominal modeling type. The underlying values are consecutive integers ranging from 1 to the number of levels that you specify in the Covering Array Factors outline. The Values that you specify in the Factors outline are the Value Labels that are assigned to the underlying integers.

Value Labels Example

You want to test an internet-based software application to detect issues arising from components in the operating environment. The four components of interest consist of a browser, the operating system, the computer's RAM, and the connection speed. To minimize testing time, you restrict yourself to testing two-way interactions.

Construct a Strength 2 covering array to test the required combinations of factor levels.

1. Select **DOE > Special Purpose > Covering Array**.
2. Select **Load Factors** from the red triangle menu.

3. Select **Help > Sample Data Library** and open Design Experiment/Software Factors.jmp.

   The factors and their levels appear in the Factors outline.

Figure A.31 Factors Outline for Software Factors

Factors

Load Design | Add Factor | Remove | Add N Factors 1 | Strength: t = 2

Name	Role	Values				
Web Browser	Categorical	Safari	IE	Firefox	Chrome	Other
Operating System	Categorical	Macintosh	Windows			
RAM	Categorical	16 MB	4 MB	8 MB		
Connection Speed	Categorical	0-1 Mbps	1-5 Mbps	>5 Mbps		

   Notice that the Role of the four factors is described as Categorical.

4. Click **Continue**.

   The Restrict Factor Level Combinations outline opens. Since all combinations of settings are possible, leave this set to **None**.

5. Click **Make Design**.

6. Click **Make Table**.

7. In the columns panel, click the asterisk next to Web Browser and select **Value Labels**.

Figure A.32 Column Info Window for Factor A

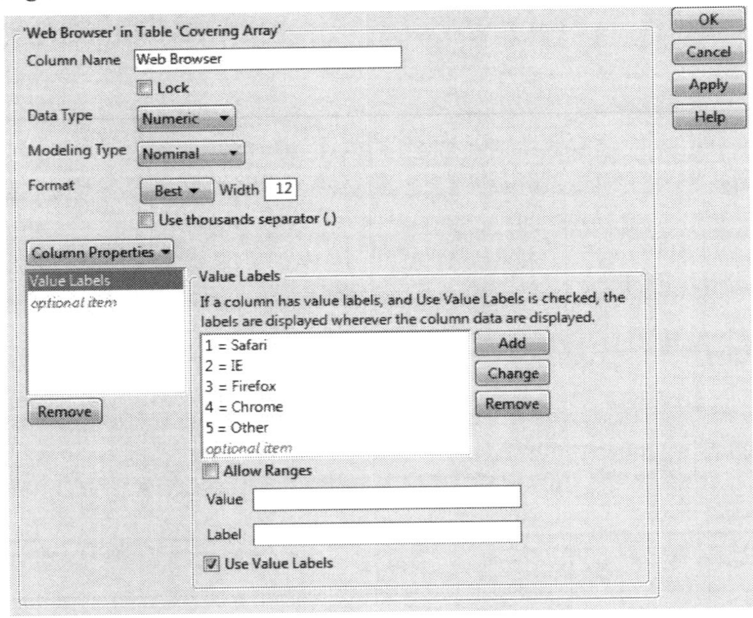

Notice that **Web Browser** has a Numeric data type and a Nominal modeling type. The underlying values of **Web Browser** are the integers 1, 2, 3, 4, and 5. These values are assigned value labels corresponding to the values that you entered when you constructed the design.

Change the value label for 2 from IE to Internet Explorer.

8. Select **2 = IE** in the Value Labels list.

Figure A.33 Value Labels Panel with Selection

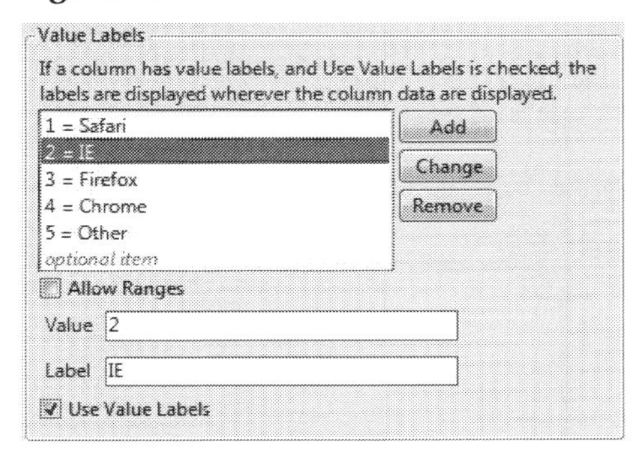

9. Type **Internet Explorer** next to Label.
10. Click **Change**.

    The change appears in the data table.

Note: To use the numeric values and not the labels, deselect **Use Value Labels**.

RunsPerBlock

When you use the DOE platforms to construct a design containing a blocking factor, the factor is assigned the Design Role column property with the value Blocking. JMP also assigns the RunsPerBlock column property to each Blocking factor. The RunsPerBlock property indicates the maximum allowable number of runs in each block. This property is used by the Evaluate Design and Augment Design platforms to indicate the blocking structure for the factor. For more details, see "Blocking" on page 88 in the "Custom Designs" chapter.

Note: The RunsPerBlock column property is assigned by JMP as part of design construction. You cannot directly assign this column property.

RunsPerBlock Example

Consider the wine tasting experiment described in “Example of a Custom Design” on page 65 in the “Custom Designs” chapter. Wine samples are tasted by five raters (Rater) and each rater tastes eight samples.

1. Select **Help > Sample Data Library** and open Design Experiment/Wine Data.jmp.
2. Click the asterisk next to Rater in the columns panel and select **Design Role**.

   Notice that the Design Role is set to Blocking.
3. Click **Cancel**.
4. Click the asterisk next to Rater in the columns panel and select **RunsPerBlock**.

Figure A.34 RunsPerBlock Column Property Panel for Rater

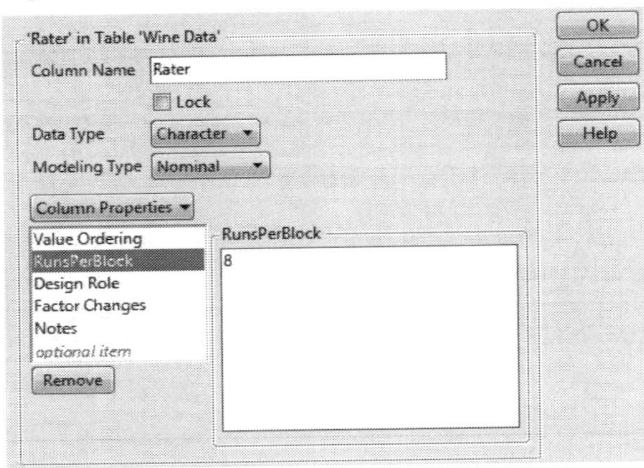

Notice that the value of RunsPerBlock is 8. The design constructed by the DOE Dialog script has 40 runs. Since there are five raters, JMP constructs a design with 40/5 = 8 runs for each rater.

ConstraintState

In the Custom and Mixture Design platforms, you can Save Constraints that you specify for a design. When you select Save Constraints, the coefficients of each linear constraint appear in a column in a data table. The value that bounds the inequality is given in the last row of the table.

Each constraint column is assigned the ConstraintState column property. This property specifies the direction of the inequality that defines the constraint. When you select Load Constraints from a design platform, the ConstraintState column property tells JMP the direction of the inequality.

Note: The ConstraintState column property is assigned by JMP as part of design construction. You cannot directly assign this column property.

ConstraintState Example

The sample data table Piepel.jmp, located in the Design Experiment folder, contains a mixture design with three continuous factors. The design is based on an experiment presented in Snee (1979) and Piepel (1988), where there are boundary constraints on each factor and three additional linear constraints. In the following example, you do the following:

1. Change one of the three additional constraints
2. Save the constraints to a table
3. Observe how the ConstraintState column property describes the direction of the inequality in the constraint

For further development of this example, see "An Extreme Vertices Example with Linear Constraints" on page 393 in the "Mixture Designs" chapter.

1. Select **Help > Sample Data Library** and open Design Experiment/Piepel.jmp.
2. In the Table panel, click the green triangle next to the **DOE Dialog** script.

   Notice the three linear constraints below the Factors outline. To make the constraints more interpretable, you want to reformulate the first constraint in terms of a "greater than or equal to" inequality.

Figure A.35 Linear Constraints beneath Factors Outline

Factors

Name	Role	Values	
X1	Mixture	0.1	0.5
X2	Mixture	0.1	0.7
X3	Mixture	0	0.7

-85	X1 +	-90	X2 +	-100	X3	≤	-90
85	X1 +	90	X2 +	100	X3	≤	95
-0.7	X1 +	0	X2 +	-1	X3	≤	-0.4

3. In the first constraint, do the following:
   - Type 85 next to X1.
   - Type 90 next to X2.
   - Type 100 next to X3.
   - Select ≥ from the inequality menu.
   - Type 90 to the right of the inequality sign.

4. Select **Save Constraints** from the Mixture Design red triangle menu.

   A table containing information about the constraints appears.

Figure A.36 Constraint Table

Untitled 17

Columns (3/0)
Constraint 1 *
Constraint 2 *
Constraint 3 *

	Constraint 1	Constraint 2	Constraint 3
1	-85	85	-0.7
2	-90	90	0
3	-100	100	-1
4	-90	95	-0.4

   Each column contains the coefficients of the factors X1, X2, and X3 in rows 1 through 3. Row 4 contains the value that appeared to the right of the inequality sign.

5. Click the asterisk next to Constraint 1.
6. Click **ConstraintState**.

Figure A.37 ConstraintState Column Property Panel

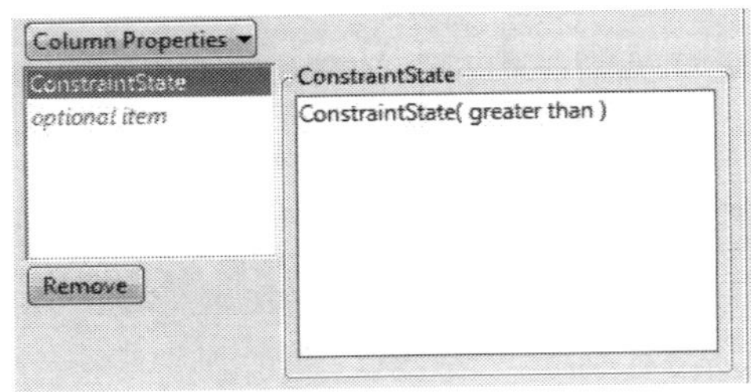

   The ConstraintState panel for X1 indicates that the direction of the inequality is “greater than” indicating greater than or equal to ≥.

7. Click **Cancel**.

Technical Details

This appendix contains information on the Alias Matrix, power calculations and the Relative Prediction Variance.

The Alias Matrix

The Alias Matrix entries represent the degree of bias imparted to model parameters by the effects that you specified in the Alias Terms outline. The Alias Matrix is also used in defining alias optimality. See "Alias Optimality" on page 127 in the "Custom Designs" chapter.

Calculations for the Alias Matrix are based on the model matrix. See "Model Matrix" on page 108 in the "Custom Designs" chapter.

Let $\mathbf{X}_1$ be the model matrix corresponding to the terms in the Model outline. Denote the matrix of model terms for the effects specified in the Alias Terms outline by $\mathbf{X}_2$.

The assumed model is given as follows:

$$\mathbf{Y} = \mathbf{X}_1\beta_1 + \varepsilon$$

Suppose that some of the alias terms are active and that the true model is given as follows:

$$\mathbf{Y} = \mathbf{X}_1\beta_1 + \mathbf{X}_2\beta_2 + \varepsilon$$

The least squares estimator of β_1 is given by:

$$\hat{\beta}_1 = (\mathbf{X}_1'\mathbf{X}_1)^{-1}\mathbf{X}_1'\mathbf{Y}$$

Under the usual regression assumptions, the expected value of $\hat{\beta}_1$ is given by:

$$E[\hat{\beta}_1] = \beta_1 + \mathbf{A}\beta_2$$

where $\mathbf{A} = (\mathbf{X}_1'\mathbf{X}_1)^{-1}\mathbf{X}_1'\mathbf{X}_2$.

The matrix $\mathbf{A}$ is called the *alias matrix*.

Designs with Hard or Very Hard Factor Changes

For designs with hard-to-change or very-hard-to-change factors, the alias matrix is given as follows:

$$\mathbf{A} = (\mathbf{X}_1'\mathbf{V}^{-1}\mathbf{X}_1)^{-1}\mathbf{X}_1'(\mathbf{V}^{-1}\mathbf{X}_2)$$

where $\mathbf{V}$ is the block diagonal covariance matrix of the responses.

Designs with If Possible Effects

For designs with If Possible effects (Bayesian D- and I-optimal designs), the alias matrix is given as follows:

$$\mathbf{A} = (\mathbf{X}_1'\mathbf{X}_1 + \mathbf{K}^2)^{-1}\mathbf{X}_1'\mathbf{X}_2$$

where **K** is a diagonal matrix with these values:

- $k = 0$ for Necessary terms
- $k = 1$ for If Possible main effects, powers, and interactions involving a categorical factor with more than two levels
- $k = 4$ for all other If Possible terms

In the Bayesian case, the alias matrix gives the aliasing of effects corresponding to a ridge regression with a prior variance of $\mathbf{K}^{-1}$. For additional detail on Bayesian designs, see "Bayesian D-Optimality" on page 124 in the "Custom Designs" chapter and "Bayesian I-Optimality" on page 126 in the "Custom Designs" chapter.

In the Custom Design platform, you can control the weights used for If Possible terms by selecting **Advanced Options > Prior Parameter Variance** from the red triangle menu. There you can set prior variances for all model terms by specifying the diagonal elements of **K**. The option updates to show the default weights when you click Make Design.

Power Calculations

The Power Analysis report gives power calculations for single parameter values and, when the design includes a categorical factor with three or more levels, for whole effects. This section describes the calculations in the two cases:

- "Power for a Single Parameter" on page 691
- "Power for a Categorical Effect" on page 692

Power for a Single Parameter

This section describes how power for the test of a single parameter is computed. Use the following notation:

X The model matrix. See the Standard Least Squares chapter in the *Fitting Linear Models* book for information on the coding for nominal effects. Also, See "Model Matrix" on page 108 in the "Custom Designs" chapter.

Note: You can view the model matrix by running Fit Model. Then select **Save Columns > Save Coding Table** from the red triangle menu for the main report.

β_i The parameter corresponding to the term of interest.

$\hat{\beta}_i$ The least squares estimator of β_i

β_i^A The Anticipated Coefficient value. The difference you want to detect is $2\beta_i^A$.

The variance of $\hat{\beta}_i$ is given by the ith diagonal entry of $\sigma^2(\mathbf{X}'\mathbf{X})^{-1}$, where σ^2 is the error variance. Denote the ith diagonal entry of $(\mathbf{X}'\mathbf{X})^{-1}$ by $(\mathbf{X}'\mathbf{X})_{ii}^{-1}$.

The error variance, σ^2, is estimated by the MSE, and has $n-p-1$ degrees of freedom, where n is the number of observations and p is the number of terms other than the intercept in the model.

The test of $H_0{:}\beta_i = 0$ is given by:

$$\frac{\hat{\beta}_i}{\sqrt{MSE(\mathbf{X}'\mathbf{X})_{ii}^{-1}}}$$

or equivalently by:

$$F_0 = \frac{\hat{\beta}_i^2}{MSE(\mathbf{X}'\mathbf{X})_{ii}^{-1}}$$

Under the null hypothesis, the test statistic F_0 has an F distribution on 1 and $n-p-1$ degrees of freedom.

If the true value of β_i is β_i^A, then F_0 has a noncentral F distribution with noncentrality parameter given by:

$$\lambda = \frac{(\beta_i^A)^2}{\sigma^2(\mathbf{X}'\mathbf{X})_{ii}^{-1}}$$

To compute the power of the test, first solve for the α-level critical value F_c:

$$\alpha = 1 - FDist(F_c, 1, n-p-1)$$

Then calculate the power as follows:

$$Power = 1 - FDist(F_c, 1, n-p-1, \lambda)$$

Power for a Categorical Effect

This section describes how power for the test for a whole categorical effect is computed. Use the following notation:

X Model matrix. See “The Alias Matrix” on page 690 in the “Technical Details” appendix.

β Vector of parameters.

$\hat{\beta}$ Least squares estimate of β.

β^A Vector of Anticipated Coefficient values.

L Matrix that defines the test for the categorical effect. The matrix **L** identifies the values of the parameters in β corresponding to the categorical effect and sets them equal to 0. The null hypothesis for the test of the categorical effect is given by:

H_0: $\mathbf{L}\beta = \mathbf{0}$

r Rank of **L**. Alternatively, r is the number of levels of the categorical effect minus one.

Note: You can view the design matrix by running Fit Model. Then select **Save Columns > Save Coding Table** from the red triangle menu for the main report.

The covariance matrix of $\hat{\beta}$ is given by $\sigma^2(\mathbf{X'X})^{-1}$, where σ^2 is the error variance.

The error variance, σ^2, is estimated by the MSE, and has $n-p-1$ degrees of freedom, where n is the number of observations and p is the number of terms other than the intercept in the model.

The test of H_0: $\mathbf{L}\beta = \mathbf{0}$ is given by:

$$F_0 = \left((\mathbf{L}\hat{\beta})'[\mathbf{L}(\mathbf{X'X})^{-1}\mathbf{L'}]^{-1}(\mathbf{L}\hat{\beta})\right)/(rMSE)$$

Under the null hypothesis, the test statistic F_0 has an F distribution on r and $n-p-1$ degrees of freedom.

If the true value of β is β^A, then F_0 has a noncentral F distribution with noncentrality parameter given by:

$$\lambda = \left((\mathbf{L}\beta^A)'[\mathbf{L}(\mathbf{X'X})^{-1}\mathbf{L'}]^{-1}(\mathbf{L}\beta^A)\right)/\sigma^2$$

To compute the power of the test, first solve for the α-level critical value F_c:

$$\alpha = 1 - FDist(F_c, r, n-p-1)$$

Then calculate the power as follows:

$$Power = 1 - FDist(F_c, r, n-p-1, \lambda)$$

Relative Prediction Variance

Consider the following notation:

X Model matrix. See "Model Matrix" on page 108 in the "Custom Designs" chapter. Custom designs provide a script that shows the model matrix. See "Save X Matrix" on page 108 in the "Custom Designs" chapter.

σ^2 Error variance.

$\hat{\beta}$ Vector of least squares estimates of the parameters.

$\mathbf{x}_i'$ The ith row of **X**.

Using this notation, the predicted response for the ith row of **X** is given by:

$$\hat{Y} = \mathbf{x}_i'\hat{\beta}$$

The relative prediction variance at the settings defined by $\mathbf{x}_i$ is given by:

$$\frac{\mathbf{x}_i' var(\hat{Y})\mathbf{x}_i}{\sigma^2} = \frac{\mathbf{x}_i' var(\mathbf{X}\hat{\beta})\mathbf{x}_i}{\sigma^2} = \mathbf{x}_i'(\mathbf{X}'\mathbf{X})^{-1}\mathbf{x}_i$$

Appendix C

References

Agresti, A., and Coull, B. A. (1998), "Approximate Is Better than 'Exact' for Interval Estimation of Binomial Proportions," *The American Statistician,* 52, 2, 119-126.

Agresti, A., and Caffo, B. N. (2000), "Simple and Effective Confidence Intervals for Proportions and Differences of Proportions Result from Adding Two Successes and Two Failures," *The American Statistician*, 54, 4, 280-288.

Atkinson, A. C. and Donev, A. N.(1992), *Optimum Experimental Designs*, Clarendon Press, Oxford p.148.

Barker (2011), C., "Power and Sample Size Calculations in JMP," SAS Institute. Retrieved January 22, 2016 from http://www.jmp.com/en_us/whitepapers/jmp/power-sample-calculations.html.

Bose, R.C., (1947), "Mathematical Theory of the Symmetrical Factorial Design" *Sankhya: The Indian Journal of Statistics,* Vol. 8, Part 2, pp. 107-166.

Box, G.E.P. (1988), "Signal–to–Noise Ratio, Performance Criteria, and Transformations," *Technometrics* 30, 1–40.

Box, G.E.P. and Behnken, D.W. (1960), "Some New Three-Level Designs for the Study of Quantitative Variables," *Technometrics* 2, 455–475.

Box, G.E.P. and Draper, N.R. (1987), *Empirical Model–Building and Response Surfaces*, New York: John Wiley and Sons.

Box, G.E.P., Hunter,W.G., and Hunter, J.S. (1978), *Statistics for Experimenters*, New York: John Wiley and Sons, Inc.

Box, G.E.P., Hunter,W.G., and Hunter, J.S. (2005), *Statistics for Experimenters: Design, Innovation, and Discovery*, 2nd edition, New York: John Wiley and Sons, Inc.

Box, G.E.P. and Meyer, R.D. (1986), "An Analysis of Unreplicated Fractional Factorials," *Technometrics* 28, 11–18.

Box, G.E.P. and Wilson, K.B. (1951), "On the Experimental Attainment of Optimum Conditions," *Journal of the Royal Statistical Society*, Series B, 13, 1-45.

Byrne, D.M. and Taguchi, G. (1986), *ASQC 40th Anniversary Quality Control Congress Transactions*, Milwaukee, WI: American Society of Quality Control, 168–177.

Chen, J., Sun, D.X., and Wu, C.F.J. (1993), "A Catalogue of Two-level and Three-Level Fractional Factorial Designs with Small Runs," *International Statistical Review*, 61, 1, p131-145, International Statistical Institute.

Clopper, C., Pearson, E. S. (1934), "The use of confidence or fiducial limits illustrated in the case of the binomial," *Biometrika* 26, 404–413.

Cochran, W.G. and Cox, G.M. (1957), *Experimental Designs*, Second Edition, New York: John Wiley and Sons.

Cohen, M.B., Gibbons, P.B., Mugridge, W.B., Colbourn, C.J. (2003), "Constructing Test Suites for Interaction Testing," *Proceedings 25th International Conference on Software Engineering*, 38–48.

Cohen, M.B., Dwyer, M.B., and Shi, J. (2007) "Interaction Testing of Highly-Configurable Systems in the Presence of Constraints," *Proceedings of the 2007 International Symposium on Software Testing and Analysis*, 129-139.

Colbourn, C. (2004), "Combinatorial Aspects of Covering Arrays," Le Matematiche, 58, 121-167.

Colbourn, C. and Dinitz, J., eds. (2010), *Handbook of Combinatorial Designs*, 2nd Edition, CRC Press.

Cornell, J.A. (1990), *Experiments with Mixtures*, Second Edition New York: John Wiley & Sons.

Cook, R.D. and Nachtsheim, C.J. (1990), "Letter to the Editor: Response to James M. Lucas," *Technometrics* 32, 363–364.

Dalal, S.R. and Mallows, C.L. (1998), "Factor-Covering Designs for Testing Software," *Technometrics* 40, 234–243.

Daniel, C. (1959), "Use of Half–normal Plots in Interpreting Factorial Two–level Experiments," *Technometrics*, 1, 311–314.

Daniel C. and Wood, F. (1980), *Fitting Equations to Data*, Revised Edition, New York: John Wiley and Sons, Inc.

Derringer, D. and Suich, R. (1980), "Simultaneous Optimization of Several Response Variables," *Journal of Quality Technology*, Oct. 1980, 12:4, 214–219.

DuMouchel, W. and Jones, B. (1994), "A Simple Bayesian Modification of *D*-Optimal Designs to Reduce Dependence on an Assumed Model," *Technometrics*, 36, 37–47.

Errore, A., Jones, B., Li, W., and Nachtsheim, C. (2016), "Using Definitive Screening Designs to Identify Active First- and Second-Order Factor Effects," forthcoming, *Journal of Quality Technology*.

Fang, K.T. and Sudfianto, A. (2006), *Design and Modeling for Computer Experiments*, Chapman and Hall CRC, Boca Ratan. Florida

Fries, A. and Hunter, W.G. (1984), "Minimum Aberration 2^(k-p) Designs," *Technometrics*, 22, 601-608.

Gotwalt, C. M., Jones, B.A., and Steinberg, D. M., (2009) "Fast Computation of Designs Robust to Parameter Uncertainty for Nonlinear Settings," *Technometrics*, 51:1, 88-95.

Goos, P. (2002) *The Optimal Design of Blocked and Split Plot Experiments*, New York: Springer.

Goos, P. and Jones, B. (2011), *Optimal Design of Experiments: A Case Study Approach*, John Wiley and Sons.

Haaland, P.D. (1989), *Experimental Design in Biotechnology*, New York: Marcel Dekker, Inc.

Hahn, G. J., Meeker, W.Q., and Feder, P. I., (1976), "The Evaluation and Comparison of Experimental Designs for Fitting Regression Relationships," *Journal of Quality Technology*, 8:3, pp. 140-157.

Hamada, M. and Wu, C.F.J. (1992), "Analysis of Designed Experiments with Complex Aliasing," *Journal of Quality Technology*, 24:3, 130-137.

Hartman, A. and Raskin, L. (2004), "Problems and Algorithms for Covering Arrays," *Discrete Mathematics*, 284, 149–156.

Hartman, A. and Raskin, L. (2004), "Problems and Algorithms for Covering Arrays," *Discrete Mathematics*, 284:1, 149-156.

Huber, J. and Zwerina, K. (1996), "The Importance of Utility Balance in Efficient Choice Designs," *Journal of Marketing Research*, 33, 307-317.

Jennrich, R.I. (1969), "Asymptotic Properties of Non-Linear Least Squares Estimators," *Annals of Mathematical Statistics*, 40, 633-643.

John, P.W.M. (1972), *Statistical Design and Analysis of Experiments*, New York: Macmillan Publishing Company, Inc.

Johnson, M.E. and Nachtsheim, C.J. (1983), "Some Guidelines for Constructing Exact D–Optimal Designs on Convex Design Spaces," *Technometrics*, 25, 271–277.

Jones, B. (1991), "An Interactive Graph For Exploring Multidimensional Response Surfaces," 1991 Joint Statistical Meetings, Atlanta, Georgia.

Jones, B. and Goos, P. (2007), "A Candidate-set-free Algorithm for Generating D-optimal Designs," *Applied Statistics*, 56, Part 3, pp 1-18. Retrieved January 22, 2016 from http://www.ncbi.nlm.nih.gov/pmc/articles/PMC3001117/.

Jones, B. and Goos, P. (2015), "Optimal Design of Blocked Experiments in the Presence of Supplementary Information about the Blocks," *Technometrics*, 47(4), 301-317.

Jones, B. and Johnson, R.T. (2009), "Design and Analysis for the Gaussian Process Model," *Quality and Reliability Engineering International*, 25, 515-524. Retrieved May 22, 2016 from http://www.stat.osu.edu/~comp_exp/jour.club/Jones_Johnson_QREI09.pdf.

Jones, B., Lin, D.K.J., and Nachtsheim, C.J. (2008), "Bayesian D-Optimal Supersaturated Designs," *Journal of Statistical Planning and Inference*, 138, 86-92.

Jones, B. and Nachtsheim, C.J. (2011), "A Class of Three-Level Designs for Definitive Screening in the Presence of Second-Order Effects," *Journal of Quality Technology*, 43, 1-15. Retrieved January 22, 2016 from http://www.jmp.com/en_us/whitepapers/jmp/class-three-level-definitive-screening.html.

Jones, B. and Nachtsheim, C.J. (2011), "Efficient Designs with Minimal Aliasing," Technometrics, 53:1, 62-71.

Jones, B. and Nachtsheim, C.J. (2013), "Definitive Screening Designs with Added Two-Level Categorical Factors," *Journal of Quality Technology*, 45, 121-129.

Jones, B. and Nachtsheim, C.J. (2016), "Blocking Schemes for Definitive Screening Designs," *Technometrics*, 58:1, 74-83.

Jones, B. and Nachtsheim, C.J. (2016), "Effective Model Selection for Definitive Screening Designs," *Technometrics*, accepted for publication.

Joseph, V.R., Gul, E., and Ba, S. (2015), "Maximum Projection Designs for Computer Experiments," *Biometrika*, 102:2, 371-380.

Kessels, R., Jones, B., Goos, P. (2011), "Bayesian Optimal Designs for Discrete Choice Experiments with Partial Profiles," *Journal of Choice Modelling*, 4(3), 52-74.

Khuri, A.I. and Cornell, J.A. (1987), *Response Surfaces: Design and Analysis*, New York: Marcel Dekker.

Kowalski, S.M., Cornell, J.A., and Vining, G.G. (2002) "Split Plot Designs and Estimation Methods for Mixture Experiments with Process Variables," *Technometrics* 44: 72-79.

Lenth, R.V. (1989), "Quick and Easy Analysis of Unreplicated Fractional Factorials," *Technometrics*, 31, 469–473.

Lekivetz, R. (2014, June 5), "What is an Alias Matrix?" Retrieved January 22, 2016 from http://blogs.sas.com/content/jmp/2014/06/05/what-is-an-alias-matrix/.

Lekivetz, R., Sitter, R., Bingham, D., Hamada, M.S., Moore, L.M., and Wendelberger, J.R. (2015), "On Algorithms for Obtaining Orthogonal and Near-Orthogonal Arrays for Main-Effects Screening," *Journal of Quality Technology* 47:1 2-13.

Lin, D. K. J. (1993), "A New Class of Supersaturated Design," *Technometrics*, 35, 28-31.

Lucas, J.M., (1990), "Letter to the Editor: Comments on Cook and Nachtsheim (1989)," *Technometrics*, 32, 363–364.

Mahalanobis, P.C. (1947), *Sankhya, The Indian Journal of Statistics,* Vol 8, Part 2, April.

Martirosyan, S. (2003), "Perfect Hash Families, Identifiable Parent Property Codes and Covering Arrays," Ph.D. thesis, Universitat Duisburg-Essen, Fakultat fur Mathematik.

Meyer, R.K. and Nachtsheim, C.J. (1995), The Coordinate Exchange Algorithm for Constructing Exact Optimal Designs," *Technometrics*, Vol 37, pp. 60-69.

Meyer, R.D., Steinberg, D.M., and Box, G.(1996), Follow-up Designs to Resolve Confounding in Multifactor Experiments, *Technometrics*, 38:4, p307.

Miller, A. and Sitter, R. R. (2005), "Using Folded-Over Nonorthogonal Designs," *Technometrics*, 47(4), 502-513.

Mitchell, T.J. (1974), "An algorithm for the Construction of D-Optimal Experimental Designs," *Technometrics*, 16:2, pp.203-210.

Montgomery, D.C. (2009), *Design and Analysis of Experiments*, Seventh Edition, Hoboken: John Wiley & Sons, Inc.

Morgan, J. (2009), "Constrained Covering Arrays: Resolving Invalid Level Combination Constraints," Technical Report, Cary NC: SAS Institute Inc., also in *Supplemental Proceedings of the 20th International Symposium on Software Reliability Engineering*.

Morgan, J. (2014), "Computing Coverage and Diversity Metrics for Constrained Covering Arrays," Technical Report, Cary NC: SAS Institute Inc.

Morris, M.D., Mitchell, T.J., and Ylvisaker, D. (1993), "Bayesian Design and Analysis of Computer Experiments: Use of Derivatives in Surface Prediction," *Technometrics* 35:2, 243-255.

Myers, R.H. (1976) *Response Surface Methodology*, Boston: Allyn and Bacon.

Myers, R.H. (1988), *Response Surface Methodology*, Virginia Polytechnic and State University.

Myers, R.H., Montgomery, D.C., and Anderson-Cook, C.M. (2009), *Response Surface Methodology, Process and Product Optimization Using Designed Experiments*, Third Edition, Hoboken: John Wiley & Sons, Inc.

Mylona, K., Goos, P., and Jones, B. (2014), "Optimal Design of Blocked and Split-plot Experiments for Fixed Effects and Variance Component Estimation," Technometrics, 56:2, 132-144.

Nayeri, P., Colbourn, C., and Konjevod, G. (2013), "Randomized Postoptimization of Covering Arrays," *European Journal of Combinatorics*, 34, 91-103.

Peipel, G.F. (1988), "Programs for Generating Extreme Vertices and Centroids of Linearly Constrained Experimental Regions," *Journal of Quality Technology* 20:2, 125-139.

Plackett, R.L. and Burman, J.P. (1947), "The Design of Optimum Multifactorial Experiments," *Biometrika*, 33, 305–325.

Sall, J. (2008), "Choice Experimental Designs Are Different," SAS Institute. Retrieved January 22, 2016, from
http://blogs.sas.com/content/jmp/2008/12/16/choice-experimental-designs-are-different/.

Santner, T.B., Williams, B., and Notz, W. (2003), *The Design and Analysis of Computer Experiments*, Springer, New York.

Schoen, E. D. (1999), "Designing Fractional Two-level Experiments with Nested Error Structures," *Journal of Applied Statistics* 26: 495-508.

Shannon, C.E. and Weaver, W. (1949), "The Mathematical Theory Of Communication," University Of Illinois Press, Urbana, Illinois.

Scheffé, H. (1958) "Experiments with Mixtures", *Journal of the Royal Statistical Society B* v20, 344–360.

Shewry, M.C. and Wynn, H.P. (1987), "Maximum Entropy Sampling," *Journal of Applied Statistics* 14, 165-170.

Snee, R.D. and Marquardt, D.W. (1974), "Extreme Vertices Designs for Linear Mixture Models," *Technometrics*, 16, 391–408.

Snee, R.D. (1975), "Experimental Designs for Quadratic Models in Constrained Mixture Spaces," *Technometrics,* 17:2, 149–159.

Snee, R.D. (1979), "Experimental Designs for Mixture Systems with Multicomponent Constraints," *Commun. Statistics: Theory and Methods,* 8(4), 303–326.

Snee, Ronald D. (1985), "Computer Aided Design of Experiments - Some Practical Experiences," *Journal of Quality Technology,* 17, 222-236.

St. John, R.C. and Draper, N.R. (1975), "D-Optimality for Regression Designs: A Review," *Technometrics*, 17, 5-23.

Taguchi, G. (1976), "An Introduction to Quality Control," Nagoya, Japan: Central Japan Quality Control Association.

Vivacqua, C.A., and Bisgaard, S. (2004), "Strip-Block Experiments for Process Improvement and Robustness,", *Quality Engineering*, 16, 495-500.

Welch, W.J. (1984), "Computer-Aided Design of Experiments for Response Estimation," *Technometrics*, 26, 217–224.

Wu, C.F. (1981), "Asymptotic Theory of Nonlinear Least Squares Estimation," *Annals of Mathematical Statistics*, 9, 501-513.

Wu, C.F.J., and Hamada, M.S. (2009), *Experiments: Planning, Analysis, and Optimization*, Second Edition, Wiley.

Xiao, L., Lin, D.K.J., and Bai, F. (2012), "Constructing Definitive Screening Designs Using Conference Matrices," *Journal of Quality Technology*, 44, 2-8.

Ye, K.Q. and Hamada, M. (2000), "Critical Values of the Lenth Method for Unreplicated Factorial Designs," *Journal of Quality Technology* 32, 57-66.

Yilmaz, C., Fouche, S., Cohen, M.B., Porter, A., Demiroz, G., and Koc, U. (2014), "Moving Forward with Combinatorial Interaction Testing," *Computer* 47, 37–45.

Zahran, A., Anderson-Cook, C.M., and Myers, R.H. (2003), "Fraction of Design Space to Assess the Prediction Capability of Response Surface Designs," *Journal of Quality Technology*, 35, 377–386.

Zhang, Z. and Zhang, J. (2011), "Characterizing Failure-Causing Parameter Interactions by Adaptive Testing," *Proceedings of the 2011 International Symposium on Software Testing and Analysis*, 331-341.

Index

Design of Experiments Guide

A

B

C

D

E

F

N

O

P

Q

R

S

T

U

V

W-Z

45444434R00395

Made in the USA
San Bernardino, CA
08 February 2017